Applied Mathematical Sciences

Volume 190

More information about this series at http://www.springer.com/series/34

Andreas Kirsch • Frank Hettlich

The Mathematical Theory of Time-Harmonic Maxwell's Equations

Expansion-, Integral-, and Variational Methods

 Springer

Andreas Kirsch
Department of Mathematics
Karlsruhe Institute of Technology (KIT)
Karlsruhe, Germany

Frank Hettlich
Department of Mathematics
Karlsruhe Institute of Technology (KIT)
Karlsruhe, Germany

ISSN 0066-5452   ISSN 2196-968X (electronic)
ISBN 978-3-319-37918-0   ISBN 978-3-319-11086-8 (eBook)
DOI 10.1007/978-3-319-11086-8
Springer Cham Heidelberg New York Dordrecht London

Mathematics Subject Classification: 31B10, 33-01, 33C55, 35A15, 35J05, 35Q61, 78-08, 78A40, 78A45

Printed on acid-free paper

Springer is part of Springer Science+Business Media (www.springer.com)

Preface

This book arose from lectures on Maxwell's equations given by the authors between 2007 and 2013. Graduate students from pure and applied mathematics, physics—including geophysics—and engineering attended these courses. We observed that the expectations of these groups of students were quite different: In geophysics expansions of the electromagnetic fields into spherical (vector-) harmonics inside and outside of balls are of particular interest. Graduate students from numerical analysis wanted to learn about the variational treatments of interior boundary value problems including an introduction to Sobolev spaces. A classical approach in scattering theory—which can be considered as a boundary value problem in the unbounded exterior of a bounded domain—uses boundary integral equation methods which are particularly helpful for deriving properties of the far field behavior of the solution. This approach is, for polygonal domains or, more generally, Lipschitz domains, also of increasing relevance from the numerical point of view because the dimension of the region to be discretized is reduced by one. In our courses we wanted to satisfy all of these wishes and designed an introduction to Maxwell's equations which covers all of these concepts—but restricted ourselves almost completely (except Sect. 4.3) to the time-harmonic case or, in other words, to the frequency domain, and to a number of model problems.

The Helmholtz equation is closely related to the Maxwell system (for time-harmonic fields). As we will see, solutions of the scalar Helmholtz equation are used to generate solutions of the Maxwell system (Hertz potentials), and every component of the electric and magnetic field satisfies an equation of Helmholtz type. Therefore, and also for didactical reasons, we will consider in each of our approaches first the simpler scalar Helmholtz equation before we turn to the technically more complicated Maxwell system. In this way one clearly sees the analogies and differences between the models.

In Chap. 1 we begin by formulating the Maxwell system in differential and integral form. We derive special cases as the E-mode and the H-mode and, in particular, the time-harmonic case. Boundary conditions and radiation conditions complement the models.

In Chap. 2 we study the particular case where the domain D is a ball. In this case we can expand the fields inside and outside of D into spherical wave functions. First we study the scalar stationary case; that is, the Laplace equation. We introduce the expansion into scalar spherical harmonics as the analogon to the Fourier expansion on circles in $\mathbb{R}^2$. This leads directly to series expansions of solutions of the Laplace equation in spherical coordinates. The extension to the Helmholtz equation requires the introduction of (spherical) Bessel- and Hankel functions. We derive the most important properties of these special functions in detail. After these preparations for the scalar Helmholtz equation we extend the analysis to the expansion of solutions of Maxwell's equations with respect to vector wave functions. None of the results of this chapter are new, of course, but we have not been able to find such a common presentation of both, the scalar and the vectorial case, in the literature. We emphasize that this chapter is completely self-contained and does not refer to any other chapter—except for the proof that the series solution of the exterior problem satisfies the radiation condition.

Chapter 3 deals with a particular scattering problem. The scattering object is of arbitrary shape but, in this chapter, with sufficiently smooth boundary ∂D. We present the classical boundary integral equation method and follow very closely the fundamental monographs [6, 7] by David Colton and Rainer Kress. In contrast to their approach we restrict ourselves to the case of smooth boundary data (as it is the usual case of the scattering by incident waves) which allows us to study the setting completely in Hölder spaces and avoids the notions of "parallel surfaces" and weak forms of the normal derivatives. For the scalar problem we restrict ourselves to the Neumann boundary condition because our main goal is the treatment for Maxwell's equations. Here we believe—as also done in [7]—that the canonical spaces on the boundary are Hölder spaces where also the surface divergence is Hölder continuous. In order to prove the necessary properties of the scalar and vector potentials a careful investigation of the differential geometric properties of the surface ∂D is needed. Parts of the technical details are moved to the appendix. We emphasize that also this chapter is self-contained and does not need any results from other chapters (except from the appendix).

As an alternative approach for studying boundary value problems for the Helmholtz equation or the Maxwell system we will study the weak or variational solution concept in Chap. 4. We restrict ourselves to the interior boundary value problem with a general source term and the homogeneous boundary condition of a perfect conductor. This makes it possible to work (almost) solely in the Sobolev spaces $H_0^1(D)$ and $H_0(\text{curl}, D)$ of functions with vanishing boundary traces or tangential boundary traces, respectively. In Sect. 4.1 we derive the basic properties of these special Sobolev spaces. The characteristic feature is that no regularity of the boundary and no trace theorems are needed. Probably the biggest difference between the scalar case of the Helmholtz equation and the vectorial case of Maxwell's equations is the fact that $H_0(\text{curl}, D)$ is not compactly imbedded in $L^2(D, \mathbb{C}^3)$ in contrast to the space $H_0^1(D)$ for the scalar problem. This makes it necessary to introduce the Helmholtz decomposition. The only proof which is beyond the scope of this elementary chapter is the proof that the subspace of $H_0(\text{curl}, D)$ consisting of divergence-free vector fields is compactly imbedded in $L^2(D, \mathbb{C}^3)$. For this part some regularity of the boundary (e.g., Lipschitz regularity) is needed. Since the proof of this fact requires more advanced properties of Sobolev spaces it is postponed to Chap. 5. We note that also this chapter is self-contained except of the before mentioned compactness property.

The final Chap. 5 presents the boundary integral equation method for Maxwell's equations on Lipschitz domains. Lipschitz domains, in particular polyhedral domains, play obviously an important role in praxis. We were encouraged by our colleagues from the numerical analysis group to include this chapter to have a reference for further studies. The investigation requires more advanced properties of Sobolev spaces than those presented in Sect. 4.1. In particular, Sobolev spaces on the boundary ∂D and the corresponding trace operators have to be introduced. Perhaps different from most of the traditional approaches we first consider the case of the cube $(-\pi, \pi)^3 \subset \mathbb{R}^3$ and introduce Sobolev spaces of periodic functions by the proper decay of the Fourier coefficients. The proofs of imbedding and trace theorems are quite elementary. Then we use, as it is quite common, the partition of unity and local maps to define the Sobolev spaces on the boundary and transfer the trace theorem to general Lipschitz domains.

We define the scalar and vector potentials in Sect. 5.2 analogously to the classical case as in Chap. 3 but have to interpret the boundary integrals as certain dual forms. The boundary operators are then defined as traces of these potentials. In this way we follow the classical approach as closely as possible. Our approach is similar but a bit more explicit than in [20], see also [12]. Once the properties of the potentials and corresponding boundary operators are known, the introduction and investigation of the boundary integral equations are almost classical. For example, for Lipschitz domains the Dirichlet boundary value problem for the scalar Helmholtz equation is solved by a (properly modified) single layer ansatz. This is preferable to a double layer ansatz because the corresponding double layer boundary operator fails to be compact (in contrast to the case of smooth boundaries). Also, the single layer boundary operator satisfies a Garding's inequality; that is, it can be decomposed into a coercive and a compact part. Analogously, the Neumann boundary value problem and the electromagnetic case are treated. In our presentation we try to show the close connection between the scalar and the vector cases.

Starting perhaps with the pioneering work of Costabel [8] many important contributions to the study of boundary integral operators in Sobolev spaces for Lipschitz boundaries have been published. It is impossible for the authors to give an overview on this subject but instead refer to the monograph [12] and the survey article [4] from which we have learned a lot. As mentioned above, our approach to introduce the Sobolev spaces, however, is different from those in, e.g., [1–3].

In the appendix we collect results from functional analysis, vector calculus, and differential geometry, in particular various forms of Green's theorem and the surface gradient and surface divergence for (smooth) functions on (smooth) surfaces.

There exist numerous monographs on electromagnetism and wave propagation as, for example, [10, 13, 15, 23, 25, 29, 30] each of which with its own scope and addressee. Perhaps different to these contributions the emphasis of our work lies in the rigorous mathematical treatment of Maxwell's equation inside or outside of bounded regions including the precise formulations of these equations and their equivalent formulations and representations in conveniently chosen function spaces which depend on the smoothness of the data. The correct choice of these function spaces makes it possible to give rigorous proofs of uniqueness and existence. In this way our style follows more the ones in the monographs [5–7, 20, 21, 24].

We want to emphasize that it was not our intention to present a comprehensive work on Maxwell's equations, not even for the time-harmonic case or any of the before mentioned subareas. As said before, this book arose from—and is intended to be—material for designing graduate courses for mathematically orientated students on electromagnetic wave propagation problems. The students should have some knowledge on vector analysis (curves, surfaces, divergence theorem) and functional analysis (normed spaces, Hilbert spaces, linear and bounded operators, dual space). The union of the topics covered in this monograph is certainly far too much for a single course. But it is very well possible to choose parts of it because the chapters are all independent of each other. For example, in the summer term 2012 (8 credit points; that is in our place, 16 weeks with 4 h per week plus exercises) one of the authors (A.K.) covered Sects. 2.1–2.6 of Chap. 2 (only interior cases), Chap. 3 without all of the proofs of the differential geometric properties of the surface and all of the jump properties of the potentials, and Chap. 4 without Sect. 4.3. Perhaps these notes can also be useful for designing courses on Special Functions (spherical harmonics, Bessel functions) or on Sobolev spaces.

One of the authors wants to dedicate this book to his father, Arnold Kirsch (1922–2013), who taught him about *simplification of problems without falsification* (as a concept of teaching mathematics in high schools). In Chap. 4 of this monograph we have picked up this concept by presenting the ideas for a special case only rather than trying to treat the most general cases. Nevertheless, we admit that other parts of the monograph (in particular of Chaps. 3 and 5) are technically rather involved.

Karlsruhe, Germany   Andreas Kirsch
Karlsruhe, Germany   Frank Hettlich
July 2014

Contents

Chapter 1
Introduction

In this introductory chapter we will explain the physical model and derive the boundary value problems which we will investigate in this monograph. We begin by formulating Maxwell's equations in differential form as our starting point. In this monograph we consider exclusively *linear* media; that is, the constitutive relations are linear. The restriction to special cases leads to the analogous equations in electrostatics or magnetostatics and, assuming periodic time dependence when going into the frequency domain, to time-harmonic fields. In the presence of media the fields have to satisfy certain continuity and boundary conditions and, if the region is unbounded, a radiation condition at infinity. We finish this chapter by introducing two model problems which we will treat in detail in the forthcoming chapters.

1.1 Maxwell's Equations

Electromagnetic wave propagation is described by four particular equations, the *Maxwell equations*, which relate five vector fields $\mathcal{E}$, $\mathcal{D}$, $\mathcal{H}$, $\mathcal{B}$, $\mathcal{J}$ and the scalar field ρ. In differential form these read as follows:

$$\frac{\partial \mathcal{B}}{\partial t} + \mathrm{curl}_x\, \mathcal{E} = 0 \qquad \text{(Faraday's Law of Induction)}$$

$$\frac{\partial \mathcal{D}}{\partial t} - \mathrm{curl}_x\, \mathcal{H} = -\mathcal{J} \quad \text{(Ampere's Law)}$$

$$\mathrm{div}_x\, \mathcal{D} = \rho \qquad \text{(Gauss' Electric Law)}$$

$$\mathrm{div}_x\, \mathcal{B} = 0 \qquad \text{(Gauss' Magnetic Law)}.$$

© Springer International Publishing Switzerland 2015
A. Kirsch, F. Hettlich, *The Mathematical Theory of Time-Harmonic Maxwell's Equations*, Applied Mathematical Sciences 190,
DOI 10.1007/978-3-319-11086-8_1

The fields $\mathcal{E}$ and $\mathcal{D}$ denote the *electric field* (in V/m) and *electric displacement* (in As/m^2), respectively, while $\mathcal{H}$ and $\mathcal{B}$ denote the *magnetic field* (in A/m) and *magnetic flux density* (in $Vs/m^2 = T =$ Tesla). Likewise, $\mathcal{J}$ and ρ denote the *current density* (in A/m^2) and *charge density* (in As/m^3) of the medium.

Here and throughout we use the *rationalized MKS-system*; that is, the fields are given with respect to the units Volt (V), Ampere (A), meter (m), and second (s). All fields depend on both, the space variable $x \in \mathbb{R}^3$ and the time variable $t \in \mathbb{R}$. We note that the differential operators are always taken with respect to the spacial variable x without indicating this in the following. The definition of the differential operators div and curl, e.g., in cartesian coordinates and basic identities are listed in the appendix.

The actual equations that govern the behavior of the electromagnetic field were first completely formulated by James Clark Maxwell (1831–1879) in *Treatise on Electricity and Magnetism* in 1873. It was the ingenious idea of Maxwell to modify Ampere's Law which was known up to that time in the form curl $\mathcal{H} = \mathcal{J}$ for stationary currents. Furthermore, he collected the four equations as a consistent theory to describe electromagnetic phenomena.

As a first observation we note that in domains where the equations are satisfied one derives from the identity div curl $\mathcal{H} = 0$ the well-known equation of continuity which combines the charge density and the current density.

Conclusion 1.1 *Gauss' Electric Law and Ampere's Law imply the* equation of continuity

$$\frac{\partial \rho}{\partial t} = \operatorname{div} \frac{\partial \mathcal{D}}{\partial t} = \operatorname{div} \left(\operatorname{curl} \mathcal{H} - \mathcal{J} \right) = -\operatorname{div} \mathcal{J}.$$

Historically, and more closely connected to the physical situation, the integral forms of Maxwell's equations should be the starting point. In order to derive these integral relations, we begin by letting S be a connected smooth surface with boundary ∂S in the interior of a region Ω_0 in $\mathbb{R}^3$ where electromagnetic waves propagate. In particular, we require that the unit normal vector $\nu(x)$ for $x \in S$ is continuous and directed always to "one side" of S, which we call the positive side of S. By $\tau(x)$ we denote a unit vector tangent to the boundary of S at $x \in \partial S$. This vector, lying in the tangent plane of S together with a second vector $n(x)$ in the tangent plane at $x \in \partial S$ and normal to ∂S is oriented such that $\{\tau, n, \nu\}$ form a mathematically positive system; that is, τ is directed counterclockwise when we "look at S from the positive side," and $n(x)$ is directed to the outside of S.

Then *Ampere's law* describing the effect of the external and the induced current on the magnetic field is of the form

$$\int\limits_{\partial S} \mathcal{H} \cdot \tau \, d\ell = \frac{d}{dt} \int\limits_{S} \mathcal{D} \cdot \nu \, ds + \int\limits_{S} \mathcal{J} \cdot \nu \, ds. \qquad (1.1)$$

It is named after André Marie Ampère (1775–1836).

Next, *Faraday's law of induction* (Michael Faraday, 1791–1867), which is

$$\int\limits_{\partial S} \mathcal{E} \cdot \tau \, d\ell = -\frac{d}{dt} \int\limits_{S} \mathcal{B} \cdot \nu \, ds, \qquad (1.2)$$

describes how a time-varying magnetic field effects the electric field.

Let us denote by $\Omega \subseteq \mathbb{R}^3$ an open set with boundary $\partial\Omega$ and outer unit normal vector $\nu(x)$ at $x \in \partial\Omega$. Finally, the equations include *Gauss' Electric Law*

$$\int\limits_{\partial\Omega} \mathcal{D} \cdot \nu \, ds = \int\limits_{\Omega} \rho \, dx \qquad (1.3)$$

describing the sources of the electric displacement, and *Gauss' Magnetic Law*

$$\int\limits_{\partial\Omega} \mathcal{B} \cdot \nu \, ds = 0 \qquad (1.4)$$

which ensures that there are no magnetic currents. Both named after Carl Friedrich Gauss (1777–1855).

In regions where the vector fields are smooth functions and μ and ε are at least continuous we can apply the following integral identities due to Stokes and Gauss for surfaces S and solids Ω lying completely in D.

$$\int\limits_{S} \operatorname{curl} \mathbf{F} \cdot \nu \, ds = \int\limits_{\partial S} \mathbf{F} \cdot \tau \, d\ell \quad \text{(Stokes)}, \qquad (1.5)$$

$$\int\limits_{\Omega} \operatorname{div} \mathbf{F} \, dx = \int\limits_{\partial\Omega} \mathbf{F} \cdot \nu \, ds \quad \text{(Gauss)}, \qquad (1.6)$$

see Appendix A.3. To derive the Maxwell's equations in differential form we choose $\mathbf{F}$ to be one of the fields $\mathcal{H}, \mathcal{E}, \mathcal{B}$, or $\mathcal{D}$. With these formulas we can eliminate the boundary integrals in (1.1)–(1.4). We then use the fact that we can vary the surface S and the solid Ω in D arbitrarily. By equating the integrands we are led to Maxwell's equations in differential form as presented in the beginning.

With Maxwell's equations many electromagnetic phenomena became explainable. For instance, they predicted the existence of electromagnetic waves as light or X-rays in vacuum. It took about 20 years after Maxwell's work when Heinrich Rudolf Hertz (1857–1894) could show experimentally the existence of electromagnetic waves, in Karlsruhe, Germany. For more details on the physical background of Maxwell's equations we refer to text books as J.D. Jackson, *Classical Electrodynamics* [14].

1.2 The Constitutive Equations

In the general setting the equations are not yet complete. Obviously, there are more unknowns than equations. The *Constitutive Equations* couple them:

$$\mathcal{D} = \mathcal{D}(\mathcal{E}, \mathcal{H}) \quad \text{and} \quad \mathcal{B} = \mathcal{B}(\mathcal{E}, \mathcal{H}).$$

The electric properties of the material, which give these relationships are complicated. In general, they depend not only on the molecular character but also on macroscopic quantities as density and temperature of the material. Also, there are time-dependent dependencies as, e.g., the hysteresis effect, i.e. the fields at time t depend also on the past.

As a first approximation one starts with representations of the form

$$\mathcal{D} = \mathcal{E} + 4\pi \mathcal{P} \quad \text{and} \quad \mathcal{B} = \mathcal{H} - 4\pi \mathcal{M}$$

where $\mathcal{P}$ denotes the electric polarization vector and $\mathcal{M}$ the magnetization of the material. These can be interpreted as mean values of microscopic effects in the material. Analogously, ρ and $\mathcal{J}$ are macroscopic mean values of the free charge and current densities in the medium.

If we ignore ferro-electric and ferro-magnetic media and if the fields are relatively small, one can model the dependencies by linear equations of the form

$$\mathcal{D} = \varepsilon \mathcal{E} \quad \text{and} \quad \mathcal{B} = \mu \mathcal{H}$$

with matrix-valued functions $\varepsilon : \mathbb{R}^3 \to \mathbb{R}^{3 \times 3}$, the *dielectric tensor*, and $\mu : \mathbb{R}^3 \to \mathbb{R}^{3 \times 3}$, the *permeability tensor*. In this case we call a medium *linear*.

The special case of an *isotropic medium* means that polarization and magnetization do not depend on the directions. Otherwise a medium is called *anisotropic*. In the isotropic case dielectricity and permeability can be modeled as just real valued functions, and we have

$$\mathcal{D} = \varepsilon \mathcal{E} \quad \text{and} \quad \mathcal{B} = \mu \mathcal{H}$$

with scalar functions $\varepsilon, \mu : \mathbb{R}^3 \to \mathbb{R}$.

In the simplest case these functions ε and μ are constant and we call such a medium *homogeneous*. It is the case, e.g., in vacuum.

We indicated already that also ρ and $\mathcal{J}$ can depend on the material and the fields. Therefore, we need a further relation. In conducting media the electric field induces a current. In a linear approximation this is described by *Ohm's Law*:

$$\mathcal{J} = \sigma \mathcal{E} + \mathcal{J}_e$$

where $\mathcal{J}_e$ is the external current density. For isotropic media the function $\sigma : \mathbb{R}^3 \to \mathbb{R}$ is called the *conductivity*. If $\sigma = 0$, then the material is called *dielectric*. In vacuum we have $\sigma = 0$ and $\varepsilon = \varepsilon_0 \approx 8.854 \cdot 10^{-12} AS/Vm$, $\mu = \mu_0 = 4\pi \cdot 10^{-7} Vs/Am$. In anisotropic media, also the function σ is matrix valued.

1.3 Special Cases

Under specific physical assumptions the Maxwell system can be reduced to elliptic second order partial differential equations. They serve often as simpler models for electromagnetic wave propagation. Also in this monograph we will always explain the approaches first for the simpler scalar wave equation.

1.3.1 Vacuum

Vacuum is a homogeneous, dielectric medium with $\varepsilon = \varepsilon_0$, $\mu = \mu_0$, and $\sigma = 0$, and no charge distributions and no external currents; that is, $\rho = 0$ and $\mathcal{J}_e = 0$. The law of induction takes the form

$$\mu_0 \frac{\partial \mathcal{H}}{\partial t} + \operatorname{curl} \mathcal{E} = 0.$$

Assuming sufficiently smooth functions a differentiation with respect to time t and an application of Ampere's Law yields

$$\varepsilon_0 \mu_0 \frac{\partial^2 \mathcal{H}}{\partial t^2} + \operatorname{curl} \operatorname{curl} \mathcal{H} = 0.$$

The term $c_0 = 1/\sqrt{\varepsilon_0 \mu_0}$ has the dimension of a velocity and is called the *speed of light*.

From the identity $\operatorname{curl}\operatorname{curl} = \nabla\operatorname{div} - \Delta$ where the vector valued Laplace operator Δ is taken componentwise it follows that the components of $\mathcal{H}$ are solutions of the linear *wave equation*

$$\frac{1}{c_0^2}\frac{\partial^2 \mathcal{H}}{\partial t^2} - \Delta\mathcal{H} = 0.$$

Analogously, one derives the same equation for the cartesian components of the electric field:

$$\frac{1}{c_0^2}\frac{\partial^2 \mathcal{E}}{\partial t^2} - \Delta\mathcal{E} = 0.$$

Therefore, a solution of the Maxwell system in vacuum can also be described by a divergence free solution of one of the two vector valued wave equations and defining the other field by Amperes Law or by Faraday's Law of Induction, respectively.

1.3.2 Electro- and Magnetostatics

Next we consider the Maxwell system in the case of stationary fields; that is, $\mathcal{E}, \mathcal{D}, \mathcal{H}, \mathcal{B}, \mathcal{J}$, and ρ are constant with respect to time. For the electric field $\mathcal{E}$ this situation in a region Ω is called *electrostatics*. The law of induction reduces to the differential equation

$$\operatorname{curl}\mathcal{E} = 0 \quad \text{in } \Omega.$$

Therefore, if Ω is simply connected, there exists a potential $u : \Omega \to \mathbb{R}$ with $\mathcal{E} = -\nabla u$ in Ω. In a homogeneous medium Gauss' Electric Law yields the *Poisson equation*

$$\rho = \operatorname{div}\mathcal{D} = -\operatorname{div}(\varepsilon_0\,\mathcal{E}) = -\varepsilon_0\,\Delta u$$

for the potential u. Thus, the electrostatics is described by the basic elliptic partial differential equation $\Delta u = -\rho/\varepsilon_0$. Mathematically, we are led to the field of *potential theory*.

Example 1.2. The most important example is the spherical symmetric electric field generated by a point charge, e.g., at the origin. For $x \in \mathbb{R}^3$ with $x \neq 0$ the function $u(x) = \frac{1}{4\pi}\frac{1}{|x|}$ is *harmonic*; that is, satisfies $\Delta u = 0$. Thus by

$$\mathcal{E}(x) = -\nabla u(x) = \frac{1}{4\pi}\frac{x}{|x|^3}, \quad x \neq 0,$$

we obtain a stationary solution, the field of an *electric monopole*.

In *magnetostatics* one considers $\mathcal{H}$ being constant in time. For the magnetic field the situation is different because by Ampere's law we have $\operatorname{curl} \mathcal{H} = \mathcal{J}$. Thus in general $\operatorname{curl} \mathcal{H}$ does not vanish. However, according to Gauss' magnetic law we have

$$\operatorname{div} \mathcal{B} = 0.$$

From this identity we conclude the existence of a *vector potential* $\mathcal{A} : \mathbb{R}^3 \to \mathbb{R}^3$ with $\mathcal{B} = -\operatorname{curl} \mathcal{A}$ in D. Substituting this into Ampere's Law yields (for homogeneous media Ω) after multiplication with μ_0 the equation

$$-\mu_0 \mathcal{J} = \operatorname{curl} \operatorname{curl} \mathcal{A} = \nabla \operatorname{div} \mathcal{A} - \Delta \mathcal{A}.$$

Since $\operatorname{curl} \nabla = 0$ we can add gradients ∇u to $\mathcal{A}$ without changing $\mathcal{B}$. We will see later that we can choose u such that the resulting potential $\mathcal{A}$ satisfies $\operatorname{div} \mathcal{A} = 0$. This choice of normalization is called *Coulomb gauge* named after Charles Augustin de Coulomb (1736–1806). With this normalization we get the Poisson equation

$$\Delta \mathcal{A} = \mu_0 \mathcal{J}$$

also in magnetostatics. We note that in this case the Laplacian is vector valued and has to be taken componentwise.

1.3.3 Time-Harmonic Fields

For our purpose of considering wave phenomena the most important situation are time-harmonic fields. Under the assumptions that the fields allow a Fourier transformation in time we set

$$E(x; \omega) = (\mathcal{F}_t \mathcal{E})(x; \omega) = \int_{\mathbb{R}} \mathcal{E}(x, t) \, e^{i\omega t} \, dt,$$

$$H(x; \omega) = (\mathcal{F}_t \mathcal{H})(x; \omega) = \int_{\mathbb{R}} \mathcal{H}(x, t) \, e^{i\omega t} \, dt,$$

etc. We note that the fields E, H, etc. are now complex valued; that is, $E(\cdot; \omega), H(\cdot; \omega) : \mathbb{R}^3 \to \mathbb{C}^3$ and also all other Fourier transformed fields. Although they are vector fields we denote them by capital Latin letters only. According to $\mathcal{F}_t(u') = -i\omega \mathcal{F}_t u$ Maxwell's equations transform into the *time-harmonic Maxwell's equations*

$$-i\omega B + \operatorname{curl} E = 0,$$

$$i\omega D + \operatorname{curl} H = \sigma E + J_e,$$

$$\operatorname{div} D = \rho,$$

$$\operatorname{div} B = 0.$$

Remark: The time-harmonic Maxwell system can also be derived from the assumption that all fields behave periodically with respect to time with the same frequency ω. Then the complex valued functions $\mathcal{E}(x,t) = e^{-i\omega t}E(x)$, $\mathcal{H}(x,t) = e^{-i\omega t}H(x)$, etc., and their real and imaginary parts satisfy the time-harmonic Maxwell system.

With the constitutive equations $D = \varepsilon E$ and $B = \mu H$ we arrive at

$$\operatorname{curl} E - i\omega\mu H = 0 \,, \tag{1.7a}$$

$$\operatorname{curl} H + (i\omega\varepsilon - \sigma)E = J_e \,, \tag{1.7b}$$

$$\operatorname{div}(\varepsilon E) = \rho \,, \tag{1.7c}$$

$$\operatorname{div}(\mu H) = 0 \,. \tag{1.7d}$$

Assuming (for simplicity only) additionally an isotropic medium we can eliminate H or E from (1.7a) and (1.7b) which yields

$$\operatorname{curl}\left(\frac{1}{i\omega\mu}\operatorname{curl} E\right) + (i\omega\varepsilon - \sigma)\,E = J_e \,. \tag{1.8}$$

and

$$\operatorname{curl}\left(\frac{1}{i\omega\varepsilon - \sigma}\operatorname{curl} H\right) + i\omega\mu\,H = \operatorname{curl}\left(\frac{1}{i\omega\varepsilon - \sigma}J_e\right), \tag{1.9}$$

respectively. Usually, one writes these equations in a slightly different way by introducing the constant values $\varepsilon_0 > 0$ and $\mu_0 > 0$ in vacuum and dimensionless, relative values $\mu_r(x), \varepsilon_r(x) \in \mathbb{R}$ and $\varepsilon_c(x) \in \mathbb{C}$, defined by

$$\mu_r = \frac{\mu}{\mu_0} \,, \qquad \varepsilon_r = \frac{\varepsilon}{\varepsilon_0} \,, \qquad \varepsilon_c = \varepsilon_r + i\,\frac{\sigma}{\omega\varepsilon_0} \,.$$

Then Eqs. (1.8) and (1.9) take the form

$$\operatorname{curl}\left(\frac{1}{\mu_r}\operatorname{curl} E\right) - k^2\varepsilon_c\,E = i\omega\mu_0 J_e \,,$$

$$\operatorname{curl}\left(\frac{1}{\varepsilon_c}\operatorname{curl} H\right) - k^2\mu_r\,H = \operatorname{curl}\left(\frac{1}{\varepsilon_c}J_e\right),$$

with the *wave number* $k = \omega\sqrt{\varepsilon_0\mu_0}$. We conclude from the second equation that $\operatorname{div}(\mu_r H) = 0$ because $\operatorname{div}\operatorname{curl}$ vanishes. For the electric field we obtain that $\operatorname{div}(\varepsilon_c E) = -\frac{i\omega\mu_0}{k^2}\operatorname{div} J_e = -\frac{i}{\omega\varepsilon_0}\operatorname{div} J_e$.

In vacuum we have $\varepsilon_c = 1$, $\mu_r = 1$ and therefore the equations reduce to

$$\operatorname{curl}\operatorname{curl} E - k^2\,E = i\omega\mu_0 J_e \,, \tag{1.10}$$

$$\operatorname{curl}\operatorname{curl} H - k^2\,H = \operatorname{curl} J_e \,. \tag{1.11}$$

Without external current density, $J_e = 0$, we obtain from $\operatorname{curl}\operatorname{curl} = \nabla\operatorname{div} - \Delta$ the *vector Helmholtz equations*

$$\Delta E + k^2 E = 0 \quad\text{and}\quad \Delta H + k^2 H = 0.$$

Obviously, the reduced problems considering E or H are symmetric and we conclude the following important lemma.

Lemma 1.3. *A vector field $E \in C^2(\Omega, \mathbb{C}^3)$ combined with $H := \frac{1}{i\omega\mu_0}\operatorname{curl}E$ provides a solution of the time-harmonic Maxwell system (1.7a)–(1.7d) for $J_e = 0$ in vacuum if and only if E is a divergence free solution of the vector Helmholtz equation; that is,*

$$\Delta E + k^2 E = 0 \quad\text{and}\quad \operatorname{div} E = 0 \quad\text{in } \Omega.$$

Analogously, a divergence free solution of the vector Helmholtz equation $H \in C^2(\Omega, \mathbb{C}^3)$ combined with $E := \frac{-1}{i\omega\varepsilon_0}\operatorname{curl}H$ leads to a solution of Maxwell's equations in vacuum.

The relationship between the Maxwell system and the vector Helmholtz equation remains true if we consider time-harmonic waves in any homogeneous medium because we only have to substitute μ_0 and ε_0 by complex valued constants μ and ε, respectively. In this complex valued case the wave number $k = \omega\sqrt{\mu\varepsilon}$ is chosen such that $\operatorname{Im} k > 0$ or $k > 0$. As an example for solutions of the Maxwell system in a homogeneous medium we consider plane waves.

Example 1.4. In the case of vacuum with $J_e = 0$ a short calculation shows that the fields

$$E(x) = p\,e^{ik\,d\cdot x} \quad\text{and}\quad H(x) = (p \times d)\,e^{ik\,d\cdot x}$$

are solutions of the homogeneous time-harmonic Maxwell equations (1.10), (1.11) provided d is a unit vector in $\mathbb{R}^3$ and $p \in \mathbb{C}^3$ with $p \cdot d = \sum_{j=1}^{3} p_j\,d_j = 0$. Such fields are called *plane time-harmonic fields* with *polarization vector* $p \in \mathbb{C}^3$ and *direction* d, because its wave fronts are planes perpendicular to d.

Additionally the following observation will be useful.

Lemma 1.5. *Let E be a divergence free solution of the vector Helmholtz equation in a domain D. Then $x \mapsto x \cdot E(x)$ is a solution of the scalar Helmholtz equation.*

Proof: By the vector Helmholtz equation and $\operatorname{div} E = 0$ we obtain

$$\Delta\big(x \cdot E(x)\big) = \sum_{j=1}^{3} \operatorname{div} \nabla\big(x_j E_j(x)\big) = \sum_{j=1}^{3} \operatorname{div}\big[E_j(x)\, e^{(j)} + x_j \nabla E_j(x)\big]$$

$$= \sum_{j=1}^{3}\left[2\frac{\partial E_j}{\partial x_j}(x) + x_j \Delta E_j(x)\right] = 2\operatorname{div} E(x) - k^2 x \cdot E(x)$$

$$= -k^2 x \cdot E(x)$$

where $e^{(j)}$ denotes the jth cartesian coordinate unit vector. □

As in the stationary situation also the time-harmonic Maxwell equations in homogeneous media can be treated with methods from potential theory. We make the assumption $\varepsilon_c = 1$, $\mu_r = 1$ and consider (1.10) and (1.11). Taking the divergence of these equations yields $\operatorname{div} H = 0$ and $k^2 \operatorname{div} E = -i\omega\mu_0 \operatorname{div} J_e$; that is, $\operatorname{div} E = -(i/\omega\varepsilon_0)\operatorname{div} J_e$. Comparing this to (1.7c) yields the time-harmonic version of the equation of continuity

$$\operatorname{div} J_e = i\omega\rho.$$

With the vector identity $\operatorname{curl}\operatorname{curl} = -\Delta + \operatorname{div}\nabla$ Eqs. (1.10) and (1.11) can be written as

$$\Delta E + k^2 E = -i\omega\mu_0 J_e + \frac{1}{\varepsilon_0}\nabla\rho, \tag{1.12}$$

$$\Delta H + k^2 H = -\operatorname{curl} J_e. \tag{1.13}$$

Let us consider the magnetic field first and introduce the *magnetic Hertz potential*: The equation $\operatorname{div} H = 0$ implies the existence of a vector potential A with $H = \operatorname{curl} A$. Thus (1.13) takes the form

$$\operatorname{curl}(\Delta A + k^2 A) = -\operatorname{curl} J_e$$

and we obtain

$$\Delta A + k^2 A = -J_e + \nabla\varphi \tag{1.14}$$

for some scalar field φ. On the other hand, if A and φ satisfy (1.14), then

$$H = \operatorname{curl} A \quad\text{and}\quad E = -\frac{1}{i\omega\varepsilon_0}(\operatorname{curl} H - J_e) = i\omega\mu_0 A - \frac{1}{i\omega\varepsilon_0}\nabla(\operatorname{div} A - \varphi)$$

satisfies the Maxwell system (1.7a)–(1.7d).

Analogously, we can introduce *electric Hertz potentials* if $J_e = 0$. Because $\operatorname{div} E = 0$ there exists a vector potential A with $E = \operatorname{curl} A$. Substituting this into (1.12) yields

$$\operatorname{curl}(\Delta A + k^2 A) = 0$$

and we obtain

$$\Delta A + k^2 A = \nabla\varphi \tag{1.15}$$

for some scalar field φ. On the other hand, if A and φ satisfy (1.15), then

$$E = \operatorname{curl} A \quad \text{and} \quad H = \frac{1}{i\omega\mu_0} \operatorname{curl} E = -i\omega\varepsilon_0 A + \frac{1}{i\omega\mu_0} \nabla(\operatorname{div} A - \varphi)$$

satisfies the Maxwell system (1.7a)–(1.7d). In any case we end up with an inhomogeneous vector Helmholtz equation.

As a particular example we may take a magnetic Hertz vector A of the form $A(x) = u(x)\,\hat{z}$ with a scalar solution u of the two-dimensional Helmholtz equation and the unit vector $\hat{z} = (0,0,1)^\top \in \mathbb{R}^3$. Then

$$H = \operatorname{curl}(u\hat{z}) = \left(\frac{\partial u}{\partial x_2}, -\frac{\partial u}{\partial x_1}, 0 \right)^\top,$$

$$E = i\omega\mu_0 \,\hat{z} + \frac{1}{-i\omega\varepsilon_0} \nabla(\partial u/\partial x_3).$$

If u is independent of x_3, then E has only a x_3-component, and the vector Helmholtz equation for A reduces to a scalar Helmholtz equation for the potential u. The situation that E has only one non-zero component is called electric mode, *E-mode*, or transverse-magnetic mode, *TM-mode*. Analogously, also the *H-mode* or *TE-mode* is considered if H consists of only one non-zero component satisfying the scalar Helmholtz equation.

1.4 Boundary and Radiation Conditions

Maxwell's equations hold only in regions with smooth parameter functions ε_r, μ_r, and σ. If we consider a situation in which a surface S separates two homogeneous media from each other, the constitutive parameters ε, μ, and σ are no longer continuous but piecewise continuous with finite jumps on S. While on both sides of S Maxwell's equations (1.7a)–(1.7d) hold, the presence of these jumps implies that the fields satisfy certain conditions on the surface.

To derive the mathematical form of this behavior, the transmission, and boundary conditions, we apply the law of induction (1.2) to a narrow rectangle-like surface R, containing the normal ν to the surface S and whose long sides C_+ and C_- are parallel to S and are on the opposite sides of it, see Fig. 1.1.

When we let the height of the narrow sides, AA' and BB', approach zero then C_+ and C_- approach a curve C on S, the surface integral $\frac{\partial}{\partial t}\int_R \mathcal{B} \cdot n\,ds$ will

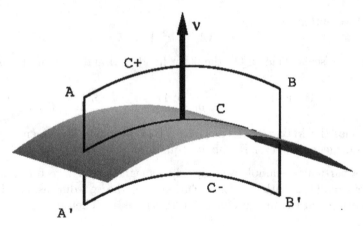

Fig. 1.1 Derivation of conditions at boundaries and interfaces from Maxwell's equations

vanish in the limit because the field remains finite. Note that the normal n is the normal to R lying in the tangential plane of S. Hence, the line integrals $\int_C \boldsymbol{\mathcal{E}}_+ \cdot \tau \, d\ell$ and $\int_C \boldsymbol{\mathcal{E}}_- \cdot \tau \, d\ell$ must be equal. Since the curve C is arbitrary the integrands $\boldsymbol{\mathcal{E}}_+ \cdot \tau$ and $\boldsymbol{\mathcal{E}}_- \cdot \tau$ coincide on every arc C; that is,

$$\nu \times \boldsymbol{\mathcal{E}}_+ \; - \; \nu \times \boldsymbol{\mathcal{E}}_- \; = \; 0 \quad \text{on } S. \tag{1.16}$$

A similar argument holds for the magnetic field in (1.1) if the current distribution $\boldsymbol{J} = \sigma \boldsymbol{\mathcal{E}} + \boldsymbol{J}_e$ remains finite. In this case, the same arguments lead to the boundary condition.

$$\nu \times \boldsymbol{\mathcal{H}}_+ \; - \; \nu \times \boldsymbol{\mathcal{H}}_- \; = \; 0 \quad \text{on } S. \tag{1.17}$$

If, however, the external current distribution is a surface current; that is, if $\boldsymbol{J}_e$ is of the form $\boldsymbol{J}_e(x + \tau\nu(x)) = \boldsymbol{J}_s(x)\delta(\tau)$ for small τ and $x \in S$ where δ denotes the delta distribution and with tangential surface field $\boldsymbol{J}_s$ and σ is finite, then the surface integral $\int_R \boldsymbol{J}_e \cdot n \, ds$ will tend to $\int_C \boldsymbol{J}_s \cdot n \, d\ell$, and so the boundary condition is

$$\nu \times \boldsymbol{\mathcal{H}}_+ \; - \; \nu \times \boldsymbol{\mathcal{H}}_- \; = \; \boldsymbol{J}_s \quad \text{on } S. \tag{1.18}$$

We will call (1.16) and (1.17) or (1.18) *transmission boundary conditions*.

A special and very important case is that of a *perfectly conducting medium* with boundary S. Such a medium is characterized by the fact that the electric field vanishes inside this medium, and (1.16) reduces to

$$\nu \times \boldsymbol{\mathcal{E}} \; = \; 0 \quad \text{on } S.$$

In realistic situations, of course, the exact form of this equation never occurs. Nevertheless, it is a common model for the case of a very large conductivity.

Another important case is the *impedance-* or *Leontovich boundary condition*

$$\nu \times \mathcal{H} = \lambda \nu \times (\mathcal{E} \times \nu) \quad \text{on } S$$

for some non-negative impedance function λ which, under appropriate conditions, may be considered as an approximation of the transmission conditions.

Of course these boundary conditions occur also in the time-harmonic case for the fields which we denote by capital Latin letters.

The situation is different for the normal components $\mathcal{E} \cdot \nu$ and $\mathcal{H} \cdot \nu$. We consider Gauss' Electric and Magnetic Laws and choose Ω to be a box which is separated by a surface S into two parts Ω_1 and Ω_2. We apply (1.3) first to all of Ω and then to Ω_1 and Ω_2 separately. The addition of the last two formulas and the comparison with the first yields that the normal component $\mathcal{D} \cdot \nu$ has to be continuous. Analogously we obtain $\mathcal{B} \cdot \nu$ to be continuous at S, if we consider (1.4). With the constitutive equations one gets

$$\nu \cdot (\varepsilon_{r,1}\mathcal{E}_1 - \varepsilon_{r,2}\mathcal{E}_2) = 0 \text{ on } S \quad \text{and} \quad \nu \cdot (\mu_{r,1}\mathcal{H}_1 - \mu_{r,2}\mathcal{H}_2) = 0 \text{ on } S.$$

Conclusion 1.6 *The normal components of $\mathcal{E}$ and/or $\mathcal{H}$ are not continuous at interfaces where ε_c and/or μ_r have jumps.*

Finally, we specify the boundary conditions to the E- and H-modes defined above (see p. 11). We assume that the surface S is an infinite cylinder in x_3-direction with constant cross section. Furthermore, we assume that the volume current density J vanishes near the boundary S and that the surface current densities take the form $J_s = j_s \hat{z}$ for the E-mode and $J_s = j_s(\nu \times \hat{z})$ for the H-mode. We use the notation $[v] := v|_+ - v|_-$ for the jump of the function v at the boundary. Then in the E-mode we obtain the transmission boundary condition

$$[u] = 0, \qquad \left[(\sigma - i\omega\varepsilon)\frac{\partial u}{\partial \nu}\right] = -j_s \quad \text{on } S,$$

and in the H-mode we get

$$[u] = j_s, \qquad \left[\mu \frac{\partial u}{\partial \nu}\right] = 0 \quad \text{on } S.$$

In scattering theory the solutions live in the unbounded exterior of a bounded domain D. In these situations the behavior of electromagnetic fields at infinity has to be taken into account. As an example we consider the fields in the case of a *Hertz dipole* at the origin.

Example 1.7. In the case of plane waves (see Example 1.4) the wave fronts are planes. Now we look for waves with *spherical wave fronts*. A direct computation shows that

$$\Phi(x) = \frac{1}{4\pi} \frac{e^{ik|x|}}{|x|}, \quad x \in \mathbb{R}^3 \setminus \{0\},$$

is a solution of the Helmholtz equation in $\mathbb{R}^3 \setminus \{0\}$. It is called the *fundamental solution* of the Helmholtz equation at the origin (see Definition 3.1) which is essential as we will see in Chap. 3. Physically it can be interpreted as the solution generated by a point source at the origin similar to the electrostatic case (see Example 1.2).

Furthermore, defining

$$H(x) = \mathrm{curl}\big[\Phi(x)\, p\big] = \frac{1}{4\pi} \mathrm{curl}\left(\frac{e^{ik|x|}}{|x|} p \right) = \frac{1}{4\pi} \nabla \frac{e^{ik|x|}}{|x|} \times p$$

for some constant vector $p \in \mathbb{C}^3$, we obtain from the identity $\mathrm{curl}\,\mathrm{curl} = \nabla \mathrm{div} - \Delta$ and the fact that Φ solves the Helmholtz equation,

$$\mathrm{curl}\, H(x) = \nabla \mathrm{div}\big[\Phi(x)\, p\big] - \Delta\big[\Phi(x)\, p\big] = \nabla \mathrm{div}\big[\Phi(x)\, p\big] + k^2\, \Phi(x)\, p$$

and thus by (A.3) $\mathrm{curl}\,\mathrm{curl}\, H = k^2\, H$. Therefore H and $E = \frac{i}{\omega\varepsilon_0} \mathrm{curl}\, H$ constitute a solution of the time-harmonic Maxwell equations in vacuum (see Lemma 1.3).

Computing the gradient

$$\nabla\Phi(x) = \frac{1}{4\pi} \left(ik - \frac{1}{|x|} \right) \frac{e^{ik|x|}}{|x|} \frac{x}{|x|} = ik\, \Phi(x) \frac{x}{|x|} - \frac{e^{ik|x|}}{4\pi|x|^2} \frac{x}{|x|} \quad (1.19)$$

we obtain by recalling the frequency $\omega > 0$ and wave number $k = \omega\sqrt{\varepsilon_0\mu_0}$ that the time-dependent magnetic field has the form

$$\mathcal{H}(x,t) = H(x)\, e^{-i\omega t} = \frac{1}{4\pi} \left(\frac{x}{|x|} \times p \right) \left(\frac{ik}{|x|} - \frac{1}{|x|^2} \right) e^{i(k|x|-\omega t)}$$

of the *Hertz dipole* centered at the origin with dipole moment $p\, e^{-i\omega t}$.

We observe that all of the functions $\mathcal{E}, \mathcal{H}, \Phi$ of this example decay as $1/|x|$ as $|x|$ tends to infinity. This asymptotic decay is not sufficient in describing a scattered field, because we also obtain solutions of the Helmholtz equation and the Maxwell equations, respectively, with this asymptotic behavior if we replace in the last example the wave number k by $-k$.

To distinguish these fields we must consider the factor $e^{i(k|x|-\omega t)}$ and compare it $e^{i(-k|x|-\omega t)}$ for the case of $-k$. In the first case we obtain "outgoing" wave fronts while in the second case where the wave number is negative we obtain "ingoing" wave fronts. For the scattering of electromagnetic waves the scattered waves have to be outgoing waves. Thus, it is required to exclude the second ones by additional conditions which are called *radiation conditions*.

From (1.19) we observe that the two cases can be distinguished by subtracting $ik\Phi$. This motivates a general characterization of radiating solutions u of the Helmholtz equation by the *Sommerfeld radiation condition*, which is

$$\lim_{|x|\to\infty} |x| \left(\frac{x}{|x|} \cdot \nabla u - iku \right) = 0.$$

Similarly, we find a condition for radiating electromagnetic fields from the behavior of the Hertz dipole. Computing

$$E(x) = \frac{i}{\omega\varepsilon_0} \operatorname{curl} H(x)$$

$$= \frac{i}{4\pi\omega\varepsilon_0} \left[k^2 \left(\frac{x}{|x|} \times p \right) \times \frac{x}{|x|} + \left(\frac{1}{|x|^2} - \frac{ik}{|x|} \right) \left(\frac{3x \cdot p}{|x|} \frac{x}{|x|} - p \right) \right] \frac{e^{ik|x|}}{|x|}$$

we conclude

$$\lim_{|x|\to 0} \left[\sqrt{\varepsilon_0} E(x) \times x + \sqrt{\mu_0} |x| H(x) \right]$$

$$= \lim_{|x|\to 0} \left[\frac{\sqrt{\mu_0}}{4\pi} \left(2 + \frac{1}{ik|x|} \right) \left(\frac{x}{|x|} \times p \right) \frac{e^{ik|x|}}{|x|} \right] = 0.$$

Analogously, we obtain $\lim_{|x|\to 0} \left[\sqrt{\mu_0} H(x) \times x - \sqrt{\varepsilon_0} |x| E(x) \right] = 0$. We note that none of the conditions are satisfied if we replace k by $-k$. The radiation condition

$$\lim_{|x|\to\infty} \left[\sqrt{\mu_0} H(x) \times x - |x| \sqrt{\varepsilon_0} E(x) \right] = 0 \tag{1.20a}$$

or

$$\lim_{|x|\to\infty} \left[\sqrt{\varepsilon_0} E(x) \times x + |x| \sqrt{\mu_0} H(x) \right] = 0, \tag{1.20b}$$

are called *Silver–Müller radiation condition* for time-harmonic electromagnetic fields.

Later (in Chap. 3) we will show that these radiation conditions are sufficient for the existence of unique solutions of scattering problems. Additionally from the representation theorem in Chap. 3 we will prove equivalent formulations and the close relationship of the Sommerfeld and the Silver–Müller radiation condition. Additionally, the limiting absorption principle will give another justification for this definition of radiating solutions.

Finally we discuss the energy of scattered waves. In general the energy density of electromagnetic fields in a linear medium is given by $\frac{1}{2}(\boldsymbol{\mathcal{E}}\cdot\boldsymbol{\mathcal{D}}+\boldsymbol{\mathcal{H}}\cdot\boldsymbol{\mathcal{B}})$. Thus, from Maxwell's equations, the identity (A.9), and the divergence theorem (Theorem A.11) in a region Ω we obtain

$$\frac{\partial}{\partial t}\left(\frac{1}{2}\int_{\Omega}\boldsymbol{\mathcal{E}}\cdot\boldsymbol{\mathcal{D}}+\boldsymbol{\mathcal{H}}\cdot\boldsymbol{\mathcal{B}}\,dx\right) = \int_{\Omega}\boldsymbol{\mathcal{E}}\cdot(\operatorname{curl}\boldsymbol{\mathcal{H}}-\boldsymbol{\mathcal{J}})-\boldsymbol{\mathcal{H}}\cdot\operatorname{curl}\boldsymbol{\mathcal{E}}\,dx$$

$$= \int_{\Omega}\operatorname{div}(\boldsymbol{\mathcal{E}}\times\boldsymbol{\mathcal{H}})-\boldsymbol{\mathcal{E}}\cdot\boldsymbol{\mathcal{J}}\,dx$$

$$= \int_{\partial\Omega}(\boldsymbol{\mathcal{E}}\times\boldsymbol{\mathcal{H}})\cdot\nu\,ds\ -\ \int_{\Omega}\boldsymbol{\mathcal{E}}\cdot\boldsymbol{\mathcal{J}}\,dx\,.$$

This conservation law for the energy of electromagnetic fields is called *Poynting's Theorem*. Physically, the right-hand side is read as the sum of the energy flux through the surface $\partial\Omega$ given by the *Poynting vector*, $\boldsymbol{\mathcal{E}}\times\boldsymbol{\mathcal{H}}$, and the electrical work of the fields with the electrical power $\boldsymbol{\mathcal{J}}\cdot\boldsymbol{\mathcal{E}}$.

Let us consider the Poynting theorem in case of time-harmonic fields with frequency $\omega>0$ in vacuum, i.e. $\boldsymbol{\mathcal{J}}=0$, $\varepsilon=\varepsilon_0$, $\mu=\mu_0$. Substituting $\boldsymbol{\mathcal{E}}(x,t)=\operatorname{Re}\left(E(x)e^{-i\omega t}\right)=\frac{1}{2}\left(E(x)e^{-i\omega t}+\overline{E(x)}e^{i\omega t}\right)$ and $\boldsymbol{\mathcal{H}}(x,t)=\frac{1}{2}\left(H(x)e^{-i\omega t}+\overline{H(x)}e^{i\omega t}\right)$ into the left-hand side of Poynting's theorem lead to

$$\frac{\partial}{\partial t}\left(\frac{1}{2}\int_{\Omega}\varepsilon_0\,|\boldsymbol{\mathcal{E}}(x,t)|^2+\mu_0\,|\boldsymbol{\mathcal{H}}(x,t)|^2\,dx\right)$$

$$= \frac{\partial}{\partial t}\left(\frac{1}{4}\int_{\Omega}\varepsilon_0\left[\operatorname{Re}\left(E(x)^2e^{-2i\omega t}\right)+|E(x)|^2\right]\right.$$

$$\left.+\ \mu_0\left[\operatorname{Re}\left(H(x)^2e^{-2i\omega t}\right)+|H(x)|^2\right]dx\right)$$

$$= -\frac{i\omega}{2}\int_{\Omega}\varepsilon_0\,\operatorname{Re}\left(E(x)^2e^{-2i\omega t}\right)+\mu_0\,\operatorname{Re}\left(H(x)^2e^{-2i\omega t}\right)dx$$

where we wrote $E(x)^2$ and $H(x)^2$ for $E(x)\cdot E(x)\in\mathbb{C}$ and $H(x)\cdot H(x)\in\mathbb{C}$, respectively. On the other hand, we compute the flux term as

$$\int_{\partial\Omega}(\boldsymbol{\mathcal{E}}\times\boldsymbol{\mathcal{H}})\cdot\nu\,ds = \frac{1}{2}\int_{\partial\Omega}\operatorname{Re}\left(E\times\overline{H}\right)\cdot\nu\,ds + \frac{1}{2}\int_{\partial\Omega}\operatorname{Re}\left(E\times H\,e^{-2i\omega t}\right)\cdot\nu\,ds\,.$$

By the Poynting Theorem the two integrals coincide for all t. Thus we conclude for the time independent term

$$\operatorname{Re}\int_{\partial\Omega}(E\times\overline{H})\cdot\nu\,ds = 0\,.$$

The vector field $E \times \overline{H}$ is called the *complex Poynting vector*. If Ω has the form $\Omega = B(0, R) \setminus D$ for a bounded domain D with sufficiently smooth boundary contained in the ball $B(0, R)$ of radius R, we observe conservation of energy in the form

$$\mathrm{Re} \int_{|x|=R} \nu \cdot (E \times \overline{H}) \, ds = \mathrm{Re} \int_{\partial D} \nu \cdot (E \times \overline{H}) \, ds \, .$$

Furthermore, from this identity we obtain

$$2\sqrt{\varepsilon_0 \mu_0} \; \mathrm{Re} \int_{\partial D} \nu \cdot (E \times \overline{H}) \, ds$$

$$= \int_{|x|=R} \varepsilon_0 \, |E|^2 + \mu_0 \, |H \times \nu|^2 \, ds \; - \int_{|x|=R} \left| \sqrt{\mu_0} \, \overline{H} \times \nu - \sqrt{\varepsilon_0} \, E \right|^2 \, ds \, .$$

If we additionally assume radiating fields, the Silver–Müller radiation condition (1.20a) or (1.20b) implies

$$\lim_{R \to \infty} \int_{|x|=R} \left| \sqrt{\mu_0} \, \overline{H} \times \nu - \sqrt{\varepsilon_0} \, E \right|^2 \, ds = 0$$

which yields boundedness of $\int_{|x|=R} |E|^2 \, ds$ and, analogously, $\int_{|x|=R} |H|^2 \, ds$ as R tends to infinity.

1.5 The Reference Problems

After this introduction into the mathematical description of electromagnetic waves the aim of the textbook becomes more obvious. In general we can distinguish (at least) three common approaches which lead to existence results of boundary value problems for linear partial differential equations: expanding solutions into spherical wave functions by separation of variables techniques, reformulation and treatment of a given boundary value problem in terms of integral equations in Banach spaces of functions on the interfaces, and the reformulation of the boundary value problem as a variational equation in Hilbert spaces of functions in the domain. It is the aim of this monograph to discuss all of these common methods in the case of time-harmonic Maxwell's equations.

We already observed the close connection of the Maxwell system with the scalar Helmholtz equation. Therefore, before we treat the more complicated situation of the Maxwell system we investigate the methods for this scalar elliptic partial differential equation in detail. Thus, the structure of all following chapters will be similar: we first discuss the technique in the case of the Helmholtz equation and then we extend it to boundary value problems for electromagnetic fields.

As we have mentioned already in the preface it is not our aim to present a collection of *all* or even some interesting boundary value problems. Instead, we present the ideas for two classical reference problems only which we introduce next.

1.5.1 Scattering by a Perfect Conductor

The first one is the scattering of electromagnetic waves in vacuum by a perfect conductor (see Fig. 1.2): Given a bounded region D with sufficiently smooth boundary ∂D and exterior unit normal vector $\nu(x)$ at $x \in \partial D$ and some solution E^{inc} and H^{inc} of the unperturbed time-harmonic Maxwell system

$$\operatorname{curl} E^{inc} - i\omega\mu_0 H^{inc} = 0 \text{ in } \mathbb{R}^3, \quad \operatorname{curl} H^{inc} + i\omega\varepsilon_0 E^{inc} = 0 \text{ in } \mathbb{R}^3,$$

the problem is to determine E, H of the Maxwell system

$$\operatorname{curl} E - i\omega\mu_0 H = 0 \text{ in } \mathbb{R}^3 \setminus \overline{D}, \quad \operatorname{curl} H + i\omega\varepsilon_0 E = 0 \text{ in } \mathbb{R}^3 \setminus \overline{D},$$

such that E satisfies the boundary condition

$$\nu \times E = 0 \quad \text{on } \partial D$$

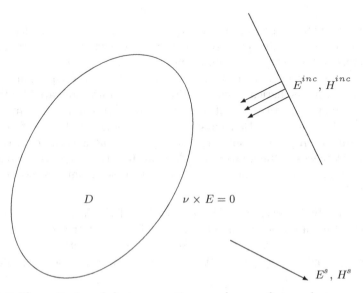

Fig. 1.2 The scattering of electromagnetic waves by a perfect conductor

and both, E and H, can be decomposed into $E = E^s + E^{inc}$ and $H = H^s + H^{inc}$ in $\mathbb{R}^3 \setminus \overline{D}$ with some scattered field E^s, H^s which satisfy the Silver–Müller radiation condition

$$\lim_{|x| \to \infty} |x| \left(\sqrt{\mu_0}\, H^s(x) \times \frac{x}{|x|} - \sqrt{\varepsilon_0}\, E^s(x) \right) = 0$$

$$\lim_{|x| \to \infty} |x| \left(\sqrt{\varepsilon_0}\, E^s(x) \times \frac{x}{|x|} + \sqrt{\mu_0}\, H^s(x) \right) = 0$$

uniformly with respect to all directions $x/|x|$.

1.5.2 A Perfectly Conducting Cavity

For the second reference problem we consider $D \subseteq \mathbb{R}^3$ to be a bounded domain with sufficiently smooth boundary ∂D and exterior unit normal vector $\nu(x)$ at $x \in \partial D$ (see Fig. 1.3). Furthermore, functions μ, ε, and σ are given on D and some source $J_e : D \to \mathbb{C}^3$. Then the problem is to determine a solution (E, H) of the time-harmonic Maxwell system

$$\operatorname{curl} E - i\omega\mu H = 0 \quad \text{in } D, \tag{1.21a}$$
$$\operatorname{curl} H + (i\omega\varepsilon - \sigma)E = J_e \quad \text{in } D, \tag{1.21b}$$

with the boundary condition

$$\nu \times E = 0 \quad \text{on } \partial D. \tag{1.21c}$$

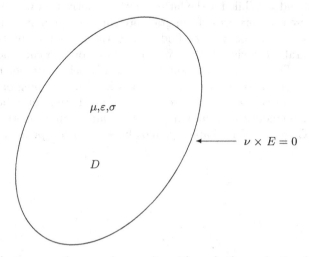

Fig. 1.3 Electromagnetic waves in a cavity with perfectly conducting boundary

Of course, for general $\mu, \varepsilon, \sigma \in L^\infty(\mathbb{R}^3)$ we have to give first a correct interpretation of the differential equations by a so-called weak formulation, which
will be presented in detail in Chap. 4.

Throughout, we will have these two reference problems in mind for the whole
presentation. We will start with constant electric parameters inside or outside a ball. Then we can expect radially symmetric solutions which can be
computed by separation of variables in terms of spherical coordinates. This
approach will be worked out in Chap. 2. It will lead us to a better understand-
ing of electromagnetic waves from its expansion into spherical wave functions.
In particular, we can see explicitly in which way boundary data are attained.

In Chap. 3 we will use the fundamental solution of the scalar Helmholtz equation to represent electromagnetic waves by integrals over the boundary ∂D
of the region D. These integral representations by boundary potentials are
the basis for deriving integral equations on ∂D. We prefer to choose the "indirect" approach; that is, to search for the solution in terms of potentials
with densities which are determined by the boundary data through a boundary integral equation. To solve this we will apply the Riesz–Fredholm theory.
In this way we have to prove the existence of unique solutions of the first
reference problem for any perfectly conducting smooth scattering obstacle.

The cavity problem will be investigated in Chap. 4. The treatment by a variational approach requires the introduction of suitable Sobolev spaces. The
Helmholtz decomposition makes it possible to transfer the ideas of the simpler scalar Helmholtz equation to the Maxwell system. The Lax–Milgram
theorem in Hilbert spaces is the essential tool to establish an existence result
for the second reference problem.

The reason for including Chap. 5 into this monograph is different than for
Chaps. 3 and 4. While for the latter ones our motivation was the teaching as-
pect (these chapters arose from graduate courses) the motivation for Chap. 5
is that we were not able to find such a thorough presentation of boundary integral methods for Maxwell's equations on Lipschitz domains in any
textbook. The integral equation methods themselves are not much different
from the classical ones on smooth boundaries. The mapping properties of the
boundary operators, however, require a detailed study of Sobolev spaces on
Lipschitz boundaries. In this Chap. 5 we use and combine methods of Chaps. 3
and 4. Therefore, this chapter cannot be studied independently of Chaps. 3
and 4.

Chapter 2
Expansion into Wave Functions

This chapter, which is totally independent of the remaining parts of this monograph,[1] studies the fact that the solutions of the scalar Helmholtz equation or the vectorial Maxwell system in balls can be expanded into certain special "wave functions." We begin by expressing the Laplacian in spherical coordinates and search for solutions of the scalar Laplace equation or Helmholtz equation by separation of the (spherical) variables. It will turn out that the spherical parts are eigensolutions of the Laplace–Beltrami operator while the radial part solves an equation of Euler type for the Laplace equation and the spherical Bessel differential equation for the case of the Helmholtz equation. The solutions of these differential equations will lead to spherical harmonics and spherical Bessel and Hankel functions. We will investigate these special functions in detail and derive many important properties. The main goal is to express the solutions of the interior and exterior boundary value problems as series of these wave functions. As always in this monograph we first present the analysis for the scalar case of the Laplace equation and the Helmholtz equation before we consider the more complicated case of Maxwell's equations.

2.1 Separation in Spherical Coordinates

The starting point of our investigation is the boundary value problem inside (or outside) the "simple" geometry of a ball in the case of a homogeneous medium. We are interested in solutions u of the Laplace equation

---

[1] Except for the proof of a radiation condition.

© Springer International Publishing Switzerland 2015
A. Kirsch, F. Hettlich, *The Mathematical Theory of Time-Harmonic Maxwell's Equations*, Applied Mathematical Sciences 190,
DOI 10.1007/978-3-319-11086-8_2

or Helmholtz equation—and later of the time-harmonic Maxwell system—which can be separated into a radial part $v : \mathbb{R}_{>0} \to \mathbb{C}$ and a spherical part $K : S^2 \to \mathbb{C}$; that is,

$$u(x) = v(r)\,K(\hat{x}), \quad r > 0,\ \hat{x} \in S^2\,,$$

where $S^2 = \{x \in \mathbb{R}^3 : |x| = 1\}$ denotes the unit sphere in $\mathbb{R}^3$. Here, $r = |x| = \sqrt{x_1^2 + x_2^2 + x_3^2}$ and $\hat{x} = \frac{x}{|x|} \in S^2$ denote the spherical coordinates; that is,

$$x = \begin{pmatrix} r\cos\varphi\sin\theta \\ r\sin\varphi\sin\theta \\ r\cos\theta \end{pmatrix} \quad \text{with} \quad r \in \mathbb{R}_{>0},\ \varphi \in [0, 2\pi),\ \theta \in [0, \pi]\,,$$

and $\hat{x} = (\cos\varphi\sin\theta,\ \sin\varphi\sin\theta,\ \cos\theta)^\top$.

In the previous chapter we have seen already the importance of the differential operator Δ in the modeling of electromagnetic waves. It occurs directly in the stationary cases and also in the differential equations for the magnetic and electric Hertz potentials. Also, solutions of the full Maxwell system solve the vector Helmholtz equation in particular cases. Thus the representation of the Laplacian in spherical polar coordinates is of essential importance.

$$\Delta = \frac{1}{r^2}\frac{\partial}{\partial r}\left(r^2\frac{\partial}{\partial r}\right) + \frac{1}{r^2\sin^2\theta}\frac{\partial^2}{\partial\varphi^2} + \frac{1}{r^2\sin\theta}\frac{\partial}{\partial\theta}\left(\sin\theta\frac{\partial}{\partial\theta}\right) \quad (2.1)$$

$$= \frac{\partial^2}{\partial r^2} + \frac{2}{r}\frac{\partial}{\partial r} + \frac{1}{r^2\sin^2\theta}\frac{\partial^2}{\partial\varphi^2} + \frac{1}{r^2\sin\theta}\frac{\partial}{\partial\theta}\left(\sin\theta\frac{\partial}{\partial\theta}\right).$$

Definition 2.1. Functions $u : \Omega \to \mathbb{R}$ which satisfy the *Laplace equation* $\Delta u = 0$ in Ω are called *harmonic functions* in Ω. Complex valued functions are harmonic if their real and imaginary parts are harmonic.

The Laplace operator separates into a radial part, which is $\frac{1}{r^2}\frac{\partial}{\partial r}(r^2\frac{\partial}{\partial r}) = \frac{\partial^2}{\partial r^2} + \frac{2}{r}\frac{\partial}{\partial r}$, and the spherical part which is called the *Laplace–Beltrami operator*, a differential operator on the unit sphere.

Definition 2.2. The differential operator $\Delta_{S^2} : C^2(S^2) \to C(S^2)$ with representation

$$\Delta_{S^2} = \frac{1}{\sin^2\theta}\frac{\partial^2}{\partial\varphi^2} + \frac{1}{\sin\theta}\frac{\partial}{\partial\theta}\left(\sin\theta\frac{\partial}{\partial\theta}\right).$$

in spherical coordinates is called *spherical Laplace–Beltrami operator* on S^2.

With the notation of the Beltrami operator and $D_r = \frac{\partial}{\partial r}$ we obtain

$$\Delta = D_r^2 + \frac{2}{r}D_r + \frac{1}{r^2}\Delta_{S^2} = \frac{1}{r^2}D_r(r^2 D_r) + \frac{1}{r^2}\Delta_{S^2}\,.$$

Assuming that a potential or a component of a time-harmonic electric or magnetic field can be separated as $u(x) = u(r\hat{x}) = v(r)K(\hat{x})$ with $r = |x|$ and $\hat{x} = x/|x| \in S^2$, a substitution into the Helmholtz equation $\Delta u + k^2 u = 0$ for some $k \in \mathbb{C}$ leads to

$$0 = \Delta u(r\hat{x}) + k^2 u(r\hat{x})$$
$$= v''(r)\,K(\hat{x}) + \frac{2}{r}v'(r)\,K(\hat{x}) + \frac{1}{r^2}v(r)\,\Delta_{S^2}K(\hat{x}) + k^2 v(r)\,K(\hat{x})\,.$$

Thus we obtain, provided $v(r)K(\hat{x}) \neq 0$,

$$\frac{v''(r) + \frac{2}{r}v'(r)}{v(r)} + \frac{1}{r^2}\frac{\Delta_{S^2}K(\hat{x})}{K(\hat{x})} + k^2 = 0\,. \tag{2.2}$$

If there is a nontrivial solution of the supposed form, it follows the existence of a constant $\lambda \in \mathbb{C}$ such that v satisfies the ordinary differential equation

$$r^2 v''(r) + 2r\,v'(r) + \left(k^2 r^2 + \lambda\right) v(r) = 0 \quad \text{for } r > 0\,, \tag{2.3}$$

and K solves the partial differential equation

$$\Delta_{S^2} K = \lambda K \quad \text{on } S^2\,. \tag{2.4}$$

In the functional analytic language the previous equation describes the problem to determine eigenfunctions K and corresponding eigenvalues λ of the spherical Laplace–Beltrami operator Δ_{S^2}. This operator is self-adjoint and non-positive with respect to the $L^2(S^2)$-norm, see Exercise 2.2. Especially, we observe that $\lambda \in \mathbb{R}_{<0}$. We will explicitly construct a complete orthonormal system of eigenfunctions although the existence of such a system follows from functional analytic arguments as well because the resolvent of Δ_{S^2} is compact.

We observe that the parameter k appears in the equation for v only. The spherical equation (2.4) is independent of k, and the system of eigenfunctions will be used for both, the Laplace and the Helmholtz equation.

Using the explicit representation of the spherical Laplace–Beltrami operator transforms (2.4) into

$$\frac{\partial^2}{\partial\varphi^2}K(\theta,\varphi) + \sin\theta\,\frac{\partial}{\partial\theta}\left(\sin\theta\,\frac{\partial}{\partial\theta}\right)K(\theta,\varphi) = \lambda \sin^2\theta\,K(\theta,\varphi)$$

where we write $K(\theta,\varphi)$ for $K(\hat{x})$. Assuming eigenfunctions of the form $K(\theta,\varphi) = y_1(\varphi)\,y_2(\theta)$ leads to

$$\frac{y_1''(\varphi)}{y_1(\varphi)} + \frac{\sin\theta\left(\sin\theta\,y_2'(\theta)\right)'}{y_2(\theta)} - \lambda\sin^2\theta = 0 \tag{2.5}$$

provided $y_1(\varphi)y_2(\theta) \neq 0$. If the decomposition is valid, there exists a constant $\mu \in \mathbb{C}$ such that $y_1'' = \mu y_1$. Since we are interested in differentiable solutions u, the function y_1 must be 2π periodic. Using a fundamental system of the linear ordinary differential equation we conclude that $\mu = -m^2$ with $m \in \mathbb{Z}$, and the general solution of $y_1'' = \mu y_1$ is given by

$$c_1 e^{im\varphi} + c_2 e^{-im\varphi} = \tilde{c}_1 \cos(m\varphi) + \tilde{c}_2 \sin(m\varphi)$$

with arbitrary constants $c_1, c_2 \in \mathbb{C}$ and $\tilde{c}_1, \tilde{c}_2 \in \mathbb{C}$, respectively.

We assume that the reader is familiar with the basics of the classical Fourier theory. In particular, we recall that every function $f \in L^2(-\pi, \pi)$ allows an expansion in the form

$$f(t) = \sum_{m \in \mathbb{Z}} f_m e^{imt}$$

with Fourier coefficients

$$f_m = \frac{1}{2\pi} \int_{-\pi}^{\pi} f(s) e^{-ims} \, ds, \quad m \in \mathbb{Z}.$$

The convergence of the series must be understood in the L^2-sense, that is,

$$\int_{-\pi}^{\pi} \left| f(t) - \sum_{m=-N}^{M} f_m e^{imt} \right|^2 dt \longrightarrow 0, \quad N, M \to \infty.$$

Thus by the previous result we conclude that the possible functions $\{e^{imt} : m \in \mathbb{Z}\}$ constitute a complete orthonormal system in $L^2(-\pi, \pi)$.

Next we consider the dependence on θ. Using $\mu = -m^2$ turns (2.5) into

$$\sin \theta \left(\sin \theta \, y_2'(\theta)\right)' - \left(m^2 + \lambda \sin^2 \theta\right) y_2(\theta) = 0$$

for y_2. With the substitution $z = \cos \theta \in (-1, 1]$ and $w(z) = y_2(\theta)$ we arrive at the *associated Legendre differential equation*

$$((1 - z^2)w'(z))' - \left(\lambda + \frac{m^2}{1 - z^2}\right) w(z) = 0. \tag{2.6}$$

An investigation of this equation will be the main task of the next section leading to the so-called *Legendre polynomials* and to a complete orthonormal system of *spherical surface harmonics*. In particular, we will see that only for $\lambda = -n(n+1)$ for $n \in \mathbb{N}_0$ there exist smooth solutions.

Obviously, the radial part of separated solutions given by the ordinary differential equation (2.3) depends on the wave number k. For $k = 0$ and

$\lambda = -n(n+1)$ we obtain the *Euler equation*

$$r^2 v''(r) \ + \ 2r\,v'(r) \ - \ n(n+1)\,v(r) \ = \ 0. \tag{2.7}$$

By the ansatz $v(r) = r^\mu$ we compute $\mu(\mu+1) = n(n+1)$, which leads to the fundamental set of solutions

$$v_{1,n}(r) = r^n \quad \text{and} \quad v_{2,n}(r) = r^{-(n+1)}.$$

If we are interested in non-singular solutions of the Laplace equation inside a ball we have to choose $v_{1,n}(r) = r^n$. In the exterior of a ball $v_{2,n}(r) = r^{-(n+1)}$ will lead to solutions which decay as $|x|$ tends to infinity.

In the case of $k \neq 0$ we rewrite the differential equation (2.3) for $\lambda = -n(n+1)$ by the substitution $z = kr$ and $v(r) = \hat{v}(kr)$ into the *spherical Bessel differential equation*

$$z^2\,\hat{v}''(z) \ + \ 2z\,\hat{v}'(z) \ + \ \left[z^2 - n(n+1)\right]\hat{v}(z) \ = \ 0. \tag{2.8}$$

In Sect. 2.5 we will discuss the Bessel functions, which solve this differential equation. Combining spherical surface harmonics and the corresponding Bessel functions will lead to solutions of the Helmholtz equation and further on also of the Maxwell system.

2.2 Legendre Polynomials

Let us consider the case of harmonic functions u, that is $\Delta u = 0$. In the previous section we saw already that a separation in spherical coordinates leads to solutions of the form $u(x) = r^n (c_1\,e^{im\varphi} + c_2\,e^{-im\varphi})\,w(\cos\theta)$ where w is determined by the associated Legendre differential equation (2.6) with $\lambda \in \mathbb{R}_{<0}$.

We discuss this real ordinary differential equation (2.6) first for the special case $m = 0$; that is,

$$\frac{d}{dz}\left[(1-z^2)\,w'(z)\right] \ - \ \lambda\,w(z) \ = \ 0, \quad -1 < z < 1. \tag{2.9}$$

This is a *differential equation of Legendre type*.

The coefficients vanish at $z = \pm 1$. We are going to determine solutions which are continuous up to the boundary, thus $w \in C^2(-1,+1) \cap C[-1,+1]$.

Theorem 2.3. *(a) For $\lambda = -n(n+1)$, $n \in \mathbb{N} \cup \{0\}$, there exists exactly one solution $w \in C^2(-1,+1) \cap C[-1,+1]$ with $w(1) = 1$. We set $P_n = w$. The* *function P_n is a polynomial of degree n and is called* Legendre polynomial *of degree n. It satisfies the differential equation*

$$\frac{d}{dx}\left[(1-x^2)\,P_n'(x)\right] + n(n+1)\,P_n(x) = 0, \quad -1 \le x \le 1.$$

(b) If there is no $n \in \mathbb{N} \cup \{0\}$ with $\lambda = -n(n+1)$, then there is no nontrivial solution of (2.9) in $C^2(-1,+1) \cap C[-1,+1]$.

Proof: We make a—first only formal—ansatz for a solution in the form of a power series:

$$w(x) = \sum_{j=0}^{\infty} a_j\, x^j$$

and substitute this into the differential equation (2.9). A simple calculation shows that the coefficients have to satisfy the recursion

$$a_{j+2} = \frac{j(j+1)+\lambda}{(j+1)(j+2)}\, a_j, \quad j = 0, 1, 2, \ldots. \tag{2.10}$$

The ratio test, applied to

$$w_1(x) = \sum_{k=0}^{\infty} a_{2k}\, x^{2k} \quad \text{and} \quad w_2(x) = \sum_{k=0}^{\infty} a_{2k+1}\, x^{2k+1}$$

separately yields that the radius of convergence is one. Therefore, for any $a_0, a_1 \in \mathbb{R}$ the function w is an analytic solution of (2.9) in $(-1,1)$.

Case 1: There is no $n \in \mathbb{N} \cup \{0\}$ with $a_j = 0$ for all $j \ge n$; that is, the series does not reduce to a finite sum. We study the behavior of $w_1(x)$ and $w_2(x)$ as x tends to ± 1. First we consider w_1 and split w_1 in the form

$$w_1(x) = \sum_{k=0}^{k_0-1} a_{2k}\, x^{2k} + \sum_{k=k_0}^{\infty} a_{2k}\, x^{2k}$$

where k_0 is chosen such that $2k_0(2k_0+1)+\lambda > 0$ and $\left[(2k_0+1) - \lambda/2\right]/[(2k_0+1)(k_0+1)] \le 1/2$. Then a_{2k} does not change its sign anymore for $k \ge k_0$. Taking the logarithm of (2.10) for $j = 2k$ yields

$$\ln|a_{2(k+1)}| = \ln\left[\frac{2k(2k+1)+\lambda}{(2k+1)(2k+2)}\right] + \ln|a_{2k}|$$

$$= \ln\left[1 - \frac{(2k+1)-\lambda/2}{(2k+1)(k+1)}\right] + \ln|a_{2k}|.$$

Now we use the elementary estimate $\ln(1-u) \ge -u - 2u^2$ for $0 \le u \le 1/2$ which yields

$$\ln |a_{2(k+1)}| \geq -\frac{(2k+1)-\lambda/2}{(2k+1)(k+1)} - \left[\frac{(2k+1)-\lambda/2}{(2k+1)(k+1)}\right]^2 + \ln |a_{2k}|$$

$$\geq -\frac{1}{k+1} - \frac{c}{k^2} + \ln |a_{2k}|$$

for some $c > 0$. Therefore, we arrive at the following estimate for $\ln |a_{2k}|$:

$$\ln |a_{2k}| \geq -\sum_{j=k_0}^{k-1} \frac{1}{j+1} - c\sum_{j=k_0}^{k-1} \frac{1}{j^2} + \ln |a_{2k_0}| \quad \text{for } k \geq k_0.$$

Using

$$\sum_{j=k_0}^{k-1} \frac{1}{j+1} \leq \sum_{j=k_0}^{k-1} \int_j^{j+1} \frac{dt}{t} = \int_{k_0}^k \frac{dt}{t} = \ln k - \ln k_0$$

yields

$$\ln |a_{2k}| \geq -\ln k + \underbrace{\left[\ln k_0 - c\sum_{j=k_0}^{\infty} \frac{1}{j^2} + \ln |a_{2k_0}|\right]}_{= \hat{c}}$$

and thus

$$|a_{2k}| \geq \frac{\exp \hat{c}}{k} \quad \text{for all } k \geq k_0.$$

We note that a_{2k}/a_{2k_0} is positive for all $k \geq k_0$ and obtain

$$\frac{w_1(x)}{a_{2k_0}} \geq \frac{a_0}{a_{2k_0}} + \sum_{k=1}^{k_0-1} \left[\frac{a_{2k}}{a_{2k_0}} - \frac{\exp \hat{c}}{k\,|a_{2k_0}|}\right] x^{2k} + \frac{\exp \hat{c}}{|a_{2k_0}|} \sum_{k=1}^{\infty} \frac{1}{k} x^{2k}$$

$$\geq \tilde{c} - \frac{\exp \hat{c}}{|a_{2k_0}|} \ln(1 - x^2).$$

From this we observe that $w_1(x) \to +\infty$ as $x \to \pm 1$ or $w_1(x) \to -\infty$ as $x \to \pm 1$ depending on the sign of a_{2k_0}. By the same arguments one shows for positive a_{2k_0+1} that $w_2(x) \to +\infty$ as $x \to +1$ and $w_2(x) \to -\infty$ as $x \to -1$. For negative a_{2k_0+1} the roles of $+\infty$ and $-\infty$ have to be interchanged. In any case, the sum $w(x) = w_1(x) + w_2(x)$ is not bounded on $[-1, 1]$ which contradicts our requirement on the solution of (2.9). Therefore, this case cannot happen.

Case 2: There is $m \in \mathbb{N}$ with $a_j = 0$ for all $j \geq m$; that is, the series reduces to a finite sum. Let m be the smallest number with this property. From the recursion formula we conclude that $\lambda = -n(n+1)$ for $n = m - 2$ and, furthermore, that $a_0 = 0$ if n is odd and $a_1 = 0$ if n is even. In particular, w is a polynomial of degree n. We can normalize w by $w(1) = 1$ because

$w(1) \neq 0$. Indeed, if $w(1) = 0$, the differential equation

$$(1 - x^2)\, w''(x) \;-\; 2x\, w'(x) \;+\; n(n+1)\, w(x) \;=\; 0$$

for the polynomial w would imply $w'(1) = 0$. Differentiating the differential equation would yield $w^{(k)}(1) = 0$ for all $k \in \mathbb{N}$, a contradiction to $w \neq 0$. $\square$

In this proof of the theorem we have proven more than stated. We collect this as a corollary.

Corollary 2.4. *The Legendre polynomials* $P_n(x) = \sum_{j=0}^{n} a_j x^j$ *have the properties*

(a) P_n *is even for even n and odd for odd n.*

(b) $a_{j+2} = \dfrac{j(j+1) - n(n+1)}{(j+1)(j+2)}\, a_j, \quad j = 0,1,\ldots,n-2.$

(c) $\displaystyle\int_{-1}^{+1} P_n(x) P_m(x)\, dx = 0 \text{ for } n \neq m.$

Proof: Only part (c) has to be shown. We multiply the differential equation (2.9) for P_n by $P_m(x)$, the differential equation (2.9) for P_m by $P_n(x)$, take the difference and integrate. This yields

$$0 = \int_{-1}^{+1} \left\{ P_m(x) \frac{d}{dx}\left[(1-x^2)\, P_n'(x) \right] - P_n(x) \frac{d}{dx}\left[(1-x^2)\, P_m'(x) \right] \right\} dx$$

$$+ \left[n(n+1) - m(m+1) \right] \int_{-1}^{+1} P_n(x)\, P_m(x)\, dx\,.$$

The first integral vanishes by partial integration. This proves part (c). $\square$

Before we return to the Laplace equation we prove some further results for the Legendre polynomials.

Lemma 2.5. *For the Legendre polynomials it holds*

$$\max_{-1 \leq x \leq 1} |P_n(x)| \;=\; 1 \quad \text{for all } n = 0, 1, 2, \ldots\,.$$

Proof: For fixed $n \in \mathbb{N}$ we define the function

$$\Phi(x) := P_n(x)^2 + \frac{1-x^2}{n(n+1)} P_n'(x)^2 \quad \text{for all } x \in [-1, +1].$$

We differentiate Φ and have

$$\Phi'(x) = 2 P_n'(x) \left[P_n(x) + \frac{1-x^2}{n(n+1)} P_n''(x) - \frac{x}{n(n+1)} P_n'(x) \right]$$

$$= \frac{2 P_n'(x)}{n(n+1)} \underbrace{\left[n(n+1) P_n(x) + \frac{d}{dx}\{(1-x^2) P_n'(x)\} + x P_n'(x) \right]}_{= 0 \text{ by the differential equation}}$$

$$= \frac{2x P_n'(x)^2}{n(n+1)}.$$

Therefore $\Phi' \geq 0$ on $[0,1]$ and $\Phi' \leq 0$ on $[-1,0]$; that is, Φ is monotonously increasing on $[0,1]$ and monotonously decreasing on $[-1,0]$. Thus, we have

$$0 \leq P_n(x)^2 \leq \Phi(x) \leq \max\{\Phi(1), \Phi(-1)\}.$$

From $\Phi(1) = \Phi(-1) = 1$ we conclude that $|P_n(x)| \leq 1$ for all $x \in [-1, +1]$. The lemma is proven by noting that $P_n(1) = 1$. $\square$

A useful representation of the Legendre polynomials is a formula first shown by B.O. Rodrigues.

Theorem 2.6. *For all $n \in \mathbb{N}_0$ the Legendre polynomials satisfy the formula of Rodrigues*

$$P_n(x) = \frac{1}{2^n\, n!} \frac{d^n}{dx^n} (x^2 - 1)^n, \quad x \in \mathbb{R}.$$

Proof: First we prove that the right-hand side solves the Legendre differential equation; that is, we show that

$$\frac{d}{dx} \left[(1 - x^2) \frac{d^{n+1}}{dx^{n+1}} (x^2 - 1)^n \right] + n(n+1) \frac{d^n}{dx^n} (x^2 - 1)^n = 0. \quad (2.11)$$

We observe that both parts are polynomials of degree n. We multiply the first part by x^j for some $j \in \{0, \dots, n\}$, integrate and use partial integration two times. This yields for $j \geq 1$

$$A_j := \int_{-1}^{1} x^j \frac{d}{dx} \left[(1 - x^2) \frac{d^{n+1}}{dx^{n+1}} (x^2 - 1)^n \right] dx$$

$$= -j \int_{-1}^{1} x^{j-1} (1 - x^2) \frac{d^{n+1}}{dx^{n+1}} (x^2 - 1)^n \, dx$$

$$= j \int_{-1}^{1} \left[(j - 1) x^{j-2} - (j + 1) x^j \right] \frac{d^n}{dx^n} (x^2 - 1)^n \, dx \, .$$

We note that no boundary contributions occur and, furthermore, that $A_0 = 0$. Now we use partial integration n times again. No boundary contributions occur either since there is always at least one factor $(x^2 - 1)$ left. For $j \leq n-1$ the integral vanishes because the n-th derivative of x^j and x^{j-2} vanish for $j < n$. For $j = n$ we have

$$A_n = -n(n + 1) (-1)^n n! \int_{-1}^{1} (x^2 - 1)^n \, dx \, .$$

Analogously, we multiply the second part of (2.11) by x^j, integrate, and apply partial integration n-times. This yields

$$B_j = n(n+1) \int_{-1}^{1} x^j \frac{d^n}{dx^n} (x^2 - 1)^n \, dx = n(n+1) (-1)^n n! \int_{-1}^{1} (x^2 - 1)^n \, dx \quad \text{if } j = n$$

and zero if $j < n$. This proves that the polynomial

$$\frac{d}{dx} \left[(1 - x^2) \frac{d^{n+1}}{dx^{n+1}} (x^2 - 1)^n \right] + n(n + 1) \frac{d^n}{dx^n} (x^2 - 1)^n$$

of degree n is orthogonal in $L^2(-1, 1)$ to all polynomials of degree at most n and, therefore, has to vanish. Furthermore,

$$\frac{d^n}{dx^n} (x^2 - 1)^n \bigg|_{x=1} = \frac{d^n}{dx^n} \left[(x + 1)^n (x - 1)^n \right] \bigg|_{x=1}$$

$$= \sum_{k=0}^{n} \binom{n}{k} \frac{d^k}{dx^k} (x + 1)^n \bigg|_{x=1} \frac{d^{n-k}}{dx^{n-k}} (x - 1)^n \bigg|_{x=1} \, .$$

From $\dfrac{d^{n-k}}{dx^{n-k}} (x - 1)^n \bigg|_{x=1} = 0$ for $k \geq 1$ we conclude

$$\frac{d^n}{dx^n}(x^2-1)^n\bigg|_{x=1} = \binom{n}{0} 2^n \frac{d^n}{dx^n}(x-1)^n\bigg|_{x=1} = 2^n\, n!$$

which proves the theorem. $\square$

As a first application of the formula of Rodrigues we can compute the norm of P_n in $L^2(-1,1)$.

Theorem 2.7. *The Legendre polynomials satisfy*

(a) $\displaystyle\int_{-1}^{+1} P_n(x)^2\, dx = \frac{2}{2n+1}$ *for all* $n = 0, 1, 2, \ldots$.

(b) $\displaystyle\int_{-1}^{+1} x\, P_n(x)\, P_{n+1}(x)\, dx = \frac{2(n+1)}{(2n+1)(2n+3)}$ *for all* $n = 0, 1, 2, \ldots$.

Proof:

(a) We use the representation of P_n by Rodrigues and n partial integrations,

$$\int_{-1}^{+1} \frac{d^n}{dx^n}(x^2-1)^n \frac{d^n}{dx^n}(x^2-1)^n\, dx$$

$$= (-1)^n \int_{-1}^{+1} \frac{d^{2n}}{dx^{2n}}\big[(x^2-1)^n\big](x^2-1)^n\, dx$$

$$= (-1)^n (2n)! \int_{-1}^{+1} (x^2-1)^n\, dx = (2n)! \int_{-1}^{+1} (1-x^2)^n\, dx.$$

It remains to compute $\displaystyle I_n := \int_{-1}^{+1} (1-x^2)^n\, dx$.

We claim:

$$I_n = 2\,\frac{2n(2n-2)\cdots 2}{(2n+1)(2n-1)\cdots 1} = 2\cdot 4^n\,\frac{(n!)^2}{(2n+1)!} \quad \text{for all } n \in \mathbb{N}.$$

The assertion is true for $n = 1$. Let it be true for $n-1$, $n \geq 2$. Then

$$I_n = I_{n-1} - \int_{-1}^{+1} x^2\,(1-x^2)^{n-1}\, dx = I_{n-1} + \int_{-1}^{+1} x\,\frac{d}{dx}\left[\frac{1}{2n}(1-x^2)^n\right] dx$$

$$= I_{n-1} - \frac{1}{2n} I_n \quad \text{and thus} \quad I_n = \frac{2n}{2n+1} I_{n-1}.$$

This proves the representation of I_n. We arrive at

$$\int_{-1}^{+1} P_n(x)^2 \, dx \;=\; \frac{1}{(2^n \, n!)^2} \, (2n)! \, 2 \cdot 4^n \, \frac{(n!)^2}{(2n+1)!} \;=\; \frac{2}{2n+1}.$$

(b) This is proven quite similarly:

$$\int_{-1}^{+1} x \, \frac{d^n}{dx^n} (x^2 - 1)^n \, \frac{dx^{n+1}}{dx^{n+1}} (x^2 - 1)^{n+1} \, dx$$

$$= (-1)^{n+1} \int_{-1}^{+1} (x^2 - 1)^{n+1} \frac{d^{n+1}}{dx^{n+1}} \left[x \frac{d^n}{dx^n} (x^2 - 1)^n \right] dx$$

$$= (-1)^{n+1} \int_{-1}^{+1} (x^2 - 1)^{n+1} \left[x \underbrace{\frac{d^{2n+1}}{dx^{2n+1}} (x^2 - 1)^n}_{=0} + (n+1) \underbrace{\frac{d^{2n}}{dx^{2n}} (x^2 - 1)^n}_{=(2n)!} \right] dx$$

$$= (n+1) \, (-1)^{n+1} (2n)! \, I_{n+1} \;=\; 2(n+1)(2n)! \, \frac{(2n+2)(2n)(2n-2) \cdots 2}{(2n+3)(2n+1)(2n-1) \cdots 1}$$

$$= (n+1) \, \frac{2(2^n n!)(2^{n+1}(n+1)!)}{(2n+1)(2n+3)},$$

which yields the assertion. $\square$

The formula of Rodrigues is also useful in proving recursion formulas for the Legendre polynomials. We prove only some of these formulas. For the remaining parts we refer to the exercises.

Theorem 2.8. *For all $x \in \mathbb{R}$ and $n \in \mathbb{N}$, $n \geq 0$, we have*

(a) $P'_{n+1}(x) \;-\; P'_{n-1}(x) \;=\; (2n+1) \, P_n(x)$,
(b) $(n+1)P_{n+1}(x) \;=\; (2n+1)x \, P_n(x) \;-\; nP_{n-1}(x)$,
(c) $P'_n(x) \;=\; nP_{n-1}(x) \;+\; xP'_{n-1}(x)$,
(d) $xP'_n(x) \;=\; nP_n(x) \;+\; P'_{n-1}(x)$,
(e) $(1 - x^2)P'_n(x) \;=\; (n+1) \big[x \, P_n(x) \;-\; P_{n+1}(x) \big]$
$\quad = -n \big[x \, P_n(x) \;-\; P_{n-1}(x) \big]$,
(f) $P'_{n+1}(x) \;+\; P'_{n-1}(x) \;=\; P_n(x) \;+\; 2x \, P'_n(x)$,

(g) $(2n+1)(1-x^2)P_n'(x) = n(n+1)[P_{n-1}(x) - P_{n+1}(x)]$,

(h) $n P_{n+1}'(x) - (2n+1)x P_n'(x) + (n+1)P_{n-1}'(x) = 0$.

In these formulas we have set $P_{-1} = 0$.

Proof:

(a) The formula is obvious for $n = 0$. Let now $n \geq 1$. We calculate, using the formula of Rodrigues,

$$P_{n+1}'(x) = \frac{1}{2^{n+1}(n+1)!} \frac{d^{n+2}}{dx^{n+2}}(x^2-1)^{n+1}$$

$$= \frac{1}{2^{n+1}(n+1)!} \frac{d^n}{dx^n}\left(2(n+1)\frac{d}{dx}\left[x(x^2-1)^n\right]\right)$$

$$= \frac{1}{2^n n!} \frac{d^n}{dx^n}\left[(x^2-1)^n + 2n x^2(x^2-1)^{n-1}\right]$$

$$= P_n(x) + \frac{1}{2^{n-1}(n-1)!} \frac{d^n}{dx^n}\left[(x^2-1)^n + (x^2-1)^{n-1}\right]$$

$$= P_n(x) + 2n P_n(x) + P_{n-1}'(x)$$

which proves formula (a).

(b) The orthogonality of the system $\{P_n : n = 0, 1, 2, \ldots\}$ implies its linear independence. Therefore, $\{P_0, \ldots, P_n\}$ forms a basis of the space $\mathcal{P}_n$ of all polynomials of degree $\leq n$. This yields existence of $\alpha_n, \beta_n \in \mathbb{R}$ and $q_{n-3} \in \mathcal{P}_{n-3}$ such that

$$P_{n+1}(x) = \alpha_n x P_n(x) + \beta_n P_{n-1}(x) + q_{n-3}.$$

The orthogonality condition implies that

$$\int_{-1}^{+1} q_{n-3} P_{n+1}\, dx = 0, \quad \int_{-1}^{+1} q_{n-3} P_{n-1}\, dx = 0, \quad \int_{-1}^{+1} x\, q_{n-3}(x) P_n(x)\, dx = 0,$$

thus

$$\int_{-1}^{+1} q_{n-3}(x)^2 dx = 0,$$

and therefore $q_{n-3} \equiv 0$. From $1 = P_{n+1}(1) = \alpha_n P_n(1) + \beta_n P_{n-1}(1) = \alpha_n + \beta_n$ we conclude that

$$P_{n+1}(x) = \alpha_n x P_n(x) + (1-\alpha_n)P_{n-1}(x).$$

We determine α_n by Theorem 2.7:

$$0 = \int_{-1}^{+1} P_{n-1}(x)\, P_{n+1}(x)\, dx$$

$$= \alpha_n \int_{-1}^{+1} x\, P_{n-1}(x)\, P_n(x)\, dx \; + \; (1 - \alpha_n) \int_{-1}^{+1} P_{n-1}(x)^2\, dx$$

$$= \alpha_n \left[\frac{2n}{(2n-1)(2n+1)} - \frac{2}{2n-1} \right] + \frac{2}{2n-1} \, ,$$

thus $\alpha_n = \dfrac{2n+1}{n+1}$. This proves part (b).

(c) The definition of P_n yields

$$P_n'(x) = \frac{1}{2^n n!} \frac{d^{n+1}}{dx^{n+1}} (x^2 - 1)^n$$

$$= \frac{1}{2^{n-1}(n-1)!} \frac{d^n}{dx^n} [x(x^2 - 1)^{n-1}]$$

$$= \frac{x}{2^{n-1}(n-1)!} \frac{d^n}{dx^n} (x^2 - 1)^{n-1} + \frac{n}{2^{n-1}(n-1)!} \frac{d^{n-1}}{dx^{n-1}} (x^2 - 1)^{n-1}$$

$$= x\, P_{n-1}'(x) \; + \; n\, P_{n-1}(x) \, .$$

(d) Differentiation of (b) and multiplication of (c) for $n+1$ instead of n by $n+1$ yields

$$(n+1)\, P_{n+1}'(x) = (2n+1)x\, P_n'(x) \; + \; (2n+1)P_n(x) \; - \; nP_{n-1}'(x),$$
$$(n+1)\, P_{n+1}'(x) = (n+1)^2 P_n(x) \; + \; (n+1)x\, P_n'(x) \, ,$$

thus by subtraction

$$0 = \left[(2n+1)-(n+1)^2\right] P_n(x) + \left[(2n+1)-(n+1)\right]x\, P_n'(x) - n\, P_{n-1}'(x)$$

which yields (d).

For the proofs of (e)–(h) we refer to the exercises.   □

Now we go back to the associated Legendre differential equation (2.6) and determine solutions in $C^2(-1,+1) \cap C[-1,+1]$ for $m \neq 0$.

Theorem 2.9. *The functions*

$$P_n^m(x) \;=\; (1-x^2)^{m/2} \frac{d^m}{dx^m} P_n(x)\,, \quad -1 < x < 1, \quad 0 \le m \le n\,,$$

are solutions of the differential equation (2.6) for $\lambda = -n(n+1)$; *that is,*

$$\frac{d}{dx}\left[(1-x^2)\frac{d}{dx}P_n^m(x)\right] + \left(n(n+1) - \frac{m^2}{1-x^2}\right) P_n^m(x) = 0\,, \quad -1 < x < 1\,,$$

for all $0 \le m \le n$. *The functions* P_n^m *are called* associated Legendre *functions.*

Proof: We compute

$$(1-x^2)\frac{d}{dx} P_n^m(x) = -m\,x\,(1-x^2)^{m/2}\frac{d^m}{dx^m}P_n(x)$$

$$+ (1-x^2)^{m/2+1}\frac{d^{m+1}}{dx^{m+1}}P_n(x)\,,$$

$$\frac{d}{dx}\left[(1-x^2)\frac{d}{dx}P_n^m(x)\right] = -m\,P_n^m(x) \;+\; m^2 x^2\,(1-x^2)^{m/2-1}\frac{d^m}{dx^m}P_n(x)$$

$$- m\,x\,(1-x^2)^{m/2}\frac{d^{m+1}}{dx^{m+1}}P_n(x)$$

$$- (m+2)\,x\,(1-x^2)^{m/2}\frac{d^{m+1}}{dx^{m+1}}P_n(x) + P_n^{m+2}(x)$$

$$= -m\,P_n^m(x) \;+\; \frac{m^2\,x^2}{1-x^2}\,P_n^m(x)$$

$$- \frac{(m+1)\,2x}{\sqrt{1-x^2}}\,P_n^{m+1}(x) \;+\; P_n^{m+2}(x)\,.$$

Differentiating equation (2.9) m times yields

$$\sum_{k=0}^{m+1}\binom{m+1}{k}\frac{d^k}{dx^k}(1-x^2)\frac{d^{m+2-k}}{dx^{m+2-k}}P_n(x) \;+\; n(n+1)\frac{d^m}{dx^m}P_n(x) \;=\; 0\,.$$

$$(2.12)$$

The sum reduces to three terms only, thus

$$(1-x^2)\frac{d^{m+2}}{dx^{m+2}}P_n(x) \;-\; (m+1)\,2x\,\frac{d^{m+1}}{dx^{m+1}}P_n(x)$$

$$+ \big[n(n+1) - m(m+1)\big]\frac{d^m}{dx^m}P_n(x) = 0\,.$$

We multiply the identity by $(1-x^2)^{m/2}$ and arrive at

$$P_n^{m+2}(x) \;-\; \frac{(m+1)\,2x}{\sqrt{1-x^2}}\,P_n^{m+1}(x) \;+\; \big[n(n+1) - m(m+1)\big]\,P_n^m(x) \;=\; 0\,.$$

Combining this with the previous equation by eliminating the terms involving $P_n^{m+1}(x)$ and $P_n^{m+2}(x)$ yields

$$\frac{d}{dx}\left[(1-x^2)\frac{d}{dx}P_n^m(x)\right]$$

$$= -m\,P_n^m(x) + \frac{m^2\,x^2}{1-x^2}\,P_n^m(x) - \left[n(n+1)-m(m+1)\right]P_n^m(x)$$

$$= \left(\frac{m^2}{1-x^2} - n(n+1)\right)P_n^m(x)$$

which proves the theorem. $\square$

2.3 Expansion into Spherical Harmonics

Now we return to the Laplace equation $\Delta u = 0$ and collect our arguments. We have shown that for $n \in \mathbb{N}$ and $0 \le m \le n$ the functions

$$h_n^m(r,\theta,\varphi) \;=\; r^n\,P_n^m(\cos\theta)\,e^{\pm i\,m\,\varphi}$$

are harmonic functions in all of $\mathbb{R}^3$. Analogously, the functions

$$v(r,\theta,\varphi) \;=\; r^{-n-1}\,P_n^m(\cos\theta)\,e^{\pm im\varphi}$$

are harmonic in $\mathbb{R}^3 \setminus \{0\}$ and decay at infinity.

It is not so obvious that the functions h_n^m are polynomials.

Theorem 2.10. *The functions* $h_n^m(r,\theta,\varphi) \;=\; r^n P_n^m(\cos\theta)\,e^{\pm im\varphi}$ *with* $0 \le m \le n$ *are homogeneous* polynomials of degree n. *The latter property means that* $h_n^m(\mu x) = \mu^n h_n^m(x)$ *for all* $x \in \mathbb{R}^3$ *and* $\mu \in \mathbb{R}$.

Proof: First we consider the case $m = 0$, that is the functions $h_n^0(r,\varphi,\theta) = r^n\,P_n(\cos\theta)$. Let n be even, that is, $n = 2\ell$ for some ℓ. Then

$$P_n(t) \;=\; \sum_{j=0}^{\ell} a_j\,t^{2j}\,, \quad t \in \mathbb{R}\,,$$

for some a_j. From $r = |x|$ and $x_3 = r\cos\theta$ we write h_n^0 in the form

$$h_n^0(x) \;=\; |x|^n\,P_n\!\left(\frac{x_3}{|x|}\right) \;=\; |x|^{2\ell}\sum_{j=0}^{\ell} a_{2j}\,\frac{x_3^{2j}}{|x|^{2j}} \;=\; \sum_{j=0}^{\ell} a_{2j}\,x_3^{2j}\,|x|^{2(\ell-j)}\,,$$

and this is obviously a polynomial of degree $2\ell = n$. The same arguments hold for odd values of n.

Now we show that also the associated functions; that is, for $m > 0$, are polynomials of degree n. We write

$$h_n^m(r, \theta, \varphi) = r^n P_n^m(\cos\theta)\, e^{im\varphi} = r^n \sin^m\theta\, \frac{d^m}{dx^m} P_n(x)\Big|_{x=\cos\theta}\, e^{im\varphi}.$$

and set $Q = \frac{d^m}{dx^m} P_n$. Let again $n = 2\ell$ be even. Then $Q(t) = \sum_{j=0}^{\ell} b_j\, t^{2j-m}$ for some b_j. Furthermore, we use the expressions

$$\cos\theta = \frac{x_3}{r}, \qquad \sin\theta = \frac{1}{r}\sqrt{x_1^2 + x_2^2},$$

as well as

$$\cos\varphi = \frac{1}{r}\frac{x_1}{\sin\theta} = \frac{x_1}{\sqrt{x_1^2 + x_2^2}}, \qquad \sin\varphi = \frac{x_2}{\sqrt{x_1^2 + x_2^2}}.$$

to obtain

$$h_n^m(x) = r^{n-m} (x_1^2 + x_2^2)^{m/2}\, Q\!\left(\frac{x_3}{r}\right) \frac{(x_1 + ix_2)^m}{(x_1^2 + x_2^2)^{m/2}}$$

$$= r^{n-m}\, Q\!\left(\frac{x_3}{r}\right) (x_1 + ix_2)^m$$

$$= \sum_{j=0}^{\ell} b_j\, r^{2(\ell-j)}\, x_3^{2j-m}\, (x_1 + ix_2)^m$$

which proves the assertion for even n. For odd n one argues analogously. It is clear from the definition that the polynomial h_n^m is homogeneous of degree n. □

Definition 2.11. Let $n \in \mathbb{N} \cup \{0\}$.

(a) Homogeneous harmonic polynomials of degree n are called *spherical harmonics of order n*.
(b) The functions $K_n : S^2 \to \mathbb{C}$, defined as restrictions of spherical harmonics of degree n to the unit sphere are called *spherical surface harmonics of order n*.

Therefore, the functions $h_n^m(r, \theta, \varphi) = r^n P_n^{|m|}(\cos\theta) e^{im\varphi}$ are spherical harmonics of order n for $-n \le m \le n$. Spherical harmonics which do not depend on φ are called *zonal*. Thus, in case of $m = 0$ the functions h_n^0 are zonal spherical harmonics.

For any spherical surface harmonic K_n of order n the function

$$H_n(x) = |x|^n K_n\left(\frac{x}{|x|}\right), \quad x \in \mathbb{R}^3,$$

is a homogeneous harmonic polynomial; that is, a spherical harmonic. We immediately have

Lemma 2.12. *If K_n denotes a surface spherical harmonic of order n, the following holds.*

(a) $K_n(-\hat{x}) = (-1)^n K_n(\hat{x})$ for all $\hat{x} \in S^2$.
(b) $\int\limits_{S^2} K_n(\hat{x}) K_m(\hat{x}) ds(\hat{x}) = 0$ for all $n \neq m$.

Proof: Part (a) follows immediately since H_n is homogeneous (set $\mu = -1$).

(b) With Green's second formula in the region $\{x \in \mathbb{R}^3 : |x| < 1\}$ we have

$$\int\limits_{S^2} \left(H_n \frac{\partial}{\partial r} H_m - H_m \frac{\partial}{\partial r} H_n \right) ds = \int\limits_{|x| \leq 1} (H_n \Delta H_m - H_m \Delta H_n)\, dx = 0.$$

Setting $f(r) = H_n(r\hat{x}) = r^n H_n(\hat{x})$ for fixed $\hat{x} \in S^2$ we have that $f'(1) = \frac{\partial}{\partial r} H_n(\hat{x}) = n H_n(\hat{x})$; that is,

$$0 = (m - n) \int\limits_{S^2} H_n H_m\, ds = (m - n) \int\limits_{S^2} K_n K_m\, ds.$$

□

Now we determine the dimension of the space of spherical harmonics for fixed order n.

Theorem 2.13. *The set of spherical surface harmonics of order n is a vector space of dimension $2n + 1$. In particular, there exists a system $\{K_n^m : -n \leq m \leq n\}$ of spherical surface harmonics of order n such that*

$$\int\limits_{S^2} K_n^m K_n^\ell\, ds = \delta_{m,\ell} = \begin{cases} 1 \text{ for } m = \ell, \\ 0 \text{ for } m \neq \ell; \end{cases}$$

that is, $\{K_n^m : -n \leq m \leq n\}$ is an orthonormal basis of this vector space.

Proof: Every homogeneous polynomial of degree n is necessarily of the form

$$H_n(x) = \sum_{j=0}^{n} A_{n-j}(x_1, x_2)\, x_3^j, \quad x \in \mathbb{R}^3, \tag{2.13}$$

where A_{n-j} are homogeneous polynomials with respect to (x_1, x_2) of degree $n - j$. Since H_n is harmonic it follows that

$$0 = \Delta H_n(x) = \sum_{j=0}^{n} x_3^j \Delta_2 A_{n-j}(x_1, x_2) + \sum_{j=2}^{n} j(j-1) x_3^{j-2} A_{n-j}(x_1, x_2),$$

where $\Delta_2 = \frac{\partial^2}{\partial x_1^2} + \frac{\partial^2}{\partial x_2^2}$ denote the two-dimensional Laplace operator. From $\Delta_2 A_0 = \Delta_2 A_1 = 0$ we conclude that

$$0 = \sum_{j=0}^{n-2} [\Delta_2 A_{n-j}(x_1, x_2) + (j+1)(j+2) A_{n-j-2}(x_1, x_2)] x_3^j,$$

and thus by comparing the coefficients

$$A_{n-j-2}(x_1, x_2) = -\frac{1}{(j+1)(j+2)} \Delta_2 A_{n-j}(x_1, x_2)$$

for all $(x_1, x_2) \in \mathbb{R}^2$, $j = 0, \ldots, n-2$; that is (replace $n - j$ by j)

$$A_{j-2} = -\frac{1}{(n-j+1)(n-j+2)} \Delta_2 A_j \quad \text{for } j = n, n-1, \ldots, 2. \quad (2.14)$$

Also, one can reverse the arguments: If A_n and A_{n-1} are homogeneous polynomials of degree n and $n-1$, respectively, then all of the functions A_j defined by (2.14) are homogeneous polynomials of degree j, and H_n is a homogeneous harmonic polynomial of order n.

Therefore, the space of all spherical harmonics of order n is isomorphic to the space

$$\{(A_n, A_{n-1}) : A_n, A_{n-1} \text{homogeneous polynomials of degree } n \text{ and } n - 1, \text{ resp.}\} .$$

From the representation $A_n(x_1, x_2) = \sum_{i=0}^{n} a_i x_1^i x_2^{n-i}$ we note that the dimension of the space of all homogeneous polynomials of degree n is just $n + 1$. Therefore, the dimension of the space of all spherical harmonics of order n is $(n + 1) + n = 2n + 1$. Finally, it is well known that any basis of this finite dimensional Euclidian space can be orthogonalized by the method of Schmidt. $\square$

Remark: The set $\{K_n^m : -n \leq m \leq n\}$ is not uniquely determined. Indeed, for any orthogonal matrix $A \in \mathbb{R}^{3 \times 3}$; that is, $A^\top A = I$, also the set $\{K_n^m(Ax) : -n \leq m \leq n\}$ is an orthonormal system. Indeed, the substitution $x = Ay$ yields

$$\int_{S^2} K_n^m(x) K_{n'}^{m'}(x) \, ds(x) = \int_{S^2} K_n^m(Ay) K_{n'}^{m'}(Ay) \, ds(y) = \delta_{n,n'} \delta_{m,m'} .$$

Now we can state that the spherical harmonics $h_n^m(x) = h_n^m(r, \theta, \varphi) = r^n P_n^{|m|}(\cos \theta)e^{im\varphi}$ with $-n \leq m \leq n$, which we have determined by separation, constitute such an orthogonal basis of the space of spherical harmonics of order n.

Theorem 2.14. *The functions $h_n^m(r, \theta, \varphi) = r^n P_n^{|m|}(\cos \theta)e^{im\varphi}$, $-n \leq m \leq n$ are spherical harmonics of order n. They are mutually orthogonal and, therefore, form an orthogonal basis of the $(2n+1)$-dimensional space of all spherical harmonics of order n.*

Proof: We already know that h_n^m are spherical harmonics of order $n \in \mathbb{N}$. Thus the theorem follows from

$$\int\limits_{|x|<1} h_n^m(x) \, \overline{h_n^\ell(x)} \, dx$$

$$= \int\limits_0^1 \int\limits_0^\pi \int\limits_0^{2\pi} r^{2n} P_n^{|m|}(\cos \theta) P_n^{|\ell|}(\cos \theta) \, e^{i(m-\ell)\varphi} \, r^2 \sin \theta \, d\varphi \, d\theta \, dr \; = \; 0$$

for $m \neq \ell$. $\square$

We want to normalize these functions. First we consider the associated Legendre functions.

Theorem 2.15. *The norm of the associated Legendre functions P_n^m in $L^2(-1, 1)$ is given by*

$$\int\limits_{-1}^{+1} P_n^m(x)^2 \, dx \; = \; \frac{2}{2n+1} \cdot \frac{(n+m)!}{(n-m)!}, \quad m = 0, \ldots, n, \quad n \in \mathbb{N} \cup \{0\}.$$

Proof: The case $m = 0$ has been proven in Theorem 2.7 already.

For $m \geq 1$ partial integration yields

$$\int\limits_{-1}^{+1} P_n^m(x)^2 \, dx = \int\limits_{-1}^{+1} (1-x^2)^m \frac{d^m}{dx^m} P_n(x) \frac{d^m}{dx^m} P_n(x) \, dx$$

$$= -\int\limits_{-1}^{+1} \frac{d}{dx}\left[(1-x^2)^m \frac{d^m}{dx^m} P_n(x)\right] \frac{d^{m-1}}{dx^{m-1}} P_n(x) \, dx \,. \quad (2.15)$$

Now we differentiate the Legendre differential equation (2.9) $(m-1)$-times (see (2.12) for m replaced by $m-1$); that is,

$$(1-x^2)\frac{d^{m+1}}{dx^{m+1}}P_n(x) - 2m\,x\,\frac{d^m}{dx^m}P_n(x) + [n(n+1) - m(m-1)]\frac{d^{m-1}}{dx^{m-1}}P_n(x) = 0,$$

thus, after multiplication by $(1-x^2)^{m-1}$,

$$\frac{d}{dx}\left[(1-x^2)^m\frac{d^m}{dx^m}P_n(x)\right] + [n(n+1)-m(m-1)](1-x^2)^{m-1}\frac{d^{m-1}}{dx^{m-1}}P_n(x) = 0.$$

Substituting this into (2.15) yields

$$\int_{-1}^{+1} P_n^m(x)^2\,dx = [n(n+1) - m(m-1)]\int_{-1}^{+1} P_n^{m-1}(x)^2\,dx$$

$$= (n+m)\,(n-m+1)\int_{-1}^{+1} P_n^{m-1}(x)^2\,dx\,.$$

This is a recursion formula for $\int_{-1}^{+1} P_n^m(x)^2\,dx$ with respect to m and yields the assertion by using the formula for $m=0$. $\quad\square$

Now we define the normalized spherical surface harmonics Y_n^m by

$$Y_n^m(\theta,\varphi) := \sqrt{\frac{(2n+1)\,(n-|m|)!}{4\pi\,(n+|m|)!}}\,P_n^{|m|}(\cos\theta)\,e^{im\varphi} \qquad (2.16)$$

with $-n \le m \le n$ and $n = 0,1,\dots.$ They form an orthonormal system in $L^2(S^2)$. We will identify $Y_n^m(x)$ with $Y_n^m(\theta,\varphi)$ for $x = (\sin\theta\cos\varphi,\ \sin\theta\sin\varphi,\ \cos\theta)^\top \in S^2$.

From (2.4) with $\lambda = -n(n+1)$ we remember that Y_n^m satisfies the differential equation

$$\frac{1}{\sin\theta}\frac{\partial}{\partial\theta}\left[\sin\theta\,\frac{\partial Y_n^m(\theta,\varphi)}{\partial\theta}\right] + \frac{1}{\sin^2\theta}\frac{\partial^2 Y_n^m(\theta,\varphi)}{\partial\varphi^2} + n(n+1)\,Y_n^m(\theta,\varphi) = 0;$$

that is,

$$\Delta_{S^2}Y_n^m + n(n+1)\,Y_n^m = 0 \qquad (2.17)$$

with the Laplace–Beltrami operator Δ_{S^2} of Definition 2.2. Thus, we see that $\lambda = -n(n+1)$ for $n \in \mathbb{N}$ are the eigenvalues of Δ_{S^2} with eigenfunctions Y_n^m, $|m| \le n$. We mentioned already (see also Exercise 2.2) that this operator Δ_{S^2} is self-adjoint and non-negative. In the setting of abstract functional analysis we note that $-\Delta$ is a densely defined and unbounded operator from $L^2(S^2)$ into itself which is self-adjoint. This observation allows the use of general functional analytic tools to prove, e.g., that the eigenfunctions $\{Y_n^m:$

$|m| \leq n, \ n = 0, 1, \ldots\}$ of $-\Delta$ form a complete orthonormal system of $L^2(S^2)$. We are going to prove this fact directly starting with the following result— which is of independent interest.

Theorem 2.16. *For any* $f \in L^2(-1, 1)$ *the following* Funk–Hecke Formula *holds.*

$$\int_{S^2} f(x \cdot y) Y_n^m(y) \, ds(y) \ = \ \lambda_n \, Y_n^m(x), \quad x \in S^2,$$

for all $n \in \mathbb{N}$ *and* $m = -n, \ldots, n$ *where* $\lambda_n = 2\pi \int_{-1}^{1} f(t) P_n(t) \, dt.$

Proof:　We keep $x \in S^2$ fixed and choose an orthogonal matrix A which depends on x such that $\hat{x} := A^{-1}x = A^{\top}x$ is the "north pole"; that is, $\hat{x} = (0, 0, 1)^{\top}$. The transformation $y = Ay'$ yields

$$\int_{S^2} f(x \cdot y) Y_n^m(y) \, ds(y) = \int_{S^2} f(x \cdot Ay') Y_n^m(Ay') \, ds(y')$$

$$= \int_{S^2} f(\hat{x} \cdot y) Y_n^m(Ay) \, ds(y).$$

The function $Y_n^m(Ay)$ is again a spherical surface harmonic of order n, thus

$$Y_n^m(Ay) \ = \ \sum_{k=-n}^{n} a_k Y_n^k(y), \quad y \in S^2, \tag{2.18}$$

where $a_k = \int_{S^2} Y_n^m(Ay) Y_n^{-k}(y) \, ds(y)$. Using this and polar coordinates $y = (\sin\theta \cos\varphi, \, \sin\theta \sin\varphi, \, \cos\theta)^{\top} \in S^2$ yields (note that $\hat{x} \cdot y = \cos\theta$)

$$\int_{S^2} f(x \cdot y) Y_n^m(y) \, ds(y) = \sum_{k=-n}^{n} a_k \int_{S^2} f(\hat{x} \cdot y) Y_n^k(y) \, ds(y)$$

$$= \sum_{k=-n}^{n} a_k \sqrt{\frac{(2n+1)(n-|k|)!}{4\pi(n+|k|)!}} \ \cdot$$

$$\cdot \int_0^{\pi} \int_0^{2\pi} f(\cos\theta) P_n^{|k|}(\cos\theta) e^{ik\varphi} \, d\varphi \, \sin\theta \, d\theta$$

$$= a_0 \sqrt{\frac{2n+1}{4\pi}} \, 2\pi \int_0^{\pi} f(\cos\theta) P_n(\cos\theta) \, \sin\theta \, d\theta$$

$$= \lambda_n \, a_0 \sqrt{\frac{2n+1}{4\pi}}$$

where we have used the substitution $t = \cos\theta$ in the last integral. Now we substitute $y = \hat{x}$ in (2.18) and have, using $Y_k^n(\hat{x}) = 0$ for $k \neq 0$,

$$Y_n^m(x) \;=\; Y_n^m(A\hat{x}) \;=\; \sum_{k=-n}^{n} a_k\, Y_n^k(\hat{x}) \;=\; a_0\,\sqrt{\frac{2n+1}{4\pi}}\, \underbrace{P_n(1)}_{=1},$$

thus

$$\int_{S^2} f(x \cdot y)\, Y_n^m(y)\, ds(y) \;=\; \lambda_n\, Y_n^m(x)$$

which proves the theorem. □

As a first application of this result we prove the addition formula. A second application will be the Jacobi–Anger expansion, see Theorem 2.32.

Theorem 2.17. *The Legendre polynomials P_n and the spherical surface harmonics defined in (2.16) satisfy the addition formula*

$$P_n(x \cdot y) \;=\; \frac{4\pi}{2n+1} \sum_{m=-n}^{n} Y_n^m(x)\, Y_n^{-m}(y) \quad \text{for all } x, y \in S^2 \,.$$

Proof: For fixed $y \in S^2$ the function $x \mapsto P_n(x \cdot y)$ is a spherical harmonic of order n and, therefore, has an expansion of the form

$$P_n(x \cdot y) \;=\; \sum_{m=-n}^{n} \int_{S^2} P_n(z \cdot y)\, Y^{-m}(z)\, ds(z)\, Y_n^m(x) \;=\; \sum_{m=-n}^{n} \lambda_n\, Y^{-m}(y)\, Y_n^m(x)$$

with

$$\lambda_n \;=\; 2\pi \int_{-1}^{1} P_n(t)\, P_n(t)\, dt \;=\; \frac{4\pi}{2n+1}$$

by Theorem 2.7 where we used the previously proven Funk–Hecke formula for $f = P_n$. □

We formulate a simple conclusion as a corollary.

Corollary 2.18.

(a) $\displaystyle \sum_{m=-n}^{n} |Y_n^m(x)|^2 \;=\; \frac{2n+1}{4\pi}$ *for all $x \in S^2$ and $n = 0, 1, \ldots$.*

(b) $\displaystyle |Y_n^m(x)| \;\leq\; \sqrt{\frac{2n+1}{4\pi}}$ *for all $x \in S^2$ and $n = 0, 1, \ldots$.*

Proof: Part (a) follows immediately from the Addition Formula of Theorem 2.17 for $y = x$. Part (b) follows directly from (a). □

After this preparation we are able to prove the completeness of the spherical surface harmonics.

Theorem 2.19. *The functions* $\{Y_n^m : -n \leq m \leq n, \; n \in \mathbb{N} \cup \{0\}\}$ *are complete in* $L^2(S^2)$; *that is, every function* $f \in L^2(S^2)$ *can be expanded into a generalized Fourier series in the form*

$$f = \sum_{n=0}^{\infty} \sum_{m=-n}^{n} (f, Y_n^m)_{L^2(S^2)} \, Y_n^m \,. \qquad (2.19a)$$

The series can also be written as

$$f(x) = \frac{1}{4\pi} \sum_{n=0}^{\infty} (2n+1) \int_{S^2} f(y) \, P_n(y \cdot x) \, ds(y) \,, \qquad x \in S^2 \,. \qquad (2.19b)$$

The convergence in (2.19a) and (2.19b) has to be understood in the L^2-*sense.*

Furthermore, on bounded sets in $C^1(S^2)$ *the series converge even uniformly; that is, for every* $M > 0$ *and* $\varepsilon > 0$ *there exists* $N_0 \in \mathbb{N}$, *depending only on* M *and* ε, *such that*

$$\left\| \sum_{n=0}^{N} \sum_{m=-n}^{n} (f, Y_n^m)_{L^2(S^2)} \, Y_n^m - f \right\|_\infty$$

$$= \max_{\hat{x} \in S^2} \left| \sum_{n=0}^{N} \sum_{m=-n}^{n} (f, Y_n^m)_{L^2(S^2)} \, Y_n^m(\hat{x}) - f(\hat{x}) \right| \leq \varepsilon$$

for all $N \geq N_0$ *and all* $f \in C^1(S^2)$ *with* $\|f\|_{1,\infty} \leq M$, *and, analogously for* (2.19b).

Here, the space $C^1(S^2)$ *consists of those functions* f *such that (with respect to spherical coordinates* θ *and* φ*) the functions* f, $\partial f/\partial \theta$, *and* $\frac{1}{\sin\theta}\partial f/\partial\varphi$ *are continuous and periodic with respect to* φ *with the norm defined by* $\|f\|_{1,\infty} = \max\{\|f\|_\infty, \|\partial f/\partial\theta\|_\infty, \|\frac{1}{\sin\theta}\partial f/\partial\varphi\|_\infty\}$.

Proof: First we prove the second part. Therefore, let $f \in C^1(S^2)$ with $\|f\|_{1,\infty} \leq M$. With the Addition Formula, see Theorem 2.17, we obtain for the partial sum

$$(S_N f)(x) = \sum_{n=0}^{N} \sum_{m=-n}^{n} (f, Y_n^m)_{L^2(S^2)} \, Y_n^m(x)$$

$$= \int_{S^2} f(y) \sum_{n=0}^{N} \sum_{m=-n}^{n} Y_n^m(x) \, Y_n^{-m}(y) \, ds(y)$$

$$= \sum_{n=0}^{N} \int_{S^2} \frac{2n+1}{4\pi} P_n(x \cdot y) \, f(y) \, ds(y) \,.$$

This yields already the equivalence of (2.19a) and (2.19b). With $(2n+1)\,P_n = P'_{n+1} - P'_{n-1}$ of Theorem 2.8 (set $P_{-1} \equiv 0$) this yields

$$(S_N f)(x) = \frac{1}{4\pi} \sum_{n=0}^{N} \int_{S^2} f(y) \left[P'_{n+1}(x \cdot y) - P'_{n-1}(x \cdot y) \right] ds(y)$$

$$= \frac{1}{4\pi} \int_{S^2} f(y) \left[P'_{N+1}(x \cdot y) + P'_N(x \cdot y) \right] ds(y).$$

Let again $z = (0,0,1)^\top$ be the north pole and choose an orthogonal matrix A (depending on x) such that $Ax = z$. Then, by the transformation formula,

$$(S_N f)(x) = \frac{1}{4\pi} \int_{S^2} f(Ay) \left[P'_{N+1}(z \cdot y) + P'_N(z \cdot y) \right] ds(y).$$

In spherical polar coordinates $y(\theta, \varphi) = (\sin\theta \cos\varphi, \sin\theta \sin\varphi, \cos\theta)^\top$ this is, defining $F(\theta) = \frac{1}{2\pi} \int_0^{2\pi} f\big(Ay(\theta,\varphi)\big)\, d\varphi$,

$$(S_N f)(x) = \frac{1}{2} \int_0^{\pi} F(\theta) \left[P'_{N+1}(\cos\theta) + P'_N(\cos\theta) \right] \sin\theta\, d\theta$$

$$= \frac{1}{2} \int_{-1}^{+1} F(\arccos t) \left[P'_{N+1}(t) + P'_N(t) \right] dt$$

$$= \underbrace{\frac{1}{2} F(\arccos t) \left[P_{N+1}(t) + P_N(t) \right] \Big|_{-1}^{+1}}_{=F(0)}$$

$$- \frac{1}{2} \int_{-1}^{+1} \frac{d}{dt} F(\arccos t) \left[P_{N+1}(t) + P_N(t) \right] dt.$$

We note that partial integration is allowed because $t \mapsto F(\arccos t)$ is continuously differentiable in $(-1, 1)$ and continuous in $[-1, 1]$. The value $\theta = 0$ corresponds to the north pole $y = z$, thus

$$F(0) = \frac{1}{2\pi} \int_0^{2\pi} f(Az)\, d\varphi = f(Az) = f(x),$$

and therefore for any $\delta \in (0, 1)$ we obtain

$$\left|(S_N f)(x) - f(x)\right| = \frac{1}{2}\left|\int_{-1}^{+1}\frac{d}{dt}F(\arccos t)\left[P_{N+1}(t) + P_N(t)\right]dt\right|$$

$$\leq \frac{1}{2}\int_{-1}^{-1+\delta} + \int_{-1+\delta}^{1-\delta}$$

$$+ \int_{1-\delta}^{1}\left|\frac{d}{dt}F(\arccos t)\right|\left(|P_{N+1}(t)| + |P_N(t)|\right)dt$$

$$= \frac{1}{2}(I_1 + I_2 + I_3).$$

We estimate these contributions separately. First we use that $|P_n(t)| \leq 1$ for all $t \in [-1, 1]$ and $n \in \mathbb{N}$ (see Lemma 2.5). From $\frac{d}{dt}\arccos t < 0$ we conclude

$$I_1 + I_3 \leq -\int_{-1}^{-1+\delta} - \int_{1-\delta}^{1}|F'(\arccos t)|\frac{d}{dt}\arccos t\, dt$$

$$\leq \|F'\|_\infty\left[\pi - \arccos(-1+\delta) + \arccos(1-\delta)\right]$$

$$\leq M\left[\pi - \arccos(-1+\delta) + \arccos(1-\delta)\right].$$

Let now $\varepsilon > 0$ be given. Choose δ such that $I_1 + I_3 \leq \frac{\varepsilon}{2}$. Then δ depends only on M and ε. With this choice of δ we consider I_2 and use the inequality of Cauchy–Schwarz.

$$I_2 \leq \underbrace{\max_{-1+\delta \leq t \leq 1-\delta}\left|\frac{d}{dt}F(\arccos t)\right|}_{=:c}\int_{-1}^{+1}1 \cdot \left(|P_{N+1}(t)| + |P_N(t)|\right)dt$$

$$\leq \sqrt{2}\,c\left(\|P_{N+1}\|_{L^2} + \|P_N\|_{L^2}\right) \leq 2\sqrt{2}\,c\sqrt{\frac{2}{2N+1}}.$$

We estimate c by

$$c \leq \|F'\|_\infty \max_{-1+\delta \leq t \leq 1-\delta}\left|\frac{d}{dt}\arccos t\right| \leq M \max_{-1+\delta \leq t \leq 1-\delta}\left|\frac{d}{dt}\arccos t\right|.$$

Now we can choose N_0, depending only on ε and M, such that $I_2 \leq \frac{\varepsilon}{2}$ for all $N \geq N_0$. Therefore, $|S_N(f)(x) - f(x)| \leq \varepsilon$ for all $N \geq N_0$. This proves uniform convergence of the Fourier series.

The first part is proven by an approximation argument. Indeed, we use the general property of orthonormal systems that $S_N f$ is the best approximation of f in the subspace span$\{Y_n^m : |m| \leq n,\ n = 0, \ldots, N\}$; that is,

$$\|S_N f - f\|_{L^2(S^2)} \leq \|g - f\|_{L^2(S^2)}$$

for all $g \in \mathrm{span}\{Y_n^m : |m| \leq n, \ n = 0, \ldots, N\}$. Let now $f \in L^2(S^2)$ and $\varepsilon > 0$ be given. Since the space $C^1(S^2)$ is dense in $L^2(S^2)$ there exists $h \in C^1(S^2)$ such that $\|h - f\|_{L^2(S^2)} \leq \varepsilon/2$. Therefore,

$$\|S_N f - f\|_{L^2(S^2)} \leq \|S_N h - f\|_{L^2(S^2)} \leq \|S_N h - h\|_{L^2(S^2)} + \|h - f\|_{L^2(S^2)}$$
$$\leq \sqrt{4\pi} \, \|S_N h - h\|_\infty + \|h - f\|_{L^2(S^2)}$$
$$\leq \sqrt{4\pi} \, \|S_N h - h\|_\infty + \frac{\varepsilon}{2} .$$

Since $S_N h$ converges uniformly to h we can find $N_0 \in \mathbb{N}$ such that the first part is less than $\varepsilon/2$ for all $N \geq N_0$ which ends the proof. $\square$

As a corollary we can prove completeness of the Legendre polynomials.

Corollary 2.20. *The polynomials $\{\sqrt{n + 1/2}\, P_n : n \in \mathbb{N}_0\}$ form a complete orthonormal system in $L^2(-1,1)$; that is, for any $f \in L^2(-1,1)$ there holds*

$$f = \sum_{n=0}^\infty f_n P_n \quad \text{with} \quad f_n = \left(n + \frac{1}{2}\right) \int_{-1}^1 f(t) P_n(t)\, dt, \ n \in \mathbb{N}_0 .$$

For $f \in C^1[-1,1]$ the series converges uniformly.

Proof: The function $g(x) = f(x_3) = f(\cos \theta)$ can be considered as a function on the sphere which is independent of φ, thus it is a zonal function. The expansion (2.19a) yields

$$f(x_3) = \sum_{n=0}^\infty \sum_{m=-n}^n a_n^m \, Y_n^m(x)$$

with

$$a_n^m = (g, Y_n^m)_{L^2(S^2)}$$
$$= \sqrt{\frac{2n+1}{4\pi} \frac{(n-m)!}{(n+m)!}} \int_0^\pi \int_0^{2\pi} f(\cos \theta) \, P_n^{|m|}(\cos \theta) \, e^{im\varphi} \sin \theta \, d\varphi \, d\theta$$
$$= \begin{cases} 0, & m \neq 0, \\ \sqrt{(2n+1)\pi} \int_0^\pi f(\cos \theta)\, P_n(\cos \theta)\, \sin \theta \, d\theta\,, & m = 0, \end{cases}$$
$$= \begin{cases} 0, & m \neq 0, \\ \sqrt{(2n+1)\pi} \int_{-1}^1 f(t)\, P_n(t)\, dt\,, & m = 0. \end{cases}$$

Thus,

$$f(x_3) \;=\; \sum_{n=0}^{\infty} \sqrt{\frac{4\pi}{2n+1}}\, f_n\, Y_n^0(x) \;=\; \sum_{n=0}^{\infty} f_n\, P_n(x_3)$$

and the proof is complete. $\square$

2.4 Laplace's Equation in the Interior and Exterior of a Ball

In the previous two sections we constructed explicitly a complete orthonormal system of functions in $L^2(S^2)$. They play exactly the role of the normalized exponential functions $\left\{\frac{1}{2\pi}\exp(in\varphi) : n \in \mathbb{Z}\right\}$ on the unit circle S^1, parametrized by $x = (\cos\varphi, \sin\varphi)^{\top}$, $\varphi \in [0, 2\pi]$; that is, the classical Fourier expansion functions. It is the aim of this section to expand solutions of the Laplace equation for balls and solve the corresponding Dirichlet boundary value problems.

Theorem 2.21. *Let $u \in C^2\big(B(0,R)\big)$ be harmonic in the ball $B(0,R)$; that is, satisfies the Laplace equation $\Delta u = 0$ in $B(0,R)$. Then there exist unique $\alpha_n^m \in \mathbb{C}$, $|m| \leq n$, $n = 0, 1, 2, \ldots$ with*

$$u(r\hat{x}) \;=\; \sum_{n=0}^{\infty} \sum_{m=-n}^{n} \alpha_n^m\, r^n\, Y_n^m(\hat{x})\,, \quad 0 \leq r < R\,, \ \hat{x} \in S^2\,. \tag{2.20}$$

The series converges uniformly with all of its derivatives in every closed ball $B[0, R']$ with $R' < R$.

Proof: For every $r \in (0, R)$ the function $\hat{x} \mapsto u(r\hat{x})$ is in $C^2(S^2)$, and, therefore, can be expanded into a series by Theorem 2.19; that is,

$$u(r\hat{x}) \;=\; \sum_{n=0}^{\infty} \sum_{m=-n}^{n} u_n^m(r)\, Y_n^m(\hat{x})\,, \quad \hat{x} \in S^2\,.$$

The coefficients are given by $u_n^m(r) = \big(u(r, \cdot), Y_n^m\big)_{L^2(S^2)}$.

We show that u_n^m satisfies a differential equation of Euler type. Using the Laplace equation in spherical coordinates for u, the self-adjoint Laplace–Beltrami operator Δ_{S^2}, and the eigenvalue equation (2.17), yields

$$\frac{d}{dr}\left(r^2 \frac{d}{dr}\right) u_n^m(r) = \int_{S^2} \frac{\partial}{\partial r}\left(r^2 \frac{\partial u(r,\hat{x})}{\partial r}\right) Y_n^{-m}(\hat{x})\, ds(\hat{x})$$

$$= -\int_{S^2} \Delta_{S^2} u(r,\hat{x})\, Y_n^{-m}(\hat{x})\, ds(\hat{x})$$

$$= -\int_{S^2} u(r,\hat{x})\, \Delta_{S^2} Y_n^{-m}(\hat{x})\, ds(\hat{x})$$

$$= n(n+1)\int_{S^2} u(r,\hat{x})\, Y_n^{-m}(\hat{x})\, ds(\hat{x}) = n(n+1)\, u_n^m(r).$$

The only smooth solution of this Euler differential equation is given by $u_n^m(r) = \alpha_n^m\, r^n$ for arbitrary α_n^m. Therefore, u has the desired form (2.20).

It remains to prove uniqueness of the expansion coefficients and uniform convergence. We fix $R' < R$ and choose $\hat{R}$ with $R' < \hat{R} < R$. Multiplying the representation of $u(\hat{R}\hat{x})$ with $Y_p^{-q}(\hat{x})$, and integrate over S^2 to obtain $(u(\hat{R},\cdot), Y_p^q)_{L^2(S^2)} = \alpha_p^q\, \hat{R}^p$ which proves uniqueness of α_n^m. Furthermore, by the addition formula of Theorem 2.17 we find the representation

$$u(r\hat{x}) = \sum_{n=0}^{\infty} \sum_{m=-n}^{n} (u(\hat{R},\cdot), Y_n^m)_{L^2(S^2)} \left(\frac{r}{\hat{R}}\right)^n Y_n^m(\hat{x})$$

$$= \sum_{n=0}^{\infty} \left(\frac{r}{\hat{R}}\right)^n \int_{S^2} \sum_{m=-n}^{n} Y_n^{-m}(\hat{y})\, Y_n^m(\hat{x})\, u(\hat{R}\hat{y})\, ds(\hat{y})$$

$$= \sum_{n=0}^{\infty} \left(\frac{r}{\hat{R}}\right)^n \frac{2n+1}{4\pi} \int_{S^2} P_n(\hat{x}\cdot\hat{y})\, u(\hat{R}\hat{y})\, ds(\hat{y})$$

for $r \le R'$ and $\hat{x} \in S^2$. From this representation and the observation that $\left|\frac{d^j P_n}{dt^j}(t)\right| \le c_j n^{2j}$ on $[-1,1]$ (see Exercise 2.6) we conclude for any differential operator $D^\ell = \partial^{|\ell|}/(\partial r^{\ell_1} \partial \theta^{\ell_2} \partial \varphi^{\ell_3})$ in spherical coordinates that the series for $D^\ell u(x)$ converges uniformly in $B[0, R']$ because, using the Cauchy–Schwarz inequality, it is dominated by the convergent series

$$c \|u(\hat{R},\cdot)\|_{L^2(S^2)} \sum_{n=0}^{\infty} (2n+1)\, n^{2\ell} \left(\frac{R'}{\hat{R}}\right)^n$$

for some $c > 0$. This ends the proof. $\square$

Now we consider the boundary value problem of Dirichlet type in $B(0, R)$; that is,

$$\Delta u = 0 \text{ in } B(0, R), \quad u = f \text{ on } \partial B(0, R), \tag{2.21}$$

for given boundary function f. We study the cases $f_R \in L^2(S^2)$ and $f_R \in C^2(S^2)$ simultaneously where we set $f_R(\hat{x}) = f(R\hat{x})$, $\hat{x} \in S^2$, here and in the following.

Theorem 2.22. (a) *For given $f_R \in L^2(S^2)$ there exists a unique solution $u \in C^2\big(B(0,R)\big)$ of $\Delta u = 0$ in $B(0,R)$ with*

$$\lim_{r \to R} \|u(r,\cdot) - f_R\|_{L^2(S^2)} = 0.$$

The solution is given by the series

$$u(r\hat{x}) = \sum_{n=0}^{\infty} \sum_{m=-n}^{n} (f_R, Y_n^m)_{L^2(S^2)} \left(\frac{r}{R}\right)^n Y_n^m(\hat{x}) \qquad (2.22a)$$

$$= \frac{1}{4\pi} \sum_{n=0}^{\infty} (2n+1) \left(\frac{r}{R}\right)^n \int_{S^2} f(R\hat{y}) \, P_n(\hat{x} \cdot \hat{y}) \, ds(\hat{y}) \qquad (2.22b)$$

for $x = r\hat{x} \in B(0,R)$. They converge uniformly on every compact ball $B[0,R']$ for any $R' < R$.

(b) *If $f_R \in C^2(S^2)$ there exists a unique solution $u \in C^2\big(B(0,R)\big) \cap C\big(B[0,R]\big)$ of (2.21) which is again given by (2.22a), (2.22b). The series converge uniformly on $B[0,R]$.*

Proof: First we note that the series coincide by the addition formula as shown in the proof of Theorem 2.21.

(a) To show uniqueness we assume that u is the difference of two solutions. Then $\Delta u = 0$ in $B(0,R)$ and $\lim_{r \to R} \|u(r,\cdot)\|_{L^2(S^2)} = 0$. By the previous theorem u can be represented as a series in the form (2.20). Let $R' < R$ and $\varepsilon > 0$ be arbitrary. Choose $R_\varepsilon \in [R', R)$ such that $\|u_\varepsilon\|_{L^2(S^2)} \leq \varepsilon$ where $u_\varepsilon(\hat{x}) = u(R_\varepsilon \hat{x})$. Multiplying the representation of $u(R_\varepsilon \hat{x})$ with $Y_p^{-q}(\hat{x})$ and integrating over S^2 yields $(u_\varepsilon, Y_p^q)_{L^2(S^2)} = \alpha_p^q R_\varepsilon^p$, thus

$$u(r\hat{x}) = \sum_{n=0}^{\infty} \sum_{m=-n}^{n} (u_\varepsilon, Y_n^m)_{L^2(S^2)} \left(\frac{r}{R_\varepsilon}\right)^n Y_n^m(\hat{x}) \quad \hat{x} \in S^2, \ r \leq R_\varepsilon.$$

Therefore, for $r \leq R'$,

$$\|u(r,\cdot)\|_{L^2(S^2)}^2 = \sum_{n=0}^{\infty} \sum_{m=-n}^{n} \left|(u_\varepsilon, Y_n^m)_{L^2(S^2)}\right|^2 \left(\frac{r}{R_\varepsilon}\right)^{2n}$$

$$\leq \sum_{n=0}^{\infty} \sum_{m=-n}^{n} \left|(u_\varepsilon, Y_n^m)_{L^2(S^2)}\right|^2$$

$$= \|u_\varepsilon\|_{L^2(S^2)}^2 \leq \varepsilon^2.$$

Since this holds for all $\varepsilon > 0$ we conclude that u has to vanish in $B(0, R')$. Since $R' < R$ was arbitrary u vanishes in $B(0, R)$.

Uniform convergence of the series and all of its derivatives on every ball $B[0, R']$ with $R' < R$ is shown as in the proof of the previous theorem. The series for $D^\ell u(x)$ is dominated by the convergent series

$$c\|f_R\|_{L^2(S^2)} \sum_{n=0}^{\infty} (2n+1)\, n^{2\ell} \left(\frac{R'}{R}\right)^n.$$

Therefore, $u \in C^\infty(B(0, R))$ solves the Laplace equation. Finally, we use the Parseval identity, applied to the series

$$u(r\hat{x}) - f(R\hat{x}) = \sum_{n=0}^{\infty} \sum_{m=-n}^{n} (f_R, Y_n^m)_{L^2(S^2)} \left[\left(\frac{r}{R}\right)^n - 1\right] Y_n^m(\hat{x});$$

that is,

$$\|u(r,\cdot) - f_R\|_{L^2(S^2)}^2 = \sum_{n=0}^{\infty} \sum_{m=-n}^{n} \left|(f_R, Y_n^m)_{L^2(S^2)}\right|^2 \left[\left(\frac{r}{R}\right)^n - 1\right]^2.$$

This term tends to zero as $r \to R$ which is shown by standard arguments: For every n the term $\left|(f_R, Y_n^m)_{L^2(S^2)}\right|^2 \left[\left(\frac{r}{R}\right)^n - 1\right]^2$ tends to zero as r tends to R. Furthermore, it is bounded by the summable term $\left|(f_R, Y_n^m)_{L^2(S^2)}\right|^2$ uniformly with respect to r.

(b) It remains to show that the series converges uniformly in $B[0, R]$. With Eq. (2.17) and the symmetry of Δ_{S^2} we express the expansion coefficient of f as

$$(f_R, Y_n^m)_{L^2(S^2)} = -\frac{1}{n(n+1)} (f_R, \Delta_{S^2} Y_n^m)_{L^2(S^2)}$$

$$= -\frac{1}{n(n+1)} (\Delta_{S^2} f_R, Y_n^m)_{L^2(S^2)}.$$

Thus, uniform convergence follows, since for $N \in \mathbb{N}$ we can estimate the remainder for any $r \leq R$ and any $\hat{x} \in S^2$ by

$$\sum_{n=N}^{\infty} \sum_{m=-n}^{n} |(f_R, Y_n^m)_{L^2(S^2)}| \left(\frac{r}{R}\right)^n |Y_n^m(\hat{x})|$$

$$\leq \sum_{n=N}^{\infty} \sum_{m=-n}^{n} |(\Delta_{S^2} f_R, Y_n^m)_{L^2(S^2)}| \frac{1}{n(n+1)} |Y_n^m(\hat{x})|$$

$$\leq \left[\sum_{n=N}^{\infty} \sum_{m=-n}^{n} |(\Delta_{S^2} f_R, Y_n^m)_{L^2(S^2)}|^2\right]^{1/2} \left[\sum_{n=N}^{\infty} \sum_{m=-n}^{n} \frac{1}{n^2(n+1)^2} |Y_n^m(\hat{x})|^2\right]^{1/2}$$

$$= \|\Delta_{S^2} f_R\|_{L^2(S^2)} \left[\sum_{n=N}^{\infty} \frac{1}{n^2(n+1)^2} \sum_{m=-n}^{n} |Y_n^m(\hat{x})|^2\right]^{1/2}$$

$$= \frac{1}{\sqrt{4\pi}} \|\Delta_{S^2} f_R\|_{L^2(S^2)} \left[\sum_{n=N}^{\infty} \frac{2n+1}{n^2(n+1)^2}\right]^{1/2}$$

where we again have used part (a) of Corollary 2.18.   □

To end this section we consider the situation in the exterior of the closed ball $B[0, R]$. We are interested in harmonic functions which tend to zero at infinity. The following theorem corresponds to Theorem 2.21.

Theorem 2.23. *Let* $u \in C^2(\mathbb{R}^3 \setminus B[0, R])$ *harmonic in the exterior of the ball* $B[0, R]$; *that is, satisfies the Laplace equation* $\Delta u = 0$ *for* $|x| > R$. *Furthermore, we assume that*

$$\lim_{r \to \infty} u(r\hat{x}) = 0 \quad \text{for every } \hat{x} \in S^2. \tag{2.23}$$

Then there exist unique coefficients $\alpha_n^m \in \mathbb{C}$, $|m| \leq n$, $n = 0, 1, 2, \ldots$ *with*

$$u(r\hat{x}) = \sum_{n=0}^{\infty} \sum_{m=-n}^{n} \alpha_n^m r^{-n-1} Y_n^m(\hat{x}), \quad r > R, \ \hat{x} \in S^2. \tag{2.24}$$

The series converges uniformly with all of its derivatives outside of every ball $B(0, R')$ *with* $R' > R$.

The proof is almost the same as the proof of Theorem 2.21. The main difference is that one has to select the solution $u_n^m(r) = \alpha_n^m r^{-n-1}$ in Euler's differential equation $\frac{d}{dr}\left(r^2 \frac{d}{dr}\right) u_n^m(r) = n(n+1) u_n^m(r)$ instead of $u_n^m(r) = \alpha_n^m r^n$. We leave the proof to the reader as Exercise 2.7.

Also, the interior boundary value problem has an exterior analog. Let again $f : \partial B(0, R) \longrightarrow \mathbb{C}$. We want to determine a function u defined in the exterior of $B(0, R)$ such that

$$\Delta u = 0 \text{ in } \mathbb{R}^3 \setminus B[0, R], \quad u = f \text{ on } \partial B(0, R), \tag{2.25}$$

and (2.23). This latter condition is needed to ensure uniqueness. Indeed, we observe that the function $u(x) = u(r, \varphi, \theta) = [(r/R)^n - (R/r)^{n+1}] P_n(\cos\theta)$ is harmonic in the exterior of $B(0, R)$ and vanishes for $r = R$.

Again, we study the cases $f_R \in L^2(S^2)$ and $f_R \in C^2(S^2)$ simultaneously where $f_R(\hat{x}) = f(R\hat{x})$, $\hat{x} \in S^2$, as before.

Theorem 2.24. (a) For given $f_R \in L^2(S^2)$ there exists a unique solution $u \in C^2(\mathbb{R}^3 \setminus B[0, R])$ of $\Delta u = 0$ in $\mathbb{R}^3 \setminus B[0, R]$ which satisfies (2.23) and

$$\lim_{r \to R} \|u(r, \cdot) - f_R\|_{L^2(S^2)} = 0.$$

The solution is given by the series

$$u(r\hat{x}) = \sum_{n=0}^{\infty} \sum_{m=-n}^{n} (f_R, Y_n^m)_{L^2(S^2)} \left(\frac{R}{r}\right)^{n+1} Y_n^m(\hat{x}) \qquad (2.26a)$$

$$= \frac{1}{4\pi} \sum_{n=0}^{\infty} (2n+1) \left(\frac{R}{r}\right)^{n+1} \int_{S^2} f(R\hat{y}) P_n(\hat{x} \cdot \hat{y}) \, ds(\hat{y}) \, (2.26b)$$

for $x = r\hat{x} \in \mathbb{R}^3 \setminus B[0, R]$. They converge uniformly for $r \geq R'$ for any $R' > R$.

(b) If $f_R \in C^2(S^2)$, there exists a unique solution $u \in C^2(\mathbb{R}^3 \setminus B[0, R]) \cap C(\mathbb{R}^3 \setminus B(0, R))$ of (2.25), (2.23) which is again given by (2.26a), (2.26b). The series converge uniformly for $r \geq R$.

The proof follows again very closely the proof of the corresponding interior case. We omit the details and refer again to Exercise 2.7.

2.5 Bessel Functions

The previous investigations for harmonic functions can be applied in electrostatics or magnetostatics (see Sect. 1.3). But for time-harmonic electromagnetic fields we will solve, as a next step towards the Maxwell system, the same Dirichlet problem for the Helmholtz equation instead of the Laplace equation. The radial functions r^n which appeared in, e.g., (2.22a), (2.22b) have to be replaced by Bessel functions. The introduction of these important functions of mathematical physics is subject of the present section.

From the separation $u(x) = v(r)K(\hat{x})$ in spherical coordinates and the eigenvalues $-n(n+1)$ of the Beltrami operator we obtained for the radial component $\hat{v}(kr) = v(r)$ with $z = kr$ the spherical Bessel differential equation

$$z^2 \hat{v}''(z) + 2z \hat{v}'(z) + \left[z^2 - n(n+1) \right] \hat{v}(z) = 0.$$

(see Eq. (2.8)). We will investigate this linear differential equation of second order for arbitrary $z \in \mathbb{C}$. For $z \neq 0$ the differential equation is equivalent to

$$\hat{v}''(z) + \frac{2}{z}\hat{v}'(z) + \left(1 - \frac{n(n+1)}{z^2} \right) \hat{v}(z) = 0 \quad \text{in } \mathbb{C} \setminus \{0\}. \qquad (2.27)$$

The coefficients of this differential equation are holomorphic in $\mathbb{C} \setminus \{0\}$ and have poles of first and second order at 0. As in the case of real z one can show that in every simply connected domain $\Omega \subseteq \mathbb{C} \setminus \{0\}$ there exist at most two linearly independent solutions of (2.27).

Lemma 2.25. *Let $\Omega \subseteq \mathbb{C} \setminus \{0\}$ be a domain. Then there exist at most two linearly independent holomorphic solutions of (2.27) in Ω.*

Proof: Let w_j, $j = 1, 2, 3$, be three solutions. The space $\mathbb{C}^2$ is of dimension 2 over the field $\mathbb{C}$. Therefore, if we fix some $x_0 \in \Omega$, there exist $\alpha_j \in \mathbb{C}$, $j = 1, 2, 3$, such that $\sum_{j=1}^{3} |\alpha_j| \neq 0$ and

$$\sum_{j=1}^{3} \alpha_j \begin{pmatrix} w_j(x_0) \\ w_j'(x_0) \end{pmatrix} = 0.$$

Set $w := \sum_{j=1}^{3} \alpha_j w_j$ in Ω. Then $w(x_0) = w'(x_0) = 0$ and thus from (2.27), also $w^{(j)}(x_0) = 0$ for all $j = 0, 1, \ldots$. Because w is holomorphic in the domain Ω we conclude by the identity theorem for holomorphic functions that w vanishes in all of Ω. Therefore, $\{w_1, w_2, w_3\}$ are linearly dependent. $\qquad \square$

Motivated by the solution $v_{1,n}(r) = r^n$ of the corresponding Eq. (2.7) for the Laplace equation and $\lambda = -n(n+1)$ we make an ansatz for a smooth solution of (2.27) as a power series in the form

$$w_1(z) = z^n \sum_{\ell=0}^{\infty} a_\ell z^\ell = \sum_{\ell=0}^{\infty} a_\ell z^{\ell+n}.$$

Substituting the ansatz into (2.27) and comparing the coefficients yields

(a) $\left[(n+\ell)(n+\ell+1) - n(n+1) \right] a_\ell = 0$ for $\ell = 0, 1$, and
(b) $\left[(n+\ell)(n+\ell+1) - n(n+1) \right] a_\ell + a_{\ell-2} = 0$ for $\ell \in \mathbb{N}$, $\ell \geq 2$.

Therefore, $a_1 = 0$, and from (b) it follows that $a_\ell = 0$ for all odd ℓ. For even ℓ we replace ℓ by 2ℓ and arrive at

$$
a_{2\ell} = -\frac{1}{(n+2\ell)(n+2\ell+1) - n(n+1)} a_{2(\ell-1)} = -\frac{1}{2\ell\,(2n+2\ell+1)} a_{2(\ell-1)}
$$

$$
= -\frac{1}{\ell}\,(n+\ell)\,\frac{1}{(2n+2\ell)(2n+2\ell+1)}
$$

and thus by induction

$$
a_{2\ell} = \frac{(2n+1)!}{n!}\frac{(-1)^\ell}{\ell!}\frac{(n+\ell)!}{(2n+2\ell+1)!} a_0 \quad \text{for all } \ell \geq 0.
$$

Altogether we have that

$$
w_1(z) = \frac{(2n+1)!}{n!} a_0\, z^n \sum_{\ell=0}^{\infty} \frac{(-1)^\ell}{\ell!}\frac{(n+\ell)!}{(2n+2\ell+1)!} z^{2\ell}.
$$

By the ratio test it is seen that the radius of convergence is infinity. Therefore, this function w_1 is a holomorphic solution of the Bessel differential equation in all of $\mathbb{C}$.

Now we determine a second solution of (2.27) which is linearly independent of w_1. Motivated by the singular solution $v_{2,n}(r) = r^{-n-1}$ of (2.7) for $\lambda = -n(n+1)$ we make an ansatz of the form

$$
w_2(z) = z^{-n-1} \sum_{\ell=0}^{\infty} a_\ell\, z^\ell = \sum_{\ell=0}^{\infty} a_\ell\, z^{\ell-n-1}.
$$

Substituting the ansatz into (2.27) and comparing the coefficients yields

(a) $\left[(\ell - n)(\ell - n - 1) - n(n+1)\right] a_\ell = 0$ for $\ell = 0, 1$, and
(b) $\left[(\ell - n)(\ell - n - 1) - n(n+1)\right] a_\ell + a_{\ell-2} = 0$ for $\ell \in \mathbb{N}$, $\ell \geq 2$.

We again set $a_\ell = 0$ for all odd ℓ. For even ℓ we replace ℓ by 2ℓ and arrive at

$$
a_{2\ell} = -\frac{1}{(2\ell - n)(2\ell - n - 1) - n(n+1)} a_{2(\ell-1)} = \frac{1}{(2\ell)\,(2n - 2\ell + 1)} a_{2(\ell-1)}
$$

and thus by induction

$$
a_{2\ell} = \frac{1}{2^\ell\,\ell!}\frac{1}{(2n - 2\ell + 1)(2n - 2\ell + 3)\cdots(2n - 1)} a_0 \quad \text{for all } \ell \geq 1.
$$

To simplify this expression we look first for $\ell \geq n$. Then

$$(2n - 2\ell + 1)(2n - 2\ell + 3) \cdots (2n - 1)$$
$$= (2n - 2\ell + 1)(2n - 2\ell + 3) \cdots (-1) \cdot 1 \cdot 3 \cdots (2n - 1)$$
$$= (-1)^{\ell-n} (2\ell - 2n - 1)!! \, (2n - 1)!!$$
$$= (-1)^{\ell-n} \frac{(2\ell - 2n)!}{2 \cdot 4 \cdots 2(\ell - n)} \frac{(2n)!}{2 \cdot 4 \cdots (2n)}$$
$$= (-1)^{\ell-n} \frac{(2\ell - 2n)!}{2^{\ell-n}(\ell - n)!} \frac{(2n)!}{2^n \, 2n!}$$

where we have used the symbol $k!! = 1 \cdot 3 \cdot 5 \cdots k$ for any odd k.
Now we consider the case $\ell < n$. Analogously we have

$$(2n-2\ell+1)(2n-2\ell+3) \cdots (2n-1) = \frac{(2n - 1)!!}{(2n - 2\ell - 1)!!} = \frac{(2n)!}{2^n \, n!} \frac{2^{n-\ell}(n - \ell)!}{(2n - 2\ell)!} .$$

Therefore, we arrive at the second solution

$$w_2(z) = a_0 \, z^{-n-1} \frac{n!}{(2n)!} \sum_{\ell=0}^{\infty} \frac{1}{\ell!} \frac{(\ell - n)!}{(2\ell - 2n)!} z^{2\ell} ,$$

where we have set

$$\frac{(\ell - n)!}{(2\ell - 2n)!} := (-1)^{n-\ell} \frac{(2n - 2\ell)!}{(n - \ell)!} \quad \text{for } \ell < n .$$

The radius of convergence of the series is again infinity. For particular normalizations the functions w_1 and w_2 are called spherical Bessel functions.

Definition 2.26. *For all $z \in \mathbb{C}$ the spherical Bessel functions of first and second kind and order $n \in \mathbb{N}_0$ are defined by*

$$j_n(z) = (2z)^n \sum_{\ell=0}^{\infty} \frac{(-1)^\ell}{\ell!} \frac{(n + \ell)!}{(2n + 2\ell + 1)!} z^{2\ell} , \quad z \in \mathbb{C} ,$$

$$y_n(z) = \frac{2 \, (-1)^{n+1}}{(2z)^{n+1}} \sum_{\ell=0}^{\infty} \frac{(-1)^\ell}{\ell!} \frac{(\ell - n)!}{(2\ell - 2n)!} z^{2\ell} , \quad z \in \mathbb{C} ,$$

where—in the definition of y_n—a quantity $\frac{(-k)!}{(-2k)!}$ for positive integers k is defined by

$$\frac{(-k)!}{(-2k)!} = (-1)^k \frac{(2k)!}{k!} , \quad k \in \mathbb{N} .$$

The functions

$$h_n^{(1)} = j_n + i \, y_n ,$$
$$h_n^{(2)} = j_n - i \, y_n ,$$

are called *Hankel functions of first and second kind and order* $n \in \mathbb{N}_0$.

For many applications the Wronskian of these functions is important.

Theorem 2.27. *For all* $n \in \mathbb{N}_0$ *and* $z \in \mathbb{C} \setminus \{0\}$ *we have*

$$W(j_n, y_n)(z) := j_n(z) \, y_n'(z) \, - \, j_n'(z) \, y_n(z) \, = \, \frac{1}{z^2}.$$

Proof: We write $W(z)$ for $W(j_n, y_n)(z)$. We multiply the spherical Bessel differential equation (2.27) for y_n by j_n and the one for j_n by y_n and subtract. This yields

$$y_n''(z) \, j_n(z) - j_n''(z) \, y_n(z) \, + \, \frac{2}{z} \left[y_n'(z) \, j_n(z) - j_n'(z) \, y_n(z) \right] \, = \, 0 \, .$$

The first term is just $W'(z)$, thus W solves the ordinary differential equation $W'(z) + \frac{2}{z} W(z) = 0$. The general solution is $W(z) = cz^{-2}$ for some $c \in \mathbb{C}$ which we determine from the leading coefficient of the Laurent series of W. By the definition of the Bessel functions we have for fixed n

$$j_n(z) = 2^n \, z^n \, \frac{n!}{(2n+1)!} \left[1 + \mathcal{O}(z^2) \right],$$

$$y_n(z) = -2^{-n} \, z^{-n-1} \, \frac{(2n)!}{n!} \left[1 + \mathcal{O}(z^2) \right],$$

$$j_n'(z) = n \, 2^n \, z^{n-1} \, \frac{n!}{(2n+1)!} \left[1 + \mathcal{O}(z^2) \right]$$

$$y_n'(z) = (n+1) \, 2^{-n} \, z^{-n-2} \, \frac{(2n)!}{n!} \left[1 + \mathcal{O}(z^2) \right]$$

as $z \to 0$. Therefore,

$$W(z) = \left[(n+1) \, z^{-2} \, \frac{(2n)!}{(2n+1)!} \, + \, n \, z^{-2} \, \frac{(2n)!}{(2n+1)!} \right] \left[1 + \mathcal{O}(z^2) \right]$$

$$= \frac{1}{z^2} \left[1 + \mathcal{O}(z^2) \right]$$

which proves that $c = 1$. $\square$

Remark: From this theorem the linear independence of $\{j_n, y_n\}$ follows immediately and thus also the linear independence of $\{j_n, h_n^{(1)}\}$. Therefore, they span the solution space of the differential equation (2.27).

The functions j_n and y_n are closely related to the sine cardinal function $\sin z / z$ and to the function $\cos z / z$. Especially, from Rayleigh's formulas below we note that $j_0(z) = \sin z / z$ and $i h_0^{(1)}(z) = \exp(iz)/z$.

Theorem 2.28 (Rayleigh's Formulas).

For any $z \in \mathbb{C}$ and $n \in \mathbb{N}_0$ we have

$$j_n(z) = (-z)^n \left(\frac{1}{z} \frac{d}{dz} \right)^n \frac{\sin z}{z},$$

$$y_n(z) = -(-z)^n \left(\frac{1}{z} \frac{d}{dz} \right)^n \frac{\cos z}{z},$$

$$h_n^{(1)}(z) = -i(-z)^n \left(\frac{1}{z} \frac{d}{dz} \right)^n \frac{\exp(iz)}{z},$$

$$h_n^{(2)}(z) = i(-z)^n \left(\frac{1}{z} \frac{d}{dz} \right)^n \frac{\exp(-iz)}{z}.$$

Proof: We prove only the form for y_n. The representation for j_n is analogous, even simpler, the ones for the Hankel functions follow immediately. We start with the power series expansion of $\cos z / z$; that is,

$$\frac{\cos z}{z} = \sum_{\ell=0}^{\infty} \frac{(-1)^\ell}{(2\ell)!} z^{2\ell-1},$$

and observe that the action of $\left(\frac{1}{z} \frac{d}{dz} \right)^n$ on a power of z is given by

$$\left(\frac{1}{z} \frac{d}{dz} \right)^n z^{2\ell-1} = (2\ell-1)(2\ell-3) \cdots (2\ell - 2n + 1) z^{2\ell-2n-1},$$

thus

$$-(-z)^n \left(\frac{1}{z} \frac{d}{dz} \right)^n \frac{\cos z}{z} = \frac{(-1)^{n+1}}{z^{n+1}} \sum_{\ell=0}^{\infty} \frac{(-1)^\ell}{(2\ell)!} (2\ell-1)(2\ell-3) \cdots (2\ell-2n+1) z^{2\ell}.$$

We discuss the term

$$q = \frac{(2\ell-1)(2\ell-3) \cdots (2\ell - 2n + 1)}{(2\ell)!}$$

separately for $\ell \geq n$ and $\ell < n$.

For $\ell \geq n$ we have, using again the notation $k!! = 1 \cdot 3 \cdots k$ for odd k,

$$q = \frac{(2\ell-1)!!}{(2\ell)!\,(2\ell-2n-1)!!} = \frac{(2\ell)!}{2^\ell \ell!} \frac{2^{\ell-n}\,(\ell-n)!}{(2\ell)!\,(2\ell-2n)!} = \frac{1}{2^n \ell!} \frac{(\ell-n)!}{(2\ell-2n)!}.$$

The case $\ell < n$ is seen analogously by splitting

$$
\begin{aligned}
q &= \frac{(2\ell - 1)(2\ell - 3)\cdots 1 \cdot (-1)\cdots(2\ell - 2n + 1)}{(2\ell)!} \\
&= \frac{(2\ell - 1)!!\,(-1)^{n-\ell}\,(2n - 2\ell - 1)!!}{(2\ell)!} \\
&= (-1)^{n-\ell}\frac{(2\ell)!}{2^\ell\,\ell!\,(2\ell)!}\frac{(2n - 2\ell)!}{2^{n-\ell}\,(n - \ell)!} = \frac{1}{2^n\,\ell!}(-1)^{n-\ell}\frac{(2n - 2\ell)!}{(n - \ell)!}\,.
\end{aligned}
$$

This proves the formula for $\cos z/z$. $\quad\square$

In the following definition our results for the Helmholtz equation are collected.

Definition 2.29. For any $n \in \mathbb{N}_0$ and $m \in \mathbb{Z}$ with $|m| \leq n$ and $k \in \mathbb{C}$ with $\mathrm{Im}\,k > 0$ or $k > 0$ the functions

$$
u_n^m(r\hat{x}) = h_n^{(1)}(kr)\,Y_n^m(\hat{x})\,,
$$

are called *spherical wave functions*. They are solutions of $\Delta u + k^2 u = 0$ in $\mathbb{R}^3 \setminus \{0\}$. The part

$$
u_n^m(r\hat{x}) = j_n(kr)\,Y_n^m(\hat{x})\,,
$$

satisfies the Helmholtz equation in all of $\mathbb{R}^3$.

The Bessel functions of the first kind j_n correspond to the functions r^n in the static case, that is $k = 0$. For the Helmholtz equation the mappings $r\hat{x} \mapsto j_n(kr)\,Y_n^m(\hat{x})$ are the expansion functions for solutions inside of balls. Furthermore the functions $h_n^{(1)}$ are used to derive expansions in the exterior of balls. From the definition we observe that they are singular at the origin of order n. Additionally, we need their asymptotic behavior for $r \to \infty$, which can be derived from the previous theorem.

Theorem 2.30. *For every $n \in \mathbb{N}$ and $z \in \mathbb{C}$ we have*

$$
h_n^{(1)}(z) = \frac{\exp[i(z - \frac{\pi}{2}(n+1))]}{z}\left[(1 + \mathcal{O}\left(\frac{1}{|z|}\right)\right] \quad \textit{for } |z| \to \infty,
$$

$$
\frac{d}{dz}h_n^{(1)}(z) = \frac{\exp[i(z - \frac{\pi}{2}n)])}{z}\left[1 + \mathcal{O}\left(\frac{1}{|z|}\right)\right] \quad \textit{for } |z| \to \infty,
$$

uniformly with respect to $z/|z|$. The corresponding formulas for j_n and y_n are derived by replacing $\exp[i(\dots)]$ by $\cos(\dots)$ and $\sin(\dots)$, respectively.

Proof: We show by induction that for every $n \in \mathbb{N}_0$ there exists a polynomial Q_n of degree at most n such that $Q_n(0) = 1$ and

$$\left(\frac{1}{z}\frac{d}{dz}\right)^n \frac{\exp(iz)}{z} = i^n \frac{\exp(iz)}{z^{n+1}} Q_n\left(\frac{1}{z}\right). \qquad (2.28)$$

This would prove the assertions.

For $n = 0$ this is obvious for the constant polynomial $Q_0 = 1$. Let it be true for n. Then

$$\left(\frac{1}{z}\frac{d}{dz}\right)^{n+1} \frac{\exp(iz)}{z} = \frac{1}{z}\frac{d}{dz}\left[i^n \frac{\exp(iz)}{z^{n+1}} Q_n\left(\frac{1}{z}\right)\right]$$

$$= \frac{i^n}{z} e^{iz}\left[\frac{i}{z^{n+1}} Q_n\left(\frac{1}{z}\right) - (n+1)\frac{1}{z^{n+2}} Q_n\left(\frac{1}{z}\right) - \frac{1}{z^{n+1}} Q_n'\left(\frac{1}{z}\right)\frac{1}{z^2}\right]$$

$$= i^{n+1} \frac{\exp(iz)}{z^{n+2}} Q_{n+1}\left(\frac{1}{z}\right)$$

with $Q_{n+1}(t) = Q_n(t) + i(n+1)\,t\,Q_n(t) + it^2\,Q_n'(t)$. Obviously, Q_{n+1} is a polynomial of order at most $n+1$ and $Q_{n+1}(0) = 1$. This proves the asymptotic forms of $h_n^{(1)}$ and its derivative.   □

For the expansion of solutions of the Helmholtz equation into spherical wave functions also the asymptotic behavior with respect to n is necessary.

Theorem 2.31. *For every $\varepsilon > 0$ and $R > 0$ we have*

$$j_n(z) = \frac{n!}{(2n+1)!}(2z)^n\left[1 + \mathcal{O}\left(\frac{1}{n}\right)\right] = \frac{1}{(2n+1)!!} z^n\left[1 + \mathcal{O}\left(\frac{1}{n}\right)\right],$$

$$y_n(z) = -\frac{(2n)!}{n!}\frac{2}{(2z)^{n+1}}\left[1 + \mathcal{O}\left(\frac{1}{n}\right)\right] = -\frac{(2n-1)!!}{z^{n+1}}\left[1 + \mathcal{O}\left(\frac{1}{n}\right)\right],$$

$$h_n^{(1)}(z) = -i\frac{(2n)!}{n!}\frac{2}{(2z)^{n+1}}\left[1 + \mathcal{O}\left(\frac{1}{n}\right)\right] = -i\frac{(2n-1)!!}{z^{n+1}}\left[1 + \mathcal{O}\left(\frac{1}{n}\right)\right]$$

for $n \to \infty$ uniformly with respect to $|z| \le R$ and uniformly with respect to $\varepsilon \le |z| \le R$, respectively. Here again, $k!! = 1 \cdot 3 \cdots k$ for odd $k \in \mathbb{N}$.

Proof: We prove only the first formula. By the definition of j_n we have

$$\left|\frac{(2n+1)!}{n!}(2z)^{-n} j_n(z) - 1\right| \le \sum_{\ell=1}^{\infty} \frac{R^{2\ell}}{\ell!}\frac{(2n+1)!\,(n+\ell)!}{n!\,(2n+2\ell+1)!}$$

$$\le \frac{1}{4(n+1)}\sum_{\ell=1}^{\infty}\frac{R^{2\ell}}{\ell!}$$

because for $\ell \ge 1$

$$\frac{(2n+1)!\,(n+\ell)!}{(2n+2\ell+1)!\,n!}$$

$$= \underbrace{\frac{1}{(2n+2)\cdots(2n+\ell+1)}}_{\leq\,1/(2n+2)}\ \underbrace{\frac{(n+1)\cdots(n+\ell)}{(2n+\ell+2)\cdots(2n+2\ell+1)}}_{\leq\,1/2}\ \leq\ \frac{1}{4(n+1)}\,.$$

□

We finish this section with a second application of the Funk–Hecke Formula of Theorem 2.16. We consider the expansion of the special solution $u(x) = \exp(ikx\cdot\hat{y})$, $x\in\mathbb{R}^3$, of the Helmholtz equation for some fixed $\hat{y}\in S^2$ which describes a plane time-harmonic electromagnetic field traveling in the direction $\hat{y}$ (see example 1.4).

Theorem 2.32. *For $\hat{x},\hat{y}\in S^2$ and $r>0$ there holds the* Jacobi–Anger expansion

$$e^{ikr\,\hat{x}\cdot\hat{y}} = 4\pi\sum_{n=0}^{\infty}\sum_{m=-n}^{n} i^n\, j_n(kr)\, Y_n^m(\hat{x})\, Y_n^{-m}(\hat{y})$$

$$= \sum_{n=0}^{\infty} i^n\,(2n+1)\, j_n(kr)\, P_n(\hat{x}\cdot\hat{y})\,.$$

For every $R>0$ the series converges uniformly with respect to $\hat{x},\hat{y}\in S^2$ and $0\leq r\leq R$.

Proof: The two representations coincide as seen from the addition formula. Uniform convergence follows from the second representation and the asymptotic behavior of j_n as $n\to\infty$ because $|P_n(t)|\leq 1$ on $[-1,1]$. We will prove the first representation and apply the Funk–Hecke formula from Theorem 2.16. Indeed, we fix $z:=kr$ and $\hat{y}\in S^2$ and expand $\hat{x}\mapsto e^{iz\,\hat{x}\cdot\hat{y}}$ into spherical harmonics; that is,

$$e^{iz\,\hat{x}\cdot\hat{y}} = \sum_{n=0}^{\infty}\sum_{m=-n}^{n} a_n^m(z,\hat{y})\, Y_n^m(\hat{x})$$

where

$$a_n^m(z,\hat{y}) = \int_{S^2} e^{iz\,x\cdot\hat{y}}\, Y_n^{-m}(x)\, ds(x) = \lambda_n(z)\, Y_n^{-m}(\hat{y})$$

with $\lambda_n(z) = 2\pi\int_{-1}^{1} e^{izt}\, P_n(t)\, dt$. It remains to show that $\lambda_n(z) = 4\pi\, i^n j_n(z)$. The function λ_n solves the Bessel differential equation. Indeed, using the differential equation (2.9) for the Legendre polynomial, we conclude that

$$\frac{d}{dz}\left(z^2 \lambda'_n(z)\right) + \left[z^2 - n(n+1)\right]\lambda_n(z)$$

$$= 2\pi \int_{-1}^{1}\left[(1-t^2)z^2 + 2izt\right]e^{izt}\,P_n(t)\,dt \;-\; n(n+1)2\pi\int_{-1}^{1} e^{izt}\,P_n(t)\,dt$$

$$= 2\pi \int_{-1}^{1}\left[(1-t^2)z^2 + 2izt\right]e^{izt}\,P_n(t)\,dt \;+\; 2\pi\int_{-1}^{1} e^{izt}\frac{d}{dt}\left[(1-t^2)P'_n(t)\right]dt$$

$$= 0$$

by two partial integrations of the second integral. Since λ_n is smooth we conclude that $\lambda_n(z) = \alpha_n j_n(z)$ for some constant $\alpha_n \in \mathbb{C}$ which we determine by the behavior at the origin. First, we use the fact that P_n is orthogonal to all polynomials of order less than n. The coefficients of the term t^n is given by $\frac{(2n-1)!}{n!\,(n-1)!\,2^{n-1}}$ which can be seen from the recursion formula (b) of Theorem 2.8 by induction (see Exercise 2.5). Therefore,

$$\frac{d^n}{dz^n}\lambda_n(0) = 2\pi\,i^n\int_{-1}^{1} t^n\,P_n(t)\,dt$$

$$= 2\pi\,i^n\frac{n!\,(n-1)!\,2^{n-1}}{(2n-1)!}\int_{-1}^{1} P_n(t)^2\,dt \;=\; 2\pi\,i^n\frac{(n!)^2\,2^{n+1}}{(2n+1)!}$$

where we have used Theorem 2.7. On the other hand, from Definition 2.26 of the Bessel function we observe that

$$\frac{d^n}{dz^n}j_n(0) = \frac{n!}{(2n+1)!}\frac{d^n}{dz^n}(2z)^n\Big|_{z=0} = \frac{(n!)^2\,2^n}{(2n+1)!}\,.$$

Comparing these two formulas yields $\alpha_n = 4\pi\,i^n$ and finishes the proof. □

2.6 The Helmholtz Equation in the Interior and Exterior of a Ball

Now we are ready to study the series expansion of solutions of the Helmholtz equation $\Delta u + k^2 u = 0$ inside and outside of balls and the corresponding boundary value problems. For the interior case we allow k to be complex valued while for the exterior case we restrict ourself to real $k > 0$. The first theorem is the analog of Theorem 2.21.

Theorem 2.33. *Let $k \in \mathbb{C} \setminus \{0\}$ with $\operatorname{Im} k \geq 0$ and $R > 0$ and $u \in C^2\big(B(0,R)\big)$ solve the Helmholtz equation $\Delta u + k^2 u = 0$ in $B(0,R)$. Then there exist unique $\alpha_n^m \in \mathbb{C}$, $|m| \leq n$, $n = 0,1,2,\ldots$ with*

$$u(r\hat{x}) = \sum_{n=0}^{\infty}\sum_{m=-n}^{n}\alpha_n^m\, j_n(kr)\,Y_n^m(\hat{x})\,, \qquad 0 \leq r < R\,,\ \hat{x} \in S^2\,.$$

The series converges uniformly with all of its derivatives in every closed ball $B[0, R']$ *with* $R' < R$.

Proof: We argue as the proof of Theorem 2.21. For every $r \in (0, R)$ the function $u(r\hat{x})$ can be expanded into a series by Theorem 2.19; that is,

$$u(r\hat{x}) = \sum_{n=0}^{\infty} \sum_{m=-n}^{n} u_n^m(r) Y_n^m(\hat{x}), \quad \hat{x} \in S^2.$$

The coefficients are given by $u_n^m(r) = \left(u(r, \cdot), Y_n^m\right)_{L^2(S^2)}$. We show that u_n^m satisfies the spherical Bessel differential equation. Using the Helmholtz equation for u in spherical polar coordinates, and that the functions Y_n^m are eigenfunctions of the self-adjoint Laplace–Beltrami operator yields for $r > R$

$$
\frac{d}{dr}\left(r^2 \frac{d}{dr}\right) u_n^m(r) + r^2 k^2 u_n^m(r)
$$

$$
= \int_{S^2} \left[\frac{\partial}{\partial r}\left(r^2 \frac{\partial u(r, \hat{x})}{\partial r}\right) + r^2 k^2 u(r, \hat{x})\right] Y_n^{-m}(\hat{x}) \, ds(\hat{x})
$$

$$
= -\int_{S^2} \Delta_{S^2} u(r, \hat{x}) \, Y_n^{-m}(\hat{x}) \, ds(\hat{x})
$$

$$
= -\int_{S^2} u(r, \hat{x}) \, \Delta_{S^2} Y_n^{-m}(\hat{x}) \, ds(\hat{x})
$$

$$
= n(n+1) \int_{S^2} u(r, \hat{x}) Y_n^{-m}(\hat{x}) \, ds(\hat{x}) \quad = \quad n(n+1) u_n^m(r).
$$

After the transformation $z = kr$ it turns into Bessel's differential equation (2.8) for the functions u_n^m. The general solution is given by $u_n^m(r) = \alpha_n^m j_n(kr) + \beta_n^m y_n(kr)$ for arbitrary α_n^m, β_n^m. Therefore, u has the form

$$u(r\hat{x}) = \sum_{n=0}^{\infty} \sum_{m=-n}^{n} \left[\alpha_n^m j_n(kr) + \beta_n^m y_n(kr)\right] Y_n^m(\hat{x}), \quad \hat{x} \in S^2, \ r > 0.$$

(2.29)

We are interested in smooth solutions in the ball $B(0, R)$. Therefore, $\beta_n^m = 0$.

To prove uniform convergence we express $u(x)$ in a different form as in the proof of Theorem 2.33. Fix $R' < R$ and choose $\hat{R}$ with $R' < \hat{R} < R$ such that $j_n(k\hat{R}) \neq 0$ for all n. This is possible because the set of all zeros of all of the analytic functions j_n is discrete. Multiplying the representation of $u(\hat{R}\hat{x})$ with $Y_p^{-q}(\hat{x})$, and integrate over S^2 to obtain $\left(u(\hat{R}, \cdot), Y_p^q\right)_{L^2(S^2)} = \alpha_p^q j_p(k\hat{R})$ which proves uniqueness of α_n^m and, furthermore, the representation of u as

$$u(r\hat{x}) = \sum_{n=0}^{\infty} \sum_{m=-n}^{n} \left(u(\hat{R}, \cdot), Y_n^m\right)_{L^2(S^2)} \frac{j_n(kr)}{j_n(k\hat{R})} Y_n^m(\hat{x})$$

$$= \sum_{n=0}^{\infty} \frac{j_n(kr)}{j_n(k\hat{R})} \int_{S^2} \sum_{m=-n}^{n} Y_n^{-m}(\hat{y}) Y_n^m(\hat{x}) u(\hat{R}\hat{y}) \, ds(\hat{y})$$

$$= \sum_{n=0}^{\infty} \frac{j_n(kr)}{j_n(k\hat{R})} \frac{2n+1}{4\pi} \int_{S^2} P_n(\hat{x} \cdot \hat{y}) u(\hat{R}\hat{y}) \, ds(\hat{y})$$

for $r \leq R'$ and $\hat{x} \in S^2$. Here we used again the Addition Formula of Theorem 2.17. So far, we followed exactly the proof of Theorem 2.21. Now we have to use again the estimate $\left|\frac{d^j P_n}{dt^j}(t)\right| \leq c_j n^{2j}$ on $[-1,1]$ (see Exercise 2.6) and the asymptotic forms of the derivatives of the Bessel functions for large orders n (see Exercise 2.9). Therefore, for any differential operator $D^\ell = \partial^{|\ell|}/(\partial r^{\ell_1} \partial \theta^{\ell_2} \partial \varphi^{\ell_3})$ the series for $D^\ell u(x)$ converges uniformly in $B[0, R']$ because it is dominated by the convergent series

$$c\|u(\hat{R}, \cdot)\|_{L^2(S^2)} \sum_{n=0}^{\infty} (2n+1) \, n^{2\ell} \left(\frac{R'}{\hat{R}}\right)^n$$

with a constant $c > 0$ depending on R'. This ends the proof. □

Now we transfer the idea of Theorem 2.22 to the *interior boundary value problem*

$$\Delta u + k^2 u = 0 \text{ in } B(0, R), \quad u = f \text{ on } \partial B(0, R). \tag{2.30}$$

We recall that we transform functions on the sphere $\{x \in \mathbb{R}^3 : |x| = R\}$ of radius R onto the unit sphere by setting $f_R(\hat{x}) = f(R\hat{x})$, $\hat{x} \in S^2$.

Theorem 2.34. *Assume that $k \in \mathbb{C}\backslash\{0\}$ and $R > 0$ are such that $j_n(kR) \neq 0$ for all $n \in \mathbb{N}_0$.*

(a) *For given $f_R \in L^2(S^2)$ there exists a unique solution $u \in C^2(B(0, R))$ of $\Delta u + k^2 u = 0$ in $B(0, R)$ with*

$$\lim_{r \to R} \|u(r, \cdot) - f_R\|_{L^2(S^2)} = 0.$$

The solution is given by

$$u(r\hat{x}) = \sum_{n=0}^{\infty} \sum_{m=-n}^{n} (f_R, Y_n^m)_{L^2(S^2)} \frac{j_n(kr)}{j_n(kR)} Y_n^m(\hat{x}) \tag{2.31a}$$

$$= \frac{1}{4\pi} \sum_{n=0}^{\infty} (2n+1) \frac{j_n(kr)}{j_n(kR)} \int_{S^2} f(R\hat{y}) P_n(\hat{x} \cdot \hat{y}) \, ds(\hat{y}). \tag{2.31b}$$

The series converge uniformly on compact subsets of $B(0, R)$.

(b) *If $f_R \in C^2(S^2)$, there exists a unique solution $u \in C^2(B(0, R)) \cap C(B[0, R])$ of (2.30) which is again given by (2.31a), (2.31b). The series converges uniformly on $B[0, R]$.*

Proof: (a) The proof of uniqueness follows the lines of the corresponding part in the proof of Theorem 2.24. Indeed, let u be a solution of the Helmholtz equation in $B(0, R)$ such that $\lim_{r \to R} \|u(r, \cdot)\|_{L^2(S^2)} = 0$. From Theorem 2.33 we conclude that u can be expanded into

$$u(r\hat{x}) = \sum_{n=0}^{\infty} \sum_{m=-n}^{n} a_n^m \, j_n(kr) \, Y_n^m(\hat{x}), \quad \hat{x} \in S^2, \, r < R.$$

Let $R_0 < R$ and $\varepsilon > 0$ be arbitrary. Choose $R_\varepsilon \in (R_0, R]$ such that $j_n(kR_\varepsilon) \neq 0$ for all n and $\|u_\varepsilon\|_{L^2(S^2)} \leq \varepsilon$ where $u_\varepsilon(\hat{x}) = u(R_\varepsilon \hat{x})$. Multiplying the representation of $u(R_\varepsilon \hat{x})$ with $Y_p^{-q}(\hat{x})$ and integrating over S^2 yields $(u_\varepsilon, Y_p^q)_{L^2(S^2)} = a_p^q \, j_p(kR_\varepsilon)$, thus

$$u(r\hat{x}) = \sum_{n=0}^{\infty} \sum_{m=-n}^{n} (u_\varepsilon, Y_n^m)_{L^2(S^2)} \frac{j_n(kr)}{j_n(kR_\varepsilon)} Y_n^m(\hat{x}), \quad \hat{x} \in S^2, \, r \leq R_\varepsilon.$$

Let now $R_1 < R_0$ be arbitrary. Since the asymptotic behavior of the Bessel function (Theorem 2.31) is uniform with respect to $R_1 \leq |z| \leq R_0$ there exists a constant $c > 0$ such that $|j_n(kr)/j_n(kR_\varepsilon)| \leq c$ for all $n \in \mathbb{N}$ and $R_1 \leq r \leq R_0$. Therefore, for $R_1 \leq r \leq R_0$,

$$\|u(r, \cdot)\|_{L^2(S^2)}^2 = \sum_{n=0}^{\infty} \sum_{m=-n}^{n} |(u_\varepsilon, Y_n^m)_{L^2(S^2)}|^2 \left| \frac{j_n(kr)}{j_n(kR_\varepsilon)} \right|^2$$

$$\leq c^2 \sum_{n=0}^{\infty} \sum_{m=-n}^{n} |(u_\varepsilon, Y_n^m)_{L^2(S^2)}|^2$$

$$= c^2 \|u_\varepsilon\|_{L^2(S^2)}^2 \leq c^2 \varepsilon^2.$$

This holds for all $\varepsilon > 0$, therefore u has to vanish in $B(0, R_0) \setminus B(0, R_1)$ and thus for $|x| < R$ because $R_1 < R_0 < R$ was arbitrary u.

It remains to show that the series provides the solution of the boundary value problem. Uniform convergence of the series and its derivatives is proven as in the previous theorem. Therefore $u \in C^\infty(B(0, R))$ and u satisfies the Helmholtz equation. Similar to the proof of Theorem 2.22 we obtain the boundary condition $\|u(r,.) - f_R\|^2_{L^2(S^2)}$, since by Theorem 2.31 the term $|\frac{j_n(kr)}{j_n(kR)} - 1|^2$ can be uniformly bounded for all $n \in \mathbb{N}$ and all $r \in [0, R]$.

(b) Also, with the uniform bound on $\left|\frac{j_n(kr)}{j_n(kR)}\right|$ for all $n \in \mathbb{N}$ and $r \in [0, R]$ we can copy the second part of the proof of Theorem 2.22 in case of the Helmholtz equation which shows part (b) of the Theorem.   □

We observe an important difference between the expansion functions $r^n Y_n^m(\hat{x})$ for the Laplace equation and $j_n(kr)Y_n^m(\hat{x})$ for the Helmholtz equation: The additional assumption $j_n(kR) \neq 0$ for all $n \in \mathbb{N}_0$ is required to obtain a unique solution of the boundary value problem. From the asymptotic form of $j_n(z)$ for real and positive z (see Theorem 2.30) we conclude that for every $n \in \mathbb{N}_0$ and fixed $R > 0$ there exist infinitely many real and positive wave numbers $k_{n,1}, k_{n,2}, \ldots$ with $j_n(k_{n,j}R) = 0$. Therefore, the functions

$$u_{n,j}^m(r\hat{x}) = j_n(k_{n,j}r)\, Y_n^m(\hat{x})\,, \quad n, j \in \mathbb{N}, \; |m| \le n\,,$$

solve the homogeneous Dirichlet boundary value problem $\Delta u_{n,j}^m + k_{n,j}^2 u_{n,j}^m = 0$ in $B[0, R]$ with $u_{n,j}^m = 0$ on the boundary $|x| = R$. The values $k_{n,j}^2$ are called Dirichlet eigenvalues of $-\Delta$ in the ball of radius R. The multiplicity is $2n+1$, and the functions $u_{n,j}^m$, $m = -n, \ldots, n$, are the corresponding eigenfunctions.

We now continue with the solutions of the Helmholtz equation in the exterior of a ball. Following the proof of Theorem 2.33 we derive again at the solution (2.29) involving the two linearly independent solutions $j_n(kr)\, Y_n^m(\hat{x})$ and $y_n(kr)\, Y_n^m(\hat{x})$. Both parts in the terms are defined for $r > 0$—and both tend to zero as r tends to infinity (see Theorem 2.30). Therefore, the requirement $u(x) \to 0$ as $|x| \to \infty$ is not sufficient to pick a unique member of the solution space. In the introduction we have motivated the physically correct choice. Here, we present a second motivation and consider the case of a conducting medium with $\varepsilon = \varepsilon_0$, $\mu = \mu_0$, and constant $\sigma > 0$. We have seen from the previous chapter that in this case k^2 has to be replaced by $\tilde{k}^2 = k^2 \varepsilon_r$ with $\varepsilon_r = 1 + i\sigma/(\omega\varepsilon_0)$. For $\tilde{k}$ itself we take the branch with $\operatorname{Re} \tilde{k} > 0$, thus also $\operatorname{Im} \tilde{k} > 0$. From Theorem 2.30 we observe that

$$j_n(\tilde{k}r)Y_n^m(\hat{x}) = \frac{f_n(\tilde{k}r)}{\tilde{k}r}\, Y_n^m(\hat{x}) \left[1 + \mathcal{O}\left(\frac{1}{r}\right)\right] \quad \text{for } r \to \infty\,,$$

where $f_n(z) = \cos(z - \frac{\pi}{2}(n+1))$. In any case we observe that for $\operatorname{Im} \tilde{k} > 0$ these functions increase exponentially as r tends to infinity. The same holds for the functions $y_n(\tilde{k}r)Y_n^m(\hat{x})$. The expansion functions $h_n^{(1)}(\tilde{k}r)Y_n^m(\hat{x})$, however, have the asymptotic forms

$$h_n^{(1)}(\tilde{k}r)Y_n^m(\hat{x}) = \frac{\exp(i\tilde{k}r)}{\tilde{k}r} Y_n^m(\hat{x}) \left[(-i)^{n+1} + \mathcal{O}\left(\frac{1}{r}\right)\right] \quad \text{for } r \to \infty,$$

and these functions decay exponentially at infinity. This is physically plausible since a conducting medium is absorbing. We note by passing that the Hankel functions $h_n^{(2)}$ of the second kind are exponentially increasing. Therefore, for conducting media the expansion functions $h_n^{(1)}(\tilde{k}r)Y_n^m(\hat{x})$ are the correct ones. In the limiting case $\sigma \to 0$ the solution should depend continuously on σ. This makes it plausible to choose these functions also for the case $\sigma = 0$. The requirement that the functions u are bounded for Im $\tilde{k} > 0$ and depend continuously on $\tilde{k}$ for Im $\tilde{k} \to 0$ on compact subsets of $\mathbb{R}^3$ is called the *limiting absorption principle*.

The remaining part we assume $k > 0$. Comparing $h_n^{(1)}$ with its derivative yields:

Lemma 2.35. *Let $k > 0$ and $n \in \mathbb{N}_0$ with Im $k \geq 0$ and $n \in \mathbb{N}_0$ and $m \in \mathbb{Z}$ with $|m| \leq n$. Then $u(r\hat{x}) = h_n^{(1)}(kr)Y_n^m(\hat{x})$ satisfies the Sommerfeld radiation condition*

$$\frac{\partial u(r\hat{x})}{\partial r} - ik\,u(r\hat{x}) = \mathcal{O}\left(\frac{1}{r^2}\right) \quad \text{for } r \to \infty, \tag{2.32}$$

uniformly with respect to $\hat{x} \in S^2$.

Proof: By Theorem 2.30 we have

$$\frac{d}{dr}h_n^{(1)}(kr) - ik\,h_n^{(1)}(kr) = \frac{\exp(ikr)}{kr}\left[(-i)^n k - ik\,(-i)^{n+1} + \mathcal{O}\left(\frac{1}{r}\right)\right]$$

$$= \frac{\exp(ikr)}{kr}\mathcal{O}\left(\frac{1}{r}\right) = \mathcal{O}\left(\frac{1}{r^2}\right) \quad \text{for } r \to \infty.$$

$\square$

We note that this radiation condition is necessary only for the case of *real* values of k. For complex values with Im $k > 0$ we note that $u(x)$ decays even exponentially as $|x|$ tends to infinity.

The importance of this radiation condition lies in the fact that it can be formulated for every function defined in the exterior of a bounded domain without making use of the Hankel functions—and still provides the correct solution. The following theorem is the analog of Theorem 2.23.

Theorem 2.36. *Let $k > 0$ and $u \in C^2(\mathbb{R}^3 \setminus B[0, R])$ satisfy $\Delta u + k^2 u = 0$ in the exterior of some ball $B[0, R]$. Furthermore, we assume that u satisfies the Sommerfeld radiation condition (2.32). Then there exist unique $\alpha_n^m \in \mathbb{C}$, $|m| \leq n$, $n = 0, 1, 2, \ldots$ with*

$$u(r\hat{x}) = \sum_{n=0}^{\infty} \sum_{m=-n}^{n} \alpha_n^m h_n^{(1)}(kr) Y_n^m(\hat{x}), \quad r > R, \ \hat{x} \in S^2. \tag{2.33}$$

The series converges uniformly with all of its derivatives on compact subsets of $\mathbb{R}^3 \setminus B[0, R]$.

Proof: We follow the arguments of the proof of Theorem 2.33 and arrive at the general form (2.29) of $u(x)$. From the radiation condition for u we conclude that every term $u_n^m(r)$ satisfies $\frac{d}{dr} u_n^m(r) - ik\, u_n^m(r) = \mathcal{O}(1/r^2)$. Since only the Hankel functions of the first kind satisfy the radiation condition in contrast to the Bessel functions we conclude that $\beta_n^m = 0$ for all $n \in \mathbb{N}_0$ and $|m| \leq n$. We continue with the arguments just as in the proof of Theorem 2.33 and omit the details. $\square$

Finally we consider the *exterior boundary value problem* to determine for given f the complex valued function u such that

$$\Delta u + k^2 u = 0 \text{ for } |x| > R, \quad u = f \text{ for } |x| = R, \tag{2.34}$$

and u satisfies Sommerfeld radiation condition (2.32).

Theorem 2.37. *Let $k > 0$ and $R > 0$.*

(a) For given $f_R \in L^2(S^2)$ there exists a unique solution $u \in C^2(\mathbb{R}^3 \setminus B[0, R])$ of $\Delta u + k^2 u = 0$ with

$$\lim_{r \to R} \|u(r, \cdot) - f_R\|_{L^2(S^2)} = 0,$$

and u satisfies Sommerfeld radiation condition (2.32). The solution is given by

$$u(r\hat{x}) = \sum_{n=0}^{\infty} \sum_{m=-n}^{n} (f_R, Y_n^m)_{L^2(S^2)} \frac{h_n^{(1)}(kr)}{h_n^{(1)}(kR)} Y_n^m(\hat{x}) \tag{2.35a}$$

$$= \frac{1}{4\pi} \sum_{n=0}^{\infty} (2n+1) \frac{h_n^{(1)}(kr)}{h_n^{(1)}(kR)} \int_{S^2} f(R\hat{y}) P_n(\hat{x} \cdot \hat{y}) \, ds(\hat{y}). \tag{2.35b}$$

The series converge uniformly on compact subsets of $\mathbb{R}^3 \setminus B[0, R]$.

(b) If $f_R \in C^2(S^2)$, there exists a unique solution $u \in C^2\big(\mathbb{R}^3 \setminus B[0,R]\big) \cap C\big(\mathbb{R}^3 \setminus B(0,R)\big)$ of (2.34) which is again given by (2.35a), (2.35b). The series converges uniformly on $B[0,R_1] \setminus B(0,R)$ for every $R_1 > R$.

Proof: The proof of uniform convergence outside of any ball $B(0,R')$ for $R' > R$ of any derivative of the series and the validity of the boundary condition follows the well-known arguments and is omitted. We just mention that for $k > 0$ the functions $h_n^{(1)}(kR)$ never vanishes—in contrast to $j_n(kR)$—because of the Wronskian of Theorem 2.27.

The proof that the function u satisfies the radiation condition is more difficult and uses a result (in the proof of Lemma 2.38) from Chap. 3.[2] The problem is that every component of the series satisfies the radiation condition but the asymptotics of $h_n(k|x|)$ as $|x|$ tends to infinity by Theorem 2.30 does not hold uniformly with respect to n. The idea to overcome this difficulty is to express u as an integral over a sphere of the form

$$u(x) = \int_{|y|=R_0} \Phi(x,y)\, g(y)\, ds(y), \quad |x| > R_0, \tag{2.36}$$

rather than a series, see Lemma 2.38 below. Here, Φ denotes the *fundamental solution* of the Helmholtz equation, defined by

$$\Phi(x,y) = \frac{\exp(ik|x-y|)}{4\pi\,|x-y|} = \frac{ik}{4\pi}\, h_0^{(1)}(k|x-y|), \quad x \neq y. \tag{2.37}$$

The fundamental solution satisfies the radiation condition (2.32) uniformly with respect to y on the compact sphere $\{y : |y| = R_0\}$ as we show in Lemma 2.39. Therefore, also u satisfies the radiation condition (2.32) and completes the proof. □

Lemma 2.38. *Let $k > 0$ and $R > 0$ and u be given by*

$$u(r\hat{x}) = \sum_{n=0}^{\infty} \sum_{m=-n}^{n} \alpha_n^m\, h_n^{(1)}(kr)\, Y_n^m(\hat{x}), \quad r > R,\ \hat{x} \in S^2. \tag{2.38}$$

We assume that the series converges uniformly on compact subsets of $\mathbb{R}^3 \setminus B[0,R]$. Then there exists $R_0 > R$ and a continuous function g on the sphere with radius R_0 such that

$$u(x) = \int_{|y|=R_0} \Phi(x,y)\, g(y)\, ds(y), \quad |x| > R_0.$$

---

[2] At least the authors do not know of any proof which uses only the elementary properties of the Hankel functions derived so far.

Here Φ denotes again the fundamental solution of the Helmholtz equation, given by (2.37).

Proof: Let $R_0 > R$ such that $j_n(kR_0) \neq 0$ for all n and fix x such that $|x| > R_0$. First, we apply Green's second identity from Theorem A.12 to the functions $w(y) = j_n(k|y|)Y_n^m(\hat{y})$ and $\Phi(x,y)$ inside the ball $B(0,R_0)$ which yields

$$j_n(kR_0) \int_{|y|=R_0} Y_n^m(\hat{y}) \frac{\partial}{\partial\nu(y)}\Phi(x,y)\,ds(y)$$

$$- k\,j_n'(kR_0) \int_{|y|=R_0} Y_n^m(\hat{y})\,\Phi(x,y)\,ds(y) = 0. \tag{2.39a}$$

Second, since $v(y) = h_n^{(1)}(k|y|)\,Y_n^m(\hat{y})$ satisfies the radiation condition (2.32) we can apply Green's representation Theorem 3.6 from the forthcoming Chap. 3 to v in the exterior of $B(0,R_0)$ which yields

$$h_n^{(1)}(k|x|)\,Y_n^m(\hat{x}) = h_n^{(1)}(kR_0) \int_{|y|=R_0} Y_n^m(\hat{y}) \frac{\partial}{\partial\nu(y)}\Phi(x,y)\,ds(y)$$

$$- k\,\frac{d}{dr}h_n^{(1)}(kR_0) \int_{|y|=R_0} Y_n^m(\hat{y})\,\Phi(x,y)\,ds(y) \tag{2.39b}$$

Equations (2.39a) and (2.39b) are two equations for the integrals. Solving for the second integral yields

$$\int_{|y|=R_0} Y_n^m(\hat{y})\,\Phi(x,y)\,ds(y) \;=\; ik\,R_0^2\,j_n(kR_0)\,h_n^{(1)}(k|x|)\,Y_n^m(\hat{x}) \tag{2.40}$$

where we have used the Wronskian of Theorem 2.27. This holds for all $|x| > R_0$. Comparing this with the form (2.38) of u yields

$$u(x) = \sum_{n=0}^{\infty}\sum_{m=-n}^{n} \frac{\alpha_n^m}{ikR_0^2 j_n(kR_0)}\,ikR_0^2 h_n^{(1)}(kr)\,j_n(kR_0)\,Y_n^m(\hat{x})$$

$$= \int_{|y|=R_0}\sum_{n=0}^{\infty}\sum_{m=-n}^{n} g_n^m\,Y_n^m(\hat{y})\,\Phi(x,y)\,ds(y)$$

for $|x| > R_0$ where

$$g_n^m \;=\; \frac{\alpha_n^m}{ik\,R_0^2\,j_n(kR_0)}. \tag{2.41}$$

Let $R' \in (R, R_0)$. The series (2.38) converges uniformly on $\{x : |x| = R'\}$, thus

$$\sum_{n=0}^{\infty} \sum_{m=-n}^{n} |a_n^m|^2 \, |h_n^{(1)}(kR')|^2 \; < \; \infty$$

and, in particular, $|a_n^m| \, |h_n^{(1)}(kR')| \leq c_1$ for all $|m| \leq n$ and $n \in \mathbb{N}$. Therefore

$$|g_n^m| \; \leq \; \frac{c_2}{|h_n^{(1)}(kR')| |j_n(kR_0)|} \; \leq \; c_3(2n+1) \left(\frac{R'}{R_0}\right)^n$$

by Theorem 2.31 which proves uniform convergence of the series

$$g(R_0 \hat{x}) \; = \; \sum_{n=0}^{\infty} \sum_{m=-n}^{n} g_n^m \, Y_n^m(\hat{x})$$

and ends the proof. □

Lemma 2.39. *For any compact set* $K \subseteq \mathbb{R}^3$ *the fundamental solution* Φ *satisfies the Sommerfeld radiation condition (2.32) uniformly with respect to* $y \in K$. *More precisely,*

$$\Phi(x, y) = \frac{\exp(ik|x|)}{4\pi|x|} e^{-ik\,\hat{x}\cdot y} \left[1 + \mathcal{O}(|x|^{-1})\right], \quad |x| \to \infty, \qquad \text{(2.42a)}$$

$$\nabla_x \Phi(x, y) = ik\,\hat{x} \, \frac{\exp(ik|x|)}{4\pi|x|} e^{-ik\,\hat{x}\cdot y} \left[1 + \mathcal{O}(|x|^{-1})\right], \quad |x| \to \infty, \text{(2.42b)}$$

uniformly with respect to $\hat{x} = x/|x| \in S^2$ *and* $y \in K$.

Proof: We set $x = r\hat{x}$ with $r = |x|$ and investigate

$$4\pi \, \Phi(r\hat{x}, y) \, r \, e^{-ikr} - e^{-ik\,\hat{x}\cdot y}$$

$$= \frac{\exp(ikr[|\hat{x} - y/r| - 1]) - |\hat{x} - y/r| \exp(-ik\,\hat{x} \cdot y)}{|\hat{x} - y/r|}$$

$$= F(1/r; \hat{x}, y)$$

with

$$F(\varepsilon; \hat{x}, y) \; = \; \frac{\exp(ik[|\hat{x} - \varepsilon y| - 1]/\varepsilon) - |\hat{x} - \varepsilon y| \exp(-ik\,\hat{x} \cdot y)}{|\hat{x} - \varepsilon y|}.$$

The function $g(\varepsilon; \hat{x}, y) = |\hat{x} - \varepsilon y|$ is analytic in a neighborhood of zero with respect to ε and $g(0; \hat{x}, y) = 1$ and $\partial g(0; \hat{x}, y)/\partial \varepsilon = -\hat{x} \cdot y$. Therefore, by the rule of l'Hospital we observe that $F(0; \hat{x}, y) = 0$, and the assertion of the first part of the lemma follows because also F is analytic with respect to ε. The second part is proven analogously. □

From formula (2.40) in the proof of Theorem 2.37 we conclude the following *addition formula* for Bessel functions.

Corollary 2.40. *Let $k > 0$. For any $x, y \in \mathbb{R}^3$ with $|x| > |y|$ we have*

$$\Phi(x, y) = \frac{ik}{4\pi} \sum_{n=0}^{\infty} (2n + 1) j_n(k|y|) h_n^{(1)}(k|x|) P_n(\hat{x} \cdot \hat{y}),$$

and the series converges uniformly for $|y| \leq R_1 < R_2 \leq |x| \leq R_3$ for any $R_1 < R_2 < R_3$.

Proof: We fix x and $R_0 < |x|$ and expand the function $\hat{y} \mapsto \Phi(x, R_0 \hat{y})$ into spherical surface harmonics; that is,

$$\Phi(x, R_0\hat{y}) = \sum_{n=0}^{\infty} \sum_{|m| \leq n} \int_{|\hat{z}|=1} \overline{Y_n^m(\hat{z})}\, \Phi(x, R_0 \hat{z})\, ds(\hat{z})\, Y_n^m(\hat{y})$$

$$= \frac{1}{R_0^2} \sum_{n=0}^{\infty} \sum_{|m| \leq n} \int_{|z|=R_0} Y_n^{-m}(\hat{z})\, \Phi(x, z)\, ds(z)\, Y_n^m(\hat{y})$$

$$= ik \sum_{n=0}^{\infty} \sum_{|m| \leq n} j_n(kR_0)\, h_n^{(1)}(k|x|)\, Y_n^{-m}(\hat{x})\, Y_n^m(\hat{y})$$

$$= \frac{ik}{4\pi} \sum_{n=0}^{\infty} (2n + 1) j_n(kR_0)\, h_n^{(1)}(k|x|)\, P_n(\hat{x} \cdot \hat{y}),$$

where we used (2.40) and the addition formula for spherical surface harmonics of Theorem 2.17. The uniform convergence follows from the asymptotic behavior of $j_n(t)$ and $h_n^{(1)}(t)$ for $n \to \infty$, uniformly with respect to t from compact subsets of $\mathbb{R}_{>0}$, see Theorem 2.31. Indeed, we have that

$$j_n(k|y|) = \frac{1}{(2n+1)!!} (k|y|)^n \left[1 + \mathcal{O}\left(\frac{1}{n}\right)\right] \quad \text{for } n \to \infty,$$

$$h_n^{(1)}(k|x|) = -i \frac{(2n-1)!!}{(k|x|)^{n+1}} \left[1 + \mathcal{O}\left(\frac{1}{n}\right)\right] \quad \text{for } n \to \infty,$$

uniformly with respect to x and y in the specified regions. Therefore,

$$|(2n+1) j_n(kR_0) h_n^{(1)}(k|x|)| = \frac{1}{k|x|} \left(\frac{|y|}{|x|}\right)^n \left[1 + \mathcal{O}\left(\frac{1}{n}\right)\right] \leq c \left(\frac{R_1}{R_2}\right)^n$$

which ends the proof. $\square$

As a corollary of Lemma 2.39 and the proof of Theorem 2.37 we have:

Corollary 2.41. *Let $k > 0$. The solution u of the exterior boundary value problem (2.34) satisfies*

$$u(r\hat{x}) = \frac{\exp(ikr)}{kr} \sum_{n=0}^{\infty} \sum_{m=-n}^{n} (f_R, Y_n^m)_{L^2(S^2)} \frac{(-i)^{n+1}}{h_n^{(1)}(kR)} Y_n^m(\hat{x}) \left[1 + \mathcal{O}\left(\frac{1}{r}\right)\right] \quad (2.43a)$$

$$= \frac{\exp(ikr)}{4\pi kr} \sum_{n=0}^{\infty} \frac{(2n+1)(-i)^{n+1}}{h_n^{(1)}(kR)} \int_{S^2} f(R\hat{y}) P_n(\hat{x} \cdot \hat{y}) ds(\hat{y}) \left[1 + \mathcal{O}\left(\frac{1}{r}\right)\right] \quad (2.43b)$$

as r tends to infinity, uniformly with respect to $\hat{x} \in S^2$.

Proof: We note that (2.43a), (2.43b) follow formally from (2.35a), (2.35b), respectively, by using the asymptotics of Theorem 2.30 for $h_n^{(1)}(kr)$. The asymptotics, however, are not uniform with respect to n, and we have to use the representation (2.36) as an integral instead. The function g has the expansion coefficients (2.41) with $\alpha_n^m = (f_R, Y_n^m)_{L^2(S^2)}/h_n^{(1)}(kR)$; that is,

$$g_n^m = \frac{(f_R, Y_n^m)_{L^2(S^2)}}{ik R_0^2 h_n^{(1)}(kR) j_n(kR_0)}.$$

Therefore, using the asymptotic form of Φ by Lemma 2.39,

$$u(x) = \int_{|y|=R_0} \Phi(x,y) g(y) ds(y)$$

$$= \frac{\exp(ikr)}{4\pi r} \int_{|y|=R_0} e^{-ik\hat{x}\cdot y} g(y) ds(y) \left[1 + \mathcal{O}\left(\frac{1}{r}\right)\right].$$

We compute with the Jacobi–Anger expansion of Theorem 2.32

$$\frac{1}{4\pi} \int_{|y|=R_0} e^{-ik\hat{x}\cdot y} g(y) ds(y)$$

$$= \frac{R_0^2}{4\pi} \int_{S^2} e^{-ikR_0 \hat{x}\cdot\hat{y}} g(R\hat{y}) ds(\hat{y}) = \frac{R_0^2}{4\pi} \left(g_R, e^{ikR_0 \hat{x}\cdot}\right)_{L^2(S^2)}$$

$$= R_0^2 \sum_{n=0}^{\infty} \sum_{|m|\leq n} g_n^m (-i)^n j_n(kR_0) Y_n^m(\hat{x})$$

$$= \sum_{n=0}^{\infty} \sum_{|m|\leq n} \frac{1}{ik} \frac{(f_R, Y_n^m)_{L^2(S^2)}}{h_n^{(1)}(kR) j_n(kR_0)} (-i)^n j_n(kR_0) Y_n^m(\hat{x})$$

$$= \frac{1}{k} \sum_{n=0}^{\infty} \sum_{|m|\leq n} \frac{(-i)^{n+1}}{h_n^{(1)}(kR)} (f_R, Y_n^m)_{L^2(S^2)} Y_n^m(\hat{x})$$

which proves the first part. For the second representation we use again the addition formula. □

This corollary implies that every radiating solution of the Helmholtz equation has an asymptotic behavior of the form

$$u(x) = \frac{\exp(ik|x|)}{r} u^\infty(\hat{x}) \left[1 + \mathcal{O}\left(\frac{1}{r}\right)\right], \quad |x| \to \infty,$$

uniformly with respect to $\hat{x} = x/|x| \in S^2$. The function $u^\infty : S^2 \to \mathbb{C}$ is called the *far field pattern* or *far field amplitude* of u and plays an important role in inverse scattering theory.

2.7 Expansion of Electromagnetic Waves

In this section we transfer the results of the previous section to the case of the time-harmonic Maxwell system in a homogeneous medium with vanishing external current and charge densities. By the close connection between the Helmholtz equation and the Maxwell system as discussed in Chap. 1 we are able to use several results from the previous section.

Thus, we consider the boundary value problems

$$\operatorname{curl} E - i\omega\mu_0 H = 0, \qquad \operatorname{curl} H + i\omega\varepsilon_0 E = 0$$

in the interior or exterior of a ball $B(0, R)$ with boundary condition

$$\nu \times E = f_R \quad \text{on } \partial B(0, R)$$

where the tangential field f_R is given. Throughout we use the constant permeability μ and dielectricity ε as μ_0 and ε_0 from vacuum. But for the interior case we allow the parameters also being complex valued. In this case we choose the branch of the wave number $k = \omega\sqrt{\varepsilon\mu}$ with $\operatorname{Im} k \geq 0$. Therefore the results will hold for any homogeneous medium.

The starting point is the observation that E, H are solutions of the source free time-harmonic Maxwell's equations in a domain $D \subseteq \mathbb{R}^3$, if and only if E (or H) is a divergence free solution of the vector Helmholtz equation; that is,

$$\Delta E + k^2 E = 0 \quad \text{and} \quad \operatorname{div} E = 0 \quad \text{in } D$$

(see Sect. 1.3, equation (1.12)). By the following lemma we can construct solutions of Maxwell's equations from scalar solutions of the Helmholtz equation.

Lemma 2.42. *Let $u \in C^\infty(D)$ satisfy $\Delta u + k^2 u = 0$ in a domain $D \subseteq \mathbb{R}^3$. Then*

$$E(x) = \operatorname{curl}(x\, u(x)) \quad and \quad H = \frac{1}{i\omega\mu} \operatorname{curl} E, \qquad x \in D, \qquad (2.44)$$

are solutions of the time-harmonic Maxwell system

$$\operatorname{curl} E - i\omega\mu H = 0 \ in \ D \quad and \quad \operatorname{curl} H + i\omega\varepsilon E = 0 \ in \ D. \qquad (2.45)$$

Furthermore, also $\tilde{E}(x) = \operatorname{curl}\operatorname{curl}(x\, u(x))$ and $\tilde{H} = \frac{1}{i\omega\mu} \operatorname{curl} \tilde{E}$ are solutions of the time-harmonic Maxwell system (2.45).

Proof: Let u satisfy the Helmholtz equation and define E and H as in the Lemma. We observe that $\operatorname{div} E = 0$. From $\operatorname{curl}\operatorname{curl} = -\Delta + \nabla \operatorname{div}$ and $\operatorname{div} E = 0$ we conclude that

$$i\omega\mu \operatorname{curl} H(x) = \operatorname{curl}\operatorname{curl} E(x) = -\Delta E(x) = -\operatorname{curl}\left[\Delta\big(x\, u(x)\big)\right]$$
$$= -\operatorname{curl}\left[2\nabla u(x) + x\,\Delta u(x)\right] = k^2 \operatorname{curl}\left[x\, u(x)\right] = k^2 E(x)$$

which proves the assertion because of $k^2 = \omega^2\varepsilon\mu$. For the proof of the second part we just take the curl of the last formula. $\quad\square$

In view of the expansion results for the Helmholtz equation of the previous section we like to consider solutions of the Helmholtz equation of the form $u(x) = j_n(k|x|)Y_n^m(\hat{x})$ where Y_n^m denotes a spherical surface harmonics of order $n \in \mathbb{N}_0$, $m = -n, \ldots, n$ and $\hat{x} = \frac{x}{|x|}$. By Lemma 2.42 we see that $E(x) = \operatorname{curl}(x\, j_n(k|x|)Y_n^m(\hat{x}))$ and $E(x) = \operatorname{curl}\operatorname{curl}(x\, j_n(k|x|)Y_n^m(\hat{x}))$ are solutions of the Maxwell system. We can also replace the Bessel function by the Hankel function, if the region D does not contain the origin, which will lead to solutions satisfying a radiation condition.

Theorem 2.43. *(a) For $n \in \mathbb{N}_0$ and $|m| \leq n$ the function $E : \mathbb{R}^3 \to \mathbb{R}^3$, defined by*

$$E(x) = \operatorname{curl}\big(x\, j_n(kr)\, Y_n^m(\hat{x})\big) = j_n(kr)\, \operatorname{Grad}_{S^2} Y_n^m(\hat{x}) \times \hat{x}, \quad (2.46a)$$

and the corresponding function

$$H(x) = \frac{1}{i\omega\mu} \operatorname{curl} E(x)$$

$$= \frac{n(n+1)}{i\omega\mu r} j_n(kr)\, Y_n^m(\hat{x})\, \hat{x}$$

$$+ \frac{1}{i\omega\mu r} \left[j_n(kr) + kr j_n'(kr)\right] \operatorname{Grad}_{S^2} Y_n^m(\hat{x}) \qquad (2.46b)$$

*satisfy the Maxwell system (2.45). Again, $r = |x|$ and $\hat{x} = x/|x|$ denote
the spherical coordinates, and Grad_{S^2} is the surface gradient on the unit
sphere (see Example A.17).*

*For the tangential components with respect to spheres of radius $r > 0$ it
holds*

$$\hat{x} \times E(x) = j_n(kr)\,\mathrm{Grad}_{S^2} Y_n^m(\hat{x})\,, \tag{2.47a}$$

$$\hat{x} \times H(x) = \frac{1}{i\omega\mu r}\left[j_n(kr) + kr j_n'(kr)\right]\hat{x} \times \mathrm{Grad}_{S^2} Y_n^m(\hat{x})\,. \tag{2.47b}$$

(b) Analogously, the vector fields $\tilde{E}(x) = \mathrm{curl\,curl}\left(x\,j_n(kr)Y_n^m(\hat{x})\right)$ and $\tilde{H} = \frac{1}{i\omega\mu}\mathrm{curl}\,\tilde{E}$ are solutions of Maxwell's equations.

*(c) The results of (a) and (b) hold in $\mathbb{R}^3 \setminus \{0\}$ if the Bessel function j_n is
replaced by the Hankel function $h_n^{(1)}$.*

Proof: (a) From Lemma 2.42 we already know that $E(x) = \mathrm{curl}\left(x\,j_n(kr)\right.$
$\left.Y_n^m(\hat{x})\right)$ generates a solution of Maxwell's equations. We have to prove
the second representation. By using the identity (A.7) and $\mathrm{curl}\,x = 0$
we find $E(x) = \nabla\left(j_n(kr)Y_n^m(\hat{x})\right) \times x$. Decomposing the gradient into its
tangential part, Grad (see (A.17)), and the radial component, $(\partial/\partial r)\,\hat{x}$,
leads to

$$E(x) = \left(\frac{1}{r}j_n(k|x|)\,\mathrm{Grad}_{S^2} Y_n^m(\hat{x}) + \frac{\partial}{\partial r}\left(j_n(k|x|)\right) Y_n^m(\hat{x})\,\hat{x}\right) \times x$$

$$= j_n(kr)\,\mathrm{Grad}_{S^2} Y_n^m(\hat{x}) \times \hat{x}$$

which proves the second representation in (2.46a).

Now we consider $\mathrm{curl}\,E$. Analogously to the proof of the previous lemma
we use $\mathrm{curl\,curl} = -\Delta + \nabla\,\mathrm{div}$ and the fact that $w(x) = j_n(kr)Y_n^m(\hat{x})$
solves the Helmholtz equation. Therefore,

$$\begin{aligned}
\mathrm{curl}\,E(x) &= -\Delta\left(xw(x)\right) + \nabla\,\mathrm{div}\left(xw(x)\right)\\
&= -2\,\nabla w(x) + k^2 x\,w(x) + 3\,\nabla w(x) + \nabla\left(x \cdot \nabla w(x)\right)\\
&= \nabla\left[\left(j_n(kr) + kr j_n'(kr)\right)Y_n^m(\hat{x})\right] + k^2 x\,j_n(kr)Y_n^m(\hat{x})\\
&= \left[\left(j_n(kr) + kr j_n'(kr)\right)' + k^2 r\,j_n(kr)\right]Y_n^m(\hat{x})\,\hat{x}\\
&\quad + \frac{1}{r}\left(j_n(kr) + kr j_n'(kr)\right)\mathrm{Grad}_{S^2} Y_n^m(\hat{x})\\
&= \frac{1}{r}\left[2kr j_n'(kr) + k^2 r^2 j_n''(kr) + k^2 r^2 j_n(kr)\right]Y_n^m(\hat{x})\,\hat{x}\\
&\quad + \frac{1}{r}\left(j_n(kr) + kr j_n'(kr)\right)\mathrm{Grad}_{S^2} Y_n^m(\hat{x})\\
&= \frac{n(n+1)}{r}j_n(kr)Y_n^m(\hat{x})\,\hat{x} + \frac{1}{r}\left(j_n(kr) + kr j_n'(kr)\right)\mathrm{Grad}_{S^2} Y_n^m(\hat{x})
\end{aligned}$$

where we used the Bessel differential equation (2.27) in the last step. Formulas (2.47a) and (2.47b) follow obviously.

(c) This is shown as in (a) and (b) by replacing j_n by $h_n^{(1)}$. $\square$

From the theorem we observe that the tangential components of the boundary values on a sphere involve the surface gradients $\mathrm{Grad}_{S^2} Y_n^m$ and $\hat{x} \times \mathrm{Grad}_{S^2} Y_n^m$. We have seen—and already used several times—that the surface gradient of $u \in C^1(S^2)$ is a tangential field, i.e. $\nu \cdot \mathrm{Grad}_{S^2} u = 0$ on S^2 with the unit normal vector $\nu(\hat{x}) = \hat{x}$ on S^2. Obviously, also $\nu \times \mathrm{Grad}_{S^2} u$ is tangential. Using the partial integration formula (A.21) we compute

$$\int_{S^2} (\hat{x} \times \mathrm{Grad}_{S^2} u) \cdot (\hat{x} \times \mathrm{Grad}_{S^2} v)\, ds = \int_{S^2} \mathrm{Grad}_{S^2} u \cdot \big((\hat{x} \times \mathrm{Grad}_{S^2} v) \times \hat{x}\big)\, ds$$

$$= \int_{S^2} \mathrm{Grad}_{S^2} u \cdot \mathrm{Grad}_{S^2} v\, ds$$

$$= -\int_{S^2} u\, \Delta_{S^2} v\, ds$$

for $u \in C^1(S^2)$ and $v \in C^2(S^2)$. Since the spherical harmonics $\{Y_n^m : -n \le m \le n,\ n \in \mathbb{N}\}$ form an orthonormal system of eigenfunctions of the spherical Laplace–Beltrami operator Δ_{S^2} with the eigenvalues $\lambda = -n(n+1)$, $n \in \mathbb{N}_0$, we conclude that the vector fields

$$U_n^m(\hat{x}) := \frac{1}{\sqrt{n(n+1)}} \mathrm{Grad}_{S^2} Y_n^m(\hat{x}) \qquad (2.48a)$$

as well as the vector fields

$$V_n^m(\hat{x}) := \hat{x} \times U_n^m(\hat{x}) = \frac{1}{\sqrt{n(n+1)}} \hat{x} \times \mathrm{Grad}_{S^2} Y_n^m(\hat{x}) \qquad (2.48b)$$

are two orthonormal systems of tangential fields on S^2. Applying partial integration again and using property (b) of the following lemma we obtain

$$\int_{S^2} (\hat{x} \times \mathrm{Grad}_{S^2} u) \cdot \mathrm{Grad}_{S^2} v\, ds = -\int_{S^2} \mathrm{Div}_{S^2}(\hat{x} \times \mathrm{Grad}_{S^2} u)\, v\, ds = 0.$$

Combining these sets of vector fields U_n^m and V_n^m yields the set of *spherical vector harmonics*. Actually, one should denote them as spherical vector surface harmonics.

The main task is to show that the spherical vector harmonics form a complete orthonormal system in the space of tangential vector fields on S^2. In preparation of the proof we need some more properties of the spherical harmonics and the surface differential operators.

Lemma 2.44.

(a) $\text{Div}_{S^2}(\hat{x} \times \text{Grad}_{S^2}u) = 0$ *for all* $u \in C^2(S^2)$ *where* Div_{S^2} *denotes the surface divergence on the unit sphere (see Example A.17).*

(b) *If a tangential field* $u \in C_t^1(S^2) = \{u \in C^1(S^2, \mathbb{C}^3) : u \cdot \nu = 0 \text{ on } S^2\}$ *satisfies*

$$\text{Div}_{S^2} u = 0 \quad \text{and} \quad \text{Div}_{S^2}(\nu \times u) = 0 \quad \text{on } S^2$$

then $u = 0$ *on* S^2.

Proof: (a) We extend u to a C^1 function in a neighborhood of S^2 and apply Corollary A.20. This yields $\text{Div}_{S^2}(\hat{x} \times \text{Grad}_{S^2}u) = \text{Div}_{S^2}(\hat{x} \times \nabla u) = -\hat{x} \cdot \text{curl}\, \nabla u = 0$ on S^2.

(b) From $\text{Div}_{S^2}(\nu \times u) = 0$ we obtain by partial integration (A.21)

$$0 = \int_{S^2} \text{Div}_{S^2}(\nu \times u)\, f\, ds = -\int_{S^2} (\nu \times u) \cdot \text{Grad}_{S^2} f\, ds \quad \text{for every } f \in C^1(S^2).$$

We express u in spherical coordinates with respect to the basis $\hat{\theta}$, $\hat{\varphi}$ of the spherical unit vectors; that is, $u = (u \cdot \hat{\theta})\hat{\theta} + (u \cdot \hat{\varphi})\hat{\varphi}$ and thus

$$\nu \times u = (u \cdot \hat{\theta})\hat{\varphi} - (u \cdot \hat{\varphi})\hat{\theta}.$$

Let $f \in C^1(S^2)$ be arbitrary. The representation of the surface gradient of f on S^2 (see Example A.17) implies by partial integration with respect to θ and φ, respectively, the identity

$$0 = \int_{S^2} (\nu \times u) \cdot \text{Grad}_{S^2} f\, ds$$

$$= \int_0^{2\pi} \int_0^{\pi} (\nu \times u) \cdot \left(\frac{\partial f}{\partial \theta} \hat{\theta} + \frac{1}{\sin\theta} \frac{\partial f}{\partial \varphi} \hat{\varphi} \right) \sin\theta\, d\theta\, d\varphi$$

$$= \int_0^{2\pi} \int_0^{\pi} \left[\frac{\partial}{\partial \theta}((u \cdot \hat{\varphi})\sin\theta) - \frac{\partial}{\partial \varphi}(u \cdot \hat{\theta}) \right] f\, d\theta\, d\varphi.$$

This leads to

$$\frac{\partial}{\partial \theta}((u \cdot \hat{\varphi})\sin\theta) = \frac{\partial}{\partial \varphi}(u \cdot \hat{\theta})$$

on S^2. Considering the anti-derivative $h(\varphi, \theta) = \int_0^\theta (u \cdot \hat{\theta})(\varphi, \theta')\, d\theta'$ we have

$$\frac{\partial h}{\partial \theta} = u \cdot \hat{\theta}$$

and

$$\frac{\partial h}{\partial \varphi}(\varphi, \theta) = \int_0^\theta \frac{\partial}{\partial \varphi}(u \cdot \hat{\theta})(\varphi, \theta')\, d\theta'$$

$$= \int_0^\theta \frac{\partial}{\partial \theta}((u \cdot \hat{\varphi})(\varphi, \theta') \sin \theta')\, d\theta' = (u \cdot \hat{\varphi}) \sin \theta.$$

Therefore we conclude that

$$u = (u \cdot \hat{\theta})\hat{\theta} + (u \cdot \hat{\varphi})\hat{\varphi} = \frac{\partial h}{\partial \theta} \hat{\theta} + \frac{1}{\sin \theta}\frac{\partial h}{\partial \varphi} \hat{\varphi} = \mathrm{Grad}_{S^2} h.$$

Finally, we use the condition $\mathrm{Div}_{S^2} u = 0$ and obtain by partial integration

$$0 = \int_{S^2} h\, \mathrm{Div}_{S^2} \bar{u}\, ds = \int_{S^2} \bar{u} \cdot \mathrm{Grad}_{S^2} h\, ds = \int_{S^2} |u|^2\, ds$$

which yields $u = 0$ on S^2. □

Remark: From the first part of the proof of part (b) we observe that $\mathrm{Div}_{S^2}(\nu \times u) = 0$ on S^2 implies the existence of a surface potential h. This motivates the common notation of the *surface curl* of a tangential field u by $\mathrm{Curl}\, u = -\mathrm{Div}(\nu \times u)$.

The following result is needed as a preparation for the expansion theorem but is also of independent interest. It proves existence and uniqueness of a solution for the equation $\Delta_{S^2} u = f$ on S^2 where $\Delta_{S^2} = \mathrm{Div}_{S^2} \mathrm{Grad}_{S^2}$ denotes again the spherical Laplace–Beltrami operator (see Definition 2.2).

Theorem 2.45. *For every $f \in C^\infty(S^2)$ with $\int_{S^2} f\, ds = 0$ there exists a unique solution $u \in C^2(S^2)$ with $\Delta_{S^2} u = -f$ on S^2 and $\int_{S^2} u\, ds = 0$. If f is given by $f = \sum_{n=1}^\infty \sum_{m=-n}^n f_n^m Y_n^m$ with $f_n^m = (f, Y_n^m)_{L^2(S^2)}$, then the solution is given by*

$$u(x) = \sum_{n=1}^\infty \sum_{m=-n}^n \frac{f_n^m}{n(n+1)} Y_n^m(x) = \frac{1}{4\pi} \sum_{n=1}^\infty \frac{2n+1}{n(n+1)} \int_{S^2} P_n(x \cdot y)\, f(y)\, ds(y).$$

Both series converge uniformly with all their derivatives. In particular, $u \in C^\infty(S^2)$.

Proof: Uniqueness is seen by Green's theorem. Indeed, if $u \in C^2(S^2)$ is the difference of two solutions, then $\Delta_{S^2} u = 0$ and thus $0 = \int_{S^2} \bar{u}\, \mathrm{Div}_{S^2} \mathrm{Grad}_{S^2} u\, ds = \int_{S^2} |\mathrm{Grad}_{S^2} u|^2 ds$; that is, $\mathrm{Grad}_{S^2} u = 0$ which implies that u is constant. The requirement $\int_{S^2} u\, ds = 0$ yields that u vanishes.

For existence it suffices to show that the series for u converges with all of its derivatives because $\Delta_{S^2} Y_n^m = -n(n+1)Y_n^m$. For partial sums we have for any $q \in \mathbb{N}$

$$(S_N u)(x) = \sum_{n=1}^{N} \sum_{m=-n}^{n} \frac{f_n^m}{n(n+1)} Y_n^m(x)$$

$$= \sum_{n=1}^{N} \frac{1}{n(n+1)} \int_{S^2} \sum_{m=-n}^{n} Y_n^m(x) \, Y_n^{-m}(y) \, f(y) \, ds(y)$$

$$= \frac{1}{4\pi} \sum_{n=1}^{N} \frac{2n+1}{n(n+1)} \int_{S^2} P_n(x \cdot y) \, f(y) \, ds(y)$$

$$= \frac{(-1)^q}{4\pi} \sum_{n=1}^{N} \frac{2n+1}{n^{q+1}(n+1)^{q+1}} \int_{S^2} \Delta_{S^2}^q P_n(x \cdot y) \, f(y) \, ds(y)$$

$$= \frac{(-1)^q}{4\pi} \sum_{n=1}^{N} \frac{2n+1}{n^{q+1}(n+1)^{q+1}} \int_{S^2} P_n(x \cdot y) \, \Delta_{S^2}^q f(y) \, ds(y)$$

where we used the addition formula of Theorem 2.17 again and $\Delta_{S^2} P_n(x \cdot y) = -n(n+1) P_n(x \cdot y)$ (because $x \mapsto P_n(x \cdot y)$ is a spherical surface harmonic) and Green's theorem q times. We consider the extension of this formula to $x \in \mathbb{R}^3$ and observe for any differential operator D^ℓ of order ℓ that

$$(D^\ell S_N u)(x) = \frac{(-1)^q}{4\pi} \sum_{n=1}^{N} \frac{2n+1}{n^{q+1}(n+1)^{q+1}} \int_{S^2} D_x^\ell P_n(x \cdot y) \, \Delta_{S^2}^q f(y) \, ds(y) \, .$$

By the chain rule and Exercise 2.6 we conclude that $|D_x^\ell P_n(x \cdot y)| \leq c n^{2\ell}$ for all $n \in \mathbb{N}$ and $x, y \in S^2$. Therefore,

$$\frac{1}{4\pi} \sum_{n=1}^{N} \frac{2n+1}{n^{q+1}(n+1)^{q+1}} \left| \int_{S^2} D_x^\ell P_n(x \cdot y) \, \Delta_{S^2}^q f(y) \, ds(y) \right|$$

$$\leq c \, \|\Delta_{S^2}^q f\|_\infty \sum_{n=1}^{N} \frac{(2n+1) \, n^{2\ell}}{n^{q+1}(n+1)^{q+1}}$$

which converges if we choose $q \geq \ell + 1$. $\square$

Theorem 2.46. *The functions*

$$U_n^m = \frac{1}{\sqrt{n(n+1)}} \, \mathrm{Grad}_{S^2} Y_n^m(\hat{x}) \quad \text{and} \quad V_n^m(\hat{x}) = \hat{x} \times U_n^m(\hat{x})$$

for $n \in \mathbb{N}$ and $-n \leq m \leq n$ constitute a complete orthonormal system in

$$L_t^2(S^2) = \{ u \in L^2(S^2) : \nu \cdot u = 0 \quad \text{on } S^2 \} \, .$$

Proof: Let $f \in C_t^\infty(S^2) = \{ u \in C^\infty(S^2, \mathbb{C}^3) : u \cdot \nu = 0 \text{ on } S^2 \}$. Then also $\mathrm{Div}_{S^2} f$, $\mathrm{Div}_{S^2}(\nu \times f) \in C^\infty(S^2)$, and by Green's theorem we have

$\int_{S^2} \text{Div}_{S^2} f \, ds = \int_{S^2} \text{Div}_{S^2} (\nu \times f) \, ds = 0$. By the previous Theorem 2.45 there exist solutions $u_1, u_2 \in C^\infty(S^2)$ with $\Delta_{S^2} u_1 = \text{Div}_{S^2} f$ and $\Delta_{S^2} u_2 = \text{Div}_{S^2} (\nu \times f)$ on S^2. Their expansions

$$u_1 = \sum_{n=1}^{\infty} \sum_{m=-n}^{n} a_n^m Y_n^m \quad \text{and} \quad u_2 = \sum_{n=1}^{\infty} \sum_{m=-n}^{n} b_n^m Y_n^m$$

converge uniformly with all of their derivatives. Set

$$g = \text{Grad}\, u_1 - \nu \times \text{Grad}\, u_2$$

$$= \sum_{n=1}^{\infty} \sum_{m=-n}^{n} \left[a_n^m \, \text{Grad}\, Y_n^m - b_n^m \, \nu \times \text{Grad}\, Y_n^m \right]$$

$$= \sum_{n=1}^{\infty} \sum_{m=-n}^{n} \left[\sqrt{n(n+1)}\, a_n^m U_n^m - \sqrt{n(n+1)}\, b_n^m V_n^m \right].$$

Then $g \in C_t^\infty(S^2)$ and, by using part (a) of Lemma 2.44, $\text{Div}_{S^2}\, g = \Delta_{S^2} u_1 = \text{Div}_{S^2} f$ and $\text{Div}_{S^2}(\nu \times g) = -\text{Div}_{S^2}\left(\nu \times (\nu \times \text{Grad}_{S^2} u_2)\right) = \text{Div}_{S^2} \text{Grad}_{S^2} u_2 = \text{Div}_{S^2}(\nu \times f)$. Application of part (b) of Lemma 2.44 yields $g = f$ which proves uniform convergence of f into vector spherical harmonics. The completeness of the set $\{U_n^m, V_n^m : -n \leq m \leq m,\ n \in \mathbb{N}\}$ in $L_t^2(S^2)$ follows again by a denseness argument just as in the proof of Theorem 2.19. $\qquad \square$

As a corollary we formulate an expansion of any vector field A in terms of the normal basis functions $\hat{x} \to Y_m^n(\hat{x})\hat{x}$ and the tangential basis functions U_m^n and V_n^m.

Corollary 2.47. *Every vector field $A \in L^2(S^2, \mathbb{C}^3)$ has an expansion of the form*

$$A(\hat{x}) = \sum_{n=0}^{\infty} \sum_{|m| \leq n} \left[a_n^m \, Y_n^m(\hat{x})\, \hat{x} + b_n^m \, U_n^m(\hat{x}) + c_n^m \, V_n^m(\hat{x}) \right] \qquad (2.49)$$

where

$$a_n^m = \int_{S^2} A(\hat{x}) \cdot \hat{x}\, Y_n^{-m}(\hat{x})\, ds(\hat{x}), \quad |m| \leq n,\ n \geq 0, \qquad (2.50a)$$

$$b_n^m = \int_{S^2} A(\hat{x}) \cdot U_n^{-m}(\hat{x})\, ds(\hat{x}), \quad |m| \leq n,\ n \geq 1,\ b_0^0 = 0, \quad (2.50b)$$

$$c_n^m = \int_{S^2} A(\hat{x}) \cdot V_n^{-m}(\hat{x})\, ds(\hat{x}), \quad |m| \leq n,\ n \geq 1,\ c_0^0 = 0. \quad (2.50c)$$

The convergence is understood in the L^2-sense. Furthermore, Parseval's equality holds; that is,

$$\sum_{n=0}^{\infty} \sum_{|m|\le n} \left[|a_m^n|^2 + |b_m^n|^2 + |c_m^n|^2 \right] \; = \; \|A\|_{L^2(S^2,\mathbb{C}^3)}^2 \,.$$

Proof: We decompose the vector field A into

$$A(\hat{x}) \; = \; \left(A(\hat{x}) \cdot \hat{x} \right)\hat{x} \; + \; \hat{x} \times \left(A(\hat{x}) \times \hat{x} \right).$$

The scalar function $\hat{x} \mapsto \hat{x} \cdot A(\hat{x})$ and the tangential vector field $\hat{x} \mapsto \hat{x} \times \left(A(\hat{x}) \times \hat{x} \right)$ can be expanded into basis functions on S^2 according to Theorem 2.19 for the scalar radial part and the previous Theorem 2.46 for the tangential part. $\square$

We apply this corollary to the vector field $\hat{x} \mapsto E(r\hat{x})$ where E satisfies Maxwell's equation.

Theorem 2.48. *Let* $E, H \in C^1\left(B(0,R), \mathbb{C}^3\right)$ *satisfy* $\operatorname{curl} E - i\omega\mu H = 0$ *and* $\operatorname{curl} H + i\omega\varepsilon E = 0$ *in* $B(0,R)$. *Then there exist unique* $\alpha_n^m, \beta_n^m \in \mathbb{C}$, $|m| \le n$, $n = 0, 1, 2, \ldots$ *such that*

$$E(x) = \sum_{n=1}^{\infty} \sum_{|m|\le n} \left[\alpha_n^m \sqrt{n(n+1)}\, \frac{j_n(kr)}{r}\, Y_n^m(\hat{x})\, \hat{x} \right.$$

$$\left. + \alpha_n^m\, \frac{\left(r j_n(kr)\right)'}{r}\, U_n^m(\hat{x}) \; + \; \beta_n^m\, j_n(kr)\, V_n^m(\hat{x}) \right] \tag{2.51a}$$

$$= \sum_{n=1}^{\infty} \sum_{|m|\le n} \frac{1}{\sqrt{n(n+1)}} \left[\alpha_n^m\, \operatorname{curl}\operatorname{curl}\left(x\, j_n(kr)\, Y_n^m(\hat{x})\right) \right.$$

$$\left. - \beta_n^m\, \operatorname{curl}\left(x\, j_n(kr)\, Y_n^m(\hat{x})\right) \right] \tag{2.51b}$$

$$H(x) = -\frac{1}{i\omega\mu} \sum_{n=1}^{\infty} \sum_{|m|\le n} \left[\beta_n^m \sqrt{n(n+1)}\, \frac{j_n(kr)}{r}\, Y_n^m(\hat{x})\, \hat{x} \right.$$

$$\left. + \beta_n^m\, \frac{\left(r j_n(kr)\right)'}{r}\, U_n^m(\hat{x}) \; + \; k^2 \alpha_n^m\, j_n(kr)\, V_n^m(\hat{x}) \right] \tag{2.51c}$$

$$= \frac{1}{i\omega\mu} \sum_{n=1}^{\infty} \sum_{|m|\le n} \frac{1}{\sqrt{n(n+1)}} \left[k^2 \alpha_n^m\, \operatorname{curl}\left(x\, j_n(kr)\, Y_n^m(\hat{x})\right) \right.$$

$$\left. - \beta_n^m\, \operatorname{curl}\operatorname{curl}\left(x\, j_n(kr)\, Y_n^m(\hat{x})\right) \right] \tag{2.51d}$$

for $x = r\hat{x}$ *and* $0 \le r < R$, $\hat{x} \in S^2$. *The series converge uniformly in every ball* $B[0, R']$ *for* $R' < R$.

Proof: Let $a_n^m(r)$, $b_n^m(r)$, and $c_n^m(r)$ be the expansion coefficients of (2.49) corresponding to the vector field given by $\hat{x} \mapsto E(r\hat{x})$. First we note that

$a_0^0(r) = \frac{1}{r^2} \int_{|x|=r} \hat{x} \cdot E(x) \, ds(x)$ vanishes by the divergence theorem because $\text{div}\, E = 0$.

Second, we consider $a_n^m(r)$ for $n \geq 1$. With

$$a_n^m(r) = \int_{S^2} \hat{x} \cdot E(r\hat{x}) \, Y_n^{-m}(\hat{x}) \, ds(\hat{x}) = \frac{1}{r} \int_{S^2} x \cdot E(x) \, Y_n^{-m}(\hat{x}) \, ds(\hat{x})$$

we observe that $r a_n^m(r)$ is the expansion coefficient of the solution $x \mapsto x \cdot E(x)$ of the scalar Helmholtz equation, see Lemma 1.5. Thus by Theorem 2.33 we obtain

$$a_n^m(r) = \frac{\tilde{a}_n^m}{r} \, j_n(kr)$$

for some $\tilde{a}_n^m$. Next, the condition $\text{div}\, E = 0$ in spherical polar coordinates reads

$$\frac{1}{r} \frac{\partial}{\partial r}(r^2 E_r(r, \hat{x})) + \text{Div}_{S^2}\, E_t(r, \hat{x}) = 0$$

(see (A.20)), and we obtain

$$\frac{1}{r}\left(r^2 a_n^m(r)\right)' = \int_{S^2} \frac{1}{r} \frac{\partial}{\partial r}(r^2 E_r(r, \hat{x})) \, Y_n^{-m}(\hat{x}) \, ds(\hat{x})$$

$$= -\int_{S^2} \text{Div}_{S^2}\, E_t(r, \hat{x}) \, Y_n^{-m}(\hat{x}) \, ds(\hat{x})$$

$$= \int_{S^2} E(r, \hat{x}) \cdot \text{Grad}_{S^2} Y_n^{-m}(\hat{x}) \, ds(\hat{x}) = \sqrt{n(n+1)}\, b_n^m(r)$$

by (A.21). Thus, with a common constant α_n^m we have

$$a_n^m(r) = \alpha_n^m \frac{\sqrt{n(n+1)}}{r} \, j_n(kr), \quad b_n^m(r) = \alpha_n^m \frac{1}{r}\left(r\, j_n(kr)\right)'. \qquad (2.52)$$

Finally, we consider $c_n^m(r)$. By partial integration it holds

$$\sqrt{n(n+1)}\, c_n^m(r) = \int_{S^2} E(r, \hat{x}) \cdot \left(\hat{x} \times \text{Grad}_{S^2} Y_n^{-m}(\hat{x})\right) ds(\hat{x})$$

$$= \int_{S^2} \left(E(r, \hat{x}) \times \hat{x}\right) \cdot \text{Grad}_{S^2} Y_n^{-m}(\hat{x}) \, ds(\hat{x})$$

$$= -\int_{S^2} \text{Div}_{S^2} \left(E(r, \hat{x}) \times \hat{x}\right) Y_n^{-m}(\hat{x}) \, ds(\hat{x})$$

$$= -r \int_{S^2} \hat{x} \cdot \text{curl}\, E(r\hat{x}) \, Y_n^{-m}(\hat{x}) \, ds(\hat{x})$$

$$= -\int_{S^2} x \cdot \text{curl}\, E(r\hat{x}) \, Y_n^{-m}(\hat{x}) \, ds(\hat{x})$$

where we have used that $\mathrm{Div}_{S^2}\big(E(r,\hat{x})\times\hat{x}\big) = \mathrm{Div}_{|x|=r}\big(E(r\hat{x})\times\hat{x}\big) = r\,\hat{x}\cdot\mathrm{curl}\,E(r\hat{x})$ (see Corollary A.20). Therefore, $\sqrt{n(n+1)}\,c_n^m(r)$ are the expansion coefficients of the scalar function $\psi(x) = x\cdot\mathrm{curl}\,E(x)$. Because ψ is a solution of the Helmholtz equation $\Delta\psi + k^2\psi = 0$, see Lemma 1.5, we again conclude from Theorem 2.33 that $c_n^m(r) = \beta_n^m\,j_n(kr)$ for some $\beta_n^m\in\mathbb{C}$. This proves the first representation (2.51a).

Furthermore, using (2.46a) and (2.46b) yields the second representation (2.51b).

It remains to show uniform convergence in $B[0,R']$ for every $R' < R$. As in the scalar case we will prove another representation of the field E. Fix $R' < R$ and choose $\hat{R}$ with $R' < \hat{R} < R$ such that $j_n(k\hat{R}) \neq 0$ and $j_n(k\hat{R}) + kj_n'(k\hat{R}) \neq 0$. The scalar function $\hat{x}\mapsto E(\hat{R}\hat{x})\cdot\hat{x}$ is smooth and we obtain from (2.51a) that

$$\alpha_n^m\,\sqrt{n(n+1)}\,\frac{j_n(k\hat{R})}{\hat{R}} = \int_{S^2}\big(E(\hat{R}\hat{y})\cdot\hat{y}\big)\,Y_n^{-m}(\hat{y})\,ds(\hat{y})\,.$$

Analogously, we obtain

$$\alpha_n^m\,\frac{j_n(k\hat{R}) + k\hat{R}j_n'(k\hat{R})}{\hat{R}} = \int_{S^2} E(\hat{R}\hat{y})\cdot U_n^{-m}(\hat{y})\,ds(\hat{y})$$

$$= -\frac{1}{\sqrt{n(n+1)}}\int_{S^2}\mathrm{Div}_{S^2}\,E(\hat{R},\hat{y})\,Y_n^{-m}(\hat{y})\,ds(\hat{y})\,,$$

and

$$\beta_n^m\,j_n(k\hat{R}) = \int_{S^2} E(\hat{R}\hat{y})\cdot V_n^{-m}(\hat{y})\,ds(\hat{y})$$

$$= \frac{1}{\sqrt{n(n+1)}}\int_{S^2} E(\hat{R},\hat{y})\cdot\big(\hat{y}\times\mathrm{Grad}_{S^2}Y_n^{-m}(\hat{y})\big)\,ds(\hat{y})$$

$$= -\frac{1}{\sqrt{n(n+1)}}\int_{S^2}\mathrm{Div}_{S^2}\big(E(\hat{R},\hat{y})\times\hat{y}\big)Y_n^{-m}(\hat{y})\,ds(\hat{y})\,.$$

Substituting these coefficients into (2.51a) and using the addition formula of Theorem 2.17 yields

$$E(r\hat{x}) = \sum_{n=1}^{\infty}\sum_{|m|\leq n}\Bigg[\frac{\hat{R}}{r}\,\frac{j_n(kr)}{j_n(k\hat{R})}\int_{S^2}\big(E(\hat{R}\hat{y})\cdot\hat{y}\big)Y_n^{-m}(\hat{y})\,ds(\hat{y})\,Y_n^m(\hat{x})\,\hat{x}$$

$$-\frac{\hat{R}}{r}\,\frac{j_n(kr) + krj_n'(kr)}{j_n(k\hat{R}) + k\hat{R}j_n'(k\hat{R})}\,\frac{1}{n(n+1)}\,\mathrm{Grad}_{S^2}\int_{S^2}\mathrm{Div}_{S^2}\,E(\hat{R},\hat{y})\,Y_n^{-m}(\hat{y})\,ds(\hat{y})\,Y_n^m(\hat{x})$$

$$-\frac{j_n(kr)}{j_n(k\hat{R})}\,\frac{1}{n(n+1)}\Big(\hat{x}\times\mathrm{Grad}_{S^2}\int_{S^2}\mathrm{Div}_{S^2}\big(E(\hat{R},\hat{y})\times\hat{y}\big)\,Y_n^{-m}(\hat{y})\,ds(\hat{y})\,Y_n^m(\hat{x})\Big)\Bigg]$$

$$\tag{2.53}$$

$$= \frac{1}{4\pi} \sum_{n=1}^{\infty} (2n+1) \left[\frac{\hat{R}}{r} \frac{j_n(kr)}{j_n(k\hat{R})} \int_{S^2} \left(E(\hat{R}\hat{y}) \cdot \hat{y} \right) P_n(\hat{x} \cdot \hat{y}) \, ds(\hat{y}) \, \hat{x} \right.$$

$$- \frac{\hat{R}}{r} \frac{j_n(kr) + krj_n'(kr)}{j_n(k\hat{R}) + k\hat{R}j_n'(k\hat{R})} \frac{1}{n(n+1)} \operatorname{Grad}_{S^2} \int_{S^2} \operatorname{Div}_{S^2} E(\hat{R}, \hat{y}) P_n(\hat{x} \cdot \hat{y}) \, ds(\hat{y})$$

$$\left. - \frac{j_n(kr)}{j_n(k\hat{R})} \frac{1}{n(n+1)} \left(\hat{x} \times \operatorname{Grad}_{S^2} \int_{S^2} \operatorname{Div}_{S^2} \left(E(\hat{R}, \hat{y}) \times \hat{y} \right) P_n(\hat{x} \cdot \hat{y}) \, ds(\hat{y}) \right) \right].$$

This has now the form of the series for the solution of the interior boundary value problem for the Helmholtz equation. The arguments used there, in the proof of Theorem 2.33, provide analogously uniform convergence of the series and its derivatives in the closed ball $B[0, R']$. □

The expansions (2.51a) and (2.51b) as well as (2.51c) and (2.51d) complement each other. The representation (2.51b) shows an expansion into vector wave functions which is closely related to the idea from Lemma 2.42 using the scalar expansion functions of the Helmholtz equation. On the other hand, (2.51a) is suitable for treating boundary value problems (see Theorem 2.49 below) and, additionally, gives some insight into the important relation of the tangential and the normal component of the electric field on surfaces.

Combining the previous results we obtain the following existence result for the interior Maxwell problem in a ball; that is, the boundary value problem for Maxwell's equation in a ball $B(0, R)$ with boundary values $\nu \times E = f_R$ on $|x| = R$. We formulate the existence result again with $f_R \in L_t^2(S^2)$ where $f_R(\hat{x}) = f(R\hat{x})$.

Theorem 2.49. *We assume that* $\varepsilon, \mu \in \mathbb{C}$ *with* $\operatorname{Im} k \geq 0$ *where again* $k = \omega\sqrt{\varepsilon\mu}$ *and* $R > 0$ *are such that* $j_n(kR) \neq 0$ *and* $j_n(kR) + kR\, j_n'(kR) \neq 0$ *for all* $n \in \mathbb{N}_0$.

(a) *For given* $f_R \in L_t^2(S^2)$ *there exists a unique solution* $E \in C^2\big(B(0, R)\big)$ *and* $H = \frac{1}{i\omega\mu} \operatorname{curl} E$ *of*

$$\operatorname{curl} E - i\omega\mu H = 0 \quad and \quad \operatorname{curl} H + i\omega\varepsilon E = 0 \quad in \ B(0, R)$$

with

$$\lim_{r \to R} \| \nu \times E(r, \cdot) - f_R \|_{L^2(S^2)} = 0.$$

The solution is given by

$$E(r\hat{x}) = \sum_{n=1}^{\infty} \sum_{m=-n}^{n} \frac{(f_R, U_n^m)_{L^2(S^2)}}{\sqrt{n(n+1)}} \frac{1}{j_n(kR)} \operatorname{curl}\left[x \, j_n(kr) \, Y_n^m(\hat{x})\right]$$

$$+ \frac{(f_R, V_n^m)_{L^2(S^2)}}{\sqrt{n(n+1)}} \frac{R}{j_n(kR) + kR \, j_n'(kR)} \operatorname{curl}\operatorname{curl}\left[x j_n(kr) \, Y_n^m(\hat{x})\right]$$

$$= \sum_{n=1}^{\infty} \sum_{m=-n}^{n} (f_R, V_n^m)_{L^2(S^2)} \frac{\sqrt{n(n+1)} \, R}{j_n(kR) + kR \, j_n'(kR)} \frac{j_n(kr)}{r} Y_n^m(\hat{x}) \, \hat{x}$$

$$+ (f_R, V_n^m)_{L^2(S^2)} \frac{R}{j_n(kR) + kR \, j_n'(kR)} \frac{j_n(kr) + kr \, j_n'(kr)}{r} U_n^m(\hat{x})$$

$$- (f_R, U_n^m)_{L^2(S^2)} \frac{j_n(kr)}{j_n(kR)} V_n^m(\hat{x}) \,,$$

$$H(r\hat{x}) = \frac{1}{i\omega\mu} \sum_{n=1}^{\infty} \sum_{m=-n}^{n} \frac{(f_R, U_n^m)_{L^2(S^2)}}{\sqrt{n(n+1)}} \frac{1}{j_n(kR)} \operatorname{curl}\operatorname{curl}\left[x \, j_n(kr) \, Y_n^m(\hat{x})\right]$$

$$+ k^2 \frac{(f_R, V_n^m)_{L^2(S^2)}}{\sqrt{n(n+1)}} \frac{R}{j_n(kR) + kR \, j_n'(kR)} \operatorname{curl}\left[x j_n(kr) \, Y_n^m(\hat{x})\right]$$

$$= \frac{1}{i\omega\mu} \sum_{n=1}^{\infty} \sum_{m=-n}^{n} (f_R, U_n^m)_{L^2(S^2)} \sqrt{n(n+1)} \frac{1}{j_n(kR)} \frac{j_n(kr)}{r} Y_n^m(\hat{x}) \, \hat{x}$$

$$+ (f_R, U_n^m)_{L^2(S^2)} \frac{1}{j_n(kR)} \frac{j_n(kr) + kr \, j_n'(kr)}{r} U_n^m(\hat{x})$$

$$- k^2 (f_R, V_n^m)_{L^2(S^2)} \frac{R}{j_n(kR) + kRj_n'(kR)} j_n(kr) \, V_n^m(\hat{x})$$

for $x = r\hat{x} \in B(0, R)$. The series converges uniformly on compact subsets of $B(0, R)$.

(b) *If $f_R \in C_t^2(S^2)$, the solution of the boundary value problem satisfies $E \in C^2(B(0, R)) \cap C(B[0, R])$ and the series converges uniformly on the closed ball $B[0, R]$.*

Proof: The proof follows the arguments as in Theorem 2.37 for the case of a boundary value problem for the Helmholtz equation.

Let us repeat the steps. To show uniqueness we assume E is a solution of the homogeneous boundary value problem, i.e. $\|\nu \times E(r\hat{x})\|_{L^2(S^2)} \to 0$ for $r \to R$ with $\nu(\hat{x}) = \hat{x}$. First we consider the expansion from Theorem 2.48 for any $R_0 < R$ and $\varepsilon > 0$ and $R_\varepsilon \in (R_0, R)$ such that $j_n(kR_\varepsilon) \neq 0$, $j_n(kR_\varepsilon) + kR_\varepsilon j_n'(kR_\varepsilon) \neq 0$ and $\|\nu \times E(R_\varepsilon \hat{x})\|_{L^2(S^2)} \leq \varepsilon$. From (2.53) we obtain

$$\hat{x} \times E(r\hat{x})$$

$$= \sum_{n=1}^{\infty} \sum_{|m|\leq n} \frac{R_\varepsilon}{r} \frac{j_n(kr) + kr\, j_n'(kr)}{j_n(kR_\varepsilon) + kR_\varepsilon\, j_n'(kR_\varepsilon)} \left(\nu \times E(R_\varepsilon\,\cdot\,), V_n^m\right)_{L^2(S^2)} U_n^m(\hat{x})$$

$$- \sum_{n=1}^{\infty} \sum_{|m|\leq n} \frac{j_n(kr)}{j_n(kR_\varepsilon)} \left(\nu \times E(R_\varepsilon\,\cdot\,), U_n^m\right)_{L^2(S^2)} V_n^m(\hat{x}).$$

By Parseval's identity we obtain for $0 < R_1 < r < R_0 < R_\varepsilon < R$

$$\|\nu \times E(r\,\cdot\,)\|_{L^2(S^2)}^2$$

$$= \sum_{n=1}^{\infty} \sum_{|m|\leq n} \left|\frac{R_\varepsilon}{r}\right|^2 \left|\frac{j_n(kr) + kr\, j_n'(kr)}{j_n(kR_\varepsilon) + kR_\varepsilon\, j_n'(kR_\varepsilon)}\right|^2 \left|\left(\nu \times E(R_\varepsilon\,\cdot\,), V_n^m\right)_{L^2(S^2)}\right|^2$$

$$+ \left|\frac{j_n(kr)}{j_n(kR_\varepsilon)}\right|^2 \left|\left(\nu \times E(R_\varepsilon\,\cdot\,), U_n^m\right)_{L^2(S^2)}\right|^2$$

$$\leq c^2 \sum_{n=1}^{\infty} \sum_{|m|\leq n} \left|\left(\nu \times E(R_\varepsilon\,\cdot\,), V_n^m\right)_{L^2(S^2)}\right|^2 + \left|\left(\nu \times E(R_\varepsilon\,\cdot\,), U_n^m\right)_{L^2(S^2)}\right|^2$$

$$\leq c^2 \|\nu \times E(R_\varepsilon)\|_{L^2(S^2)} \leq c^2 \varepsilon^2. \tag{2.54}$$

Here we have used again as in the proof of 2.34 a uniform bound for all $n \in \mathbb{N}$ of $\left|\frac{j_n(kr)}{j_n(kR_\varepsilon)}\right| \leq c$ and of $\left|\frac{R_\varepsilon}{r} \frac{j_n(kr)+kr\, j_n'(kr)}{j_n(kR_\varepsilon)+kR_\varepsilon\, j_n'(kR_\varepsilon)}\right| \leq c$ for $r \in [R_1, R_0]$, where the last estimate follows from Theorem 2.31 by using the identity $j_n(z) + zj_n'(z) = zj_{n-1}(z) - nj_n(z)$ (see Exercise 2.8).

Since (2.54) holds for any $\varepsilon > 0$ we obtain by the expansion of Theorem 2.48 that $\alpha_n^m (rj_n(kr))' = 0$ and $\beta_n^m j_n(kr) = 0$ for all $r \in (R_1, R_0)$ and $|m| \leq n$, $n \in \mathbb{N}$. Thus it follows $\alpha_n^m = \beta_n^m = 0$, and the expansion leads to $E = 0$ in $B(0, R)$.

Next we show that the function $E(r\hat{x})$ given in the Theorem satisfies the boundary condition in L^2 sense. For $r < R$ we compare the representation (2.51b) with the given expansion and define

$$\alpha_n^m = \frac{R}{j_n(kR) + kRj_n'(kR)}(f_R, V_n^m)_{L^2} \quad \text{and} \quad \beta_n^m = -\frac{1}{j_n(kR)}(f_R, U_n^m)_{L^2}.$$

Thus, we obtain from (2.51a)

$$\hat{x} \times E(r\hat{x}) = \sum_{n=1}^{\infty} \sum_{m=-n}^{n} \frac{\alpha_n^m}{r}(rj_n(kr))' \,\hat{x} \times U_n^m(\hat{x}) + \beta_n^m j_n(kr)\,\hat{x} \times V_n^m(\hat{x})$$

$$= \sum_{n=1}^{\infty} \sum_{m=-n}^{n} \frac{\alpha_n^m}{r}(rj_n(kr))'\, V_n^m(\hat{x}) - \beta_n^m j_n(kr)\, U_n^m(\hat{x}).$$

With the definition of α_n^m and β_n^m it follows by Parseval's identity

$$\|\nu \times E(r\,.) - f_R\|_{L^2} = \sum_{n=1}^{\infty} \sum_{m=-n}^{n} \left| \frac{j_n(kr)}{j_n(kR)} - 1 \right|^2 |(f_r, U_n^m)_{L^2}|^2$$

$$+ \sum_{n=1}^{\infty} \sum_{m=-n}^{n} \left| \frac{R}{r} \frac{j_n(kr) + krj_n'(kr)}{j_n(kR) + krj_n'(kR)} - 1 \right|^2 |(f_r, V_n^m)_{L^2}|^2 .$$

As in Theorem 2.34 we can adapt the arguments of Theorem 2.22 in the case of the Laplacian, since by Theorem 2.31 the terms $\left| \frac{j_n(kr)}{j_n(kR)} - 1 \right|$ and $\left| \frac{R}{r} \frac{j_n(kr) + krj_n'(kr)}{j_n(kR) + kRj_n'(kR)} - 1 \right|$ are uniformly bounded for $r \in [0, R]$ and converge to zero as r tends to R for every n. Thus we conclude that $\|\nu \times E(r, \cdot) - f_R\|_{L^2} \to 0$ if $r \to R$.

Furthermore, the second part of the proof of Theorem 2.22 can be transferred to the electric fields by the uniform bounds on the ratios $\left| \frac{j_n(kr)}{j_n(kR)} \right|$ and $\left| \frac{R}{r} \frac{j_n(kr) + krj_n'(kr)}{j_n(kR) + kRj_n'(kR)} \right|$ and we conclude part (b) of the Theorem. $\square$

The last part in this chapter is devoted to the Maxwell problem in the exterior $\mathbb{R}^3 \backslash \overline{B(0, R)}$ of a ball. Here we restrict ourself to real $\varepsilon_0, \mu_0 > 0$ that is $k = \omega\sqrt{\varepsilon_0\mu_0} > 0$. First we observe that replacing the Bessel functions by the Hankel functions of the first kind in the constructed expansion functions will lead to solutions satisfying a radiation condition. At this point we find it very convenient to use the scalar Sommerfeld radiation condition (2.32) for the solutions $x \cdot E(x)$ and $x \cdot H(x)$ of the Helmholtz equation. These conditions imply in particular that the radial components of the electric and magnetic fields decay as $1/r^2$.; that is, the fields E and H are "almost" tangential fields on large spheres. Later (see Corollary 2.53 and Remark 3.31) we will show that this form of the radiation condition is equivalent to the better known Silver–Müller radiation condition.

Theorem 2.50. Let $E \in C^2(\mathbb{R}^3 \setminus B[0, R])$ and $H = \frac{1}{i\omega\mu} \operatorname{curl} E$ be solutions of the time-harmonic Maxwell equations

$$\operatorname{curl} E - i\omega\mu H = 0 \text{ in } D \quad \text{and} \quad \operatorname{curl} H + i\omega\varepsilon E = 0 \text{ in } \mathbb{R}^3 \setminus B[0, R].$$

Furthermore, let $x \mapsto x \cdot E(x)$ and $x \mapsto x \cdot H(x)$ satisfy the Sommerfeld radiation condition (2.32). Then there exist unique $\alpha_n^m, \beta_n^m \in \mathbb{C}$, $|m| \le n$, $n = 0, 1, 2, \ldots$ such that

$$E(x) = \sum_{n=1}^{\infty} \sum_{|m| \leq n} \left[\frac{\alpha_n^m \sqrt{n(n+1)}}{r} h_n^{(1)}(kr) Y_n^m(\hat{x}) \hat{x} \right.$$

$$\left. + \frac{\alpha_n^m}{r} \left(r h_n^{(1)}(kr)\right)' U_n^m(\hat{x}) + \beta_n^m h_n^{(1)}(kr) V_n^m(\hat{x}) \right] \quad (2.55a)$$

$$= \sum_{n=1}^{\infty} \sum_{|m| \leq n} \frac{1}{\sqrt{n(n+1)}} \left[\alpha_n^m \operatorname{curl} \operatorname{curl} \left(x \, h_n^{(1)}(kr) Y_n^m(\hat{x})\right) \right.$$

$$\left. - \beta_n^m \operatorname{curl} \left(x \, h_n^{(1)}(kr) Y_n^m(\hat{x})\right) \right] \quad (2.55b)$$

for $x = r\hat{x}$ and $r > R$, $\hat{x} \in S^2$. The series converges uniformly with all of its derivatives in compact subsets of $\mathbb{R}^3 \setminus B[0, R]$.

The field H has the forms (2.51c), (2.51d) with $h_n^{(1)}$ replacing j_n.

Proof: The proof uses exactly the same arguments as the proof of Theorem 2.48. First we note that $r a_n^m(r) = \int_{S^2} x \cdot E(x) Y_n^{-m}(\hat{x}) \, ds(\hat{x})$ is the expansion coefficient of the solution $u(x) = x \cdot E(x)$ of the Helmholtz equation with respect to $\{Y_n^m : |m| \leq n, \ n \in \mathbb{N}\}$. By assumption u is radiating, therefore $r a_n^m(r) = \tilde{a}_n^m h^{(1)}(kr)$ for some coefficient $\tilde{a}_n^m$. Therefore, (2.52) holds for j_n replaced by $h_n^{(1)}$. Analogously, $c_n^m(r)$ is given by $c_n^m(r) = \beta_n^m h_n^{(1)}(kr)$ for some β_n^m because also $x \mapsto x \cdot \operatorname{curl} E(x)$ is radiating. This yields the form (2.55a) for $E(x)$ and from this also (2.55b) by using (2.46a), (2.46b) for $h_n^{(1)}$ instead of j_n. Uniform convergence of this series is shown as in the proof of Theorem 2.48. □

In contrast to the interior problem and similar to the Helmholtz equation the exterior Maxwell problem for a ball is uniquely solvable for any $k \in \mathbb{C}$ with $\operatorname{Im} k \geq 0$ and every $R > 0$.

Theorem 2.51. Let $k \in \mathbb{C} \setminus \{0\}$ with $\operatorname{Im} k \geq 0$ and let $R > 0$.

(a) For given $f_R \in L_t^2(S^2)$ there exist unique solutions $E, H \in C^2(\mathbb{R}^3 \setminus B[0, R])$ of

$$\operatorname{curl} E - i\omega\mu H = 0 \quad \text{and} \quad \operatorname{curl} H + i\omega\varepsilon E = 0 \quad \text{in } \mathbb{R}^3 \setminus B[0, R]$$

with

$$\lim_{r \to R} \|\nu \times E(r, \cdot) - f_R\|_{L^2(S^2)} = 0,$$

such that $x \mapsto x \cdot E(x)$ and $x \mapsto x \cdot H(x)$ satisfy the Sommerfeld radiation condition (2.32). The solution is given by

$$E(r\hat{x}) = \sum_{n=1}^{\infty} \sum_{m=-n}^{n} \frac{(f_R, U_n^m)_{L^2(S^2)}}{\sqrt{n(n+1)}} \frac{1}{h_n^{(1)}(kR)} \operatorname{curl}\left[x\, h_n^{(1)}(kr)\, Y_n^m(\hat{x})\right]$$

$$+ \frac{(f_R, V_n^m)_{L^2(S^2)}}{\sqrt{n(n+1)}} \frac{R}{h_n^{(1)}(kR) + kR\, h_n^{(1)'}(kR)} \operatorname{curl}\operatorname{curl}\left[x h_n^{(1)}(kr)\, Y_n^m(\hat{x})\right]$$

$$= \sum_{n=1}^{\infty} \sum_{m=-n}^{n} (f_R, V_n^m)_{L^2(S^2)} \frac{\sqrt{n(n+1)}\, R}{h_n^{(1)}(kR) + kR\, h_n^{(1)'}(kR)} \frac{h_n^{(1)}(kr)}{r} Y_n^m(\hat{x})\, \hat{x}$$

$$+ (f_R, V_n^m)_{L^2(S^2)} \frac{R}{h_n^{(1)}(kR) + kR\, h_n^{(1)'}(kR)} \frac{h_n^{(1)}(kr) + kr\, h_n^{(1)'}(kr)}{r} U_n^m(\hat{x})$$

$$- (f_R, U_n^m)_{L^2(S^2)} \frac{h_n^{(1)}(kr)}{h_n^{(1)}(kR)} V_n^m(\hat{x})\, .$$

The series converges uniformly on compact subsets of $\mathbb{R}^3 \setminus B[0, R]$.

The field H has the forms as in Theorem 2.49 with $h_n^{(1)}$ replacing j_n.

(b) *If $f_R \in L_t^2(S^2) \cap C^2(S^2)$, the solution E is continuous in all of $\mathbb{R}^3 \setminus B(0, R)$, and its series representation converges uniformly on $B[0, R_1] \setminus B(0, R)$ for every $R_1 > R$.*

Proof: We omit the proofs of uniqueness and the fact that the series solves the boundary value problem because the arguments are almost the same as in the proof of the corresponding Theorem 2.49 for the interior boundary value problem. Again, we only mention that $h_n^{(1)}(kR)$ and $h_n^{(1)}(kR) + kRh_n^{(1)'}(kR)$ do not vanish for any kR which follows again by Theorem 2.27. To show the radiation condition we multiply the second representation of $E(x)$ by $x = r\hat{x}$ and arrive at

$$x \cdot E(x) = \sum_{n=0}^{\infty} \sum_{m=-n}^{n} (f_R, V_n^m)_{L^2(S^2)} \sqrt{n(n+1)} \frac{Rh_n^{(1)}(kr)}{h_n^{(1)}(kR) + kRh_n^{(1)'}(kR)} Y_n^m(\hat{x})\, .$$

This scalar solution of the Helmholtz equation has just the form of (2.38). Application of Lemmas 2.38 and 2.39 yields that $x \mapsto x \cdot E(x)$ satisfies the Sommerfeld radiation condition.

The corresponding proof for $x \cdot H(x)$ uses the same arguments for the representation of $H = \frac{1}{i\omega\mu} \operatorname{curl} E$. $\square$

For later use we formulate and prove the analog of Lemma 2.38.

Lemma 2.52. *Let*

$$E(x) = \sum_{n=1}^{\infty} \sum_{|m|\leq n} \left[\alpha_n^m \operatorname{curl}\left(x\, h_n^{(1)}(kr)\, Y_n^m(\hat{x})\right) + \beta_n^m \operatorname{curl}\operatorname{curl}\left(x\, h_n^{(1)}(kr)\, Y_n^m(\hat{x})\right)\right]$$

converge uniformly on compact subsets of $\mathbb{R}^3 \setminus B[0, R]$. *Then there exists* $R_0 > R$ *and scalar continuous functions* g *and* h *such that*

$$E(x) = \text{curl} \left[\int_{|y|=R_0} g(y)\, \Phi(x, y)\, ds(y)\, x \right]$$

$$+ \text{curl}^2 \left[\int_{|y|=R_0} h(y)\, \Phi(x, y)\, ds(y)\, x \right] \tag{2.56a}$$

$$H(x) = \frac{1}{i\omega\mu} \text{curl}\, E(x) = \frac{1}{i\omega\mu} \text{curl}^2 \left[\int_{|y|=R_0} g(y)\, \Phi(x, y)\, ds(y)\, x \right]$$

$$+ i\omega\varepsilon\, \text{curl} \left[\int_{|y|=R_0} h(y)\, \Phi(x, y)\, ds(y)\, x \right] \tag{2.56b}$$

for $|x| > R_0$. *Here,* Φ *denotes again the fundamental solution of the Helmholtz equation, given by (2.37). In particular, the solution of the boundary value problem of the previous theorem can be represented in this form.*

Proof: We follow the proof of Lemma 2.38. Let $R_0 > R$ such that $j_n(kR_0) \neq 0$ for all n and substitute the right-hand side of (2.40) into the series for $E(x)$. This yields

$$E(x) = \sum_{n=1}^{\infty} \sum_{|m| \leq n} \frac{a_n^m}{ikR_0^2 j_n(kR_0)} \text{curl} \left[\int_{|y|=R_0} Y_n^m(\hat{y})\, \Phi(x, y)\, ds(y)\, x \right]$$

$$+ \frac{\beta_n^m}{ikR_0^2 j_n(kR_0)} \text{curl}\,\text{curl} \left[\int_{|y|=R_0} Y_n^m(\hat{y})\, \Phi(x, y)\, ds(y)\, x \right]$$

$$= \text{curl} \left[\int_{|y|=R_0} g(y)\, \Phi(x, y)\, ds(y)\, x \right]$$

$$+ \text{curl}\,\text{curl} \left[\int_{|y|=R_0} h(y)\, \Phi(x, y)\, ds(y)\, x \right]$$

with

$$g(y) = \sum_{n=1}^{\infty} \sum_{|m| \leq n} \frac{a_n^m}{ikR_0^2 j_n(kR_0)} Y_n^m(\hat{y}),$$

$$h(y) = \sum_{n=1}^{\infty} \sum_{|m| \leq n} \frac{\beta_n^m}{ikR_0^2 j_n(kR_0)} Y_n^m(\hat{y}).$$

The uniform convergence of the series for g and h is shown in the same way as in the proof of Lemma 2.38. The form of H is obvious by $\text{curl}^2 = -\Delta + \nabla \text{div}$ and $k^2 = \omega^2 \mu\varepsilon$. $\square$

We finish this chapter by the following corollary which states that any pair (E, H) of solutions of the Maxwell system which satisfy the scalar radiation conditions of Theorems 2.50 and 2.51 satisfy also the Silver–Müller radiation condition (1.20a), (1.20b); or even,

$$\sqrt{\mu_0}\, H(x) \times \hat{x} \; - \; \sqrt{\varepsilon_0}\, E(x) \; = \; \mathcal{O}(|x|^{-2}) \qquad\qquad (2.57a)$$

and

$$\sqrt{\varepsilon_0}\, E(x) \times \hat{x} \; + \; \sqrt{\mu_0}\, H(x) \; = \; \mathcal{O}(|x|^{-2}) \qquad\qquad (2.57b)$$

uniformly in $x/|x| \in S^2$. As mentioned above, the scalar radiation conditions of Theorems 2.50 and 2.51 imply in particular that the radial components $\hat{x} \cdot E(x)$ and $\hat{x} \cdot H(x)$ decay faster than $\mathcal{O}(|x|^{-1})$ to zero as $|x|$ tends to infinity; that is, $E(x)$ and $H(x)$ are "almost" tangential fields on large spheres. On the other hand, the Silver–Müller radiation condition implies that the combination $\sqrt{\mu_0} H(x) \times \hat{x} - \sqrt{\varepsilon} E(x)$ decays faster than $\mathcal{O}(|x|^{-1})$ to zero as $|x|$ tends to infinity; that is, $E(x)$ and $H(x)$ are "almost" orthogonal to each other.

Corollary 2.53. *Let $E \in C^2(\mathbb{R}^3 \setminus B[0, R])$ and $H = \frac{1}{i\omega\mu}$ curl E be solutions of the time-harmonic Maxwell equations*

$$\text{curl } E - i\omega\mu H = 0 \text{ in } D \quad \text{and} \quad \text{curl } H + i\omega\varepsilon E = 0 \text{ in } \mathbb{R}^3 \setminus B[0, R]$$

such that $x \mapsto x \cdot E(x)$ and $x \mapsto x \cdot H(x)$ satisfy the Sommerfeld radiation condition (2.32). Then E and H satisfy the Silver–Müller radiation condition (2.57a), (2.57b).

Proof: From Theorem 2.50 and Lemma 2.52 we conclude that a representation of the form (2.56a) holds. Both terms on the right-hand side satisfy the Silver–Müller radiation condition, see Lemma 3.29 of the next chapter. □.

2.8 Exercises

Exercise 2.1. (a) Prove that the Laplacian Δ is given in spherical polar coordinates by (2.1).

(b) Prove that the Laplacian Δ is given in planar polar coordinates $x = \begin{pmatrix} r\cos\varphi \\ r\sin\varphi \end{pmatrix}$ by

$$\Delta \; = \; \frac{1}{r}\frac{\partial}{\partial r}\left(r\frac{\partial}{\partial r}\right) + \frac{1}{r}\frac{\partial^2}{\partial\varphi^2}\,.$$

Exercise 2.2. Show that the Laplace–Beltrami operator Δ_{S^2} is self-adjoint and non-positive; that is,

$$\int_{S^2} f(\hat{x}) \, \Delta_{S^2} g(\hat{x}) \, ds(\hat{x}) = \int_{S^2} g(\hat{x}) \, \Delta_{S^2} f(\hat{x}) \, ds(\hat{x}),$$

$$\int_{S^2} f(\hat{x}) \, \Delta_{S^2} f(\hat{x}) \, ds(\hat{x}) \leq 0$$

for all $f, g \in C^2(S^2)$.

Exercise 2.3. Construct harmonic functions in $\mathbb{R}^2$ or $\mathbb{R}^2 \setminus \{0\}$ by separation of variables in plane polar coordinates.

Exercise 2.4. Prove parts (e)–(h) of Lemma 2.8. Hints:

(e): Show by using the differential equation for P_n and part (c) that the derivatives of both sides of the first equation coincide. For the second equation use (b).
(f) This follows from (c) and (a).
(g) This follows using both equations of (e).
(h) This follows from (b) and (a).

Exercise 2.5. Show that the leading coefficient of P_n is given by $\frac{(2n-1)!}{n! \, (n-1)! \, 2^{n-1}}$; that is,

$$P_n(t) = \frac{(2n-1)!}{n! \, (n-1)! \, 2^{n-1}} t^n + Q_{n-1}(t)$$

for some polynomial Q_{n-1} of order less than n. Hint: Use Theorem 2.8.

Exercise 2.6. Use property (c) of Theorem 2.8 to show for every $\ell \in \mathbb{N}$ the existence of $c_\ell > 0$ such that

$$\|P_n^{(\ell)}\|_\infty = \max_{|t| \leq 1} \left| \frac{d^\ell P_n(t)}{dt^\ell} \right| \leq c_\ell \, n^{2\ell} \quad \text{for all } n \in \mathbb{N}.$$

Exercise 2.7. Try to prove Theorem 2.23.

Exercise 2.8. Prove the following representations of the derivative by using Rayleigh's Formulas from Theorem 2.28

$$f_n'(z) = \frac{n}{z} f_n(z) - f_{n+1}(z), \quad f_n'(z) = -f_{n-1}(z) - \frac{n+1}{z} f_n(z), \quad z \neq 0,$$

where f_n is any of the functions j_n, y_n, $h_n^{(1)}$, or $h_n^{(2)}$.

Exercise 2.9. Prove the following asymptotics for the derivatives by modifying the proof of Theorem 2.31. For any $r \in \mathbb{N}$ and $\varepsilon > 0$ and $R > \varepsilon$:

$$\frac{d^r}{dz^r} j_n(z) = \frac{1}{(2n+1)!!} \frac{d^r}{dz^r} z^n \left[1 + \mathcal{O}\left(\frac{1}{n}\right)\right] \quad \text{for } n \to \infty,$$

$$\frac{d^r}{dz^r} h_n^{(1)}(z) = -i \, (2n-1)!! \frac{d^r}{dz^r} \frac{1}{z^{n+1}} \left[1 + \mathcal{O}\left(\frac{1}{n}\right)\right] \quad \text{for } n \to \infty,$$

uniformly with respect to $|z| \leq R$ and uniformly with respect to $\varepsilon \leq |z| \leq R$, respectively.

Exercise 2.10. Try to formulate and prove the corresponding theorems of Sect. 2.7 for the boundary condition $\nu \times H = f$ on $\partial B(0, R)$.

Chapter 3
Scattering from a Perfect Conductor

In Sect. 2.6 of the previous chapter we have studied the scattering of plane waves by balls. In this chapter we investigate the same problem for arbitrary shapes. To treat this boundary value problem we introduce the boundary integral equation method which reformulates the boundary value problem in terms of an integral equation on the boundary of the region. For showing existence of a solution of this integral equation we will apply the Riesz–Fredholm theory. Again, as in the previous chapter, we consider first the simpler scattering problem for the scalar Helmholtz equation in Sect. 3.1 before we turn to Maxwell's equations in Sect. 3.2.

3.1 A Scattering Problem for the Helmholtz Equation

For this first part the *scattering problem* we are going to solve is the following one:

Given an incident field u^{inc}; that is, a solution u^{inc} of the Helmholtz equation $\Delta u^{\text{inc}} + k^2 u^{\text{inc}} = 0$ in all of $\mathbb{R}^3$, find the total field $u \in C^2(\mathbb{R}^3 \setminus \overline{D}) \cap C^1(\mathbb{R}^3 \setminus D)$ such that

$$\Delta u + k^2 u = 0 \quad \text{in } \mathbb{R}^3 \setminus \overline{D}, \qquad \frac{\partial u}{\partial \nu} = 0 \quad \text{on } \partial D, \qquad (3.1)$$

and such that the scattered field $u^s = u - u^{\text{inc}}$ satisfies the Sommerfeld radiation condition (2.32); that is,

$$\frac{\partial u^s(r\hat{x})}{\partial r} - ik\, u^s(r\hat{x}) = \mathcal{O}\left(\frac{1}{r^2}\right) \quad \text{for } r \to \infty, \qquad (3.2)$$

© Springer International Publishing Switzerland 2015
A. Kirsch, F. Hettlich, *The Mathematical Theory of Time-Harmonic Maxwell's Equations*, Applied Mathematical Sciences 190,
DOI 10.1007/978-3-319-11086-8_3

uniformly with respect to $\hat{x} \in S^2$. For smooth fields $u \in C^1(\mathbb{R}^3 \setminus D)$ and smooth boundaries the normal derivative is given by $\partial u / \partial \nu = \nabla u \cdot \nu$ where $\nu = \nu(x)$ denotes the exterior unit normal vector at $x \in \partial D$.

In this section we restrict ourselves to the Neumann boundary problem as the model problem. The Dirichlet boundary value problem will be treated in Chap. 5.

Since we want to consider the classical situation of scattering in homogeneous media, throughout this chapter we make the following assumptions.

Assumption: Let the wave number k be real and strictly positive unless stated otherwise. Let $D \subseteq \mathbb{R}^3$ be a finite union of bounded domains D_j such that $\overline{D}_j \cap \overline{D}_\ell = \emptyset$ for $j \neq \ell$. Furthermore, we assume that the boundary ∂D is C^2-smooth (see Definition A.7 of the Appendix) and that the complement $\mathbb{R}^3 \setminus \overline{D}$ is connected.

Before we investigate uniqueness and existence of a solution of this scattering problem we study general properties of solutions of the Helmholtz equation $\Delta u + k^2 u = 0$ in bounded and unbounded domains.

3.1.1 Representation Theorems

We begin with the (really!) fundamental solution of the Helmholtz equation, compare (2.37).

Lemma 3.1. *For $k \in \mathbb{C}$ the function $\Phi_k : \{(x,y) \in \mathbb{R}^3 \times \mathbb{R}^3 : x \neq y\} \to \mathbb{C}$, defined by*

$$\Phi_k(x,y) = \frac{e^{ik|x-y|}}{4\pi|x-y|}, \quad x \neq y,$$

is called the fundamental solution *of the Helmholtz equation, i.e. it holds that*

$$\Delta_x \Phi_k(x,y) + k^2 \Phi_k(x,y) = 0 \quad \text{for } x \neq y.$$

Proof: This is easy to check. $\square$

We often suppress the index k; that is, write Φ for Φ_k. Since the fundamental solution and/or its derivatives will occur later on as kernels of various integral operators, we begin with the investigation of certain integrals.

Lemma 3.2. *(a) Let* $K : \{(x, y) \in \mathbb{R}^3 \times \overline{D} : x \neq y\} \to \mathbb{C}$ *be continuous. Assume that there exists* $c > 0$ *and* $\beta \in (0, 1]$ *such that*

$$|K(x, y)| \leq \frac{c}{|x - y|^{3-\beta}} \quad \text{for all } x \in \mathbb{R}^3 \text{ and } y \in D \text{ with } x \neq y.$$

Then the integral $\int_D K(x, y) \, dy$ *exists in the sense of Lebesgue and there exists* $c_\beta > 0$ *with*

$$\int_{D \setminus B(x, \tau)} |K(x, y)| \, dy \leq c_\beta \quad \text{for all } x \in \mathbb{R}^3 \text{ and all } \tau > 0, \quad (3.3a)$$

$$\int_{D \cap B(x, \tau)} |K(x, y)| \, dy \leq c_\beta \tau^\beta \quad \text{for all } x \in \mathbb{R}^3 \text{ and all } \tau > 0. \quad (3.3b)$$

(b) Let $K : \{(x, y) \in \partial D \times \partial D : x \neq y\} \to \mathbb{C}$ *be continuous. Assume that there exists* $c > 0$ *and* $\beta \in (0, 1]$ *such that*

$$|K(x, y)| \leq \frac{c}{|x - y|^{2-\beta}} \quad \text{for all } x, y \in \partial D \text{ with } x \neq y.$$

Then $\int_{\partial D} K(x, y) \, ds(y)$ *exists and there exists* $c_\beta > 0$ *with*

$$\int_{\partial D \setminus B(x, \tau)} |K(x, y)| \, ds(y) \leq c_\beta \quad \text{for all } x \in \partial D \text{ and } \tau > 0, \quad (3.3c)$$

$$\int_{\partial D \cap B(x, \tau)} |K(x, y)| \, ds(y) \leq c_\beta \tau^\beta \quad \text{for all } x \in \partial D \text{ and } \tau > 0. \quad (3.3d)$$

(c) Let $K : \{(x, y) \in \partial D \times \partial D : x \neq y\} \to \mathbb{C}$ *be continuous. Assume that there exists* $c > 0$ *and* $\beta \in (0, 1)$ *such that*

$$|K(x, y)| \leq \frac{c}{|x - y|^{3-\beta}} \quad \text{for all } x, y \in \partial D \text{ with } x \neq y.$$

Then there exists $c_\beta > 0$ *with*

$$\int_{\partial D \setminus B(x, \tau)} |K(x, y)| \, ds(y) \leq c_\beta \tau^{\beta-1} \quad \text{for all } x \in \partial D \text{ and } \tau > 0.$$

$$(3.3e)$$

Proof: (a) Fix $x \in \mathbb{R}^3$ and choose $R > 0$ such that $\overline{D} \subseteq B(0, R)$. First case: $|x| \leq 2R$. Then $D \subseteq B(x, 3R)$ and thus, using spherical polar coordinates with respect to x,

$$\int_{D\setminus B(x,\tau)} \frac{1}{|x-y|^{3-\beta}}\,dy \leq \int_{\tau<|y-x|<3R} \frac{1}{|x-y|^{3-\beta}}\,dy = 4\pi \int_{\tau}^{3R} \frac{1}{r^{3-\beta}}\,r^2\,dr$$

$$= \frac{4\pi}{\beta}\left[(3R)^\beta - \tau^\beta\right] \leq \frac{4\pi}{\beta}(3R)^\beta.$$

Second case: $|x| > 2R$. Then $|x-y| \geq |x|-|y| \geq R$ for $y \in D$ and thus

$$\int_{D\setminus B(x,\tau)} \frac{1}{|x-y|^{3-\beta}}\,dy \leq \frac{1}{R^{3-\beta}} \int_{B(0,R)} dy = \frac{1}{R^{3-\beta}}\frac{4\pi}{3}R^3.$$

This proves (3.3a). For (3.3b) we compute

$$\int_{D\cap B(x,\tau)} |K(x,y)|\,dy \leq c\int_{|x-y|<\tau} \frac{1}{|x-y|^{3-\beta}}\,dy$$

$$= 4\pi \int_0^\tau \frac{1}{r^{3-\beta}}\,r^2\,dr = \frac{4\pi c}{\beta}\tau^\beta.$$

(b) We choose a local coordinate system as in Definition A.7; that is, a covering of ∂D by cylinders of the form $U_j = R_j C_j + z^{(j)}$ where $R_j \in \mathbb{R}^{3\times 3}$ are rotations and $z^{(j)} \in \mathbb{R}^3$ and $C_j = B_2(0,\alpha_j) \times (-2\rho_j, 2\rho_j)$, and where $\partial D \cap U_j$ is expressed as $\partial D \cap U_j = \{R_j x + z^{(j)} : (x_1, x_2) \in B_2(0,\alpha_j), \ x_3 = \xi_j(x_1, x_2)\}$ for some (smooth) function $\xi_j : B_2(0,\alpha_j) \to (-\rho_j, \rho_j)$. Furthermore, we choose a corresponding *partition of unity* (see Theorem A.9); that is, a family of functions $\phi_j \in C^\infty(\mathbb{R}^3)$, $j = 1,\ldots,m$, with

- $0 \leq \phi_j \leq 1$ in $\mathbb{R}^3$,
- the support $S_j := \operatorname{supp}(\phi_j)$ is contained in U_j, and
- $\sum_{j=1}^m \phi_j(x) = 1$ for all $x \in \partial D$.

Then, obviously, $\partial D \subseteq \bigcup_{j=1}^m S_j$. Furthermore, there exists $\delta > 0$ such that $|x-y| \geq \delta$ for all $(x,y) \in S_j \times \bigcup_{\ell \neq j}(U_\ell \setminus \overline{U_j})$ for every j. Indeed, otherwise there would exist some j and a sequence $(x_k, y_k) \in S_j \times \bigcup_{\ell \neq j}(U_\ell \setminus \overline{U_j})$ with $|x_k - y_k| \to 0$. There exist convergent subsequences $x_k \to z$ and $y_k \to z$. Then $z \in S_j$ and $z \in \bigcup_{\ell \neq j}(\overline{U_\ell} \setminus U_j)$ which is a contradiction because $S_j \subseteq U_j$.

The integral is decomposed as

$$\int_{\substack{\partial D \\ |x-y|>\tau}} \frac{ds(y)}{|x-y|^{2-\beta}} = \sum_{\ell=1}^m \int_{\substack{\partial D \cap U_\ell \\ |x-y|>\tau}} \frac{\phi_j(y)}{|x-y|^{2-\beta}}\,ds(y).$$

We fix $x \in U_j \cap \partial D$. For $y \notin U_j \cap \partial D$ we have that $|x-y| \geq \delta$. Therefore, the integrals over $\partial D \cap U_\ell$ for $\ell \neq j$ are easily estimated.

For the integral over $\partial D \cap U_j$ let $x = R_j u + z^{(j)}$ and $y = R_j v + z^{(j)}$ for some $u, v \in B_2(0, \alpha_j) \times (-\rho_j, \rho_j)$ with $u_3 = \xi_j(u_1, u_2)$ and $v_3 = \xi_j(v_1, v_2)$. We set $\tilde{u} = (u_1, u_2)$ and $\tilde{v} = (v_1, v_2)$ and certainly have an estimate of the form

$$c_1 |x - y| \leq |\tilde{u} - \tilde{v}| \leq c_2 |x - y| \quad \text{for all } x, y \in U_j \cap \partial D.$$

Using polar coordinates with respect to $\tilde{u}$ we estimate

$$\int\limits_{\substack{\partial D \cap U_j \\ |x-y|>\tau}} \frac{ds(y)}{|x - y|^{2-\beta}} \leq c_2^{2-\beta} \int\limits_{B_2(0,\alpha_j) \backslash B_2(\tilde{u}, c_1 \tau)} \frac{1}{|\tilde{u} - \tilde{v}|^{2-\beta}} \sqrt{1 + |\nabla \xi_j(\tilde{v})|^2}\, d\tilde{v}$$

$$\leq \hat{c} \int\limits_{c_1 \tau < |\tilde{v} - \tilde{u}| < 2\alpha_j} \frac{d\tilde{v}}{|\tilde{u} - \tilde{v}|^{2-\beta}} = 2\pi \hat{c} \int_{c_1 \tau}^{2\alpha_j} \frac{r}{r^{2-\beta}}\, dr$$

$$\leq 2\pi \hat{c}\, \frac{(2\alpha_j)^\beta}{\beta}.$$

The proofs of (3.3d) and part (c) follow the same lines. $\square$

The following representation theorem implies that any solution of the Helmholtz equation is already determined by its Dirichlet- and Neumann data on the boundary. This theorem is totally equivalent to Cauchy's integral representation formula for holomorphic functions.

Theorem 3.3. *(Green's Representation Theorem in the Interior of D)*

For any $k \in \mathbb{C}$ and $u \in C^2(D) \cap C^1(\overline{D})$ with $\Delta u \in L(D)$ we have the representation

$$\int_D \Phi(x, y) [\Delta u(y) + k^2 u(y)]\, dy + \int_{\partial D} \left\{ u(y) \frac{\partial \Phi}{\partial \nu(y)}(x, y) - \Phi(x, y) \frac{\partial u}{\partial \nu}(y) \right\} ds(y)$$

$$= \begin{cases} -u(x), & x \in D, \\ -\frac{1}{2} u(x), & x \in \partial D, \\ 0, & x \notin \overline{D}. \end{cases}$$

Remarks:

- This theorem tells us that, for $x \in D$, any function u can be expressed as a sum of three potentials:

$$(\tilde{S}\varphi)(x) = \int_{\partial D} \varphi(y)\,\Phi(x,y)\,ds(y)\,, \quad x \notin \partial D\,, \tag{3.4a}$$

$$(\tilde{D}\varphi)(x) = \int_{\partial D} \varphi(y)\,\frac{\partial \Phi}{\partial \nu(y)}(x,y)\,ds(y)\,, \quad x \notin \partial D\,, \tag{3.4b}$$

$$(\mathcal{V}\varphi)(x) = \int_{D} \varphi(y)\,\Phi(x,y)\,dy\,, \quad x \in \mathbb{R}^3\,, \tag{3.4c}$$

which are called *single layer potential, double layer potential,* and *volume potential,* respectively, with density φ. We will investigate these potential in detail in Sect. 3.1.2 below.

- The one-dimensional analogon is (for $x \in D = (a,b) \subseteq \mathbb{R}$)

$$u(x) = \frac{1}{2ik} \int_a^b e^{ik|x-y|}\big[u''(y) + k^2 u(y)\big]\,dy$$

$$+ \frac{1}{2ik}\left[u(y)\,\frac{d}{dy}e^{ik|x-y|} - u'(y)\,e^{ik|x-y|}\right]_a^b\,.$$

Therefore, the one-dimensional fundamental solution is $\Phi(x,y) = -\exp(ik|x-y|)/(2ik)$, see Exercise 3.1.

Proof of Theorem 3.3: First we fix $x \in D$ and a small closed ball $B[x,r] \subseteq D$ centered at x with radius $r > 0$. For $y \in \partial B(x,r)$ the normal vector $\nu(y) = \frac{x-y}{|y-x|} = (x-y)/r$ is directed into the interior of $B(x,r)$. We apply Green's second identity to u and $v(y) := \Phi(x,y)$ in the domain $D_r := D \setminus B[x,r]$. Then

$$\int_{\partial D}\left\{u(y)\,\frac{\partial \Phi}{\partial \nu(y)}(x,y) - \Phi(x,y)\,\frac{\partial u}{\partial \nu}(y)\right\}ds(y) \tag{3.5a}$$

$$+ \int_{\partial B(x,r)}\left\{u(y)\,\frac{\partial \Phi}{\partial \nu(y)}(x,y) - \Phi(x,y)\,\frac{\partial u}{\partial \nu}(y)\right\}ds(y) \tag{3.5b}$$

$$= \int_{D_r}\left\{u(y)\,\Delta_y\Phi(x,y) - \Phi(x,y)\,\Delta u(y)\right\}dy$$

$$= -\int_{D_r}\Phi(x,y)\big[\Delta u(y) + k^2 u(y)\big]\,dy\,,$$

if one uses the Helmholtz equation for Φ. We compute the integral (3.5b). We observe that

$$\nabla_y\Phi(x,y) = \frac{\exp(ik|x-y|)}{4\pi|x-y|}\left(ik - \frac{1}{|x-y|}\right)\frac{y-x}{|x-y|}$$

and thus for $|y - x| = r$:

$$\Phi(x, y) = \frac{\exp(ikr)}{4\pi r},$$

$$\frac{\partial \Phi}{\partial \nu(y)}(x, y) = \frac{x - y}{r} \cdot \nabla_y \Phi(x, y) = -\frac{\exp(ikr)}{4\pi r}\left(ik - \frac{1}{r}\right).$$

Therefore, we compute the integral (3.5b) as

$$\int_{\partial B(x,r)} \left\{ u(y) \frac{\partial \Phi}{\partial \nu(y)}(x, y) - \Phi(x, y)\frac{\partial u}{\partial \nu}(y) \right\} ds(y)$$

$$= \frac{\exp(ikr)}{4\pi r} \int_{|y-x|=r} \left\{ u(y)\left(\frac{1}{r} - ik\right) - \frac{\partial u}{\partial \nu}(y) \right\} ds$$

$$= \frac{\exp(ikr)}{4\pi r^2} \int_{|y-x|=r} u(y)\, ds - \frac{\exp(ikr)}{4\pi r} \int_{|y-x|=r} \left\{ ik\, u(y) + \frac{\partial u}{\partial \nu}(y) \right\} ds.$$

For $r \to 0$ the first term tends to $u(x)$, because the surface area of $\partial B(x, r)$ is just $4\pi r^2$ and

$$\frac{\exp(ikr)}{4\pi r^2} \int_{|y-x|=r} u(y)\, ds = \frac{\exp(ikr)}{4\pi r^2} \int_{|y-x|=r} [u(y) - u(x)]\, ds$$
$$+ \exp(ikr)\, u(x),$$

where we obtain by continuity

$$\left| \frac{1}{4\pi r^2} \int_{|y-x|=r} [u(y) - u(x)]\, ds \right| \leq \sup_{y \in B(x,r)} |u(y) - u(x)| \to 0, \quad r \to 0.$$

Similarly the second term tends to zero. Therefore, also the limit of the volume integral exists as $r \to 0$ and yields the desired formula for $x \in D$.

Let now $x \in \partial D$. Then we proceed in the same way. The domains of integration in (3.5a) and (3.5b) have to be replaced by $\partial D \backslash B(x, r)$ and $\partial B(x, r) \cap D$, respectively. In the computation the region $\{y \in \mathbb{R}^3 : |y - x| = r\}$ has to be replaced by $\{y \in D : |y - x| = r\}$. By Lemma A.10 its surface area is $2\pi r^2 + \mathcal{O}(r^3)$ which gives the factor $1/2$ of $u(x)$.

For $x \notin \overline{D}$ the functions u and $v = \Phi(x, \cdot)$ are both solutions of the Helmholtz equation in all of D. Application of Green's second identity in D yields the assertion. □

We note that the volume integral vanishes if u is a solution of the Helmholtz equation $\Delta u + k^2 u = 0$ in D. In this case the function u can be expressed solely as a combination of a single and a double layer surface potential. This observation is a first hint on a reformulation of the boundary value problem in terms of an integral equation on the surface ∂D.

From the proof we note that our assumptions on the smoothness of the boundary ∂D are too strong. For the representation theorem the domain D has to satisfy exactly the assumptions which are needed for Green's theorems to hold.

As a corollary from the representation theorem we obtain the following conclusion.

Corollary 3.4. *Let $u \in C^2(D)$ be a real- or complex valued solution of the Helmholtz equation in D. Then u is analytic, i.e. one can locally expand u into a power series. That is, for every $z \in D$ there exists $r > 0$ such that u has the form*

$$u(x) \;=\; \sum_{n \in \mathbb{N}^3} a_n (x_1 - z_1)^{n_1} (x_2 - z_2)^{n_2} (x_3 - z_3)^{n_3} \quad \text{for} \quad \sum_{j=1}^{3} |x_j - z_j|^2 < r^2,$$

where we use the notation $\mathbb{N} = \mathbb{Z}_{\geq 0} = \{0, 1, 2, \dots\}$.

Proof: From the previous representation of $u(x)$ as a difference of a single and a double layer potential and the smoothness of the kernels $x \mapsto \Phi(x, y)$ and $x \mapsto \partial \Phi(x, y) / \partial \nu(y)$ for $x \neq y$ it follows immediately that $u \in C^\infty(D)$. The proof of analyticity is technically not easy if one avoids methods from complex analysis.[1] If one uses these methods, then one can argue as follows: Fix $\hat{x} \in D$ and choose $r > 0$ such that $B[\hat{x}, r] \subseteq D$. Define the region $R \subseteq \mathbb{C}^3$ and the function $v : R \to \mathbb{C}$ by

$$R = \left\{ z \in \mathbb{C}^3 : |\operatorname{Re} z - \hat{x}| < r/2, \ |\operatorname{Im} z| < r/2 \right\},$$

$$v(z) = \int_{\partial D} \left[\frac{\exp\left[ik\sqrt{\sum_{j=1}^{3}(z_j - y_j)^2}\right]}{4\pi \sqrt{\sum_{j=1}^{3}(z_j - y_j)^2}} \frac{\partial u}{\partial \nu}(y) \right.$$
$$\left. - u(y) \frac{\partial}{\partial \nu(y)} \frac{\exp\left[ik\sqrt{\sum_{j=1}^{3}(z_j - y_j)^2}\right]}{4\pi \sqrt{\sum_{j=1}^{3}(z_j - y_j)^2}} \right] ds(y)$$

for $z \in R$. Taking the principal value with cut along the negative real axis of the square root of the complex number $\sum_{j=1}^{3}(z_j - y_j)^2$ is not a problem because $\operatorname{Re} \sum_{j=1}^{3}(z_j - y_j)^2 = \sum_{j=1}^{3}(\operatorname{Re} z_j - y_j)^2 - (\operatorname{Im} z_j)^2 = |\operatorname{Re} z - y|^2 - |\operatorname{Im} z|^2 > 0$ because of $|\operatorname{Re} z - y| \geq |y - \hat{x}| - |\hat{x} - \operatorname{Re} z| > r - r/2 = r/2$ and $|\operatorname{Im} z| < r/2$. Obviously, the function v is holomorphic in R and thus (complex) analytic. □

---

[1] We refer to [19, Sect. 2.4], for a proof.

As a second corollary we can easily prove the following version of *Holmgren's uniqueness theorem* for the Helmholtz equation.

Theorem 3.5. *(Holmgren's uniqueness theorem)*
Let D be a domain with C^2-boundary and $u \in C^1(\overline{D}) \cap C^2(D)$ be a solution of the Helmholtz equation $\Delta u + k^2 u = 0$ in D. Furthermore, let U be an open set such that $U \cap \partial D \neq \emptyset$. If $u = 0$ and $\partial u / \partial \nu = 0$ on $U \cap \partial D$, then u vanishes in all of D.

Proof: Let $z \in U \cap \partial D$ and $B \subseteq U$ a ball centered at z such that $\Gamma = D \cap \partial B \neq \emptyset$. Then $\partial(B \cap D) = \Gamma \cup (B \cap \partial D)$. The reader should sketch the situation. We define v by

$$v(x) = \int_\Gamma \left\{ \Phi(x,y) \frac{\partial u}{\partial \nu}(y) - u(y) \frac{\partial \Phi}{\partial \nu(y)}(x,y) \right\} ds(y), \quad x \in B.$$

Then v satisfies the Helmholtz equation in B. Application of Green's representation formula of Theorem 3.3 to u in $B \cap D$ yields[2]

$$u(x) = \int_{\partial(B \cap D)} \left\{ \Phi(x,y) \frac{\partial u}{\partial \nu}(y) - u(y) \frac{\partial \Phi}{\partial \nu(y)}(x,y) \right\} ds(y) = v(x), \quad x \in B \cap D,$$

because the integral vanishes on $\partial D \cap B$. By the same theorem we conclude that $v(x) = 0$ for $x \in B \setminus \overline{D}$. Since v is analytic by the previous corollary we conclude that v vanishes in all of B. In particular, u vanishes in $B \cap D$. Again, u is analytic in D and D is connected, thus also u vanishes in all of D. □

For radiating solutions of the Helmholtz equation we have the following version of Green's representation theorem which we prove even for complex valued k.

Theorem 3.6. *(Green's Representation Theorem in the Exterior of D)*

Let $k \in \mathbb{C} \setminus \{0\}$ with $k > 0$ or $\mathrm{Im}\, k > 0$ and $u \in C^2(\mathbb{R}^3 \setminus \overline{D}) \cap C^1(\mathbb{R}^3 \setminus D)$ be a solution of the Helmholtz equation $\Delta u + k^2 u = 0$ in $\mathbb{R}^3 \setminus \overline{D}$. Furthermore, let u satisfy the Sommerfeld radiation condition (3.2). Then we have the representation

$$\int_{\partial D} \left\{ u(y) \frac{\partial \Phi}{\partial \nu(y)}(x,y) - \Phi(x,y) \frac{\partial u}{\partial \nu}(y) \right\} ds(y) = \begin{cases} u(x), & x \notin \overline{D}, \\ \frac{1}{2} u(x), & x \in \partial D, \\ 0, & x \in D. \end{cases}$$

The domain integral as well as the surface integral (for $x \in \partial D$) exists.

---

[2] In this case the region $B \cap D$ does not meet the smoothness assumptions of the beginning of this section. The representation theorem still holds by the remark following Theorem 3.3.

Proof: Let first $x \notin \overline{D}$. We choose $R > |x|$ such that $\overline{D} \subseteq B(0, R)$ and apply Green's representation Theorem 3.3 in the annular region $B(0, R) \setminus \overline{D}$. Noting that $\Delta u + k^2 u = 0$ and that $\nu(y)$ for $y \in \partial D$ is directed into the interior of $B(0, R) \setminus \overline{D}$ yields

$$u(x) = \int_{\partial D} \left\{ u(y) \frac{\partial \Phi}{\partial \nu(y)}(x, y) - \Phi(x, y) \frac{\partial u}{\partial \nu}(y) \right\} ds(y)$$

$$- \int_{|y|=R} \left\{ u(y) \frac{\partial \Phi}{\partial \nu(y)}(x, y) - \Phi(x, y) \frac{\partial u}{\partial \nu}(y) \right\} ds(y).$$

We show that the surface integral over $\partial B(0, R)$ tends to zero as R tends to infinity. We write this surface integral (for fixed x) as

$$I_R = \int_{|y|=R} u(y) \left[\frac{\partial \Phi}{\partial \nu(y)}(x, y) - ik\, \Phi(x, y) \right] ds(y)$$

$$- \int_{|y|=R} \Phi(x, y) \left[\frac{\partial u}{\partial \nu}(y) - ik\, u(y) \right] ds(y)$$

and use the Cauchy–Schwarz inequality:

$$|I_R|^2 \leq \int_{|y|=R} |u|^2 ds \int_{|y|=R} \left| \frac{\partial \Phi}{\partial \nu(y)}(x, y) - ik\, \Phi(x, y) \right|^2 ds(y)$$

$$+ \int_{|y|=R} |\Phi(x, y)|^2 ds(y) \int_{|y|=R} \left| \frac{\partial u}{\partial \nu} - ik\, u \right|^2 ds(y)$$

From the radiation conditions of $\Phi(x, \cdot)$ for fixed x with respect to y and of u we conclude that the integrands of the second and fourth integral behave as $\mathcal{O}(1/R^4)$ as $R \to \infty$. Since the surface area of $\partial B(0, R)$ is equal to $4\pi R^2$ we conclude that the second and fourth integral tend to zero as R tends to infinity. Furthermore, the integrand of the third integral behaves as $\mathcal{O}(1/R^2)$ as $R \to \infty$. Therefore, the third integral is bounded. It remains to show that also $\int_{|y|=R} |u|^2 ds$ is bounded. This follows again from the radiation condition. Indeed, from the radiation condition we have

$$\mathcal{O}\left(\frac{1}{R^2} \right) = \int_{|x|=R} \left| \frac{\partial u}{\partial r} - iku \right|^2 ds$$

$$= \int_{|x|=R} \left\{ \left| \frac{\partial u}{\partial r} \right|^2 + |ku|^2 \right\} ds + 2\, \text{Im} \left[k \int_{|x|=R} u \frac{\partial \overline{u}}{\partial r} ds \right].$$

Green's theorem, applied in $B(0, R) \setminus D$ to the function u yields

$$\int\limits_{|x|=R} u\, \frac{\partial \overline{u}}{\partial r}\, ds \;=\; \int\limits_{\partial D} u\, \frac{\partial \overline{u}}{\partial \nu}\, ds \;+\; \int\limits_{B(0,R)\setminus D} \left[|\nabla u|^2 - \overline{k}^2 |u|^2 \right] dx\,.$$

We multiply by k and take the imaginary part. This yields

$$\mathrm{Im} \left[k \int\limits_{|x|=R} u\, \frac{\partial \overline{u}}{\partial r}\, ds \right] = \mathrm{Im} \left[k \int\limits_{\partial D} u\, \frac{\partial \overline{u}}{\partial \nu}\, ds \right]$$

$$+\; \mathrm{Im}\,(k) \int\limits_{B(0,R)\setminus D} \left[|\nabla u|^2 + |k|^2 |u|^2 \right] dx$$

$$\geq \mathrm{Im} \left[k \int\limits_{\partial D} u\, \frac{\partial \overline{u}}{\partial r}\, ds \right]$$

and thus

$$\int\limits_{|x|=R} \left| \frac{\partial u}{\partial r} \right|^2 + |ku|^2\, ds \;\leq\; -2\,\mathrm{Im} \left[k \int\limits_{\partial D} u\, \frac{\partial \overline{u}}{\partial r}\, ds \right] + \mathcal{O}\!\left(\frac{1}{R^2} \right).$$

This implies, in particular, that $\int_{|y|=R} |u|^2 ds$ is bounded. Altogether, we have shown that I_R tends to zero as R tends to infinity. □

3.1.2 Volume and Surface Potentials

We have seen in the preceding section that any function can be represented by a combination of volume and surface potentials. The integral equation method for solving boundary value problems for the Helmholtz equation and Maxwell's equations rely heavily on the smoothness properties of these potentials. This subsection is concerned with the investigation of these potentials. The analysis is quite technical and uses some tools from differential geometry (see Sect. A.3).

In this section we allow k to be an arbitrary complex number $k \in \mathbb{C}$ with $\mathrm{Im}\, k \geq 0$. We recall the fundamental solution for $k \in \mathbb{C}$; that is,

$$\Phi(x, y) \;=\; \frac{e^{ik|x-y|}}{4\pi |x - y|}\,, \qquad x \neq y\,, \tag{3.6}$$

and begin with the *volume potential*

$$w(x) = \int_D \varphi(y)\, \Phi(x,y)\, dy\,, \quad x \in \mathbb{R}^3\,, \tag{3.7}$$

where $D \subseteq \mathbb{R}^3$ is any open and bounded set.

Lemma 3.7. *Let $\varphi \in L^\infty(D)$ be any complex-valued function. Then $w \in C^1(\mathbb{R}^3)$ and*

$$\frac{\partial w}{\partial x_j}(x) = \int_D \varphi(y)\, \frac{\partial \Phi}{\partial x_j}(x,y)\, dy\,, \quad x \in \mathbb{R}^3\,, \ j = 1,2,3\,. \tag{3.8}$$

Proof: We fix $j \in \{1,2,3\}$ and a real valued function $\eta \in C^1(\mathbb{R})$ with $0 \le \eta(t) \le 1$ and $\eta(t) = 0$ for $t \le 1$ and $\eta(t) = 1$ for $t \ge 2$. We set

$$v(x) = \int_D \varphi(y)\, \frac{\partial \Phi}{\partial x_j}(x,y)\, dy\,, \quad x \in \mathbb{R}^3\,,$$

and note that the integral exists by Lemma 3.2 because $\left|\frac{\partial \Phi}{\partial x_j}(x,y)\right| \le c_0\,(|x - y|)^{-2}$ for some $c_0 > 0$. Furthermore, set

$$w_\varepsilon(x) = \int_D \varphi(y)\, \Phi(x,y)\, \eta\big(|x-y|/\varepsilon\big)\, dy\,, \quad x \in \mathbb{R}^3\,.$$

Then $w_\varepsilon \in C^1(\mathbb{R}^3)$ and

$$v(x) - \frac{\partial w_\varepsilon}{\partial x_j}(x) = \int_{|y-x|\le 2\varepsilon} \varphi(y)\, \frac{\partial}{\partial x_j}\big\{\Phi(x,y)\,[1 - \eta(|x-y|/\varepsilon)]\big\}\, dy\,.$$

Thus, we have

$$\left|v(x) - \frac{\partial w_\varepsilon}{\partial x_j}(x)\right| \le \|\varphi\|_\infty \int_{|y-x|\le 2\varepsilon} \left\{\left|\frac{\partial \Phi}{\partial x_j}(x,y)\right| + \frac{\|\eta'\|_\infty}{\varepsilon}\,|\Phi(x,y)|\right\} dy$$

$$\le c_1 \int_{|y-x|\le 2\varepsilon} \left[\frac{1}{|x-y|^2} + \frac{1}{\varepsilon\,|x-y|}\right] dy$$

$$= c_1 \int_0^{2\pi}\int_0^\pi\int_0^{2\varepsilon} \left[\frac{1}{r^2} + \frac{1}{\varepsilon\,r}\right] r^2 \sin\theta\, dr\, d\theta\, d\varphi = 16\pi\, c_1\, \varepsilon\,.$$

Therefore, $\partial w_\varepsilon/\partial x_j \to v$ uniformly in $\mathbb{R}^3$, which shows $w \in C^1(\mathbb{R}^3)$ and $\partial w/\partial x_j = v$. $\square$

This regularity result is not sufficient in view of second order differential equations. Higher regularity is obtained for Hölder-continuous densities.

Definition 3.8. For a set $T \subseteq \mathbb{R}^3$ and $\alpha \in (0,1]$ we define the space $C^{0,\alpha}(T)$ of bounded, uniformly *Hölder-continuous functions* by

$$C^{0,\alpha}(T) := \left\{ v \in C(T) : v \text{ bounded and } \sup_{x,y \in T,\ x \neq y} \frac{|v(x) - v(y)|}{|x - y|^\alpha} < \infty \right\}.$$

Note that any Hölder-continuous function is uniformly continuous and, therefore, has a continuous extension to $\overline{T}$. The space $C^{0,\alpha}(T)$ is a Banach space with norm

$$\|v\|_{C^{0,\alpha}(T)} := \underbrace{\sup_{x \in T} |v(x)|}_{= \|v\|_\infty} + \sup_{x,y \in T,\ x \neq y} \frac{|v(x) - v(y)|}{|x - y|^\alpha}. \tag{3.9}$$

Now we can complement the previous Lemma and obtain that volume potentials with Hölder-continuous densities are two times differentiable.

Theorem 3.9. *Let $\varphi \in C^{0,\alpha}(D)$ and let w be the volume potential. Then $w \in C^2(D) \cap C^\infty(\mathbb{R}^3 \setminus \overline{D})$ and*

$$\Delta w + k^2 w = \begin{cases} -\varphi, & \text{in } D, \\ 0, & \text{in } \mathbb{R}^3 \setminus \overline{D}. \end{cases}$$

Furthermore, if D_0 is any C^2-smooth domain with $D \subseteq D_0$, then

$$\frac{\partial^2 w}{\partial x_i \partial x_j}(x) = \int_{D_0} \frac{\partial^2 \Phi}{\partial x_i \partial x_j}(x,y) \left[\varphi(y) - \varphi(x)\right] dy$$

$$- \varphi(x) \int_{\partial D_0} \frac{\partial \Phi}{\partial x_i}(x,y)\, \nu_j(y)\, ds(y) \tag{3.10}$$

for $x \in D$ and $i,j \in \{1,2,3\}$ where we have extended φ by zero in $D_0 \setminus D$. If $\varphi = 0$ on ∂D, then $w \in C^2(\mathbb{R}^3)$.

Proof: First we note that the volume integral in the last formula exists. Indeed, we fix $x \in D$ and split the region of integration into $D_0 = D \cup (D_0 \setminus D)$. The integral over $D_0 \setminus D$ exists because the integrand is smooth. The integral over D exists again by Lemma 3.2 because $|\varphi(y) - \varphi(x)||\partial^2 \Phi/(\partial x_i \partial x_j)(x,y)| \leq c|x - y|^{\alpha - 3}$ for $y \in D$. The existence of the surface integral is obvious because $x \notin \partial D_0$.
We fix $i,j \in \{1,2,3\}$ and the same function $\eta \in C^1(\mathbb{R})$ as in the previous lemma. We define $v := \partial w / \partial x_i$,

$$u(x) := \int_{D_0} \frac{\partial^2 \Phi}{\partial x_i \partial x_j}(x,y) \left[\varphi(y) - \varphi(x)\right] dy - \varphi(x) \int_{\partial D_0} \frac{\partial \Phi}{\partial x_i}(x,y)\, \nu_j(y)\, ds(y),$$

for $x \in D$, and

$$v_\varepsilon(x) := \int_D \varphi(y)\, \eta\big(|x-y|/\varepsilon\big)\, \frac{\partial \Phi}{\partial x_i}(x,y)\, dy\,, \quad x \in \mathbb{R}^3\,.$$

Then $v_\varepsilon \in C^1(D)$ and for $x \in D$

$$\begin{aligned}
\frac{\partial v_\varepsilon}{\partial x_j}(x) &= \int_D \varphi(y)\, \frac{\partial}{\partial x_j}\left\{ \eta\big(|x-y|/\varepsilon\big)\, \frac{\partial \Phi}{\partial x_i}(x,y) \right\} dy \\
&= \int_{D_0} \big[\varphi(y) - \varphi(x)\big]\, \frac{\partial}{\partial x_j}\left\{ \eta\big(|x-y|/\varepsilon\big)\, \frac{\partial \Phi}{\partial x_i}(x,y) \right\} dy \\
&\quad + \varphi(x) \int_{D_0} \frac{\partial}{\partial x_j}\left\{ \eta\big(|x-y|/\varepsilon\big)\, \frac{\partial \Phi}{\partial x_i}(x,y) \right\} dy \\
&= \int_{D_0} \big[\varphi(y) - \varphi(x)\big]\, \frac{\partial}{\partial x_j}\left\{ \eta\big(|x-y|/\varepsilon\big)\, \frac{\partial \Phi}{\partial x_i}(x,y) \right\} dy \\
&\quad - \varphi(x) \int_{\partial D_0} \frac{\partial \Phi}{\partial x_i}(x,y)\, \nu_j(y)\, ds(y)
\end{aligned}$$

provided $2\varepsilon \leq d(x, \partial D_0)$. In the last step we used the Divergence Theorem. Therefore, for $x \in D$,

$$\begin{aligned}
\left| u(x) - \frac{\partial v_\varepsilon}{\partial x_j}(x) \right| & \\
&\leq \int_{|y-x|\leq 2\varepsilon} |\varphi(y) - \varphi(x)|\, \left| \frac{\partial}{\partial x_j}\left\{ \big(1 - \eta(|x-y|/\varepsilon)\big)\, \frac{\partial \Phi}{\partial x_i}(x,y) \right\} \right| dy \\
&\leq c_1 \int_{|y-x|\leq 2\varepsilon} \left(\frac{1}{|y-x|^3} + \frac{\|\eta'\|_\infty}{\varepsilon\, |y-x|^2} \right) |y-x|^\alpha\, dy \\
&= c_2 \int_0^{2\varepsilon} \left(\frac{1}{r^{1-\alpha}} + \frac{\|\eta'\|_\infty}{\varepsilon}\, r^\alpha \right) dr \\
&= c_2 \left[\frac{(2\varepsilon)^\alpha}{\alpha} + \frac{(2\varepsilon)^{1+\alpha}}{(1+\alpha)\,\varepsilon} \right] \leq c_3\, \varepsilon^\alpha
\end{aligned}$$

provided $2\varepsilon \leq \mathrm{dist}(x, \partial D)$. Therefore, $\partial v_\varepsilon/\partial x_j \to u$ uniformly on compact subsets of D. Also, $v_\varepsilon \to v$ uniformly on compact subsets of D and thus $w \in C^2(D)$ and $u = \partial^2 w/(\partial x_i \partial x_j)$. This proves (3.10). Finally, we fix $z \in D$ and set $D_0 = B(z, R)$ where R is chosen such that $D \subseteq B(z, R)$. For $x \in D$ we have

$$\Delta w(x) = -k^2 \int_{B(z,R)} \Phi(x,y) \left[\varphi(y) - \varphi(x)\right] dy$$

$$- \varphi(x) \int_{|y-z|=R} \sum_{j=1}^{3} \frac{y_j - z_j}{R} \left[\frac{\exp(ikR)}{4\pi R}\left(ik - 1/R\right)\frac{x_j - y_j}{R}\right] ds(y).$$

This holds for all $x \in D$. For $x = z$ we conclude

$$\Delta w(z) = -k^2 w(z) + k^2 \varphi(z) \int_{B(z,R)} \frac{\exp(ik|z-y|)}{4\pi |z-y|} dy - \varphi(z) e^{ikR}(1 - ikR),$$

i.e.

$$\Delta w(z) + k^2 w(z) = -\varphi(z) \underbrace{\left[e^{ikR}(1 - ikR) - k^2 \int_{B(z,R)} \frac{\exp(ik|z-y|)}{4\pi |z-y|} dy\right]}_{= 1}$$

because

$$k^2 \int_{B(z,R)} \frac{\exp(ik|z-y|)}{4\pi |z-y|} dy = k^2 \int_0^R r\, e^{ikr} dr = -ikR\, e^{ikR} + e^{ikR} - 1.$$

If φ vanishes on ∂D, then the extension of φ by zero is in $C^{0,\alpha}(\mathbb{R}^3)$ (see Exercise 3.5). Therefore, we can apply the result above to any ball $D = D_0 = B(0, R)$ which yields that the second derivatives of w are continuous in $\mathbb{R}^3$. $\square$

The following corollary shows that the volume integral is bounded when considered as an integral operator between suitable spaces.

Corollary 3.10. *Let D be C^2-smooth and $A \subseteq \mathbb{R}^3$ be a closed set with $A \subseteq D$ or $A \subseteq \mathbb{R}^3 \setminus \overline{D}$. Furthermore, let w be the volume integral with density $\varphi \in C^{0,\alpha}(D)$. Then there exists $c > 0$ with*

$$\|w\|_{C^1(\mathbb{R}^3)} \leq c\|\varphi\|_\infty \quad \text{and} \quad \|w\|_{C^2(A)} \leq c\|\varphi\|_{C^{0,\alpha}(D)} \qquad (3.11)$$

for all $\varphi \in C^{0,\alpha}(D)$.

Proof: We estimate

$$|\Phi(x,y)| = \frac{1}{4\pi|x-y|},$$

$$\left|\frac{\partial \Phi}{\partial x_j}(x,y)\right| \leq c_1 \left[\frac{1}{|x-y|^2} + \frac{1}{|x-y|}\right],$$

$$\left|\frac{\partial^2 \Phi}{\partial x_i \partial x_j}(x,y)\right| \leq c_2 \left[\frac{1}{|x-y|^3} + \frac{1}{|x-y|^2} + \frac{1}{|x-y|}\right].$$

Thus for $x \in \mathbb{R}^3$ we can apply (3.8) and Lemma 3.2 above and obtain

$$|w(x)| \leq \|\varphi\|_\infty \int_D \frac{1}{4\pi|x-y|}\, dy \; \leq \; c\,\|\varphi\|_\infty \,,$$

$$\left|\frac{\partial w}{\partial x_j}(x)\right| \leq \|\varphi\|_\infty \int_D \left|\frac{\partial \Phi}{\partial x_j}(x,y)\right| dy \; \leq \; c\,\|\varphi\|_\infty \,,$$

where the constant $c > 0$ can be chosen independent of x and φ. This proves the first estimate of (3.11). Let now $x \in A$. If $A \subseteq D$, then there exists $\delta > 0$ with $|x-y| \geq \delta$ for all $x \in A$ and $y \in \partial D$. By (3.10) for $D = D_0$ we have

$$\left|\frac{\partial^2 w}{\partial x_i \partial x_j}(x)\right| \leq \|\varphi\|_{C^{0,\alpha}(D)} \int_D \left|\frac{\partial^2 \Phi}{\partial x_i \partial x_j}(x,y)\right| |x-y|^\alpha\, dy$$

$$+ \|\varphi\|_\infty \int_{\partial D} \left|\frac{\partial \Phi}{\partial x_i}(x,y)\right| ds(y)$$

$$\leq c_3\,\|\varphi\|_{C^{0,\alpha}(D)} \int_D \frac{dy}{|x-y|^{3-\alpha}} + c_4\,\|\varphi\|_\infty \int_{\partial D} \frac{ds(y)}{|x-y|^2}$$

$$\leq c_5\,\|\varphi\|_{C^{0,\alpha}(D)} + \frac{c_4}{\delta^2}\|\varphi\|_\infty \int_{\partial D} ds$$

$$\leq c\,\|\varphi\|_{C^{0,\alpha}(D)} \,.$$

If $A \subseteq \mathbb{R}^3 \setminus \overline{D}$, then there exists $\delta > 0$ with $|x-y| \geq \delta$ for all $x \in A$ and $y \in D$. Therefore, we can estimate

$$\left|\frac{\partial^2 w}{\partial x_i \partial x_j}(x)\right| \leq \|\varphi\|_\infty \int_D \left|\frac{\partial^2 \Phi}{\partial x_i \partial x_j}(x,y)\right| dy \leq \|\varphi\|_\infty \frac{c}{\delta^3} \int_D dy \leq c\,\|\varphi\|_{C^{0,\alpha}(D)} \,.$$

$\square$

We continue with the *single layer surface potential*; that is, the function

$$v(x) \; = \; \tilde{S}\varphi(x) \; = \; \int_{\partial D} \varphi(y)\,\Phi(x,y)\, ds(y)\,, \qquad x \in \mathbb{R}^3 \qquad (3.12)$$

The investigation of this potential requires some elementary facts from differential geometry which we have collected in Sect. A.3 of Chap. A. First we note that for continuous densities φ the integral exists by Lemma 3.2 above even for $x \in \partial D$ because Φ has a singularity of the form $|\Phi(x,y)| = 1/(4\pi|x-y|)$. Before we prove continuity of v we make the following general remark.

Remark 3.11. A function $v : \mathbb{R}^3 \to \mathbb{C}$ is Hölder-continuous if

(a) v is bounded,
(b) v is Hölder-continuous in some H_ρ,
(c) v is Lipschitz continuous in $\mathbb{R}^3 \setminus U_\delta$ for every $\delta > 0$.

Here, H_ρ is a strip around ∂D of thickness ρ and U_δ is the neighborhood of ∂D with thickness δ as defined in Lemma A.10; that is,

$$H_\rho := \big\{ z + t\nu(z) : z \in \partial D,\ |t| < \rho \big\},$$
$$U_\delta := \big\{ x \in \mathbb{R}^3 : \inf_{z \in \partial D} |x - z| < \delta \big\}.$$

Proof: Choose $\delta > 0$ such that $U_{3\delta} \subseteq H_\rho$.

1st case: $|x_1 - x_2| < \delta$ and $x_1 \in U_{2\delta}$. Then $x_1, x_2 \in U_{3\delta} \subseteq H_\rho$ and Hölder-continuity follows from (b).

2nd case: $|x_1 - x_2| < \delta$ and $x_1 \notin U_{2\delta}$. Then $x_1, x_2 \notin U_\delta$ and thus from (c):

$$\big| v(x_1) - v(x_2) \big| \ \leq \ c|x_1 - x_2| \ \leq \ c\delta^{1-\alpha} |x_1 - x_2|^\alpha.$$

3rd case: $|x_1 - x_2| \geq \delta$. Then, by (a),

$$\big| v(x_1) - v(x_2) \big| \ \leq \ 2\|v\|_\infty \ \leq \ \frac{2\|v\|_\infty}{\delta^\alpha} |x_1 - x_2|^\alpha.$$

$\square$

Using this remark we show that the single layer potential with continuous density is Hölder-continuous.

Theorem 3.12. *The single layer potential v from (3.12) with continuous density φ is uniformly Hölder-continuous in all of $\mathbb{R}^3$, and for every $\alpha \in (0,1)$ there exists $c > 0$ (independent of φ) with*

$$\|v\|_{C^\alpha(\mathbb{R}^3)} \ \leq \ c\|\varphi\|_\infty. \tag{3.13}$$

Proof: We check the conditions (a)–(c) of Remark 3.11. Boundedness follows from Lemma 3.2 because $\big| \Phi(x,y) \big| = 1/(4\pi|x - y|)$. For (b) and (c) we write

$$\big| v(x_1) - v(x_2) \big| \ \leq \ \|\varphi\|_\infty \int_{\partial D} \big| \Phi(x_1, y) - \Phi(x_2, y) \big|\, ds(y) \tag{3.14}$$

and estimate

$$
\begin{aligned}
\left|\Phi(x_1, y) - \Phi(x_2, y)\right| &\leq \frac{1}{4\pi}\left|\frac{1}{|x_1 - y|} - \frac{1}{|x_2 - y|}\right| \\
&\quad + \frac{1}{4\pi|x_1 - y|}\left|e^{ik|x_1 - y|} - e^{ik|x_2 - y|}\right| \\
&\leq \frac{|x_1 - x_2|}{4\pi\,|x_1 - y||x_2 - y|} + \frac{k\,|x_1 - x_2|}{4\pi\,|x_1 - y|}
\end{aligned} \tag{3.15}
$$

because $\left|\exp(it) - \exp(is)\right| \leq |t - s|$ for all $t, s \in \mathbb{R}$. Now (c) follows because $|x_j - y| \geq \delta$ for $x_j \notin U_\delta$.

To show (b); that is, Hölder-continuity in H_ρ, let $x_1, x_2 \in H_\rho$, where $\rho < \rho_0$ is chosen as in Lemma A.10 such that there is the unique representation of x_j in the form $x_j = z_j + t_j\nu(z_j)$ with $z_j \in \partial D$ and $|t_j| < \rho_0$. We set $\Gamma_{z,r} = \{y \in \partial D : |y - z| < r\}$ and split the domain of integration into $\Gamma_{z_1,r}$ and $\partial D \setminus \Gamma_{z_1,r}$ where we set $r = 3|x_1 - x_2|$. The integral over $\Gamma_{z_1,r}$ is simply estimated by

$$
\begin{aligned}
\int_{\Gamma_{z_1,r}} \left|\Phi(x_1, y) - \Phi(x_2, y)\right| ds(y) &\leq \frac{1}{4\pi}\int_{\Gamma_{z_1,r}} \frac{ds(y)}{|x_1 - y|} + \frac{1}{4\pi}\int_{\Gamma_{z_1,r}} \frac{ds(y)}{|x_2 - y|} \\
&\leq \frac{1}{2\pi}\int_{\Gamma_{z_1,r}} \frac{ds(y)}{|z_1 - y|} + \frac{1}{2\pi}\int_{\Gamma_{z_2,2r}} \frac{ds(y)}{|z_2 - y|}.
\end{aligned}
$$

By Lemma A.10 we conclude that $|z_j - y| \leq 2|x_j - y|$ for $j = 1, 2$ and $\Gamma_{z_1,r} \subseteq \Gamma_{z_2,2r}$ because $|y - z_2| \leq |y - z_1| + |z_1 - z_2| \leq |y - z_1| + 2|x_1 - x_2| \leq |y - z_1| + r$. The estimate

$$
\int_{\Gamma_{z,\rho}} \frac{ds(y)}{|z - y|} \leq c_1\,\rho
$$

for some c_1 independent of z and ρ has been proven in (3.3d). Therefore, we have shown that

$$
\int_{\Gamma_{z_1,r}} \left|\Phi(x_1, y) - \Phi(x_2, y)\right| ds(y) \leq \frac{c}{2\pi}(r + 2r) = \frac{9c}{2\pi}|x_1 - x_2| \leq \tilde{c}|x_1 - x_2|^\alpha
$$

where $\tilde{c}$ is independent of x_j.

Now we continue with the integral over $\partial D \setminus \Gamma_{z_1,r}$. For $y \in \partial D \setminus \Gamma_{z_1,r}$ we have $3|x_1 - x_2| = r \leq |y - z_1| \leq 2|y - x_1|$, thus $|x_2 - y| \geq |x_1 - y| - |x_1 - x_2| \geq (1 - 2/3)|x_1 - y| = |x_1 - y|/3$ and therefore

$$
\left|\Phi(x_1, y) - \Phi(x_2, y)\right| \leq \frac{3|x_1 - x_2|}{4\pi|x_1 - y|^2} + \frac{k|x_1 - x_2|}{4\pi|x_1 - y|} \leq \frac{3|x_1 - x_2|}{\pi|z_1 - y|^2} + \frac{k|x_1 - x_2|}{2\pi|z_1 - y|}
$$

because $|z_1 - y| \leq 2|x_1 - y|$. Then we estimate

$$
\int\limits_{\partial D \setminus \Gamma_{z_1,r}} \left| \Phi(x_1, y) - \Phi(x_2, y) \right| ds(y)
$$

$$
\leq \frac{|x_1 - x_2|}{\pi} \int\limits_{\partial D \setminus \Gamma_{z_1,r}} \left[\frac{3}{|z_1 - y|^2} + \frac{k}{2|z_1 - y|} \right] ds(y)
$$

$$
= \frac{|x_1 - x_2|^\alpha}{\pi} \int\limits_{\partial D \setminus \Gamma_{z_1,r}} (r/3)^{1-\alpha} \left[\frac{3}{|z_1 - y|^2} + \frac{k}{2|z_1 - y|} \right] ds(y)
$$

$$
\leq \frac{|x_1 - x_2|^\alpha}{\pi \, 3^{1-\alpha}} \int\limits_{\partial D \setminus \Gamma_{z_1,r}} \left[\frac{3}{|z_1 - y|^{2-(1-\alpha)}} + \frac{k}{2 \, |z_1 - y|^{1-(1-\alpha)}} \right] ds(y)
$$

because $r^{1-\alpha} \leq |y - z_1|^{1-\alpha}$ for $y \in \partial D \setminus \Gamma_{z_1,r}$. Therefore,

$$
\int\limits_{\partial D \setminus \Gamma_{z_1,r}} \left| \Phi(x_1, y) - \Phi(x_2, y) \right| ds(y) \leq \frac{1}{\pi \, 3^{1-\alpha}} |x_1 - x_2|^\alpha \left[3\hat{c}_{1-\alpha} + \frac{k}{2} \hat{c}_{2-\alpha} \right]
$$

with the constants $\hat{c}_\beta$ from Lemma 3.2, part (b). Altogether we have shown the existence of $c > 0$ with

$$
\left| v(x_1) - v(x_2) \right| \leq c \|\varphi\|_\infty |x_1 - x_2|^\alpha \quad \text{for all } x_1, x_2 \in H_{\rho_0}
$$

which ends the proof. $\square$

In preparation of the investigation of the double layer surface potential we prove an other auxiliary result which will be used often.

Lemma 3.13. *For $\varphi \in C^{0,\alpha}(\partial D)$ and $a \in C(\partial D, \mathbb{C}^3)$ define*

$$
w(x) = \int_{\partial D} [\varphi(y) - \varphi(z)] \, a(y) \cdot \nabla_y \Phi(x, y) \, ds(y), \quad x \in H_{\rho_0},
$$

where $x = z + t\nu(z) \in H_{\rho_0}$ with $|t| < \rho_0$ and $z \in \partial D$. Then the integral exists for $x \in \partial D$ and w is Hölder-continuous in H_{ρ_0} for any exponent $\beta < \alpha$. Furthermore, there exists $c > 0$ with $|w(x)| \leq c \|\varphi\|_{C^{0,\beta}(\partial D)}$ for all $x \in H_{\rho_0}$, and the constant c does not depend on x and φ, but it may depend on α.

Proof: For $x_j = z_j + t_j \nu(z_j) \in H_{\rho_0}$, $j = 1, 2$, we have to estimate

$$
\begin{aligned}
w(x_1) - w(x_2) &= \int_{\partial D} \big[\varphi(y) - \varphi(z_1)\big]\, a(y) \cdot \nabla_y \Phi(x_1, y) \\
&\quad - \big[\varphi(y) - \varphi(z_2)\big]\, a(y) \cdot \nabla_y \Phi(x_2, y)\, ds(y) \qquad (3.16) \\
&= \big[\varphi(z_2) - \varphi(z_1)\big] \int_{\partial D} a(y) \cdot \nabla_y \Phi(x_1, y)\, ds(y) \\
&\quad + \int_{\partial D} \big[\varphi(y) - \varphi(z_2)\big]\, a(y) \cdot \big[\nabla_y \Phi(x_1, y) - \nabla_y \Phi(x_2, y)\big]\, ds(y)
\end{aligned}
$$

and thus

$$
\begin{aligned}
\big|w(x_1) - w(x_2)\big| &\leq \big|\varphi(z_2) - \varphi(z_1)\big| \left| \int_{\partial D} a(y) \cdot \nabla_y \Phi(x_1, y)\, ds(y) \right| \\
&\quad + \|a\|_\infty \int_{\partial D} \big|\varphi(y) - \varphi(z_2)\big|\, \big|\nabla_y \Phi(x_1, y) - \nabla_y \Phi(x_2, y)\big|\, ds(y). \qquad (3.17)
\end{aligned}
$$

We need the following estimates of $\nabla_y \Phi$: For $x, x_l \in H_{\rho_0}$ and $y \in \partial D$ there exists $c > 0$ with

$$
\big|\nabla_y \Phi(x, y)\big| \leq \frac{c}{|x - y|^2}, \quad x \neq y,
$$

$$
\big|\nabla_y \Phi(x_1, y) - \nabla_y \Phi(x_2, y)\big| \leq c \frac{|x_1 - x_2|}{|x_1 - y|^3}, \quad \text{with } 0 < |x_1 - y| \leq 3|x_2 - y|.
$$

Proof of these estimates: The first one is obvious. For the second one we observe that $\Phi(x, y) = \phi(|x - y|)$ with $\phi(t) = \exp(ikt)/(4\pi t)$, thus $\phi'(t) = \phi(t)(ik - 1/t)$ and $\phi''(t) = \phi'(t)(ik - 1/t) + \phi(t)/t^2 = \phi(t)\big[(ik - 1/t)^2 + 1/t^2\big]$ and therefore $|\phi'(t)| \leq c_1/t^2$ and $|\phi''(t)| \leq c_2/t^3$ for $0 < t \leq 1$. For $t \leq 3s$ we have

$$
\begin{aligned}
\big|\phi'(t) - \phi'(s)\big| = \left| \int_s^t \phi''(\tau)\, d\tau \right| &\leq c_2 \left| \int_s^t \frac{d\tau}{\tau^3} \right| = \frac{c_2}{2} \big|t^{-2} - s^{-2}\big| \\
&= \frac{c_2}{2} \frac{|t^2 - s^2|}{t^2 s^2} \leq \frac{c_2}{2} |t - s| \left(\frac{1}{t s^2} + \frac{1}{t^2 s} \right) \\
&\leq \frac{c_2}{2} |t - s| \left(\frac{9}{t^3} + \frac{3}{t^3} \right) = 6 c_2 \frac{|t - s|}{t^3}.
\end{aligned}
$$

Setting $t = |x_1 - y|$ and $s = |x_2 - y|$ and observing that $|t - s| \leq |x_1 - x_2|$ yields the estimate

$$\left|\nabla_y \Phi(x_1, y) - \nabla_y \Phi(x_2, y)\right| = \left|\phi'(|x_1 - y|)\frac{y - x_1}{|y - x_1|} - \phi'(|x_2 - y|)\frac{y - x_2}{|y - x_2|}\right|$$

$$\leq \left|\phi'(|x_1 - y|) - \phi'(|x_2 - y|)\right|$$

$$+ \left|\phi'(|x_2 - y|)\right|\left|\frac{y - x_1}{|y - x_1|} - \frac{y - x_2}{|y - x_2|}\right|$$

$$\leq 6c_2\frac{|x_2 - x_1|}{|y - x_1|^3} + 2\left|\phi'(|x_2 - y|)\right|\frac{|x_2 - x_1|}{|y - x_1|}$$

$$\leq 6c_2\frac{|x_2 - x_1|}{|y - x_1|^3} + 2c_1\frac{|x_2 - x_1|}{|y - x_1||y - x_2|^2}$$

$$\leq (6c_2 + 18c_1)\frac{|x_2 - x_1|}{|y - x_1|^3} \, .$$

This yields the second estimate.

Now we split the region of integration again into $\Gamma_{z_1,r}$ and $\partial D \setminus \Gamma_{z_1,r}$ with $r = 3|x_1 - x_2|$ where $\Gamma_{z,r} = \{y \in \partial D : |y - z| < r\}$. Let $c > 0$ denote a generic constant which may differ from line to line. The integral (in the form (3.16)) over $\Gamma_{z_1,r}$ is estimated by

$$\int_{\Gamma_{z_1,r}} \left|\left[\varphi(y) - \varphi(z_1)\right] a(y) \cdot \nabla_y \Phi(x_1, y)\right.$$

$$\left. - \left[\varphi(y) - \varphi(z_2)\right] a(y) \cdot \nabla_y \Phi(x_2, y)\right| ds(y)$$

$$\leq c \int_{\Gamma_{z_1,r}} \left[|y - z_1|^\alpha \frac{1}{|y - x_1|^2} + |y - z_2|^\alpha \frac{1}{|y - x_2|^2}\right] ds(y) \qquad (3.18)$$

$$\leq c \int_{\Gamma_{z_1,r}} |y - z_1|^\alpha \frac{1}{|y - z_1|^2} ds(y) + c \int_{\Gamma_{z_2,2r}} |y - z_2|^\alpha \frac{1}{|y - z_2|^2} ds(y)$$

because $\Gamma_{z_1,r} \subseteq \Gamma_{z_2,2r}$ and $|x_j - y| \geq |z_j - y|/2$. Therefore, using (3.3d), this term behaves as $r^\alpha = 3^\alpha |x_1 - x_2|^\alpha \leq c|x_1 - x_2|^\beta$.

We finally consider the integral over $\partial D \setminus \Gamma_{z_1,r}$ and use the form (3.17):

$$I := \left|\varphi(z_2) - \varphi(z_1)\right|\left|\int_{\partial D \setminus \Gamma_{z_1,r}} a(y) \cdot \nabla_y \Phi(x_1, y) ds(y)\right| +$$

$$+ \|a\|_\infty \int_{\partial D \setminus \Gamma_{z_1,r}} |\varphi(y) - \varphi(z_2)| \left|\nabla_y \Phi(x_1, y) - \nabla_y \Phi(x_2, y)\right| ds(y)$$

$$\leq c \int_{\partial D \setminus \Gamma_{z_1,r}} \left[|z_2 - z_1|^\alpha \frac{1}{|x_1 - y|^2} + |y - z_2|^\alpha \frac{|x_1 - x_2|}{|x_1 - y|^3}\right] ds(y)$$

Since $y \in \partial D \setminus \Gamma_{z_1,r}$ we can use the estimates

$$|x_1 - x_2| \;=\; \frac{r}{3} \;\leq\; \frac{1}{3}|y - z_1| \;\leq\; \frac{2}{3}|y - x_1| \;<\; |y - x_1|$$

and

$$|y - z_2| \;\leq\; 2|y - x_2| \;\leq\; 2|y - x_1| + 2|x_1 - x_2| \;\leq\; 4|y - x_1|\,.$$

Thus, we obtain

$$I \leq c \int\limits_{\partial D \setminus \Gamma_{z_1,r}} \left[|x_2 - x_1|^{\alpha} \frac{1}{|z_1 - y|^2} + \frac{|x_1 - x_2|}{|x_1 - y|^{3-\alpha}} \right] ds(y)$$

$$\leq c|x_1 - x_2|^{\beta} \int\limits_{\partial D \setminus \Gamma_{z_1,r}} \left[|x_2 - x_1|^{\alpha - \beta} \frac{1}{|z_1 - y|^2} + \frac{|x_1 - x_2|^{1-\beta}}{|z_1 - y|^{3-\alpha}} \right] ds(y)$$

$$\leq c|x_1 - x_2|^{\beta} \int\limits_{\partial D \setminus \Gamma_{z_1,r}} \frac{1}{|z_1 - y|^{2-(\alpha - \beta)}} \, ds(y) \;\leq\; c\,\hat{c}_{\alpha - \beta}|x_1 - x_2|^{\beta}$$

with a constant $c > 0$ and the constant $\hat{c}_{\alpha - \beta}$ from Lemma 3.2. This, together with (3.18) proves the Hölder-continuity of w. The proof of the estimate $|w(x)| \leq v\|\varphi\|_{C^{0,\beta}(\partial D)}$ for $x \in H_{\rho_0}$ is simpler and left to the reader. $\quad\square$

Next we consider the *double layer surface potential*

$$v(x) \;=\; \tilde{D}\varphi(x) \;=\; \int_{\partial D} \varphi(y) \frac{\partial \Phi_k}{\partial \nu(y)}(x,y)\, ds(y)\,, \qquad x \in \mathbb{R}^3 \setminus \partial D\,, \qquad (3.19)$$

for Hölder-continuous densities. Here we indicate the dependence on $k \geq 0$ by writing Φ_k.

Theorem 3.14. *The double layer potential v from (3.19) with Hölder-continuous density $\varphi \in C^{0,\alpha}(\partial D)$ can be continuously extended from D to $\overline{D}$ and from $\mathbb{R}^3 \setminus \overline{D}$ to $\mathbb{R}^3 \setminus D$ with limiting values*

$$\lim_{\substack{x \to x_0 \\ x \in D}} v(x) = -\frac{1}{2}\,\varphi(x_0) + \int_{\partial D} \varphi(y) \frac{\partial \Phi_k}{\partial \nu(y)}(x_0, y)\, ds(y)\,, \qquad x_0 \in \partial D, \quad (3.20a)$$

$$\lim_{\substack{x \to x_0 \\ x \notin \overline{D}}} v(x) = +\frac{1}{2}\,\varphi(x_0) + \int_{\partial D} \varphi(y) \frac{\partial \Phi_k}{\partial \nu(y)}(x_0, y)\, ds(y)\,, \qquad x_0 \in \partial D. \quad (3.20b)$$

v is Hölder-continuous in D and in $\mathbb{R}^3 \setminus \overline{D}$ with exponent β for every $\beta < \alpha$.

Proof: First we note that the integrals exist because for $x_0, y \in \partial D$ we can estimate

$$\left| \frac{\partial \Phi_k}{\partial \nu(y)}(x_0, y) \right| = \left| \frac{\exp(ik|x_0 - y|)}{4\pi|x_0 - y|} \left(ik - \frac{1}{|x_0 - y|} \right) \right| \frac{|\nu(y) \cdot (y - x_0)|}{|y - x_0|}$$
$$\leq \frac{c}{|x_0 - y|}.$$

by Lemma A.10. Furthermore, v has a decomposition into $v = v_0 + v_1$ where

$$v_0(x) = \int_{\partial D} \varphi(y) \frac{\partial \Phi_0}{\partial \nu(y)}(x, y)\, ds(y)\,, \quad x \in \mathbb{R}^3 \setminus \partial D\,,$$
$$v_1(x) = \int_{\partial D} \varphi(y) \frac{\partial (\Phi_k - \Phi_0)}{\partial \nu(y)}(x, y)\, ds(y)\,, \quad x \in \mathbb{R}^3 \setminus \partial D\,.$$

The kernel $K = \nabla_y(\Phi_k - \Phi_0)$ has the form $K(x, y) = A(|x-y|^2) + \frac{y-x}{|y-x|} B(|x - y|^2)$, $(x, y) \in \mathbb{R}^3 \times \partial D$, with analytic functions A and B and is similar to the kernel of the single layer potential. From this it follows that v_1 is Hölder-continuous in all of $\mathbb{R}^3$.

We continue with the analysis of v_0 and note that we have again to prove estimates of the form (a) and (b) of Remark 3.11.

First, for $x \in U_{\delta/4}$ we have $x = z + t\nu(z)$ with $|t| < \delta/4$ and $z \in \partial D$. We write $v_0(x)$ in the form

$$v_0(x) = \underbrace{\int_{\partial D} [\varphi(y) - \varphi(z)] \frac{\partial \Phi_0}{\partial \nu(y)}(x, y)\, ds(y)}_{= \,\tilde{v}_0(x)} + \varphi(z) \int_{\partial D} \frac{\partial \Phi_0}{\partial \nu(y)}(x, y)\, ds(y)\,.$$

The function $\tilde{v}_0$ is Hölder-continuous in $U_{\delta/4}$ with exponent $\beta < \alpha$ by Lemma 3.13 (take $a(y) = \nu(y)$). This proves the estimate (a) and (b) of Remark 3.11 for $\tilde{v}_0$.

Now we consider the decomposition

$$v_0(x) = \tilde{v}_0(x) + \varphi(z) \int_{\partial D} \frac{\partial \Phi_0}{\partial \nu(y)}(x, y)\, ds(y)\,.$$

By Green's representation (Theorem 3.3) for $k = 0$ and $u = 1$ we observe that

$$\int_{\partial D} \frac{\partial \Phi_0}{\partial \nu(y)}(x, y)\, ds(y) = \begin{cases} -1, & x \in D\,, \\ -1/2, & x \in \partial D\,, \\ 0, & x \notin \overline{D}\,. \end{cases}$$

Therefore,

$$\lim_{x \to x_0, \ x \in D} v_0(x) = \tilde{v}_0(x_0) - \varphi(x_0) = v_0(x_0) + \frac{1}{2}\varphi(x_0) - \varphi(x_0) = -\frac{1}{2}\varphi(x_0),$$

$$\lim_{x \to x_0, \ x \notin \overline{D}} v_0(x) = \tilde{v}_0(x_0) = v_0(x_0) + \frac{1}{2}\varphi(x_0).$$

This ends the proof. $\square$

The next step is an investigation of the *derivative of the single layer potential* (3.12); that is,

$$v(x) = \int_{\partial D} \varphi(y)\, \Phi(x,y)\, ds(y), \quad x \in \mathbb{R}^3.$$

First we show the following auxiliary result.

Lemma 3.15. *(a) There exists $c > 0$ with*

$$\left| \int_{\partial D \setminus B(x,\tau)} \nabla_x \Phi(x,y)\, ds(y) \right| \le c \quad \text{for all } x \in \mathbb{R}^3, \ \tau > 0,$$

(b)

$$\lim_{\tau \to 0} \int_{\partial D \setminus B(x,\tau)} \nabla_x \Phi(x,y)\, ds(y) = \int_{\partial D} H(y)\, \Phi(x,y)\, ds(y)$$
$$- \int_{\partial D} \nu(y)\, \frac{\partial \Phi}{\partial \nu(y)}(x,y)\, ds(y)$$

for $x \in \mathbb{R}^3$ where $H(y) = \left(\operatorname{Div} \hat{e}_t^1(y), \operatorname{Div} \hat{e}_t^2(y), \operatorname{Div} \hat{e}_t^3(y)\right)^\top \in \mathbb{R}^3$ and $\hat{e}_t^j(y) = \nu(y) \times \left(\hat{e}^j \times \nu(y)\right)$, $j = 1, 2, 3$, the tangential components of the unit vectors $\hat{e}^j$.

Proof: For any $x \in \mathbb{R}^3$ we have

$$\int_{\partial D \setminus B(x,\tau)} \nabla_x \Phi(x,y)\, ds(y) = - \int_{\partial D \setminus B(x,\tau)} \nabla_y \Phi(x,y)\, ds(y)$$
$$= - \int_{\partial D \setminus B(x,\tau)} \operatorname{Grad}_y \Phi(x,y)\, ds(y)$$
$$- \int_{\partial D \setminus B(x,\tau)} \frac{\partial \Phi}{\partial \nu(y)}(x,y)\, \nu(y)\, ds(y)$$

and thus for any fixed vector $a \in \mathbb{C}^3$ by the previous theorem, again with $\Gamma(x, \tau) = \partial D \cap B(x, \tau)$ and $a_t(y) = \nu(y) \times (a \times \nu(y))$, and Lemma A.18 we obtain

$$\int_{\partial D \backslash B(x,\tau)} a \cdot \nabla_x \Phi(x, y)\, ds(y) = \int_{\partial D \backslash B(x,\tau)} \mathrm{Div}\, a_t(y)\, \Phi(x, y)\, ds(y)$$

$$+ \int_{\partial \Gamma(x,\tau)} a_t(y) \cdot (\tau(y) \times \nu(y))\, \Phi(x, y)\, ds(y)$$

$$+ \int_{\partial D \backslash B(x,\tau)} \frac{\partial \Phi}{\partial \nu(y)}(x, y)\, a \cdot \nu(y)\, ds(y).$$

The first and third integrals converge uniformly with respect to $x \in \partial D$ when τ tends to zero because the integrands are weakly singular. For the second integral we note that

$$\left| \int_{\partial \Gamma(x,\tau)} a_t(y) \cdot \nu_0(y)\, \Phi(x, y)\, ds(y) \right| = \frac{1}{4\pi\tau} \left| \int_{\partial \Gamma(x,\tau)} a_t(y) \cdot (\tau(y) \times \nu(y))\, ds(y) \right|$$

$$= \frac{1}{4\pi\tau} \left| \int_{\Gamma(x,\tau)} \mathrm{Div}\, a_t(y)\, ds(y) \right|$$

and this tends to zero uniformly with respect to $x \in \partial D$ when τ tends to zero. The conclusion follows if we take for a the unit coordinate vectors $\hat{e}^{(j)}$.
$\square$

As an abbreviation we write in the following

$$v(x_0)\big|_- = \lim_{\substack{x \to x_0 \\ x \in D}} v(x), \quad v(x_0)\big|_+ = \lim_{\substack{x \to x_0 \\ x \notin \overline{D}}} v(x) \quad \text{and}$$

$$\frac{\partial v}{\partial \nu}(x_0)\bigg|_- = \lim_{\substack{x \to x_0 \\ x \in D}} \nu(x_0) \cdot \nabla v(x) \text{ and } \frac{\partial v}{\partial \nu}(x_0)\bigg|_+ = \lim_{\substack{x \to x_0 \\ x \notin \overline{D}}} \nu(x_0) \cdot \nabla v(x),$$

and obtain the following relations for the traces of the derivatives of the single layer surface potential.

Theorem 3.16. *The derivative of the single layer potential v from (3.12) with Hölder-continuous density $\varphi \in C^{0,\alpha}(\partial D)$ can be continuously extended from D to $\overline{D}$ and from $\mathbb{R}^3 \setminus \overline{D}$ to $\mathbb{R}^3 \setminus D$. The tangential component is continuous, i.e. $\mathrm{Grad}\, v|_- = \mathrm{Grad}\, v|_+$, and the limiting values of the normal derivatives are*

$$\frac{\partial v}{\partial \nu}(x)\Big|_{\pm} = \mp \frac{1}{2}\varphi(x) + \int_{\partial D} \varphi(y)\frac{\partial \Phi}{\partial \nu(x)}(x,y)\,ds(y), \quad x \in \partial D. \quad (3.21)$$

Proof: We note that the integral exists (see proof of Theorem 3.14). First we consider the density 1, i.e. we set

$$v_1(x) = \int_{\partial D} \Phi(x,y)\,ds(y), \quad x \in \mathbb{R}^3.$$

By the previous lemma we have that

$$\nabla v_1(x) = -\int_{\partial D} \nu(y)\frac{\partial \Phi}{\partial \nu(y)}(x,y)\,ds(y) + \int_{\partial D} H(y)\,\Phi(x,y)\,ds(y), \quad x \notin \partial D,$$
$$(3.22)$$

where again $H(y) = \big(\mathrm{Div}\,\hat{e}_t^1(y), \mathrm{Div}\,\hat{e}_t^2(y), \mathrm{Div}\,\hat{e}_t^3(y)\big)^{\top} \in \mathbb{R}^3$.

The right-hand side is the sum of a double and a single layer potential. By Theorems 3.12 and 3.14 it has a continuous extension to the boundary from the inside and the outside with limiting values

$$\nabla v_1(x)|_{\pm} = \mp\frac{1}{2}\nu(x) - \int_{\partial D} \nu(y)\frac{\partial \Phi}{\partial \nu(y)}(x,y)\,ds(y) + \int_{\partial D} H(y)\,\Phi(x,y)\,ds(y)$$
$$= \mp\frac{1}{2}\nu(x) + \int_{\partial D} \nabla_x\Phi(x,y)\,ds(y)$$

for $x \in \partial D$. The last integral has to be interpreted as a Cauchy principal value as in part (b) of the previous Lemma 3.15. In particular, the tangential component is continuous and the normal derivative jumps, and we have

$$\frac{\partial v_1}{\partial \nu}(x)\Big|_{\pm} = \mp\frac{1}{2} + \int_{\partial D} \frac{\partial \Phi}{\partial \nu(x)}(x,y)\,ds(y), \quad x \in \partial D.$$

Now we consider v and have for $x = z + t\nu(z) \in H_{\rho_0} \setminus \partial D$ (that is, $t \neq 0$)

$$\nabla v(x) = \int_{\partial D} \varphi(y)\,\nabla_x\Phi(x,y)\,ds(y)$$
$$= \underbrace{\int_{\partial D} \nabla_x\Phi(x,y)\,[\varphi(y) - \varphi(z)]\,ds(y)}_{=\,\tilde{v}(x)} + \varphi(z)\,\nabla v_1(x).$$

Application of Lemma 3.13 yields that $\tilde{v}$ is Hölder-continuous in all of H_{ρ_0} with limiting value

$$\tilde{v}(x) = \int_{\partial D} \nabla_x\Phi(x,y)\,[\varphi(y) - \varphi(x)]\,ds(y) \quad \text{for } x \in \partial D.$$

For further use we formulate the result derived so far.

$$\nabla v(x)\big|_{\pm} = \int_{\partial D} \nabla_x \Phi(x,y)\left[\varphi(y) - \varphi(x)\right] ds(y) \; + \; \varphi(x)\left[\mp \frac{1}{2}\nu(x)\right.$$

$$\left. - \int_{\partial D} \nu(y)\,\frac{\partial \Phi}{\partial \nu(y)}(x,y)\,ds(y) + \int_{\partial D} H(y)\,\Phi(x,y)\,ds(y)\right].$$

$$(3.23)$$

This proves that the gradient has continuous extensions from the inside and outside of D and on ∂D we have

$$\frac{\partial v}{\partial \nu}(x)\bigg|_{\pm} = \mp \frac{1}{2}\varphi(x) + \varphi(x)\int_{\partial D}\frac{\partial \Phi}{\partial \nu(x)}(x,y)\,ds(y)$$

$$+ \int_{\partial D}\frac{\partial \Phi}{\partial \nu(x)}(x,y)\left[\varphi(y) - \varphi(x)\right]ds(y)$$

$$= \mp\frac{1}{2}\varphi(x) \; + \; \int_{\partial D}\varphi(y)\,\frac{\partial \Phi}{\partial \nu(x)}(x,y)\,ds(y)\,,$$

$$\mathrm{Grad}\,v(x)\big|_{\pm} = \int_{\partial D}\mathrm{Grad}\,_x\Phi(x,y)\left[\varphi(y) - \varphi(z)\right]ds(y)$$

$$+ \varphi(z)\,\mathrm{Grad}\,v_1(x) \qquad\qquad (3.24)$$

for $x \in \partial D$. □

3.1.3 Boundary Integral Operators

It is the aim of this subsection to investigate the mapping properties of the traces of the single and double layer potentials on the boundary ∂D. We start with a general theorem on boundary integral operators with singular kernels.

Theorem 3.17. *Let* $\Lambda = \left\{(x,y) \in \partial D \times \partial D : x \neq y\right\}$ *and* $K \in C(\Lambda)$.

(a) Let there exist $c > 0$ *and* $\alpha \in (0,1)$ *such that*

$$\left|K(x,y)\right| \leq \frac{c}{|x-y|^{2-\alpha}} \quad \textit{for all } (x,y) \in \Lambda, \qquad (3.25a)$$

$$\left|K(x_1,y) - K(x_2,y)\right| \leq c\,\frac{|x_1 - x_2|}{|x_1 - y|^{3-\alpha}} \quad \textit{for all } (x_1,y),(x_2,y) \in \Lambda$$

$$(3.25b)$$

$$\textit{with } |x_1 - y| \geq 3|x_1 - x_2|\,.$$

Then the operator $\mathcal{K}_1 : C(\partial D) \to C^{0,\alpha}(\partial D)$, defined by

$$(\mathcal{K}_1\varphi)(x) = \int_{\partial D} K(x,y)\,\varphi(y)\,ds(y)\,, \quad x \in \partial D\,,$$

is well defined and bounded.

(b) Let there exist $c > 0$ such that:

$$|K(x,y)| \le \frac{c}{|x-y|^2} \quad \text{for all } (x,y) \in \Lambda, \qquad (3.25c)$$

$$|K(x_1,y) - K(x_2,y)| \le c\frac{|x_1 - x_2|}{|x_1 - y|^3} \quad \text{for all } (x_1,y),(x_2,y) \in \Lambda$$
$$(3.25d)$$

$$\text{with } |x_1 - y| \ge 3|x_1 - x_2|\,,$$

$$\left|\int_{\partial D \setminus B(x,r)} K(x,y)\,ds(y)\right| \le c \quad \text{for all } x \in \partial D \text{ and } r > 0\,. \qquad (3.25e)$$

Then the operator $\mathcal{K}_2 : C^{0,\alpha}(\partial D) \to C^{0,\alpha}(\partial D)$, defined by

$$(\mathcal{K}_2\varphi)(x) = \int_{\partial D} K(x,y)\,\big[\varphi(y) - \varphi(x)\big]\,ds(y)\,, \quad x \in \partial D\,,$$

is well defined and bounded.

Proof: (a) We follow the idea of the proof of Theorem 3.12 and write

$$\big|(\mathcal{K}_1\varphi)(x_1) - (\mathcal{K}_1\varphi(x_2)\big| \le \|\varphi\|_\infty \int_{\partial D} \big|K(x_1,y) - K(x_2,y)\big|\,ds(y)\,.$$

We split the region of integration again into $\Gamma_{x_1,r}$ and $\partial D \setminus \Gamma_{x_1,r}$ where again $\Gamma_{x_1,r} = \{y \in \partial D : |y - x_1| < r\}$ and set $r = 3|x_1 - x_2|$. The integral over $\Gamma_{x_1,r}$ can be estimated with (3.3d) of Lemma 3.2 by

$$\int_{\Gamma_{x_1,r}} \big|K(x_1,y) - K(x_2,y)\big|\,ds(y) \le c \int_{\Gamma_{x_1,r}} \frac{ds(y)}{|x_1 - y|^{2-\alpha}} + c \int_{\Gamma_{x_2,2r}} \frac{ds(y)}{|x_2 - y|^{2-\alpha}}$$

$$\le c'r^\alpha = (c'3^\alpha)\,|x_1 - x_2|^\alpha$$

because $\Gamma_{x_1,r} \subseteq \Gamma_{x_2,2r}$. Here we used formula (3.3d) of Lemma 3.2.

For the integral over $\partial D \setminus \Gamma_{x_1,r}$ we note that $|y - x_1| \geq r = 3|x_1 - x_2|$ for $y \in \partial D \setminus \Gamma_{x_1,r}$, thus

$$\int_{\partial D \setminus \Gamma_{x_1,r}} |K(x_1,y) - K(x_2,y)|\, ds(y) \leq c\,|x_1 - x_2| \int_{\partial D \setminus \Gamma_{x_1,r}} \frac{ds(y)}{|x_1 - y|^{3-\alpha}}$$

$$\leq c'|x_1 - x_2|\,r^{\alpha-1} = c'3^{\alpha-1}\,|x_1 - x_2|^{\alpha}$$

where we used estimate (3.3e). The proof of $|(\mathcal{K}_1\varphi)(x)| \leq c\|\varphi\|_\infty$ is similar, even simpler, and is left to the reader.

For part (b) we follow the ideas of the proof of Lemma 3.13. We write

$$|(\mathcal{K}_2\varphi)(x_1) - (\mathcal{K}_2\varphi(x_2)|$$

$$\leq \int_{\Gamma_{x_1,r}} \Big| K(x_1,y)[\varphi(y) - \varphi(x_1)] - K(x_2,y)[\varphi(y) - \varphi(x_2)] \Big|\, ds(y)$$

$$+ |\varphi(x_2) - \varphi(x_1)| \left| \int_{\partial D \setminus \Gamma_{x_1,r}} K(x_1,y)\, ds(y) \right|$$

$$+ \int_{\partial D \setminus \Gamma_{x_1,r}} |\varphi(y) - \varphi(x_2)|\,|K(x_1,y) - K(x_2,y)|\, ds(y)$$

$$\leq c\|\varphi\|_{C^{0,\alpha}(\partial D)} \left[\int_{\Gamma_{x_1,r}} \frac{ds(y)}{|y - x_1|^{2-\alpha}} + \int_{\Gamma_{x_2,2r}} \frac{ds(y)}{|y - x_2|^{2-\alpha}} \right]$$

$$+ c\|\varphi\|_{C^{0,\alpha}(\partial D)}|x_1 - x_2|^{\alpha}$$

$$+ c\|\varphi\|_{C^{0,\alpha}(\partial D)} \int_{\partial D \setminus \Gamma_{x_1,r}} |y - x_2|^{\alpha}\frac{|x_1 - x_2|}{|y - x_1|^{3}}\, ds(y)$$

$$\leq c\|\varphi\|_{C^{0,\alpha}(\partial D)} \left[r^{\alpha} + |x_1 - x_2|^{\alpha} + |x_1 - x_2| \int_{\partial D \setminus \Gamma_{x_1,r}} \frac{ds(y)}{|y - x_1|^{3-\alpha}} \right]$$

because $|y - x_2| \leq |y - x_1| + |x_1 - x_2| = |y - x_1| + r/3 \leq 2|y - x_1|$. The last integral has been estimated by $r^{\alpha-1}$, see (3.3e). This proves that

$$|(\mathcal{K}_2\varphi)(x_1) - (\mathcal{K}_2\varphi(x_2)| \leq c\|\varphi\|_{C^{0,\alpha}(\partial D)}\,|x_1 - x_2|^{\alpha}.$$

The proof of $|(\mathcal{K}_2\varphi)(x)| \leq c\|\varphi\|_{C^{0,\alpha}(\partial D)}$ is again simpler and is left to the reader. $\square$

One of the essential properties of the boundary operators is their compactness in Hölder spaces. This follows from the previous theorem and the following compactness result.

Lemma 3.18. *The embedding* $C^{0,\alpha}(\partial D) \to C(\partial D)$ *is compact for every* $\alpha \in (0,1)$.

Proof: We have to prove that the unit ball $B = \{\varphi \in C^{0,\alpha}(\partial D) : \|\varphi\|_{C^{0,\alpha}(\partial D)} \leq 1\}$ is relatively compact, i.e. its closure is compact, in $C(\partial D)$. This follows directly by the theorem of Arcela–Ascoli (see, e.g., [11], Appendix C.7). Indeed, B is equi-continuous because

$$|\varphi(x_1) - \varphi(x_2)| \leq \|\varphi\|_{C^{0,\alpha}(\partial D)}|x_1 - x_2|^\alpha \leq |x_1 - x_2|^\alpha$$

for all $x_1, x_2 \in \partial D$. Furthermore, B is bounded. $\square$

Thus from the last theorem we immediately conclude the following corollary.

Corollary 3.19. *Under the assumptions of Theorem 3.17 the operator* $\mathcal{K}_1$ *is compact from* $C^{0,\alpha}(\partial D)$ *into itself for every* $\alpha \in (0,1)$.

Proof: The operator $\mathcal{K}_1$ is bounded from $C(\partial D)$ into $C^{0,\alpha}(\partial D)$ and the embedding $C^{0,\alpha}(\partial D)$ into $C(\partial D)$ is compact, which implies compactness of $\mathcal{K}_1 : C^{0,\alpha}(\partial D) \to C^{0,\alpha}(\partial D)$. $\square$

We apply this result to the *boundary integral operators* which appear in the traces of the single and double layer potentials of Theorems 3.12, 3.14, and 3.16.

Theorem 3.20. *Let* $k \in \mathbb{C}$ *with* $\operatorname{Im} k \geq 0$. *The operators* $\mathcal{S}, \mathcal{D}, \mathcal{D}' : C^{0,\alpha}(\partial D) \to C^{0,\alpha}(\partial D)$, *defined by*

$$(\mathcal{S}\varphi)(x) = \int_{\partial D} \varphi(y)\, \Phi_k(x,y)\, ds(y)\,, \quad x \in \partial D\,, \tag{3.26a}$$

$$(\mathcal{D}\varphi)(x) = \int_{\partial D} \varphi(y)\, \frac{\partial \Phi_k}{\partial \nu(y)}(x,y)\, ds(y)\,, \quad x \in \partial D\,, \tag{3.26b}$$

$$(\mathcal{D}'\varphi)(x) = \int_{\partial D} \varphi(y)\, \frac{\partial \Phi_k}{\partial \nu(x)}(x,y)\, ds(y)\,, \quad x \in \partial D\,, \tag{3.26c}$$

are well defined and compact. Additionally, the operator $\mathcal{S}$ *is bounded from* $C^{0,\alpha}(\partial D)$ *into* $C^{1,\alpha}(\partial D)$. *Here,* $C^{1,\alpha}(\partial D) = \{u \in C^{0,\alpha}(\partial D) : \operatorname{Grad} u \in C^{0,\alpha}(\partial D, \mathbb{C}^3)\}$ *equipped with its canonical norm.*

Proof: We have to check the assumptions (3.25a) and (3.25b) of Theorem 3.17. For $x, y \in \partial D$ we have by the definition of the fundamental solution Φ and part (a) of Lemma A.10 that

$$\left|\Phi(x,y)\right| = \frac{1}{4\pi|x-y|},$$

$$\left|\frac{\partial}{\partial\nu(y)}\Phi(x,y)\right| = \frac{1}{4\pi|x-y|}\left|ik - \frac{1}{|x-y|}\right|\frac{|(y-x)\cdot\nu(y)|}{|x-y|}$$

$$\leq \frac{c}{4\pi|x-y|}\left|ik - \frac{1}{|x-y|}\right||x-y|$$

$$\leq \frac{c}{4\pi|x-y|}\left[k|x-y|+1\right] \leq \frac{c(kd+1)}{4\pi|x-y|}$$

where $d = \sup\{|x-y| : x, y \in \partial D\}$. The same estimate holds for $\partial\Phi(x,y)/\partial\nu(x)$. This proves (3.25a) with $\alpha = 1$. Furthermore, we will prove (3.25b) with $\alpha = 1$. Let $x_1, x_2, y \in \partial D$ such that $|x_1 - y| \geq 3|x_1 - x_2|$. Then, for any $t \in [0,1]$, we conclude that $|x_1 + t(x_2 - x_1) - y| \geq |x_1 - y| - |x_2 - x_1| \geq |x_1 - y| - |x_1 - y|/3 = 2|x_1 - y|/3$.
First we consider Φ and apply the mean value theorem:

$$\left|\Phi(x_1,y) - \Phi(x_2,y)\right| \leq |x_1 - x_2| \sup_{0\leq t\leq 1} \left|\nabla_x\Phi(x_1 + t(x_2 - x_1),y)\right|$$

$$\leq c \sup_{0\leq t\leq 1} \frac{|x_1 - x_2|}{|x_1 + t(x_2 - x_1) - y|^2} \leq c\frac{9}{4}\frac{|x_1 - x_2|}{|x_1 - y|^2}.$$

To show the corresponding estimate for the normal derivative of the fundamental solution we can restrict ourselves to the case $k = 0$ as in the proof of Theorem 3.14.

Furthermore, we assume again $x_1, x_2, y \in \partial D$ such that $|x_1 - y| \geq 3|x_1 - x_2|$. Then

$$\left|\frac{\partial}{\partial\nu(y)}\Phi_0(x_1,y) - \frac{\partial}{\partial\nu(y)}\Phi_0(x_2,y)\right|$$

$$\leq \frac{1}{4\pi|x_1 - y|^3}\underbrace{\left|\nu(y)\cdot(y-x_1) - \nu(y)\cdot(y-x_2)\right|}_{=\nu(y)\cdot(x_2-x_1)}$$

$$+ \frac{1}{4\pi}\left|\frac{1}{|x_1 - y|^3} - \frac{1}{|x_2 - y|^3}\right|\left|\nu(y)\cdot(y-x_2)\right|$$

$$\leq \frac{1}{4\pi|x_1 - y|^3}\left[\left|(\nu(y) - \nu(x_1))\cdot(x_2 - x_1)\right| + \left|\nu(x_1)\cdot(x_2 - x_1)\right|\right]$$

$$+ \frac{1}{4\pi}\left|\frac{1}{|x_1 - y|^3} - \frac{1}{|x_2 - y|^3}\right|\left|\nu(y)\cdot(y-x_2)\right|$$

Now we use the estimates (a) and (b) of Lemma A.10 for the first term and the mean value theorem for the second term. By $|x_1 + t(x_2 - x_1) - y| \geq 2|x_1 - y|/3$ we have

$$\left| \frac{\partial}{\partial \nu(y)} \Phi_0(x_1, y) - \frac{\partial}{\partial \nu(y)} \Phi_0(x_2, y) \right|$$

$$\leq c \frac{|y - x_1||x_2 - x_1| + |x_1 - x_2|^2}{|x_1 - y|^3} + c \frac{|y - x_2|^2|x_1 - x_2|}{|x_1 - y|^4}.$$

Estimate (3.25b) follows from $|x_1 - x_2| \leq |x_1 - y|/3$ and $|y - x_2| \leq |y - x_1| + |x_1 - x_2| \leq 4|y - x_1|/3$.

The same arguments hold for the normal derivative with respect to x.

Finally, we have to show that $\mathrm{Grad}\, \mathcal{S}$ is bounded from $C^{0,\alpha}(\partial D)$ into $C^{0,\alpha}(\partial D, \mathbb{C}^3)$. But this is given from the representation (3.24). $\quad\Box$

3.1.4 Uniqueness and Existence

Now we come back to the scattering problem (3.1) and (3.2) from the beginning of the section. We first study the question of uniqueness. For absorbing media; that is Im $k > 0$, uniqueness can be seen directly from an application of Greens formulas and the radiation condition. But in scattering theory we are interested in the case of a real and positive wave number k. The following lemma is fundamental for proving uniqueness and tells us that a solution of the Helmholtz equation $\Delta u + k^2 u = 0$ for $k > 0$ cannot decay faster than $1/|x|$ as x tends to infinity. We will give two proofs of this result. The first—and shorter—one uses the expansion arguments from the previous chapter. In particular, properties of the spherical Bessel- and Hankel functions are used. The second proof which goes back to the original work by Rellich (see [28]) avoids the use of these special functions but is far more technical and also needs a stronger assumption on the field. For completeness, we present both versions. We begin with the first form.

Lemma 3.21 (Rellich's Lemma, First Form). *Let $u \in C^2(\mathbb{R}^3 \setminus B[0, R_0])$ be a solution of the Helmholtz equation $\Delta u + k^2 u = 0$ for $|x| > R_0$ and wave number $k \in \mathbb{R}_{>0}$ such that*

$$\lim_{R \to \infty} \int_{|x|=R} |u|^2 \, ds = 0.$$

Then u vanishes for $|x| > R_0$.

Proof: The general solution of the Helmholtz equation in the exterior of $B(0, R_0)$ is given by (2.29); that is,

$$u(r\hat{x}) = \sum_{n=0}^{\infty} \sum_{m=-n}^{n} \left[a_n^m h_n^{(1)}(kr) + b_n^m j_n(kr) \right] Y_n^m(\hat{x}), \quad \hat{x} \in S^2, \; r > R_0,$$

for some $a_n^m, b_n^m \in \mathbb{C}$. The spherical harmonics $\{Y_n^m : |m| \le n, \ n \in \mathbb{N}_0\}$ form an orthogonal system. Therefore, Parseval's theorem yields

$$\sum_{n=0}^{\infty} \sum_{m=-n}^{n} \left| a_n^m \, h_n^{(1)}(kr) + b_n^m \, j_n(kr) \right|^2 = \int_{S^2} |u(r\hat{x})|^2 ds(\hat{x}),$$

and from the assumption on u we note that $r^2 \int_{S^2} |u(r\hat{x})|^2 ds(\hat{x})$ tends to zero as r tends to infinity. Especially, for every fixed $n \in \mathbb{N}_0$ and m with $|m| \le n$ we conclude that

$$r^2 \left| a_n^m \, h_n^{(1)}(kr) + b_n^m \, j_n(kr) \right|^2 \longrightarrow 0$$

as r tends to infinity. Defining $c_n^m = a_n^m + b_n^m$ we can write it as $(kr) \, i \, a_n^m \, y_n(kr) + (kr) \, c_n^m \, j_n(kr) \to 0$ for $r \to \infty$. Now we use the asymptotic behavior of $j_n(kr)$ and $y_n(kr)$ as r tends to infinity. By Theorem 2.30 we conclude that

$$i \, a_n^m \, \text{Im} \left[e^{ikr} (-i)^{n+1} \right] + c_n^m \, \text{Re} \left[e^{ikr} (-i)^{n+1} \right] \longrightarrow 0, \quad r \to \infty.$$

The term $(-i)^{n+1}$ can take the values ± 1 and $\pm i$. Therefore, depending on n, we have that

$$i \, a_n^m \sin(kr) + c_n^m \cos(kr) \longrightarrow 0 \quad \text{or} \quad i \, a_n^m \cos(kr) - c_n^m \sin(kr) \longrightarrow 0.$$

In any case, a_n^m and c_n^m have to vanish by taking particular sequences $r_j \to \infty$. This shows that also $b_n^m = 0$. Since it holds for all n and m we conclude that u vanishes. $\square$

As mentioned above, the second proof avoids the use of the Bessel and Hankel functions but needs, however, a stronger assumption on u.

Lemma 3.22 (Rellich's Lemma, Second Form). *Let* $u \in C^2(\mathbb{R}^3 \setminus B[0, R_0])$ *be a solution of the Helmholtz equation* $\Delta u + k^2 u = 0$ *for* $|x| > R_0$ *with wave number* $k \in \mathbb{R}_{>0}$ *such that*

$$\lim_{r \to \infty} \int_{|x|=r} |u|^2 \, ds = 0 \quad \text{and} \quad \lim_{r \to \infty} \int_{|x|=r} \left| \frac{x}{|x|} \cdot \nabla u(x) \right|^2 ds = 0.$$

Then u *vanishes for* $|x| > R_0$.

Proof: The proof, which is taken from the monograph [18, Section VII.3] is lengthy, and we will structure it. Without loss of generality we assume that u is real valued because we can consider real and imaginary parts separately.

1st step: Transforming the integral onto the unit sphere $S^2 = \{x \in \mathbb{R}^3 : |x| = 1\}$ we conclude that

$$\int_{|\hat{x}|=1} |u(r\hat{x})|^2 \, r^2 \, ds(\hat{x}) = \int_{|x|=r} |u(x)|^2 \, ds(x) \quad \text{and} \quad \int_{|\hat{x}|=1} \left|\frac{\partial u}{\partial r}(r\hat{x})\right|^2 r^2 \, ds(\hat{x})$$

$$(3.27)$$

tend to zero as r tends to infinity. We transform the partial differential equation into an ordinary differential equation (not quite!) for the function $v(r, \hat{x}) = r\, u(r, \hat{x})$ with respect to r. We write $v(r)$ and $v'(r)$ and $v''(r)$ for $v(r, \cdot)$ and $\partial v(r, \cdot)/\partial r$ and $\partial^2 v(r, \cdot)/\partial r^2$, respectively. Then (3.27) yields that $\|v(r)\|_{L^2(S^2)} \to 0$ and $\|v'(r)\|_{L^2(S^2)} \to 0$ as $r \to \infty$. The latter follows from $\frac{\partial}{\partial r}\left(ru(r, \cdot)\right) = \frac{1}{r}\left(ru(r, \cdot)\right) + r\frac{\partial u}{\partial r}(r, \cdot)$ and the triangle inequality. We observe that $u = \frac{1}{r}v$, thus $r^2\frac{\partial u}{\partial r} = -v + r\frac{\partial v}{\partial r}$ and $\frac{\partial}{\partial r}\left(r^2\frac{\partial u}{\partial r}\right) = r\frac{\partial^2 v}{\partial r^2}$, thus

$$\begin{aligned}
0 &= \frac{1}{r^2}\frac{\partial}{\partial r}\left(r^2\frac{\partial u}{\partial r}(r, \theta, \phi)\right) + \frac{1}{r^2}\Delta_{S^2}u(r, \theta, \phi) + k^2 u(r, \theta, \phi) \\
&= \frac{1}{r}\left[\frac{\partial^2 v}{\partial r^2}(r, \theta, \phi) + k^2 v(r, \theta, \phi) + \frac{1}{r^2}\Delta_{S^2}v(r, \theta, \phi)\right],
\end{aligned}$$

i.e.

$$v''(r) + k^2 v(r) + \frac{1}{r^2}\Delta_{S^2}v(r) = 0 \quad \text{for } r \geq R_0, \tag{3.28}$$

where again $\Delta_{S^2} = \operatorname{Div}\operatorname{Grad}$ denotes the Laplace–Beltrami operator on the unit sphere; that is, in polar coordinates $\hat{x} = (\sin\theta\cos\phi, \sin\theta\sin\phi, \cos\theta)^\top$

$$(\Delta_{S^2}w)(\theta, \phi) = \frac{1}{\sin\theta}\frac{\partial}{\partial\theta}\left(\sin\theta\frac{\partial w}{\partial\theta}(\theta, \phi)\right) + \frac{1}{\sin^2\theta}\frac{\partial^2 w}{\partial\phi^2}(\theta, \phi)$$

for any $w \in C^2(S^2)$. It is easily seen either by direct integration or by application of Theorem A.11 that Δ_{S^2} is self-adjoint and negative definite, i.e.

$$\left(\Delta_{S^2}v, w\right)_{L^2(S^2)} = \left(v, \Delta_{S^2}w\right)_{L^2(S^2)} \quad \text{and} \quad \left(\Delta_{S^2}v, v\right)_{L^2(S^2)} \leq 0$$

for all $v, w \in C^2(S^2)$.

2nd step: We introduce the functions E, v_m and F by

$$E(r) := \|v'(r)\|^2_{L^2(S^2)} + k^2\|v(r)\|^2_{L^2(S^2)} + \frac{1}{r^2}\left(\Delta_{S^2}v(r), v(r)\right)_{L^2(S^2)},$$

$$v_m(r) := r^m v(r),$$

$$F(r, m, c) := \|v'_m(r)\|^2_{L^2(S^2)} + \left(k^2 + \frac{m(m+1)}{r^2} - \frac{2c}{r}\right)\|v_m(r)\|^2_{L^2(S^2)}$$

$$+ \frac{1}{r^2}\left(\Delta_{S^2}v_m(r), v_m(r)\right)_{L^2(S^2)},$$

for $r \geq R_0$, $m \in \mathbb{N}$, $c \geq 0$. In the following we write $\|\cdot\|$ and $(\cdot, \cdot)$ for $\|\cdot\|_{L^2(S^2)}$ and $(\cdot, \cdot)_{L^2(S^2)}$, respectively. We show:

(a) E satisfies $E'(r) \geq 0$ for all $r \geq R_0$.

(b) The functions v_m solve the differential equation

$$v_m''(r) - \frac{2m}{r} v_m'(r) + \left(\frac{m(m+1)}{r^2} + k^2\right) v_m(r) + \frac{1}{r^2} \Delta_{S^2} v_m(r) = 0.$$
(3.29)

(c) For every $c > 0$ there exist $r_0 = r_0(c) \geq R_0$ and $m_0 = m_0(c) \in \mathbb{N}$ such that

$$\frac{\partial}{\partial r}[r^2 F(r, m, c)] \geq 0 \quad \text{for all } r \geq r_0, \ m \geq m_0.$$

(d) Expressed in terms of v the function F has the forms

$$F(r, m, c) = r^{2m} \left\{ \left\| v'(r) + \frac{m}{r} v(r) \right\|^2 + \left(k^2 + \frac{m(m+1)}{r^2} - \frac{2c}{r}\right) \|v(r)\|^2 \right.$$
$$\left. + \frac{1}{r^2} \left(\Delta_{S^2} v(r), v(r)\right) \right\}$$
(3.30a)

$$= r^{2m} \left\{ E(r) + \frac{2m}{r} \left(v(r), v'(r)\right) \right.$$
$$\left. + \left(\frac{m(2m+1)}{r^2} - \frac{2c}{r}\right) \|v(r)\|^2 \right\}.$$
(3.30b)

Proof of these statements:

(a) We just differentiate E and substitute the second derivative from (3.28). Note that $\frac{d}{dr}\|v(r)\|^2 = 2(v, v')$ and $\frac{d}{dr}(\Delta_{S^2} v, v) = 2(\Delta_{S^2} v, v')$:

$$E'(r) = 2\big(v'(r), v''(r)\big) + 2k^2\big(v(r), v'(r)\big) - \frac{1}{r^3}\left(\Delta_{S^2} v(r), v(r)\right)$$
$$+ \frac{2}{r^2}\left(\Delta_{S^2} v(r), v'(r)\right)$$
$$= 2\left(v'(r), \ \left[v''(r) + k^2 v(r) + \frac{1}{r^2}\Delta_{S^2} v(r)\right]\right) - \frac{1}{r^3}\left(\Delta_{S^2} v(r), v(r)\right)$$
$$= -\frac{1}{r^3}\left(\Delta_{S^2} v(r), v(r)\right) \geq 0.$$

(b) We substitute $v(r) = r^{-m} v_m(r)$ into (3.28) and obtain directly (3.29). We omit the calculation.

(c) Again we differentiate $r^2 F(r, m, c)$ with respect to r, substitute the form of v_m'' from (3.29) and obtain

$$\frac{\partial}{\partial r}[r^2 F(r, m, c)]$$

$$= 2r\|v_m'(r)\|^2 + 2r^2(v_m'(r), v_m''(r)) + 2(k^2 r - c)\|v_m(r)\|^2$$

$$+ 2r^2\left(k^2 + \frac{m(m+1)}{r^2} - \frac{2c}{r}\right)(v_m(r), v_m'(r)) + 2(\Delta_{S^2} v_m(r), v_m'(r))$$

$$= \cdots = 2r(1+2m)\|v_m'(r)\|^2 - 4cr(v_m'(r), v_m(r)) + 2(k^2 r - c)\|v_m(r)\|^2$$

$$= 2r\left[\left\|\sqrt{1+2m}\, v_m'(r) - \frac{c}{\sqrt{1+2m}}\, v_m(r)\right\|^2\right.$$

$$\left. + \left(k^2 - \frac{c}{r} - \frac{c^2}{1+2m}\right)\|v_m(r)\|^2\right].$$

From this the assertion (c) follows if r_0 and m_0 are chosen such that the bracket $(\cdots)$ is positive.

(d) The first equation is easy to see by just inserting the form of v_m. For the second form one uses simply the binomial theorem for the first term and the definition of $E(r)$.

3rd step: We begin with the actual proof of the lemma and show first that there exists $R_1 \geq R_0$ such that $\|v(r)\| = 0$ for all $r \geq R_1$. Assume, on the contrary, that this is not the case. Then, for every $R \geq R_0$ there exists $\hat{r} \geq R$ such that $\|v(\hat{r})\| > 0$.
We choose the constants $\hat{c} > 0$, r_0, m_0, r_1, m_1 in the following order:

- Choose $\hat{c} > 0$ with $k^2 - \frac{2\hat{c}}{R_0} > 0$.

- Choose $r_0 = r_0(\hat{c}) \geq R_0$ and $m_0 = m_0(\hat{c}) \in \mathbb{N}$ according to property (c) above, i.e. such that $\frac{\partial}{\partial r}[r^2 F(r, m, \hat{c})] \geq 0$ for all $r \geq r_0$ and $m \geq m_0$.

- Choose $r_1 > r_0$ such that $\|v(r_1)\| > 0$.

- Choose $m_1 \geq m_0$ such that $m_1(m_1+1)\|v(r_1)\|^2 + (\Delta_{S^2} v(r_1), v(r_1)) > 0$.

Then, by (3.30a) and because $k^2 - \frac{2\hat{c}}{r_1} \geq k^2 - \frac{2\hat{c}}{R_0} > 0$, it follows that $F(r_1, m_1, \hat{c}) > 0$ and thus, by the monotonicity of $r \mapsto r^2 F(r, m_1, \hat{c})$ that also $F(r, m_1, \hat{c}) > 0$ for all $r \geq r_1$. Therefore, from (3.30b) we conclude that, for $r \geq r_1$,

$$0 < r^{-2m_1} F(r, m_1, \hat{c})$$

$$= E(r) + \frac{2m_1}{r} \left(v(r), v'(r) \right) + \left(\frac{m_1(2m_1+1)}{r^2} - \frac{2\hat{c}}{r} \right) \|v(r)\|^2$$

$$= E(r) + \frac{m_1}{r} \frac{d}{dr} \|v(r)\|^2 + \frac{1}{r} \left(\frac{m_1(2m_1+1)}{r} - 2\hat{c} \right) \|v(r)\|^2 .$$

Choose now $r_2 \geq r_1$ such that $\frac{m_1(2m_1+1)}{r_2} - 2\hat{c} < 0$. Finally, choose $\hat{r} \geq r_2$ such that $\frac{d}{dr} \|v(\hat{r})\|^2 \leq 0$. (This is possible because $\|v(r)\|^2 \to 0$ as $r \to \infty$.) We finally have

$$0 < p := \hat{r}^{-2m_1} F(\hat{r}, m_1, \hat{c}) \leq E(\hat{r}).$$

By the monotonicity of E we conclude that $E(r) \geq p$ for all $r \geq \hat{r}$. On the other hand, by the definition of $E(r)$ we have that $E(r) \leq \|v'(r)\|^2 + k^2 \|v(r)\|^2$ and this tends to zero as r tends to infinity. This is a contradiction. Therefore, there exists $R_1 \geq R_0$ with $v(r) = 0$ for all $r \geq R_1$. $\square$

Applying the Rellich Lemma we can now prove uniqueness of the scattering problem.

Theorem 3.23. Let $k > 0$. For any incident field u^{inc} there exists at most one solution $u \in C^2(\mathbb{R}^3 \setminus \overline{D}) \cap C^1(\mathbb{R}^3 \setminus D)$ of the scattering problem (3.1) and (3.2).

Proof: Let u be the difference of two solutions. Then u satisfies (3.1) and also the radiation condition (3.2). From the radiation condition we conclude that

$$\int\limits_{|x|=R} \left| \frac{\partial u}{\partial r} - iku \right|^2 ds = \int\limits_{|x|=R} \left\{ \left| \frac{\partial u}{\partial r} \right|^2 + k^2 |u|^2 \right\} ds + 2k \,\mathrm{Im} \int\limits_{|x|=R} u \frac{\partial \overline{u}}{\partial r} ds$$

tends to zero as R tends to infinity. Green's theorem, applied in $B_R \setminus D$ to the function u yields that

$$\int\limits_{|x|=R} u \frac{\partial \overline{u}}{\partial r} ds = \int\limits_{\partial D} u \frac{\partial \overline{u}}{\partial r} ds + \int\limits_{B_R \setminus D} \left[|\nabla u|^2 - k^2 |u|^2 \right] dx$$

$$= \int\limits_{B_R \setminus D} \left[|\nabla u|^2 - k^2 |u|^2 \right] dx$$

because the surface integral over ∂D vanishes by the boundary condition. The volume integral is real valued. Therefore, its imaginary part vanishes and we conclude that

$$\int_{|x|=R} \left|\frac{\partial u}{\partial r}\right|^2 + k^2 |u|^2 ds \to 0$$

as R tends to infinity. Rellich's lemma (in the form Lemma 3.21 or 3.22) implies that u vanishes outside of every ball which encloses ∂D. Finally, we note that u is an analytic function in the exterior of D (see Corollary 3.4). Since the exterior of D is connected we conclude that u vanishes in $\mathbb{R}^3 \setminus D$. □

We turn to the question of *existence* of a solution and choose the integral equation method for its treatment. We follow the approach of [7, Chapter 3], but prefer to work in the space $C^{0,\alpha}(\partial D)$ of Hölder-continuous functions rather than in the space of merely continuous functions. This avoids the necessity to introduce the class of continuous functions for which the normal derivatives exist "in the uniform sense along the normal."

Let us recall the notion of the single layer surface potential of (3.4a), see also (3.12), and make the ansatz for the scattered field in the form of a single layer potential; that is,

$$u^s(x) = \int_{\partial D} \varphi(y)\, \Phi(x,y)\, ds(y)\,, \quad x \in \mathbb{R}^3 \setminus \partial D\,, \tag{3.31}$$

where again $\Phi(x,y) = \exp(ik|x-y|)/(4\pi|x-y|)$ denotes the fundamental solution of the Helmholtz equation, and $\varphi \in C^{0,\alpha}(\partial D)$ is some density to be determined. We remark already here that we will face some difficulties with this ansatz but we begin with this for didactical reasons. First we note that u^s solves the Helmholtz equation in the exterior of $\overline{D}$ and also the radiation condition. This follows from the corresponding properties of the fundamental solution $\Phi(\cdot, y)$, uniformly with respect to y on the compact surface ∂D. Furthermore, by Theorems 3.12 and 3.16 the function u^s and its derivatives can be extended continuously from the exterior into $\mathbb{R}^3 \setminus D$ with limiting values

$$u^s(x)\big|_+ = \int_{\partial D} \varphi(y)\, \Phi(x,y)\, ds(y) = (\mathcal{S}\varphi)(x)\,, \quad x \in \partial D\,, \tag{3.32a}$$

$$\frac{\partial u^s}{\partial \nu}(x)\bigg|_+ = -\frac{1}{2}\varphi(x) + \int_{\partial D} \varphi(y)\, \frac{\partial}{\partial \nu(x)}\Phi(x,y)\, ds(y)$$

$$= -\frac{1}{2}\varphi(x) + (\mathcal{D}'\varphi)(x)\,, \quad x \in \partial D\,, \tag{3.32b}$$

where we used the notations of the boundary integral operators from Theorem 3.20. Therefore, in order that $u = u^{inc} + u^s$ satisfies the boundary condition $\partial u/\partial \nu = 0$ on ∂D the density φ has to satisfy the boundary integral equation

$$-\frac{1}{2}\varphi + \mathcal{D}'\varphi = -\frac{\partial u^{inc}}{\partial \nu} \quad \text{in } C^{0,\alpha}(\partial D). \tag{3.33}$$

By Theorem 3.20 the operator $\mathcal{D}'$ is compact. Therefore, we can apply the Riesz–Fredholm theory. By Theorem A.2, existence follows from uniqueness. To prove uniqueness we assume that $\varphi \in C^{0,\alpha}(\partial D)$ satisfies the homogeneous equation $-\frac{1}{2}\varphi + \mathcal{D}'\varphi = 0$. Define v to be the single layer potential with density φ just as in (3.31), but for arbitrary $x \notin \partial D$. Then, again from the jump conditions of the normal derivative of the single layer, $\partial v/\partial \nu|_+ = -\frac{1}{2}\varphi + \mathcal{D}'\varphi = 0$. Therefore, v is the solution of the exterior Neumann problem with vanishing boundary data. The uniqueness result of Theorem 3.23 yields that v vanishes in the exterior of D. Furthermore, v is continuous in $\mathbb{R}^3$, thus v is a solution of the Helmholtz equation in D with vanishing boundary data. At this point we wish to conclude that v vanishes also in D, because then we could conclude by the jump of the normal derivatives of the potential at ∂D that φ vanishes. However, this is not always the case. Indeed, there are nontrivial solutions in D if, and only if, k^2 is an eigenvalue of $-\Delta$ in D with respect to Dirichlet boundary conditions (see Theorem 2.34).

This is the reason why it is necessary to modify the ansatz (3.31). There are several ways how to do it, see the discussion in [6, Chapters 3 and 4]. We choose a modification which we have not found in the literature. It avoids the use of double layer potentials. On the other hand, however, it results in a system of two equations which increases the numerical effort considerably. We assume for simplicity that D is connected although this is not necessary as one observes from the following arguments.

We choose an open ball $B = B(z, \rho)$ with boundary Γ such that $\Gamma \subseteq D$ and such that k^2 is not an eigenvalue of $-\Delta$ inside B with respect to Dirichlet boundary conditions. By Theorem 2.34 from the previous chapter we observe that we have to choose the radius ρ of B such that $k\rho$ is not a zero of any of the Bessel functions j_n.

Now we make an ansatz for u^s as a sum of two single layer potentials in the form

$$u^s(x) = (\tilde{S}_{\partial D}\varphi)(x) + (\tilde{S}_\Gamma \psi)(x)$$
$$= \int_{\partial D} \varphi(y)\, \Phi(x,y)\, ds(y) + \int_\Gamma \psi(y)\, \Phi(x,y)\, ds(y), \quad x \notin \overline{D}, \tag{3.34}$$

where $\phi \in C^{0,\alpha}(\partial D)$ and $\psi \in C^{0,\alpha}(\Gamma)$ are two densities to be determined from the system of two boundary integral equations

$$-\frac{1}{2}\varphi + \mathcal{D}'_{\partial D}\varphi + \frac{\partial}{\partial \nu}\tilde{S}_{\Gamma}\psi = -\frac{\partial u^{\text{inc}}}{\partial \nu} \quad \text{on } \partial D, \tag{3.35a}$$

$$\left(\frac{\partial}{\partial \nu} + ik\right)\tilde{S}_{\partial D}\varphi - \frac{1}{2}\psi + \mathcal{D}'_{\Gamma}\psi + ik\,S_{\Gamma}\psi = 0 \quad \text{on } \Gamma. \tag{3.35b}$$

The operators S_{Γ} and $\mathcal{D}'_{\Gamma}$ denote the boundary operators S and $\mathcal{D}'$, respectively, on the boundary Γ instead of ∂D. These two equations can be written in matrix form as

$$-\frac{1}{2}\begin{pmatrix}\varphi\\\psi\end{pmatrix} + \begin{pmatrix}\mathcal{D}'_{\partial D} & \partial\tilde{S}_{\Gamma}/\partial\nu\\(\partial/\partial\nu + ik)\tilde{S}_{\partial D} & \mathcal{D}'_{\Gamma} + ik\,S_{\Gamma}\end{pmatrix}\begin{pmatrix}\varphi\\\psi\end{pmatrix} = -\begin{pmatrix}\partial u^{\text{inc}}/\partial\nu\\0\end{pmatrix}$$

in $C^{0,\alpha}(\partial D) \times C^{0,\alpha}(\Gamma)$. The operators $\mathcal{D}'_{\partial D}$, $\mathcal{D}'_{\Gamma} + ik\,S_{\Gamma}$, $\partial\tilde{S}_{\Gamma}/\partial\nu$, and $(\partial/\partial\nu+ik)\tilde{S}_{\partial D}$ are all compact. Therefore, we can apply the Riesz–Fredholm theory to this system. By Theorem A.2 existence is assured if the homogeneous system admits only the trivial solution $\varphi = 0$ and $\psi = 0$. Therefore, let $(\varphi, \psi) \in C^{0,\alpha}(\partial D) \times C^{0,\alpha}(\Gamma)$ be a solution of the homogeneous system and define the v as the sum of the single layers with densities φ and ψ for all x in $\mathbb{R}^3\backslash(\partial D\cup\Gamma)$. From the jump condition for the normal derivative and the first homogeneous integral equation we conclude—just in the above case of only one single layer potential—that $\partial v/\partial v\big|_{+} = -\frac{1}{2}\varphi + \mathcal{D}'_{\partial D}\varphi + \partial\tilde{S}_{\Gamma}\psi/\partial\nu = 0$. Again, v is a solution of the exterior Neumann problem with vanishing boundary data. Therefore, by the uniqueness theorem, v vanishes in the exterior of D. Furthermore, v is continuous in $\mathbb{R}^3$ and satisfies also the Helmholtz equation in $D\backslash\overline{B}$. From the jump conditions on the boundary Γ we conclude that

$$\frac{\partial v}{\partial\nu}\bigg|_{+} + ikv = \left(\frac{\partial}{\partial\nu} + ik\right)\tilde{S}_{\partial D}\varphi - \frac{1}{2}\psi + \mathcal{D}'_{\Gamma}\psi + ik\,S_{\Gamma}\psi = 0 \quad \text{on } \Gamma.$$

Therefore, $v = 0$ on ∂D and $\partial v/\partial\nu\big|_{+} + ikv = 0$ on Γ. Application of Green's first theorem in $D \backslash \overline{B}$ yields

$$\int_{D\backslash\overline{B}} [|\nabla v|^2 - k^2|v|^2]\, dx = \int_{\partial D} \overline{v}\frac{\partial v}{\partial\nu}\,ds - \int_{\Gamma}\overline{v}\frac{\partial v}{\partial\nu}\bigg|_{+} ds = ik\int_{\Gamma}|v|^2\,ds.$$

Taking the imaginary part yields that v vanishes on Γ and therefore also $\partial v/\partial\nu\big|_{+} = 0$ on Γ. Holmgren's uniqueness Theorem 3.5 implies that v vanishes in all of $D \backslash B$. The jump conditions for the normal derivatives on ∂D yield

$$0 = \frac{\partial v}{\partial\nu}\bigg|_{-} - \frac{\partial v}{\partial\nu}\bigg|_{+} = \varphi \quad \text{on } \partial D.$$

Therefore, v is a single layer potential on Γ with density ψ and vanishes on Γ. The wave number k^2 is not a Dirichlet eigenvalue of $-\Delta$ in B by the choice of the radius of B. Therefore, v vanishes also in B. The jump conditions on Γ yield

$$0 = \left.\frac{\partial v}{\partial \nu}\right|_{-} - \left.\frac{\partial v}{\partial \nu}\right|_{+} = \psi \quad \text{on } \Gamma.$$

Therefore, $\varphi = 0$ on ∂D and $\psi = 0$ on Γ and we have shown injectivity for the system of integral equations.

If D consists of several components $D = \bigcup_{m=1}^{M} D_m$, then one has to choose balls B_m in each of the domains D_m and make an ansatz as a sum of single layers on ∂D and ∂B_m for $m = 1, \ldots, M$.

Application of the Riesz–Fredholm theory to (3.35a) and (3.35b) yields the desired existence result.

Theorem 3.24. *There exists a unique solution $u \in C^2(\mathbb{R}^3 \setminus \overline{D}) \cap C^1(\mathbb{R}^3 \setminus D)$ of the scattering problem (3.1) and (3.2).*

This approach by using an ansatz with some density (or pair of densities) which has no physical meaning is sometimes called the *dual* integral equation method or indirect approach in contrast to those which arise from the representation theorem. We do not present this approach here but refer to Remark 3.37 below in the case of Maxwell's equations and [6, Section 3.9].

3.2 A Scattering Problem for the Maxwell System

In the second part of this chapter we focus on our main task, the scattering by electromagnetic waves. Here the idea of the integral equation method as shown for the scalar Helmholtz equation will be extended to the following *scattering problem* for the Maxwell system.

Given a solution $(E^{\text{inc}}, H^{\text{inc}})$ of the Maxwell system

$$\text{curl } E^{\text{inc}} - i\omega\mu_0 H^{\text{inc}} = 0, \quad \text{curl } H^{\text{inc}} + i\omega\varepsilon_0 E^{\text{inc}} = 0$$

in some neighborhood of D, determine the total fields $E, H \in C^1(\mathbb{R}^3 \setminus \overline{D}, \mathbb{C}^3) \cap C(\mathbb{R}^3 \setminus D, \mathbb{C}^3)$ such that

$$\text{curl } E - i\omega\mu_0 H = 0 \quad \text{and} \quad \text{curl } H + i\omega\varepsilon_0 E = 0 \quad \text{in } \mathbb{R}^3 \setminus \overline{D}, \quad (3.36a)$$

E satisfies the boundary condition

$$\nu \times E = 0 \quad \text{on } \partial D, \quad (3.36b)$$

and the radiating parts $E^s = E - E^{\text{inc}}$ and $H^s = H - H^{\text{inc}}$ satisfy the *Silver–Müller radiation conditions*

$$\sqrt{\varepsilon_0}\, E^s(x) - \sqrt{\mu_0}\, H^s(x) \times \frac{x}{|x|} = \mathcal{O}\left(\frac{1}{|x|^2}\right), \tag{3.37a}$$

and

$$\sqrt{\mu_0}\, H^s(x) + \sqrt{\varepsilon_0}\, E^s(x) \times \frac{x}{|x|} = \mathcal{O}\left(\frac{1}{|x|^2}\right), \tag{3.37b}$$

uniformly with respect to $x/|x|$.

Again throughout the section we fix some assumptions, in view of a classical formulation of the scattering problem in vacuum.

Assumption: Let the wave number be given by $k = \omega\sqrt{\varepsilon_0\mu_0} > 0$ with constants $\varepsilon_0, \mu_0 > 0$ and the obstacle $D \subseteq \mathbb{R}^3$ be bounded and C^2-smooth such that the complement $\mathbb{R}^3 \setminus \overline{D}$ is connected.

The case of a (homogeneous) conducting medium; that is, $\sigma > 0$, can be treated as well without any difficulty. In this case we just have $k \in \mathbb{C}$ with $\text{Im } k > 0$.

Clearly, after renaming the unknown fields, the scattering problem is a special case of the following *Exterior Boundary Value Problem:*

Given a tangential field $f \in C^{0,\alpha}(\partial D, \mathbb{C}^3)$, i.e. $\nu(x) \cdot f(x) = 0$ on ∂D, such that $\text{Div } f \in C^{0,\alpha}(\partial D)$ determine radiating solutions $E^s, H^s \in C^1(\mathbb{R}^3 \setminus \overline{D}, \mathbb{C}^3) \cap C(\mathbb{R}^3 \setminus D, \mathbb{C}^3)$; that is, E^s, H^s satisfy the radiating conditions (3.37a) and (3.37b), of the system

$$\text{curl } E^s - i\omega\mu_0 H^s = 0 \quad \text{and} \quad \text{curl } H^s + i\omega\varepsilon_0 E^s = 0 \quad \text{in } \mathbb{R}^3 \setminus \overline{D}, \tag{3.38a}$$

such that E^s satisfies the boundary condition

$$\nu \times E^s = f \quad \text{on } \partial D. \tag{3.38b}$$

We note that the assumption on the surface divergence of f is necessary by Corollary A.20.

Additionally, we note that in the previous chapter we had assumed a different kind of radiation condition. We had made the assumption that the scalar fields $x \mapsto x \cdot E(x)$ and $x \mapsto x \cdot H(x)$—which are solutions of the scalar Helmholtz equation by Lemma 1.5—satisfy Sommerfeld's radiation condition 3.2. Later (in Remark 3.31) we will discuss the equivalence of both radiation conditions.

3.2.1 Representation Theorems

We have seen in the previous section (Theorem 3.3) that every sufficiently smooth function u can be written as a sum of a volume potential with density $\Delta u + k^2 u$, a single layer potential with density $\partial u / \partial \nu$, and a double layer potential with potential u. Additionally we know from Lemma 1.3 that each component of solutions E and H of the homogeneous Maxwell system in a domain D satisfies the Helmholtz equation. Thus we obtain that these components are analytic and can be represented by surface potentials. But more appropriate for Maxwell's equations is a representation in terms of vector potentials which we will discuss next.

Theorem 3.25. *Let $k \in \mathbb{C}$ and $E \in C^1(D, \mathbb{C}^3) \cap C(\overline{D}, \mathbb{C}^3)$ such that $\operatorname{curl} E \in C(\overline{D}, \mathbb{C}^3)$ and $\operatorname{div} E \in C(\overline{D})$. Then we have for $x \in D$:*

$$
E(x) = \operatorname{curl} \int_D \Phi(x, y) \operatorname{curl} E(y) \, dy \; - \; \nabla \int_D \Phi(x, y) \operatorname{div} E(y) \, dy
$$

$$
- \, k^2 \int_D E(y) \, \Phi(x, y) \, dy
$$

$$
- \, \operatorname{curl} \int_{\partial D} \big[\nu(y) \times E(y) \big] \, \Phi(x, y) \, ds(y)
$$

$$
+ \, \nabla \int_{\partial D} \big[\nu(y) \cdot E(y) \big] \, \Phi(x, y) \, ds(y) \, .
$$

Furthermore, the right-hand side of this equation vanishes for $x \notin \overline{D}$ and is equal to $\frac{1}{2} E(x)$ for $x \in \partial D$.

Proof: Fix $z \in D$, choose $r > 0$ such that $B[z, r] \subseteq D$, and set $D_r = D \setminus B[z, r]$. For $x \in B(z, r)$ we set

$$
I_r(x) := \operatorname{curl} \int_{D_r} \Phi(x, y) \operatorname{curl} E(y) \, dy \; - \; \nabla \int_{D_r} \Phi(x, y) \operatorname{div} E(y) \, dy
$$

$$
- \, k^2 \int_{D_r} E(y) \, \Phi(x, y) \, dy - \operatorname{curl} \int_{\partial D_r} \big[\nu(y) \times E(y) \big] \, \Phi(x, y) \, ds(y)
$$

$$
+ \, \nabla \int_{\partial D_r} \big[\nu(y) \cdot E(y) \big] \, \Phi(x, y) \, ds(y)
$$

$$
= \int_{D_r} \nabla_x \Phi(x, y) \times \operatorname{curl} E(y) \, dy - \int_{D_r} \nabla_x \Phi(x, y) \operatorname{div} E(y) \, dy
$$

$$- k^2 \int_{D_r} E(y)\, \Phi(x,y)\, dy - \operatorname{curl} \int_{\partial D_r} \left[\nu(y) \times E(y) \right] \Phi(x,y)\, ds(y)$$

$$+ \nabla \int_{\partial D_r} \left[\nu(y) \cdot E(y) \right] \Phi(x,y)\, ds(y)\,,$$

where we have applied the identity A.6. We will show that $I_r(x)$ vanishes. Indeed, we can interchange differentiation and integration and write

$$I_r(x) = \operatorname{curl} \left[\int_{D_r} \Phi(x,y)\, \operatorname{curl} E(y)\, dy \; - \; \int_{\partial D_r} \left[\nu(y) \times E(y) \right] \Phi(x,y)\, ds(y) \right]$$

$$- \nabla \left[\int_{D_r} \Phi(x,y)\, \operatorname{div} E(y)\, dy \; - \; \int_{\partial D_r} \left[\nu(y) \cdot E(y) \right] \Phi(x,y)\, ds(y) \right]$$

$$- k^2 \int_{D_r} E(y)\, \Phi(x,y)\, dy$$

$$= \operatorname{curl} \left[\int_{D_r} \left\{ \operatorname{curl}_y \left[E\, \Phi(x,\cdot) \right] - \nabla_y \Phi(x,\cdot) \times E \right\} dy \right.$$

$$\left. - \int_{\partial D_r} \nu \times \left[E\, \Phi(x,\cdot) \right] ds \right]$$

$$- \nabla \left[\int_{D_r} \left\{ \operatorname{div}_y \left[E\, \Phi(x,\cdot) \right] - \nabla_y \Phi(x,\cdot) \cdot E \right\} dy \right.$$

$$\left. - \int_{\partial D_r} \nu \cdot \left[E\, \Phi(x,\cdot) \right] ds \right]$$

$$- k^2 \int_{D_r} E\, \Phi(x,\cdot)\, dy\,.$$

Now we use the divergence theorem in the forms:

$$\int_{D_r} \operatorname{div} F\, dx = \int_{\partial D_r} \nu \cdot F\, ds\,,$$

$$\int_{D_r} \operatorname{curl} F\, dx = \int_{D_r} \begin{pmatrix} \operatorname{div}(0, F_3, -F_2) \\ \vdots \end{pmatrix} dx \; = \; \int_{\partial D_r} \nu \times F\, ds\,. \tag{3.39}$$

Therefore,

$$I_r(x) = - \operatorname{curl} \int_{D_r} \nabla_y \Phi(x,\cdot) \times E\, dy \; + \; \nabla \int_{D_r} \nabla_y \Phi(x,\cdot) \cdot E\, dy$$

$$- k^2 \int_{D_r} E\, \Phi(x,\cdot)\, dy$$

$$= \int_{D_r} \left\{ \nabla_x \left[E \cdot \nabla_y \Phi(x,\cdot) \right] - \operatorname{curl}_x \left[\nabla_y \Phi(x,\cdot) \times E \right] \right\} dy$$

$$-k^2 \int_{D_r} E\,\Phi(x,\cdot)\,dy$$

$$= -\int_{D_r} E\left[\Delta_y\Phi(x,\cdot) + k^2\Phi(x,\cdot)\right]dy = 0.$$

Here we used the formula

$$\operatorname{curl}_x\left[\nabla_y\Phi(x,\cdot)\times E\right] = -E\operatorname{div}_x\nabla_y\Phi(x,\cdot) + (E\cdot\nabla_x)\nabla_y\Phi(x,\cdot)$$
$$= E\,\Delta_y\Phi(x,\cdot) + \nabla_x\left[E\cdot\nabla_y\Phi(x,\cdot)\right].$$

Therefore, $I_r(x) = 0$ for all $x \in B(z,r)$, i.e.

$$0 = \int_{D_r} \nabla_x\Phi(x,y)\times\operatorname{curl}E(y)\,dy - \int_{D_r} \nabla_x\Phi(x,y)\operatorname{div}E(y)\,dy$$

$$-k^2\int_{D_r} E(y)\,\Phi(x,y)\,dy$$

$$-\operatorname{curl}\int_{\partial D}\left[\nu(y)\times E(y)\right]\Phi(x,y)\,ds(y)$$

$$+\nabla\int_{\partial D}\left[\nu(y)\cdot E(y)\right]\Phi(x,y)\,ds(y)$$

$$-\int_{|y-z|=r}\nabla_x\Phi(x,y)\times\left[\nu(y)\times E(y)\right]ds(y)$$

$$+\int_{|y-z|=r}\nabla_x\Phi(x,y)\left[\nu(y)\cdot E(y)\right]ds(y).$$

We set $x = z$ and compute the last two surface integrals explicitly. We recall that

$$\nabla_z\Phi(z,y) = \frac{\exp(ik|z-y|)}{4\pi|z-y|}\left(ik - \frac{1}{|z-y|}\right)\frac{z-y}{|z-y|}$$

and thus for $|z-y| = r$:

$$-\int_{|y-z|=r}\nabla_z\Phi(z,y)\times\left[\nu(y)\times E(y)\right]ds(y) + \int_{|y-z|=r}\nabla_z\Phi(z,y)\left[\nu(y)\cdot E(y)\right]ds(y)$$

$$= \frac{\exp(ikr)}{4\pi r}\left(ik - \frac{1}{r}\right)$$

$$\times\underbrace{\int_{|y-z|=r}\left[\frac{z-y}{|z-y|}\left(\frac{z-y}{|z-y|}\cdot E(y)\right) - \frac{z-y}{|z-y|}\times\left(\frac{z-y}{|z-y|}\times E(y)\right)\right]ds(y)}_{= E(y)}$$

$$= \frac{\exp(ikr)}{4\pi r} \left(ik - \frac{1}{r}\right) \int\limits_{|y-z|=r} E(y)\, ds(y)$$

$$= -e^{ikr} E(z) + ik \frac{\exp(ikr)}{4\pi r} \int\limits_{|y-z|=r} E(y)\, ds(y)$$

$$+ \frac{\exp(ikr)}{4\pi r^2} \int\limits_{|y-z|=r} \left[E(z) - E(y)\right] ds(y).$$

This term converges to $-E(z)$ as r tends to zero. This proves the formula for $x \in D$.

The same arguments (replacing D_r by D) lead to $I_r(x) = 0$ and yield that the expression vanishes if $x \notin \overline{D}$. The formula for $x \in \partial D$ follows from the same arguments as in the proof of Theorem 3.3. □

This representation holds for any (smooth) vector field E. The relationship with Maxwell's equations becomes clear with the following identity.

Lemma 3.26. *Let $H \in C^1(D, \mathbb{C}^3) \cap C(\overline{D}, \mathbb{C}^3)$ such that $\operatorname{curl} H \in C(\overline{D}, \mathbb{C}^3)$. Then, for $x \in D$,*

$$\operatorname{curl} \int\limits_D \Phi(x, y)\, H(y)\, dy \;-\; \int\limits_D \Phi(x, y)\, \operatorname{curl} H(y)\, dy$$

$$+ \int\limits_{\partial D} \left[\nu(y) \times H(y)\right] \Phi(x, y)\, ds(y) \;=\; 0.$$

Proof: The volume potential $\int_D \Phi(x, y) H(y)\, dy$ is continuously differentiable by Lemma 3.7, therefore

$$\operatorname{curl} \int\limits_D \Phi(x, y)\, H(y)\, dy \;-\; \int\limits_D \Phi(x, y)\, \operatorname{curl} H(y)\, dy$$

$$= \int\limits_D \left[\nabla_x \Phi(x, y) \times H(y) - \Phi(x, y)\, \operatorname{curl} H(y)\right] dy$$

$$= -\int\limits_D \left[\nabla_y \Phi(x, y) \times H(y) + \Phi(x, y)\, \operatorname{curl} H(y)\right] dy$$

$$= -\int\limits_D \operatorname{curl}_y \left[\Phi(x, y)\, H(y)\right] dy$$

where we used (A.7). Now we choose $r > 0$ such that $B[x,r] \subseteq D$ and apply Green's formula (A.16a) in $D_r = D \setminus B[x,r]$. This yields

$$\int_{D_r} \text{curl}_y \left[\Phi(x,y) \, H(y) \right] dy = \int_{\partial D_r} \left[\nu(y) \times H(y) \right] \Phi(x,y) \, ds(y)$$

The boundary contribution

$$\int_{|y-x|=r} \left[\nu(y) \times H(y) \right] \Phi(x,y) \, ds(y)$$

tends to zero as $r \to 0$ which yields the desired result. $\square$

Subtracting the last identity multiplied by $i\omega\mu_0$ from the representation of E in Theorem 3.25 shows that if E and H solve the Maxwell system the volume integrals annihilate and we obtain a representation only by surface integrals. This leads to the well-known Stratton–Chu formula.

Theorem 3.27 (Stratton–Chu Formula).
Let $k = \omega\sqrt{\varepsilon_0\mu_0} > 0$ and $E, H \in C^1(D,\mathbb{C}^3) \cap C(\overline{D},\mathbb{C}^3)$ satisfy Maxwell's equations

$$\text{curl}\, E - i\omega\mu_0 H = 0 \text{ in } D, \quad \text{curl}\, H + i\omega\varepsilon_0 E = 0 \text{ in } D.$$

Then we have for $x \in D$:

$$E(x) = -\text{curl} \int_{\partial D} \left[\nu(y) \times E(y) \right] \Phi(x,y) \, ds(y)$$
$$+\nabla \int_{\partial D} \left[\nu(y) \cdot E(y) \right] \Phi(x,y) \, ds(y)$$
$$- i\omega\mu_0 \int_{\partial D} \left[\nu(y) \times H(y) \right] \Phi(x,y) \, ds(y)$$
$$= -\text{curl} \int_{\partial D} \left[\nu(y) \times E(y) \right] \Phi(x,y) \, ds(y)$$
$$+ \frac{1}{i\omega\varepsilon_0} \text{curl}^2 \int_{\partial D} \left[\nu(y) \times H(y) \right] \Phi(x,y) \, ds(y),$$
$$H(x) = -\text{curl} \int_{\partial D} \left[\nu(y) \times H(y) \right] \Phi(x,y) \, ds(y)$$
$$- \frac{1}{i\omega\mu_0} \text{curl}^2 \int_{\partial D} \left[\nu(y) \times E(y) \right] \Phi(x,y) \, ds(y).$$

For $x \notin \overline{D}$ both integral expressions on the right-hand side vanish.

Proof: The second term in the representation of E in Theorem 3.25 vanishes because of div $E = 0$. As mentioned above, subtraction of the identity from Lemma 3.26, multiplied by $i\omega\mu_0$, proves the first formula. For the second one we set $D_r = D \setminus B[x, r]$ for $r > 0$ such that $B[x, r] \subseteq D$ and compute

$$\int_{\partial D_r} [\nu(y) \cdot E(y)]\, \Phi(x, y)\, ds(y) = -\frac{1}{i\omega\varepsilon_0} \int_{\partial D_r} [\nu(y) \cdot \operatorname{curl} H(y)]\, \Phi(x, y)\, ds(y)$$

$$= -\frac{1}{i\omega\varepsilon_0} \underbrace{\int_{\partial D_r} \nu(y) \cdot \operatorname{curl}[H(y)\, \Phi(x, y)]\, ds(y)}_{= 0 \text{ by the divergence theorem}}$$

$$+ \frac{1}{i\omega\varepsilon_0} \int_{\partial D_r} \nu(y) \cdot [\nabla_y \Phi(x, y) \times H(y)]\, ds(y)$$

$$= \frac{1}{i\omega\varepsilon_0} \int_{\partial D_r} \nabla_y \Phi(x, y) \cdot [H(y) \times \nu(y)]\, ds(y)\,.$$

Now we let r tend to zero. We observe that the integral $\int_{|y-x|=r} \nabla_y \Phi(x, y) \cdot [H(y) \times \nu(y)]\, ds(y)$ vanishes because $\nabla_y \Phi(x, y)$ and $\nu(y)$ are parallel. Therefore,

$$\int_{\partial D} [\nu(y) \cdot E(y)]\, \Phi(x, y)\, ds(y) = \frac{1}{i\omega\varepsilon_0} \int_{\partial D} \nabla_y \Phi(x, y) \cdot [H(y) \times \nu(y)]\, ds(y)$$

$$= \frac{1}{i\omega\varepsilon_0} \operatorname{div} \int_{\partial D} \Phi(x, y)\, [\nu(y) \times H(y)]\, ds(y)\,.$$

Taking the gradient and using $\operatorname{curl} \operatorname{curl} = \nabla \operatorname{div} - \Delta$ yields

$$\nabla \int_{\partial D} [\nu(y) \cdot E(y)]\, \Phi(x, y)\, ds(y)$$

$$= \frac{1}{i\omega\varepsilon_0} \operatorname{curl} \operatorname{curl} \int_{\partial D} \Phi(x, y)\, [\nu(y) \times H(y)]\, ds(y)$$

$$- \frac{k^2}{i\omega\varepsilon_0} \int_{\partial D} \Phi(x, y)\, [\nu(y) \times H(y)]\, ds(y)\,.$$

This ends the proof for E. The representation for $H(x)$ follows directly by $H = \frac{1}{i\omega\mu_0} \operatorname{curl} E$ and $\operatorname{curl} \operatorname{curl} \operatorname{curl} = -\operatorname{curl} \Delta$. $\quad\square$

Similar to the scalar case we are also interested in a representation formula in the exterior of a domain D. As an example of specific solutions to the homogeneous Maxwell equations which exist in $\mathbb{R}^3$ except of one source point we already introduced dipoles in Sect. 1.7.

Definition 3.28. For $k = \omega\sqrt{\varepsilon_0\mu_0} > 0$ vector fields of the form

$$E_{md}(x) = \operatorname{curl}\left[p\,\Phi(x,y)\right],$$

$$H_{md}(x) = \frac{1}{i\omega\mu_0}\operatorname{curl} E_{md}(x) = \frac{1}{i\omega\mu_0}\operatorname{curl}\operatorname{curl}\left[p\,\Phi(x,y)\right],$$

$$H_{ed}(x) = \operatorname{curl}\left[p\,\Phi(x,y)\right],$$

$$E_{ed}(x) = -\frac{1}{i\omega\varepsilon_0}\operatorname{curl} H_{ed}(x) = -\frac{1}{i\omega\varepsilon_0}\operatorname{curl}\operatorname{curl}\left[p\,\Phi(x,y)\right],$$

for some $y \in \mathbb{R}^3$ and $p \in \mathbb{C}^3$ are called *magnetic* and *electric dipols*, respectively, at y with polarization p.

As seen from the introduction of Chap. 1 the behavior of these fields for $|x| \to \infty$ motivates the radiation condition. More specific we have the following results.

Lemma 3.29. *Let* $k = \omega\sqrt{\varepsilon_0\mu_0} > 0$.

(a) *The electromagnetic fields* E_{md}, H_{md} *and* E_{ed}, H_{ed} *of a magnetic or electric dipol, respectively, satisfy the Silver–Müller radiation condition (3.37a), (3.37b); that is,*

$$\sqrt{\varepsilon_0}\,E(x) - \sqrt{\mu_0}\,H(x) \times \frac{x}{|x|} = \mathcal{O}\left(\frac{1}{|x|^2}\right), \tag{3.40a}$$

and

$$\sqrt{\mu_0}\,H(x) + \sqrt{\varepsilon_0}\,E(x) \times \frac{x}{|x|} = \mathcal{O}\left(\frac{1}{|x|^2}\right), \tag{3.40b}$$

uniformly with respect to $x/|x| \in S^2$ *and* (p,y) *in compact subsets of* $\mathbb{C}^3 \times \mathbb{R}^3$.

(b) *The fields* $x \mapsto x \cdot \operatorname{curl}\left[p\,\Phi(x,y)\right]$ *and* $x \mapsto x \cdot \operatorname{curl}\operatorname{curl}\left[p\,\Phi(x,y)\right]$ *satisfies the scalar Sommerfeld radiation condition (3.2) uniformly with respect to* (p,y) *in compact subsets of* $\mathbb{C}^3 \times \mathbb{R}^3$.

(c) *The fields* $E(x) = \operatorname{curl}\left[x\,\Phi(x,y)\right]$ *and* $H = \frac{1}{i\omega\mu_0}\operatorname{curl} E$ *are radiating solution of the Maxwell system.*

Proof: (a) Direct computation yields

$$E_{md}(x) = \operatorname{curl}\left[p\,\Phi(x,y)\right] = \Phi(x,y)\left(ik - \frac{1}{|x-y|}\right)\left(\frac{x-y}{|x-y|} \times p\right)$$

$$= ik\,\Phi(x,y)\left(\frac{x-y}{|x-y|} \times p\right) + \mathcal{O}\left(\frac{1}{|x|^2}\right),$$

$$H_{md}(x) = \frac{1}{i\omega\mu_0}\operatorname{curl}\operatorname{curl}\left[p\,\Phi(x,y)\right] = \frac{1}{i\omega\mu_0}(-\Delta + \nabla\operatorname{div})\left[p\,\Phi(x,y)\right]$$

$$= \cdots = \frac{k^2}{i\omega\mu_0}\Phi(x,y)\left[p - \frac{x-y}{|x-y|}\frac{(x-y)\cdot p}{|x-y|}\right] + \mathcal{O}\left(\frac{1}{|x|^2}\right).$$

for $|x| \to \infty$. With $\frac{x-y}{|x-y|} = \frac{x}{|x|} + \mathcal{O}(1/|x|^2)$ and $k = \omega\sqrt{\mu_0\varepsilon_0}$ the first assertion follows. Analogously, the second assertion can be proven.

(b) We prove the assertion only for $u(x) = x \cdot \mathrm{curl}\left[p\,\Phi(x,y)\right]$ which we write as

$$u(x) = x \cdot \left[\nabla_x\Phi(x,y) \times p\right] = \left(ik - \frac{1}{|x-y|}\right)\Phi(x,y)\frac{x\cdot[(x-y)\times p]}{|x-y|}$$

$$= \left(ik - \frac{1}{|x-y|}\right)\Phi(x,y)\frac{p\cdot(y\times x)}{|x-y|} = ik\,\Phi(x,y)\frac{p\cdot(y\times x)}{|x-y|} + \mathcal{O}(|x|^2)$$

and thus by differentiating this expression

$$\hat{x}\cdot\nabla u(x) = -k^2\Phi(x,y)\frac{\hat{x}\cdot(x-y)}{|x-y|}\frac{p\cdot(y\times x)}{|x-y|} + \mathcal{O}(|x|^2)$$

$$= -k^2\Phi(x,y)\frac{p\cdot(y\times x)}{|x-y|} + \mathcal{O}(|x|^2)$$

which proves the assertion by noting that the $\mathcal{O}$-terms hold uniformly with respect to (p,y) in compact subsets of $\mathbb{C}^3 \times \mathbb{R}^3$.

(c) By Lemma 2.42 it suffices to show that E and H satisfy the Silver–Müller radiation condition. We compute

$$E(x) = \mathrm{curl}\left[x\,\Phi(x,y)\right] = \nabla_x\Phi(x,y) \times x$$

$$= \Phi(x,y)\left[ik - \frac{1}{|x-y|}\right]\frac{x-y}{|x-y|} \times x$$

$$= \Phi(x,y)\left[ik - \frac{1}{|x-y|}\right]\frac{x\times y}{|x-y|}$$

$$= ik\,\Phi(x,y)\,(\hat{x}\times y) + \mathcal{O}\left(\frac{1}{|x|^2}\right)$$

and

$$\mathrm{curl}\,E(x) = \nabla\left(\Phi(x,y)\left[ik - \frac{1}{|x-y|}\right]\right) \times \frac{x\times y}{|x-y|}$$

$$+ \Phi(x,y)\left[ik - \frac{1}{|x-y|}\right]\mathrm{curl}\,\frac{x\times y}{|x-y|}$$

$$= ik\, \nabla_x \Phi(x,y) \times \frac{x \times y}{|x-y|} - \nabla_x \frac{\Phi(x,y)}{|x-y|} \times \frac{x \times y}{|x-y|} + \mathcal{O}\left(\frac{1}{|x|^2}\right)$$

$$= ik\, \nabla_x \Phi(x,y) \times \frac{x \times y}{|x-y|} + \mathcal{O}\left(\frac{1}{|x|^2}\right)$$

because

$$\operatorname{curl} \frac{x \times y}{|x-y|} = \nabla_x \frac{1}{|x-y|} \times (x \times y) + \frac{1}{|x-y|} \operatorname{curl}_x(x \times y) = \mathcal{O}\left(\frac{1}{|x|}\right).$$

and

$$\nabla_x \frac{\Phi(x,y)}{|x-y|} = \mathcal{O}\left(\frac{1}{|x|^2}\right).$$

Finally, we have

$$ik\, \nabla_x \Phi(x,y) \times \frac{x \times y}{|x-y|} = ik\, \Phi(x,y)\left[ik - \frac{1}{|x-y|}\right] \frac{x-y}{|x-y|} \times \frac{x \times y}{|x-y|}$$

$$= -k^2 \Phi(x,y)\left[\frac{(x-y)\cdot y}{|x-y|^2} x - \frac{(x-y)\cdot x}{|x-y|^2} y\right]$$

and thus

$$\hat{x} \times \operatorname{curl} E(x) = k^2 \Phi(x,y)\, (\hat{x} \times y) + \mathcal{O}\left(\frac{1}{|x|^2}\right).$$

Substituting the form of H and $k = \omega\sqrt{\varepsilon_0 \mu_0}$ ends the proof. $\square$

We note from the lemma that for the electric and magnetic dipoles both, the Silver–Müller radiation condition for the pair (E, H) and the Sommerfeld radiation condition for the scalar functions $e(x) = x\cdot E(x)$ and $h(x) = x\cdot H(x)$ hold. The following representation theorem will show that these two kinds of radiation conditions are indeed equivalent (see Remark 3.31).

Theorem 3.30. *(Stratton–Chu Formula in Exterior Domains)*
Let $k = \omega\sqrt{\varepsilon_0 \mu_0} > 0$ and $E, H \in C^1(\mathbb{R}^3 \setminus \overline{D}, \mathbb{C}^3) \cap C(\mathbb{R}^3 \setminus D, \mathbb{C}^3)$ *solutions of the homogeneous Maxwell's equations*

$$\operatorname{curl} E - i\omega\mu_0 H = 0, \quad \operatorname{curl} H + i\omega\varepsilon_0 E = 0$$

in $\mathbb{R}^3 \setminus \overline{D}$ *which satisfy also one of the Silver–Müller radiation conditions (3.40a) or (3.40b). Then*

$$\operatorname{curl} \int_{\partial D} \left[\nu(y) \times E(y) \right] \Phi(x, y) \, ds(y)$$

$$- \frac{1}{i\omega\varepsilon_0} \operatorname{curl} \operatorname{curl} \int_{\partial D} \left[\nu(y) \times H(y) \right] \Phi(x, y) \, ds(y)$$

$$= \begin{cases} 0, & x \in D, \\ E(x), & x \notin \overline{D}, \end{cases}$$

and

$$\operatorname{curl} \int_{\partial D} \left[\nu(y) \times H(y) \right] \Phi(x, y) \, ds(y)$$

$$+ \frac{1}{i\omega\mu_0} \operatorname{curl} \operatorname{curl} \int_{\partial D} \left[\nu(y) \times E(y) \right] \Phi(x, y) \, ds(y)$$

$$= \begin{cases} 0, & x \in D, \\ H(x), & x \notin \overline{D}. \end{cases}$$

Proof: Let us first assume the radiation condition (3.40a). Fix $x \notin \partial D$. We apply Theorem 3.27 in the region $D_R = \{y \notin \overline{D} : |y| < R\}$ for large values of R (such that $R > |x|$). Then the assertion follows if one can show that

$$I_R := \operatorname{curl} \int_{|y|=R} \left[\nu(y) \times E(y) \right] \Phi(x, y) \, ds(y)$$

$$- \frac{1}{i\omega\varepsilon_0} \operatorname{curl} \operatorname{curl} \int_{|y|=R} \left[\nu(y) \times H(y) \right] \Phi(x, y) \, ds(y)$$

tends to zero as $R \to \infty$. To do this we first prove that $\int_{|y|=R} |E|^2 ds$ is bounded with respect to R. The binomial theorem yields

$$\int_{|y|=R} |\sqrt{\varepsilon_0} \, E - \sqrt{\mu_0} \, H \times \nu|^2 ds = \varepsilon_0 \int_{|y|=R} |E|^2 ds + \mu_0 \int_{|y|=R} |H \times \nu|^2 ds$$

$$-2\sqrt{\varepsilon_0\mu_0} \operatorname{Re} \int_{|y|=R} E \cdot (\overline{H} \times \nu) \, ds.$$

We have by the divergence theorem

$$\int_{|y|=R} E \cdot (\overline{H} \times \nu) \, ds = \int_{\partial D} E \cdot (\overline{H} \times \nu) \, ds + \int_{D_R} \operatorname{div}(E \times \overline{H}) \, dx$$

$$= \int_{\partial D} E \cdot (\overline{H} \times \nu) \, ds + \int_{D_R} \left[\overline{H} \right] \cdot \operatorname{curl} E - E \cdot \operatorname{curl} \overline{H} \right] dx$$

$$= \int_{\partial D} E \cdot (\overline{H} \times \nu) \, ds + \int_{D_R} \left[i\omega\mu_0 |H|^2 - i\omega\varepsilon_0 |E|^2 \right] dx.$$

This term is purely imaginary, thus

$$\int_{|y|=R} |\sqrt{\varepsilon_0}\, E - \sqrt{\mu_0}\, H \times \nu|^2 ds = \varepsilon_0 \int_{|y|=R} |E|^2 ds + \mu_0 \int_{|y|=R} |H \times \nu|^2 ds$$
$$- 2\sqrt{\varepsilon_0 \mu_0}\, \mathrm{Re} \int_{\partial D} E \cdot (\overline{H} \times \nu)\, ds.$$

From this the boundedness of $\int_{|y|=R} |E|^2 ds$ follows because the left-hand side tends to zero by the radiation condition (3.40b).
Now we write I_R in the form

$$I_R = \mathrm{curl} \int_{|y|=R} \left\{ [\nu(y) \times E(y)]\, \Phi(x,y) + \frac{1}{ik} E(y) \times \nabla_y \Phi(x,y) \right\} ds(y)$$
$$- \frac{1}{i\omega\varepsilon_0}\, \mathrm{curl} \int_{|y|=R} \left([\nu(y) \times H(y)] + \sqrt{\frac{\varepsilon_0}{\mu_0}}\, E(y) \right) \times \nabla_y \Phi(x,y)\, ds(y).$$

Let us first consider the second term. The bracket $(\cdots)$ tends to zero as $1/R^2$ by the radiation condition (3.40a). Taking the curl of the integral results in second order differentiations of Φ. Since Φ and all derivatives decay as $1/R$ the total integrand decays as $1/R^3$. Therefore, this second term tends to zero because the surface area is only $4\pi R^2$.
For the first term we observe that

$$[\nu(y) \times E(y)]\, \Phi(x,y) + \frac{1}{ik} E(y) \times \nabla_y \Phi(x,y)$$
$$= E(y) \times \left[-\frac{y}{R}\, \Phi(x,y) + \frac{1}{ik} \frac{y-x}{|y-x|}\, \Phi(x,y) \left(ik - \frac{1}{|y-x|} \right) \right].$$

For fixed x and arbitrary y with $|y| = R$ the bracket $[\cdots]$ tends to zero of order $1/R^2$. The same holds true for all of the partial derivatives with respect to x. Therefore, the first term can be estimated by the inequality of Cauchy–Schwartz

$$\frac{c}{R^2} \int_{|y|=R} |E(y)|\, ds \leq \frac{c}{R^2} \sqrt{\int_{|y|=R} 1^2\, ds} \sqrt{\int_{|y|=R} |E(y)|^2\, ds}$$
$$= \frac{c\sqrt{4\pi}}{R} \sqrt{\int_{|y|=R} |E(y)|^2\, ds}$$

and this tends also to zero.
This proves the representation of E by using the first radiation condition (3.40a). The representation of $H(x)$ follows again by computing $H = \frac{1}{i\omega\mu_0}\, \mathrm{curl}\, E$. If the second radiation condition (3.40b) is assumed, one can argue as before and derive the representation of H first. $\square$

We draw the following conclusions from this result.

Remark 3.31.

(a) If E, H are solutions of Maxwell's equations in $\mathbb{R}^3 \setminus \overline{D}$, then each of the radiation conditions (3.40a) and (3.40b) implies the other one.

(b) The Silver–Müller radiation condition for solutions E, H of the Maxwell system is equivalent to the Sommerfeld radiation condition (3.2) for every component of E and H. This follows from the fact that the fundamental solution Φ and every derivative of Φ satisfies the Sommerfeld radiation condition.

(c) The Silver–Müller radiation condition for solutions E, H of the Maxwell system is equivalent to the Sommerfeld radiation condition (3.2) for the scalar functions $e(x) = x \cdot E(x)$ and $h(x) = x \cdot H(x)$. This follows from the fact that the Silver–Müller radiation condition implies a representation of the form of the previous theorem which implies the scalar Sommerfeld radiation conditions for $e(x)$ and $h(x)$ by Lemma 3.29, and the scalar Sommerfeld radiation conditions for $e(x)$ and $h(x)$ imply the representation (2.56a) which satisfies the Silver–Müller radiation conditions again by Lemma 3.29.

(d) The asymptotic behavior of Φ yields

$$E(x) = \mathcal{O}\left(\frac{1}{|x|}\right) \quad \text{and} \quad H(x) = \mathcal{O}\left(\frac{1}{|x|}\right)$$

for $|x| \to \infty$ uniformly with respect to all directions $x/|x|$.

Sometimes it is convenient to eliminate one of the fields E or H from the Maxwell system and work with only one of them. If we eliminate H, then E solves the second order equation

$$\operatorname{curl}^2 E - k^2 E = 0 \tag{3.41}$$

where again $k = \omega\sqrt{\mu_0\varepsilon_0}$ denotes the wave number. If, on the other hand, E satisfies (3.41), then E and $H = \frac{1}{i\omega\mu_0}E$ solve the Maxwell system. Indeed, the first Maxwell equation is satisfied by the definition of H. Also the second Maxwell equation is satisfied because

$$\operatorname{curl} H = \frac{1}{i\omega\mu_0}\operatorname{curl}^2 E = \frac{1}{i\omega\mu_0}\big[\nabla\underbrace{\operatorname{div} E}_{=0} - \underbrace{\Delta E}_{=-k^2 E}\big] = \frac{k^2}{i\omega\mu_0}E = -i\omega\varepsilon_0 E.$$

The Silver–Müller radiation condition (3.37a) or (3.37b) turns into

$$\operatorname{curl} E(x) \times \hat{x} - ik\,E(x) = \mathcal{O}\big(|x|^{-2}\big), \quad |x| \to \infty, \tag{3.42}$$

uniformly with respect to $\hat{x} = x/|x|$.

3.2.2 Vector Potentials and Boundary Integral Operators

In Sect. 3.2.3 we will prove existence of solutions of the scattering problem by a boundary integral equation method. Analogously to the scalar case we have to introduce *vector potentials*. Motivated by the Stratton–Chu formulas we have to consider the curl and the double-curl of the single layer potential

$$v(x) = \int_{\partial D} a(y)\, \Phi(x,y)\, ds(y), \quad x \in \mathbb{R}^3, \tag{3.43}$$

where $a \in C^{0,\alpha}(\partial D, \mathbb{C}^3)$ is a tangential field; that is, $a(y) \cdot \nu(y) = 0$ for all $y \in \partial D$.

Lemma 3.32. *Let v be defined by (3.43). Then $E = \operatorname{curl} v$ satisfies (3.41) in all of $\mathbb{R}^3 \setminus \partial D$ and also the radiation condition (3.42).*

The **proof** follows immediately from Lemma 3.29. □

In the next theorem we study the behavior of E at the boundary.

Theorem 3.33. *The curl of the potential v from (3.43) with Hölder-continuous tangential field $a \in C^{0,\alpha}(\partial D, \mathbb{C}^3)$ can be continuously extended from D to $\overline{D}$ and from $\mathbb{R}^3 \setminus \overline{D}$ to $\mathbb{R}^3 \setminus D$. The limiting values of the tangential components are*

$$\nu(x) \times \operatorname{curl} v(x)\big|_{\pm} = \pm\frac{1}{2} a(x) + \nu(x) \times \int_{\partial D} \operatorname{curl}_x\big[a(y)\Phi(x,y)\big]\, ds(y), \quad x \in \partial D. \tag{3.44}$$

If, in addition, the surface divergence $\operatorname{Div} a$ (see Sect. A.5) is continuous, then $\operatorname{div} v$ is continuous in all of $\mathbb{R}^3$ with limiting values

$$\operatorname{div} v(x)\big|_{\pm} = \int_{\partial D} \Phi(x,y)\, \operatorname{Div} a(y)\, ds(y), \quad x \in \partial D,$$

and also for the normal component of $\operatorname{curl} v$ it holds

$$\nu \cdot \operatorname{curl} v\big|_{+} = \nu \cdot \operatorname{curl} v\big|_{-} \quad \text{on } \partial D.$$

If, furthermore, $\operatorname{Div} a \in C^{0,\alpha}(\partial D)$, then $\operatorname{curl} \operatorname{curl} v$ can be continuously extended from D to $\overline{D}$ and from $\mathbb{R}^3 \setminus \overline{D}$ to $\mathbb{R}^3 \setminus D$.

The limiting values are

$$\nu \times \operatorname{curl} \operatorname{curl} v\big|_{+} = \nu \times \operatorname{curl} \operatorname{curl} v\big|_{-} \quad \text{on } \partial D, \tag{3.45}$$

and

$$\nu(x) \cdot \mathrm{curl}\,\mathrm{curl}\,v(x)\big|_{\pm}$$

$$= \mp \frac{1}{2}\,\mathrm{Div}\,a(x) + \int_{\partial D}\left[\mathrm{Div}\,a(y)\,\frac{\partial \Phi}{\partial \nu(x)}(x,y) + k^2 \nu(x) \cdot a(y)\,\Phi(x,y)\right]ds(y),$$

for $x \in \partial D$.

Proof: The components of $\mathrm{curl}\,v$ are combinations of partial derivatives of the single layer potential. Therefore, by Theorem 3.16 the field $\mathrm{curl}\,v$ has continuous extensions to ∂D from both sides. It remains to show the representation of the tangential components of these extensions on the boundary. We recall the special neighborhoods H_ρ of ∂D from Sect. A.3 and write $x \in H_{\rho_0}$ in the form $x = z + t\nu(z)$, $z \in \partial D$, $0 < |t| < \rho_0$. Then we have

$$\nu(z) \times \mathrm{curl}\,v(x) = \int_{\partial D} \nu(z) \times \left[\nabla_x \Phi(x,y) \times a(y)\right]ds(y) \qquad (3.46)$$

$$= \int_{\partial D}\left[\nu(z) - \nu(y)\right] \times \left[\nabla_x \Phi(x,y) \times a(y)\right]ds(y)$$

$$+ \int_{\partial D} a(y)\,\frac{\partial \Phi}{\partial \nu(y)}(x,y)\,ds(y).$$

The first term is continuous in all of $\mathbb{R}^3$ by Lemma 3.13, the second in $\mathbb{R}^3 \setminus \partial D$ because it is a double layer potential. The limiting values are

$$\nu(x) \times \mathrm{curl}\,v(x)\big|_{\pm} = \int_{\partial D}\left[\nu(x) - \nu(y)\right] \times \left[\nabla_x \Phi(x,y) \times a(y)\right]ds(y)$$

$$\pm \frac{1}{2}\,a(x) + \int_{\partial D} a(y)\,\frac{\partial \Phi}{\partial \nu(y)}(x,y)\,ds(y)$$

$$= \pm \frac{1}{2}\,a(x) + \int_{\partial D} \nu(x) \times \left[\nabla_x \Phi(x,y) \times a(y)\right]ds(y)$$

which has the desired form.

Similarly, for the normal component we obtain

$$\nu(z) \cdot \mathrm{curl}\,v(x) = \int_{\partial D}\left[\nu(z) - \nu(y)\right] \cdot \left[\nabla_x \Phi(x,y) \times a(y)\right]ds(y)$$

$$+ \int_{\partial D} \nu(y)\left[\nabla_x \Phi(x,y) \times a(y)\right]ds(y)$$

$$= \int_{\partial D}\left[a(y) \times (\nu(z) - \nu(y))\right] \cdot \nabla_x \Phi(x,y)\,ds(y)$$

$$- \int_{\partial D} a(y)\left[\nu(y) \times \nabla_y \Phi(x,y)\right]ds(y)$$

$$= \int_{\partial D} \left[a(y) \times (\nu(z) - \nu(y)) \right] \cdot \nabla_x \Phi(x,y) \, ds(y)$$

$$- \int_{\partial D} \mathrm{Grad}\,_y \Phi(x,y) \cdot \left[a(y) \times \nu(y) \right] ds(y)$$

$$= \int_{\partial D} \left[a(y) \times (\nu(z) - \nu(y)) \right] \cdot \nabla_x \Phi(x,y) \, ds(y)$$

$$- \int_{\partial D} \Phi(x,y) \, \mathrm{Div}\left[a(y) \times \nu(y) \right] ds(y) \,,$$

where the first integral is continuous at ∂D as above and the second is a single layer potential and therefore also continuous. Thus taking the limit we obtain

$$\nu \cdot \mathrm{curl}\, v \big|_+ = \nu \cdot \mathrm{curl}\, v \big|_- \,.$$

For the divergence we write

$$\mathrm{div}\, v(x) = \int_{\partial D} a(y) \cdot \nabla_x \Phi(x,y) \, ds(y) = - \int_{\partial D} a(y) \cdot \nabla_y \Phi(x,y) \, ds(y)$$

$$= - \int_{\partial D} a(y) \cdot \mathrm{Grad}\,_y \Phi(x,y) \, ds(y) = \int_{\partial D} \Phi(x,y) \, \mathrm{Div}\, a(y) \, ds(y) \,.$$

This, again, is a single layer potential and thus continuous.

Finally, because $\mathrm{curl}\,\mathrm{curl} = \nabla\,\mathrm{div} - \Delta$ and $\Delta_x \Phi(x,y) = -k^2 \Phi(x,y)$, we conclude that

$$\mathrm{curl}\,\mathrm{curl}\, v(x) = \nabla\,\mathrm{div} \int_{\partial D} a(y)\, \Phi(x,y)\, ds(y) + k^2 \int_{\partial D} a(y)\, \Phi(x,y)\, ds(y)$$

$$= \nabla \int_{\partial D} \Phi(x,y) \, \mathrm{Div}\, a(y) \, ds(y) + k^2 \int_{\partial D} a(y)\, \Phi(x,y)\, ds(y)$$

from which the assertion for the boundary values of $\mathrm{curl}\,\mathrm{curl}\, v$ follows by Theorem 3.16. □

The continuity properties of the derivatives of v give rise to corresponding boundary integral operators. It is convenient to not only define the spaces $C_t(\partial D)$ and $C_t^{0,\alpha}(\partial D)$ of continuous and Hölder-continuous tangential fields, respectively, but also of Hölder-continuous tangential fields such that the surface divergence is also Hölder-continuous. Therefore, we define:

$$C_t(\partial D) = \left\{ a \in C(\partial D, \mathbb{C}^3) : a(y) \cdot \nu(y) = 0 \text{ on } \partial D \right\},$$

$$C_t^{0,\alpha}(\partial D) = C_t(\partial D) \cap C^{0,\alpha}(\partial D, \mathbb{C}^3),$$

$$C_{Div}^{0,\alpha}(\partial D) = \left\{ a \in C_t^{0,\alpha}(\partial D) : \mathrm{Div}\, a \in C^{0,\alpha}(\partial D) \right\}.$$

We equip $C_t(\partial D)$ and $C_t^{0,\alpha}(\partial D)$ with the ordinary norms of $C(\partial D, \mathbb{C}^3)$ and $C^{0,\alpha}(\partial D, \mathbb{C}^3)$, respectively, and $C_{Div}^{0,\alpha}(\partial D)$ with the norm $\|a\|_{C_{Div}^{0,\alpha}(\partial D)} = \|a\|_{C^{0,\alpha}(\partial D)} + \|\operatorname{Div} a\|_{C^{0,\alpha}(\partial D)}$. Then we can prove:

Theorem 3.34. *(a) The boundary operator* $\mathcal{M} : C_t(\partial D) \to C_t^{0,\alpha}(\partial D)$, *defined by*

$$(\mathcal{M}a)(x) = \nu(x) \times \int_{\partial D} \operatorname{curl}_x [a(y)\,\Phi(x,y)]\, ds(y), \quad x \in \partial D, \quad (3.47)$$

is well defined and bounded.
(b) $\mathcal{M}$ *is well defined and compact from* $C_{Div}^{0,\alpha}(\partial D)$ *into itself.*
(c) The boundary operator $\mathcal{L} : C_{Div}^{0,\alpha}(\partial D) \to C_{Div}^{0,\alpha}(\partial D)$, *defined by*

$$(\mathcal{L}a)(x) = \nu(x) \times \operatorname{curl}\operatorname{curl} \int_{\partial D} a(y)\,\Phi(x,y)\, ds(y), \quad x \in \partial D, \quad (3.48)$$

is well defined and bounded. Here, the right-hand side is the trace of $\operatorname{curl}^2 v$ *with* v *from (3.43) which exists by the previous theorem, see (3.45).*

Proof: (a) We see from (3.46) that the kernel of this integral operator has the form

$$G(x,y) = \nabla_x \Phi(x,y) [\nu(x) - \nu(y)]^\top - \frac{\partial \Phi}{\partial \nu(x)}(x,y)\, I\,.$$

The component $g_{j,\ell}$ of the first term is given by

$$g_{j,\ell}(x,y) = \frac{\partial \Phi}{\partial x_j}(x,y) [\nu_\ell(x) - \nu_\ell(y)]\,,$$

and this satisfies certainly the first assumption of part (a) of Theorem 3.17 for $\alpha = 1$. For the second assumption we write

$$g_{j,\ell}(x_1,y) - g_{j,\ell}(x_2,y) = [\nu_\ell(x_1) - \nu_\ell(x_2)] \frac{\partial \Phi}{\partial x_j}(x_1,y)$$

$$+ [\nu_\ell(x_2) - \nu_\ell(y)] \left[\frac{\partial \Phi}{\partial x_j}(x_1,y) - \frac{\partial \Phi}{\partial x_j}(x_2,y) \right]$$

and thus by the same arguments as in the proof Lemma 3.13

$$|g_{j,\ell}(x_1,y) - g_{j,\ell}(x_2,y)| \leq c\frac{|x_1 - x_2|}{|x_1 - y|^2} + c|x_2 - y|\frac{|x_1 - x_2|}{|x_1 - y|^3} \leq c'\frac{|x_1 - x_2|}{|x_1 - y|^2}\,.$$

This settles the first term of G. For the second term we observe that

$$\frac{\partial \Phi}{\partial \nu(x)}(x,y) = -\frac{\partial \Phi}{\partial \nu(y)}(x,y) + [\nu(x) - \nu(y)] \cdot \nabla_x \Phi(x,y).$$

The first term is just the kernel of the double layer operator treated in the previous theorem. For the second we can apply the first part again because it is just $\sum_{\ell=1}^{3} g_{\ell,\ell}(x,y)$.

(b) We note that the space $C_{\mathrm{Div}}^{0,\alpha}(\partial D)$ is a subspace of $C_t^{0,\alpha}(\partial D)$ with bounded embedding and, furthermore, the space $C_t^{0,\alpha}(\partial D)$ is compactly embedded in $C_t(\partial D)$ by Lemma 3.18. Therefore, $\mathcal{M}$ is compact from $C_{\mathrm{Div}}^{0,\alpha}(\partial D)$ into $C_t^{0,\alpha}(\partial D)$. It remains to show that $\mathrm{Div}\,\mathcal{M}$ is compact from $C_{\mathrm{Div}}^{0,\alpha}(\partial D)$ into $C^{0,\alpha}(\partial D)$.

For $a \in C_{\mathrm{Div}}^{0,\alpha}(\partial D)$ we define the potential v by

$$v(x) = \int_{\partial D} a(y)\,\Phi(x,y)\,ds(y), \quad x \in D.$$

Then $\mathcal{M}a = \nu \times \mathrm{curl}\,v|_- + \frac{1}{2}a$ by Theorem 3.33. Furthermore, by Theorem 3.33 again we conclude that for $x \notin \partial D$

$$\mathrm{curl}\,\mathrm{curl}\,v(x) = (\nabla\,\mathrm{div} - \Delta)\int_{\partial D} a(y)\,\Phi(x,y)\,ds(y)$$

$$= \nabla \int_{\partial D} \mathrm{Div}\,a(y)\,\Phi(x,y)\,ds(y) + k^2 \int_{\partial D} a(y)\,\Phi(x,y)\,ds(y)$$

and thus by Corollary A.20 and the jump condition of the derivative of the single layer (Theorem 3.16)

$$\mathrm{Div}(\mathcal{M}a)(x) = -\nu(x) \cdot \mathrm{curl}\,\mathrm{curl}\,v(x)|_- + \frac{1}{2}\,\mathrm{Div}\,a(x)$$

$$= -\int_{\partial D}\left[\mathrm{Div}\,a(y)\frac{\partial \Phi}{\partial \nu(x)}(x,y) + k^2\nu(x) \cdot a(y)\Phi(x,y)\right] ds(y)$$

$$= -\mathcal{D}'(\mathrm{Div}\,a) - k^2\,\nu \cdot \mathcal{S}a.$$

The assertion follows because $\mathcal{D}' \circ \mathrm{Div}$ and $\nu \cdot \mathcal{S}$ are both compact from $C_{\mathrm{Div}}^{0,\alpha}(\partial D)$ into $C^{0,\alpha}(\partial D)$.

(c) Define

$$w(x) = \mathrm{curl}\,\mathrm{curl} \int_{\partial D} a(y)\,\Phi(x,y)\,ds(y), \quad x \notin \partial D.$$

Writing again $\operatorname{curl}^2 = \nabla \operatorname{div} - \Delta$ yields for $x = z + t\nu(z) \in H_\rho \setminus \partial D$:

$$w(x) = \nabla \operatorname{div} \int_{\partial D} \Phi(x,y)\, a(y)\, ds(y) \; + \; k^2 \int_{\partial D} a(y)\, \Phi(x,y)\, ds(y)$$

$$= \int_{\partial D} \nabla_x \Phi(x,y)\, \operatorname{Div} a(y)\, ds(y) \; + \; k^2 \int_{\partial D} a(y)\, \Phi(x,y)\, ds(y)$$

$$= \int_{\partial D} \nabla_x \Phi(x,y) \big[\operatorname{Div} a(y) - \operatorname{Div} a(z)\big]\, ds(y)$$

$$+ \; \operatorname{Div} a(z) \int_{\partial D} \nabla_x \Phi(x,y)\, ds(y) \; + \; k^2 \int_{\partial D} a(y)\, \Phi(x,y)\, ds(y)$$

We use Lemma 3.15 and arrive at

$$w(x) = \int_{\partial D} \nabla_x \Phi(x,y) \big[\operatorname{Div} a(y) - \operatorname{Div} a(z)\big]\, ds(y)$$

$$+ \; k^2 \int_{\partial D} a(y)\, \Phi(x,y)\, ds(y) + \; \operatorname{Div} a(z) \int_{\partial D} H(y)\, \Phi(x,y)\, ds(y)$$

$$- \; \operatorname{Div} a(z) \int_{\partial D} \nu(y)\, \frac{\partial \Phi}{\partial \nu(y)}(x,y)\, ds(y)\,.$$

Therefore, by Lemma 3.13, Theorems 3.12 and 3.14 the tangential component of w is continuous, thus $\mathcal{L}a$ is given by

$$(\mathcal{L}a)(x) = \nu(x) \times w(x) = \nu(x) \times \int_{\partial D} \nabla_x \Phi(x,y)\big[\operatorname{Div} a(y) - \operatorname{Div} a(x)\big]\, ds(y)$$

$$+ \; \operatorname{Div} a(x)\, \nu(x) \times \int_{\partial D} H(y)\, \Phi(x,y)\, ds(y)$$

$$- \; \operatorname{Div} a(x)\, \nu(x) \times \int_{\partial D} \nu(y)\, \frac{\partial \Phi}{\partial \nu(y)}(x,y)\, ds(y)$$

$$+ \; k^2\, \nu(x) \times \int_{\partial D} a(y)\, \Phi(x,y)\, ds(y)\,.$$

The boundedness of the last three terms as operators from $C^{0,\alpha}_{Div}(\partial D)$ into $C^{0,\alpha}_t(\partial D)$ follow from the boundedness of the single and double layer boundary operators $\mathcal{S}$ and $\mathcal{D}$. For the boundedness of the first term we apply part (b) of Theorem 3.17. The assumptions

$$\big|\nu(x) \times \nabla_x \Phi(x,y)\big| \; \leq \; \frac{c}{|x-y|^2} \quad \text{for all } x,y \in \partial D,\; x \neq y, \quad \text{and}$$

$$\big|\nu(x_1) \times \nabla_x \Phi(x_1,y) - \nu(x_2) \times \nabla_x \Phi(x_2,y)\big| \; \leq \; c\, \frac{|x_1 - x_2|}{|x_1 - y|^3}$$

for all $x_1, x_2, y \in \partial D$ with $|x_1 - y| \geq 3|x_1 - x_2|$ are simple to prove (cf. proof of Lemma 3.13). For the third assumption, namely

$$\left| \int_{\partial D \setminus B(x,r)} \nu(x) \times \nabla_x \Phi(x,y)\, ds(y) \right| \leq \left| \int_{\partial D \setminus B(x,r)} \nabla_x \Phi(x,y)\, ds(y) \right| \leq c$$

(3.49)

for all $x \in \partial D$ and $r > 0$ we refer to Lemma 3.15. This proves boundedness of $\mathcal{L}$ from $C^{0,\alpha}_{Div}(\partial D)$ into $C^{0,\alpha}_t(\partial D)$. We consider now the surface divergence $\mathrm{Div}\,\mathcal{L}a$. By Corollary A.20 and the form of w we conclude that $\mathrm{Div}\,\mathcal{L}a = \mathrm{Div}(\nu \times w) = -\nu \cdot \mathrm{curl}\,w$. For $x = z + t\nu(z) \in H_\rho \setminus \partial D$ we compute

$$\mathrm{curl}\,w(x) = \mathrm{curl}^3 \int_{\partial D} a(y)\,\Phi(x,y)\,ds(y) = k^2 \int_{\partial D} \nabla_x \Phi(x,y) \times a(y)\,ds(y)$$

$$= -k^2 \int_{\partial D} \nabla_y \Phi(x,y) \times a(y)\,ds(y)$$

$$= k^2 \int_{\partial D} \nabla_y \Phi(x,y) \times \big[a(z) - a(y)\big]\,ds(y)$$

$$\quad - k^2 \int_{\partial D} \nabla_y \Phi(x,y)\,ds(y) \times a(z)$$

$$= k^2 \int_{\partial D} \nabla_y \Phi(x,y) \times \big[a(z) - a(y)\big]\,ds(y)$$

$$\quad - k^2 \int_{\partial D} \mathrm{Grad}_{\,y} \Phi(x,y)\,ds(y) \times a(z)$$

$$\quad - k^2 \int_{\partial D} \nu(y)\, \frac{\partial \Phi}{\partial \nu(y)}(x,y)\,ds(y) \times a(z)$$

$$= k^2 \int_{\partial D} \nabla_y \Phi(x,y) \times \big[a(z) - a(y)\big]\,ds(y)$$

$$\quad + k^2 \int_{\partial D} H(y)\,\Phi(x,y)\,ds(y) \times a(z)$$

$$\quad - k^2 \int_{\partial D} \nu(y)\, \frac{\partial \Phi}{\partial \nu(y)}(x,y)\,ds(y) \times a(z)\,.$$

The normal component is bounded by the same arguments as above. □

3.2.3 Uniqueness and Existence

Since we now have collected the integral operators with their mapping properties which will be applied in case of Maxwell's equations, we continue with the investigation of the scattering problem. As a next step we prove that there exists at most one solution of the exterior boundary value problem (3.38a), (3.38b), (3.40a), (3.40b) and therefore also to the scattering problem (3.36a), (3.36b),(3.40a), (3.40b).

Theorem 3.35. *For every tangential field $f \in C_{Div}^{0,\alpha}(\partial D, \mathbb{C}^3)$ there exists at most one radiating solution $E^s, H^s \in C^1(\mathbb{R}^3 \setminus \overline{D}, \mathbb{C}^3) \cap C(\mathbb{R}^3 \setminus D, \mathbb{C}^3)$ of the exterior boundary value problem (3.38a), (3.38b), (3.40a), (3.40b).*

Proof: Let E_j^s, H_j^s for $j = 1, 2$ be two solutions of the boundary value problem. Then the difference $E^s = E_1^s - E_2^s$ and $H^s = H_1^s - H_2^s$ solve the exterior boundary value problem for boundary data $f = 0$.
In the proof of the Stratton–Chu formula (Theorem 3.30) we have derived the following formula:

$$
\int_{|y|=R} |\sqrt{\varepsilon_0}\, E^s - \sqrt{\mu_0}\, H^s \times \nu|^2 ds
$$

$$
= \varepsilon_0 \int_{|y|=R} |E^s|^2 ds + \mu_0 \int_{|y|=R} |H^s \times \nu|^2 ds
$$

$$
- 2\sqrt{\varepsilon_0 \mu_0}\, \mathrm{Re} \int_{\partial D} E^s \cdot (\overline{H^s} \times \nu)\, ds .
$$

The integral $\int_{\partial D} E^s \cdot (\overline{H^s} \times \nu)\, ds = \int_{\partial D} \overline{H^s} \cdot (\nu \times E^s)\, ds$ vanishes by the boundary condition. Since the left-hand side tends to zero by the radiation condition we conclude that $\int_{|y|=R} |E^s|^2 ds$ tends to zero as $R \to \infty$. Furthermore, by the Stratton–Chu formula (Theorem 3.30) in the exterior of $\overline{D}$ we can represent $E^s(x)$ in the form

$$
E^s(x) = \mathrm{curl} \int_{\partial D} [\nu(y) \times E^s(y)]\, \Phi(x, y)\, ds(y)
$$

$$
- \frac{1}{i\omega\varepsilon_0}\, \mathrm{curl}\,\mathrm{curl} \int_{\partial D} [\nu(y) \times H^s(y)]\, \Phi(x, y)\, ds(y)
$$

for $x \notin \overline{D}$. From these two facts we observe that every component $u = E_j^s$ satisfies the Helmholtz equation $\Delta u + k^2 u = 0$ in the exterior of $\overline{D}$ and $\lim_{R \to 0} \int_{|x|=R} |u| ds = 0$. Furthermore, we recall that $\Phi(x, y)$ satisfies the Sommerfeld radiation condition (3.2) and therefore also $u = E_j^s$. The triangle inequality in the form $|z| \leq |z - w| + |w|$, thus $|z|^2 \leq 2|z - w|^2 + 2|w|^2$ yields

$$
\int_{|x|=R} \left| \frac{x}{|x|} \cdot \nabla u(x) \right|^2 ds(x)
$$

$$
\leq 2 \int_{|x|=R} \left| \frac{x}{|x|} \cdot \nabla u(x) - iku(x) \right|^2 ds(x) + 2k^2 \int_{|x|=R} |u(x)|^2 ds(x) ,
$$

and this tends to zero as R tends to zero. We are now in the position to apply Rellich's Lemma 3.21 (or Lemma 3.22). This yields $u = 0$ in the exterior of $\overline{D}$ and ends the proof. $\square$

We turn to the question of *existence* of the scattering problem (3.36a), (3.36b), (3.40a), (3.40b) or, more generally, the exterior boundary value problem (3.38a), (3.38b), (3.40a), (3.40b). As in the scalar case we first prove an existence result which is not optimal but rather serves as a preliminary result to motivate a more complicated approach.

Theorem 3.36. *Assume that $w = 0$ is the only solution of the following interior boundary value problem:*

$$\operatorname{curl}\operatorname{curl} w - k^2 w = 0 \text{ in } D, \quad \nu \times \operatorname{curl} w = 0 \text{ on } \partial D. \qquad (3.50)$$

Then, for every $f \in C^{0,\alpha}_{\mathrm{Div}}(\partial D)$, there exists a unique solution $E^s, H^s \in C^1(\mathbb{R}^3 \setminus \overline{D}, \mathbb{C}^3) \cap C(\mathbb{R}^3 \setminus D, \mathbb{C}^3)$ of the exterior boundary value problem (3.38a), (3.38b), (3.40a), (3.40b). In particular, under this assumption, the scattering problem (3.36a), (3.36b), (3.40a), (3.40b) has a unique solution for every incident field. The solution has the form

$$E^s(x) = \operatorname{curl} \int_{\partial D} a(y)\, \Phi(x,y)\, ds(y), \quad H^s(x) = \frac{1}{i\omega\mu_0} \operatorname{curl} E^s(x), \quad x \notin D,$$
$$(3.51)$$

for $a \in C^{0,\alpha}_{\mathrm{Div}}(\partial D)$ which is the unique solution of the boundary integral equation

$$\frac{1}{2} a(x) + \nu(x) \times \operatorname{curl} \int_{\partial D} a(y)\, \Phi(x,y)\, ds(y) = f(x), \quad x \in \partial D. \qquad (3.52)$$

Proof: First we note that, by the jump condition of Theorem 3.33, E^s and H^s from (3.51) solve the boundary value problem if a solves (3.52); that is with the operator $\mathcal{M}$,

$$\frac{1}{2} a + \mathcal{M}a = f.$$

Since $\mathcal{M}$ is compact we can apply the Riesz–Fredholm theory (Theorem A.2) to this equation; that is, existence is assured if uniqueness holds. Therefore, let $a \in C^{0,\alpha}_{\mathrm{Div}}(\partial D)$ satisfy $\frac{1}{2} a + \mathcal{M}a = 0$ on ∂D. Define E^s and H^s as in (3.51) in all of $\mathbb{R}^3 \setminus \partial D$. Then E^s, H^s satisfy the Maxwell system and $\nu \times E^s(x)\big|_+ = \frac{1}{2} a + \mathcal{M}a = 0$ on ∂D. The uniqueness result of Theorem 3.35 yields $E^s = 0$ in $\mathbb{R}^3 \setminus D$. Application of Theorem 3.33 again yields that $\nu \times \operatorname{curl} E^s$ is continuous on ∂D, thus $\nu \times \operatorname{curl} E^s\big|_- = 0$ on ∂D. From our assumption we conclude that E^s vanishes also inside of D. Now we apply Theorem 3.33 a third time and have that $a = \nu \times E^s\big|_- - \nu \times E^s\big|_+ = 0$. Thus, the homogeneous integral equation admits only $a = 0$ as a solution and, therefore, there exists a unique solution of the inhomogeneous equation for every right-hand side $f \in C_{\mathrm{Div}}^{0,\alpha}(\partial D)$.   □

Remark 3.37. This integral equation approach for proving existence is some-times called the *dual integral equation method* in contrast to the—perhaps more natural—equation which arises from the Stratton–Chu formula of The-orem 3.30. Indeed, if E and H solve (3.38a), (3.38b) and (3.40a) or (3.40b) then, by the representation of H of Theorem 3.30,

$$
\begin{aligned}
H(x) = \mathrm{curl} &\int_{\partial D} \left[\nu(y) \times H(y) \right] \Phi(x,y)\, ds(y) \\
&+ \frac{1}{i\omega\mu_0}\, \mathrm{curl}\,\mathrm{curl} \int_{\partial D} f(y)\, \Phi(x,y)\, ds(y)
\end{aligned}
\tag{3.53}
$$

for $x \notin \overline{D}$. Taking the trace and using the jump conditions of Theorem 3.33 yields the boundary integral equation

$$
\frac{1}{2} b \;-\; \mathcal{M}b \;=\; \frac{1}{i\omega\mu_0}\, \mathcal{L}f \quad \text{on } \partial D
\tag{3.54}
$$

for $b = \nu \times H$. The operator $-\mathcal{M}$ is the adjoint of the operator $\mathcal{M}$ with respect to the dual form $\langle a, b \rangle = \int_{\partial D} a \cdot (\nu \times b)\, ds$ which can be seen by interchanging the orders of integration (see Lemma 5.61 for the analogous result for $\mathcal{L}$). Therefore, the operator $\frac{1}{2}I + \mathcal{M}$ of (3.52) is the adjoint of the operator $\frac{1}{2}I - \mathcal{M}$ which appears in the *primal integral equation* (3.54). Under the assumptions of the previous theorem also (3.54) is uniquely solvable. On the other hand, the solution b of (3.54) provides the solution of the exterior boundary value problem. Indeed, let $b \in C_{\mathrm{Div}}^{0,\alpha}(\partial D)$ be a solution of (3.54) and define H in all of $\mathbb{R}^3 \setminus \partial D$ by (3.53) where we replace $\nu \times H$ by b. Then $\nu \times H|_+ = \frac{1}{2}b + \mathcal{M}b + \frac{1}{i\omega\mu_0}\mathcal{L}f = b$ and thus $\nu \times H|_+ = b = \nu \times H|_+ - \nu \times H|_-$, that is, $\nu \times H|_- = 0$. Therefore,

$$
\begin{aligned}
E(x) &= -\frac{1}{i\omega\varepsilon_0}\, \mathrm{curl}\, H(x) \\
&= -\frac{1}{i\omega\varepsilon_0}\, \mathrm{curl}^2 \int_{\partial D} \left[\nu(y) \times H(y) \right] \Phi(x,y)\, ds(y) \\
&\quad + \mathrm{curl} \int_{\partial D} f(y)\, \Phi(x,y)\, ds(y)
\end{aligned}
$$

satisfies $\nu \times \mathrm{curl}\, E|_- = 0$ on ∂D which implies that E vanishes in D by as-sumption. Therefore, $f = \nu \times E|_+ - \nu \times E|_- = \nu \times E|_+$ which shows that H and E solve (3.38a), (3.38b).

We note that by the general existence theorem below (Theorem 3.40) there al-ways exists a solution of (3.54), independent of the choice of the wave number.

The solution is, however, not always unique, and the general Theorem A.2 is only applicable under the assumption that k^2 is not an eigenvalue of (3.50).

We continue with the problem of finding an appropriate ansatz for the solution of (3.38a), (3.38b), (3.40a), (3.40b). The problem of the simple ansatz (3.51) is similar to the one for the scalar case. The eigenvalues of the problem (3.50) are "artificially" introduced by the form (3.51) of the ansatz. In general, there exist eigenvalues of the problem (3.50). To construct an example we apply Lemma 2.42. Thus we know, if u solves the scalar Helmholtz equation $\Delta u + k^2 u = 0$ in the unit ball $B(0,1)$, then $v(x) = \mathrm{curl}[u(x)x]$ solves $\mathrm{curl}\,\mathrm{curl}\,v - k^2 v = 0$ in $B(0,1)$. Furthermore, we have $\nu(x) \times v(x) = x \times [\nabla u(x) \times x] = \mathrm{Grad}\,u(x)$ on S^2.

Example 3.38. We claim that the field

$$w(r, \theta, \phi) = \mathrm{curl}\,\mathrm{curl}\left[\sin\theta\left(k\cos(kr) - \frac{\sin(kr)}{r}\right)\hat{r}\right]$$

satisfies $\mathrm{curl}\,\mathrm{curl}\,w - k^2 w = 0$ in $B(0,1)$ and $\nu(x) \times \mathrm{curl}\,w(x) = k^2\cos\theta$ $[k\cos k - \sin k]\,\hat{\theta}$ on $S^2 = \partial B(0,1)$.

Indeed, we can write $w(x) = \mathrm{curl}\,\mathrm{curl}(u(x)x)$ with

$$u(r, \theta, \phi) = \sin\theta\left(k\frac{\cos(kr)}{r} - \frac{\sin(kr)}{r^2}\right) = \sin\theta\,\frac{d}{dr}\frac{\sin(kr)}{r}.$$

By direct evaluation of the Laplace operator in spherical coordinates we see that u satisfies the Helmholtz equation $\Delta u + k^2 u = 0$ in all of $\mathbb{R}^3$, because $\frac{\sin(kr)}{r}$ actually is analytic in $\mathbb{R}$. By Lemma 2.42 we have that $v(x) = \mathrm{curl}(u(x)x)$ satisfies $\mathrm{curl}\,\mathrm{curl}\,v - k^2 v = 0$, and thus also w satisfies $\mathrm{curl}\,\mathrm{curl}\,w - k^2 w = 0$. We compute $\mathrm{curl}\,w$ as $\mathrm{curl}\,w(x) = \mathrm{curl}[\nabla\,\mathrm{div} - \Delta](u(x)x) = k^2\,\mathrm{curl}(u(x)x) = k^2 v(x)$ and thus, using $v(x) = \nabla u(x) \times x$, $\nu(x) \times \mathrm{curl}\,w(x) = k^2\,\mathrm{Grad}\,u(x) = k^2\cos\theta[k\cos k - \sin k]\,\hat{\theta}$ on S^2.

Therefore, if $k > 0$ is any zero of $\psi(k) = k\cos k - \sin k$, then the corresponding field w satisfies (3.50).

Thus again as in the scalar case the insufficient ansatz (3.51) has to be modified. We propose a modification of the form

$$E^s(x) = \mathrm{curl}\int_{\partial D} a(y)\,\Phi(x, y)\,ds(y) \tag{3.55a}$$

$$+\,\eta\,\mathrm{curl}\,\mathrm{curl}\int_{\partial D}\left[\nu(y) \times (\hat{S}_i^2 a)(y)\right]\Phi(x, y)\,ds(y), \tag{3.55b}$$

$$H^s = \frac{1}{i\omega\mu_0}\,\mathrm{curl}\,E^s \tag{3.55c}$$

for some constant $\eta \in \mathbb{C}$, some $a \in C_{Div}^{0,\alpha}(\partial D)$, and where the bounded operator $\hat{S}_i : C^{0,\alpha}(\partial D, \mathbb{C}^3) \to C^{1,\alpha}(\partial D, \mathbb{C}^3)$ is the single layer surface operator with the value $k = i$, considered componentwise; that is, $\hat{S}_i a = (S_i a_1, S_i a_2, S_i a_3)^\top$ for $a = (a_1, a_2, a_3)^\top \in C^{0,\alpha}(\partial D, \mathbb{C}^3)$. We note that $\hat{S}_i$ is bounded from $C^{0,\alpha}(\partial D, \mathbb{C}^3)$ into $C^{1,\alpha}(\partial D, \mathbb{C}^3)$ by Theorem 3.20. Therefore, the operator $\mathcal{K} : a \mapsto \nu \times \hat{S}_i^2 a$ is compact from $C_{Div}^{0,\alpha}(\partial D, \mathbb{C}^3)$ into itself. We need the following additional result for the single layer boundary operator S_i for wavenumber $k = i$.

Lemma 3.39. *The operator S_i is self-adjoint with respect to $\langle \varphi, \psi \rangle_{\partial D} = \int_{\partial D} \varphi \psi \, ds$; that is*

$$\langle S_i \varphi, \psi \rangle_{\partial D} = \langle \varphi, S_i \psi \rangle_{\partial D} \quad \text{for all } \varphi, \psi \in C^{0,\alpha}(\partial D)$$

and one-to-one.

Proof: Let $\varphi, \psi \in C^{0,\alpha}(\partial D)$ and define $u = \tilde{S}_i \varphi$ and $v = \tilde{S}_i \psi$ as the single layer potentials with densities φ and ψ, respectively. Then u and v are solutions of the Helmholtz equation $\Delta u - u = 0$ and $\Delta v - v = 0$ in $\mathbb{R}^3 \setminus \partial D$. Furthermore, u and v and their derivatives decay exponentially as $|x|$ tends to infinity. u and v are continuous in all of $\mathbb{R}^3$ and $\partial v / \partial \nu|_- - \partial v / \partial \nu|_+ = \psi$. By Green's formula we have

$$\langle S_i \varphi, \psi \rangle_{\partial D} = \int_{\partial D} (S_i \varphi) \, \psi \, ds = \int_{\partial D} u \left(\frac{\partial v}{\partial \nu}\bigg|_- - \frac{\partial v}{\partial \nu}\bigg|_+ \right) ds$$

$$= \int_D (\nabla u \cdot \nabla v + u \, v) \, dx + \int_{B(0,R) \setminus \overline{D}} (\nabla u \cdot \nabla v + u \, v) \, dx$$

$$- \int_{|x|=R} u \frac{\partial v}{\partial r} \, ds \, .$$

The integral over the sphere $\{x : |x| = R\}$ tends to zero and thus

$$\langle S_i \varphi, \psi \rangle_{\partial D} = \int_{\mathbb{R}^3} (\nabla u \cdot \nabla v + u \, v) \, dx \, .$$

This term is symmetric with respect to u and v.

Furthermore, if $S_i \varphi = 0$ we conclude for $\psi = \overline{\varphi}$ that u vanishes in $\mathbb{R}^3$. The jump condition $\partial u / \partial \nu|_- - \partial u / \partial \nu|_+ = \varphi$ implies that φ vanishes which shows injectivity of S_i. $\square$

Analogously to the beginning of the previous section we observe with the help of Theorem 3.33 that the ansatz (3.55b), (3.55c) solves the exterior boundary value problem if $a \in C^{0,\alpha}_{Div}(\partial D)$ solves the equation

$$\frac{1}{2} a + \mathcal{M}a + \eta \mathcal{L} \mathcal{K} a = c \quad \text{on } \partial D \tag{3.56}$$

where again $\mathcal{K}a = \nu \times \hat{S}_i^2 a$. Finally, we can prove the general existence theorem.

Theorem 3.40. *For every $f \in C^{0,\alpha}_{Div}(\partial D)$, there exists a unique radiating solution $E^s, H^s \in C^1(\mathbb{R}^3 \setminus \overline{D}) \cap C(\mathbb{R}^3 \setminus D)$ of the exterior boundary value problem (3.38a), (3.38b), (3.40a), (3.40b). In particular, under this assumption, the scattering problem has a unique solution for every incident field. The solution has the form of (3.55b), (3.55c) for any $\eta \in \mathbb{C} \setminus \mathbb{R}$ and some $a \in C^{0,\alpha}_{Div}(\partial D)$ which is the unique solution of the boundary equation (3.56).*

Proof: We make the ansatz (3.55b), (3.55c) and have to discuss the boundary equation (3.56). The compactness of $\mathcal{M}$ and $\mathcal{K}$ and the boundedness of $\mathcal{L}$ yields compactness of the composition $\mathcal{L} \mathcal{K}$. Therefore, the Riesz–Fredholm theory is applicable to (3.56), in particular Theorem A.2.

To show uniqueness, let a be a solution of the homogeneous equation. Then, with the ansatz (3.55b), (3.55c), we conclude that $\nu \times E^s|_+ = 0$ and thus $E^s = 0$ in $\mathbb{R}^3 \setminus D$ by the uniqueness result. From the jump conditions of Theorem 3.33 we conclude that (note that $\text{curl}^3 \int_{\partial D} a(y) \, \Phi(x,y) \, ds(y) = -\text{curl}\, \Delta \int_{\partial D} a(y) \, \Phi(x,y) \, ds(y) = k^2 \, \text{curl} \int_{\partial D} a(y) \, \Phi(x,y) \, ds(y)$)

$$\nu \times E^s|_- = \nu \times E^s|_- - \nu \times E^s|_+ = -a \,,$$
$$\nu \times \text{curl}\, E^s|_- = \nu \times \text{curl}\, E^s|_- - \nu \times \text{curl}\, E^s|_+ = -\eta \, k^2 \mathcal{K}a \,.$$

and thus

$$\int_{\partial D} (\nu \times \text{curl}\, E^s|_-) \cdot \overline{E^s} \, ds = \eta k^2 \int_{\partial D} (\mathcal{K}a) \cdot (\overline{a} \times \nu) \, ds$$

$$= \eta k^2 \int_{\partial D} (\nu \times \hat{S}_i^2 a) \cdot (\overline{a} \times \nu) \, ds$$

$$= -\eta k^2 \int_{\partial D} \hat{S}_i^2 a \cdot \overline{a} \, ds$$

$$= -\eta k^2 \int_{\partial D} |\hat{S}_i a|^2 \, ds \,.$$

The left-hand side is real valued by Green's theorem applied in D. Also the integral on the right-hand side is real. Because η is not real we conclude that both integrals vanish and thus also a by the injectivity of $\hat{S}_i$. $\square$

3.3 Exercises

Exercise 3.1. Prove the one-dimensional form of the representation theorem (Theorem 3.3) with fundamental solution $\Phi(x, y) = \frac{i}{2k} \exp(ik|x - y|)$ for $x, y \in \mathbb{R}$, $x \neq y$.

Exercise 3.2. Use (2.40) to prove that, for $|x| > |y|$,

$$\Phi(x, y) = ik \sum_{n=0}^{\infty} \sum_{m=-n}^{n} j_n(k|y|) \, h_n^{(1)}(k|x|) \, Y_n^m(\hat{x}) \, Y_n^{-m}(\hat{y})$$

$$= \frac{ik}{4\pi} \sum_{n=0}^{\infty} (2n + 1) \, j_n(k|y|) \, h_n^{(1)}(k|x|) \, P_n(\hat{x} \cdot \hat{y})$$

where again $\hat{x} = x/|x|$, $\hat{y} = y/|y|$. Show that the series converge uniformly on $\{(x, y) \in \mathbb{R}^3 \times \mathbb{R}^3 : |y| \leq R_1 < R_2 \leq |x| \leq R_3\}$ for all $R_1 < R_2 < R_3$.

Exercise 3.3. Show for the example of the unit ball $D = B(0, 1)$ that in general the interior Neumann boundary value problem

$$\Delta u + k^2 u = 0 \text{ in } D, \qquad \frac{\partial u}{\partial \nu} = f \text{ on } \partial D, \tag{3.57}$$

is not uniquely solvable.

Exercise 3.4. Show that the interior Neumann boundary value problem (3.57) can be solved by a single layer ansatz on ∂D provided uniqueness holds. Here $D \subseteq \mathbb{R}^3$ is any bounded domain with sufficiently smooth boundary such that the exterior is connected.

Exercise 3.5. Let $\varphi \in C^{0,\alpha}(D)$ with $\varphi = 0$ on ∂D. Then the extension of φ by zero outside of D is in $C^{0,\alpha}(\mathbb{R}^3)$.

The following Exercises 3.6–3.9 study the potential theoretic case; that is, for $k = 0$, in $\mathbb{R}^3$. Again, $D \subseteq \mathbb{R}^3$ is a bounded domain with sufficiently smooth boundary such that the exterior is connected.

Exercise 3.6. Show that the exterior Neumann boundary value problem

$$\Delta u = 0 \text{ in } \mathbb{R}^3 \setminus \overline{D}, \qquad \frac{\partial}{\partial \nu} u = f \text{ on } \partial D, \tag{3.58a}$$

has at most one solution $u \in C^2(\mathbb{R}^3 \setminus \overline{D}) \cap C^1(\mathbb{R}^3 \setminus D)$ which satisfies the decay conditions

$$u(x) = \mathcal{O}(|x|^{-1}), \quad \nabla u(x) = \mathcal{O}(|x|^{-2}), \quad |x| \to \infty, \qquad (3.58b)$$

uniformly with respect to $x/|x| \in S^2$.

Exercise 3.7. Show that for any $f \in C^{0,\alpha}(\partial D)$ the exterior Neumann boundary value problem (3.58a), (3.58b) has a unique solution as a single layer ansatz on ∂D.

Exercise 3.8. Show by using Green's theorem that the homogeneous interior Neumann boundary value problem; that is, (3.58a) in D for $f = 0$ is only solved by constant functions. Show that the condition $\int_{\partial D} f(x)\, ds = 0$ is a necessary condition for the inhomogeneous interior Neumann boundary value problem to be solvable.

Exercise 3.9. (a) Show that the double layer boundary operator (3.26b) for $k = 0$ maps the subspace $\{\varphi \in C^{0,\alpha}(\partial D) : \int_{\partial D} \varphi(x)ds = 0\}$ into itself. *Hint:* Study the single layer ansatz with density φ in D and make use of the jump conditions.
(b) Prove existence of the interior Neumann boundary value problem for any $f \in \{\varphi \in C^{0,\alpha}(\partial D) : \int_{\partial D} \varphi(x)ds = 0\}$ by a single layer ansatz with a density in this subspace.

Chapter 4

The Variational Approach to the Cavity Problem

In this chapter we want to introduce the reader to a second powerful approach for solving boundary value problems for the Maxwell system (or more general partial differential equations) which is the basis of, e.g., the Finite Element technique. We introduce this idea for the cavity problem as our reference problem which has been formulated in the introduction, see (1.21a)–(1.21c). The problem is to determine vector fields E and H with

$$\operatorname{curl} E - i\omega\mu H = 0 \quad \text{in } D, \tag{4.1a}$$

$$\operatorname{curl} H + (i\omega\varepsilon - \sigma)E = J_e \quad \text{in } D, \tag{4.1b}$$

$$\nu \times E = 0 \quad \text{on } \partial D. \tag{4.1c}$$

Here, J_e is a given vector field on D which describes the source which we assume to be in $L^2(D, \mathbb{C}^3)$. It is natural to search for L^2-solutions E and H. But then it follows from (4.1a) and (4.1b) that also $\operatorname{curl} E, \operatorname{curl} H \in L^2(D, \mathbb{C}^3)$.

Assuming scalar valued electric parameter μ, ε, σ and a sufficiently smooth domain D, we first multiply (4.1b) by some smooth vector field ψ with vanishing tangential components on ∂D, then integrate over D, use partial integration and (4.1a). This yields

$$
\begin{aligned}
\int_D J_e \cdot \psi \, dx &= \int_D \left[\operatorname{curl} H \cdot \psi + (i\omega\varepsilon - \sigma)E \cdot \psi \right] dx \\
&= \int_D \left[H \cdot \operatorname{curl} \psi + (i\omega\varepsilon - \sigma)E \cdot \psi \right] dx \\
&= \int_D \left(\frac{1}{i\omega\mu} \operatorname{curl} E \cdot \operatorname{curl} \psi + (i\omega\varepsilon - \sigma)E \cdot \psi \right) dx.
\end{aligned}
$$

© Springer International Publishing Switzerland 2015
A. Kirsch, F. Hettlich, *The Mathematical Theory of Time-Harmonic Maxwell's Equations*, Applied Mathematical Sciences 190,
DOI 10.1007/978-3-319-11086-8_4

The boundary contribution vanishes because of the boundary condition for ψ. Multiplication with $i\omega$ yields

$$\int_D \left[\frac{1}{\mu} \operatorname{curl} E \cdot \operatorname{curl} \psi - \omega^2 \left(\varepsilon + i \frac{\sigma}{\omega} \right) E \cdot \psi \right] dx = i\omega \int_D J_e \cdot \psi \, dx . \quad (4.2)$$

Now the idea of the approach is to consider this equation in a Hilbert space setting, where the right-hand side is treated as a linear bounded functional. Considering the left-hand side as a bilinear form in such a Hilbert space we can hope for an application of the Riesz representation theorem or, more generally, the Lax–Milgram theorem (see Theorem A.6) to obtain existence results.

Thus the first crucial step is to find appropriate Hilbert spaces which, in particular, incorporate the boundary condition (4.1c) and allows a rigorous treatment of the idea. These requirements lead to the introduction of Sobolev spaces. We begin with Sobolev spaces of scalar functions in Sect. 4.1.1 before we continue with vector valued functions in Sect. 4.1.2, which will be used in Sect. 4.2 for the cavity problem.

4.1 Sobolev Spaces

4.1.1 Basic Properties of Sobolev Spaces of Scalar Functions

We are going to present the definition of the Sobolev space $H_0^1(D)$ and its basic properties which are needed to treat the boundary value problem $\Delta u + k^2 u = f$ in D and $u = 0$ on ∂D. The definitions and proofs are all elementary—mainly because we do not need any smoothness assumptions on the set $D \subseteq \mathbb{R}^3$ which is in this section always an open set. Some of the analogous properties of the space $H^1(D)$ where no boundary condition is incorporated are needed for the Helmholtz decomposition of Sect. 4.1.3 and will be postponed to that subsection.

We assume that the reader is familiar with the following basic function spaces. We have used some of them before already.

$$C^k(\overline{D}) = \left\{ u : \overline{D} \to \mathbb{C} : \begin{array}{l} u \text{ is } k \text{ times continuously differentiable in } D \\ \text{and all derivatives can be continuously} \\ \text{extended to } \overline{D} \end{array} \right\},$$

$$C^k(\overline{D}, \mathbb{C}^3) = \left\{ u : \overline{D} \to \mathbb{C}^3 : u_j \in C^k(\overline{D}) \text{ for } j = 1, 2, 3 \right\},$$

$$C_0^k(D) = \left\{ u \in C^k(\overline{D}) : \text{supp}(u) \text{ is compact and } \text{supp}(u) \subseteq D \right\},$$

$$C_0^k(D, \mathbb{C}^3) = \left\{ u \in C^k(\overline{D}, \mathbb{C}^3) : u_j \in C_0^k(D) \text{ for } j = 1, 2, 3 \right\},$$

$$L^p(D) = \left\{ u : D \to \mathbb{C} : u \text{ is Lebesgue-measurable and } \int_D |u|^p dx < \infty \right\},$$

$$L^\infty(D) = \left\{ u : D \to \mathbb{C} : \exists\, c > 0 : |u(x)| \le c \text{ a.e.} \right\},$$

$$L^p(D, \mathbb{C}^3) = \left\{ u : D \to \mathbb{C}^3 : u_j \in L^p(D) \text{ for } j = 1, 2, 3 \right\},$$

with $p \in [1, \infty)$, and

$$\mathcal{S} = \left\{ u \in C^\infty(\mathbb{R}^3) : \sup_{x \in \mathbb{R}^3} \left[|x|^p |D^q u(x)| \right] < \infty \text{ for all } p \in \mathbb{N}_0, q \in \mathbb{N}_0^3 \right\}.$$

Here, the support of a measurable function u is defined as

$$\text{supp}(u) = \bigcap \left\{ K \subseteq \overline{D} : K \text{ closed and } u(x) = 0 \text{ a.e. on } D \setminus K \right\}.$$

Note that in general $\text{supp}(u) \subseteq \overline{D}$, see also Exercise 4.1. The norms in the spaces $C^k(\overline{D}, \mathbb{C}^3)$ (for bounded D) and $L^p(D, \mathbb{C}^3)$ for $p \in [1, \infty)$ are canonical, the norm in $L^\infty(D)$ is given by

$$\|u\|_\infty := \inf \left\{ c > 0 : |u(x)| \le c \text{ a.e. on } D \right\}.$$

The differential operator D^q for $q = (q_1, q_2, q_3) \in \mathbb{N}^3$ is defined by

$$D^q = \frac{\partial^{q_1 + q_2 + q_3}}{\partial x_1^{q_1} \partial x_2^{q_2} \partial x_3^{q_3}}.$$

Definition 4.1. Let $D \subseteq \mathbb{R}^3$ be an open set. A function $u \in L^2(D)$ possesses a variational gradient in $L^2(D)$ if there exists $F \in L^2(D, \mathbb{C}^3)$ with

$$\int_D u \nabla \psi \, dx = -\int_D F \psi \, dx \quad \text{for all } \psi \in C_0^\infty(D). \tag{4.3}$$

We write $\nabla u = F$. Using the denseness of $C_0^\infty(\mathbb{R}^3)$ in $L^2(\mathbb{R}^3)$ (see Lemma 4.9) it is easy to show that F is unique—if it exists (see Exercise 4.5).

Definition 4.2. We define the Sobolev space $H^1(D)$ by

$$H^1(D) = \left\{ u \in L^2(D) : u \text{ possesses a variational gradient } \nabla u \in L^2(D, \mathbb{C}^3) \right\}$$

and equip $H^1(D)$ with the inner product

$$(u,v)_{H^1(D)} = (u,v)_{L^2(D)} + (\nabla u, \nabla v)_{L^2(D)} = \int_D \left[u\overline{v} + \nabla u \cdot \nabla \overline{v} \right] dx.$$

Theorem 4.3. *The space $H^1(D)$ is a Hilbert space.*

Proof: Only completeness has to be shown. Let (u_n) be a Cauchy sequence in $H^1(D)$. Then (u_n) and (∇u_n) are Cauchy sequences in $L^2(D)$ and $L^2(D, \mathbb{C}^3)$, respectively, and thus convergent: $u_n \to u$ and $\nabla u_n \to F$ for some $u \in L^2(D)$ and $F \in L^2(D, \mathbb{C}^3)$. We show that $F = \nabla u$: For $\psi \in C_0^\infty(D)$ we conclude that

$$\int_D u_n \nabla \psi \, dx = - \int_D \nabla u_n \, \psi \, dx$$

by the definition of the variational gradient. The left-hand side converges to $\int_D u \nabla \psi \, dx$, the right-hand side to $-\int_D F \psi \, dx$, and thus $F = \nabla u$. $\quad\square$

The proof of the following simple lemma uses only the definition of the variational derivative.

Lemma 4.4. *Let $D \subseteq \mathbb{R}^3$ be an open, bounded set.*

(a) *The space $C^1(\overline{D})$ is contained in $H^1(D)$, and the embedding is bounded. For $u \in C^1(\overline{D})$ the classical derivatives $\partial u / \partial x_j$ coincide with the variational derivatives.*

(b) *Let $u \in H^1(D)$ and $\phi \in C_0^\infty(D)$. Let v be the extension of $u\phi$ by zero into $\mathbb{R}^3$. Then $v \in H_0^1(\mathbb{R}^3)$ and the product rule holds; that is,*

$$\nabla v = \begin{cases} \phi \nabla u + u \nabla \phi & \text{in } D, \\ 0 & \text{in } \mathbb{R}^3 \setminus D. \end{cases} \tag{4.4}$$

Proof: (a) For $u \in C^1(\overline{D})$ and $\psi \in C_0^\infty(D)$ we note that the extension v of the product $u\psi$ by zero is in $C^1(\mathbb{R}^3)$. Therefore, choosing a box $Q = (-R, R)^3$ containing $\overline{D}$ we obtain by the product rule and the fundamental theorem of calculus

$$\int_D [\psi \nabla u + u \nabla \psi] \, dx = \int_D \nabla(u\psi) \, dx = \int_Q \nabla v \, dx = 0.$$

Therefore, the classical derivative is also the variational derivative and thus belongs to $L^2(D, \mathbb{C}^3)$.

(b) Let $\psi \in C_0^\infty(\mathbb{R}^3)$. Then $\phi\psi \in C_0^\infty(D)$. Let $g \in L^2(\mathbb{R}^3, \mathbb{C}^3)$ be the right-hand side of (4.4). Then

$$\int_{\mathbb{R}^3} [\psi\, g + v\, \nabla \psi]\, dx = \int_D [\psi\phi\, \nabla u + \psi u\, \nabla \phi + u\phi\, \nabla \psi]\, dx$$

$$= \int_D [(\psi\phi)\, \nabla u + u\, \nabla(\psi\phi)]\, dx = 0$$

by the definition of the variational derivative of u. $\square$

We continue by showing some denseness results. The proofs will rely on the technique of mollifying the given function, the Fourier transform and the convolution of functions. We just collect and formulate the corresponding results and refer to, e.g., [17], for the proofs.

The Fourier transform $\mathcal{F}$ is defined by

$$(\mathcal{F}u)(x) = \hat{u}(x) = \frac{1}{(2\pi)^{3/2}} \int_{\mathbb{R}^3} u(y)\, e^{-i\, x \cdot y} dy, \quad x \in \mathbb{R}^3.$$

$\mathcal{F}$ is well defined as an operator from S into itself and also from $L^1(\mathbb{R}^3)$ into the space $C_b(\mathbb{R}^3)$ of bounded continuous functions on $\mathbb{R}^3$. Furthermore, $\mathcal{F}$ is unitary with respect to the L^2-norm; that is,

$$(u, v)_{L^2(\mathbb{R}^3)} = (\hat{u}, \hat{v})_{L^2(\mathbb{R}^3)} \quad \text{for all } u, v \in S. \tag{4.5}$$

Now we use the following result from the theory of Lebesgue integration.

Lemma 4.5. *The space S is dense in $L^2(\mathbb{R}^3)$.*

Proof: We only sketch the arguments. First we use without proof that the space of step functions with compact support is dense in $L^2(\mathbb{R}^3)$. Every step function is a finite linear combination of functions of the form $u(x) = 1$ on U and $u(x) = 0$ outside of U where U is some open bounded set. We leave the constructive proof that functions of this type can be approximated by smooth functions to the reader (see Exercise 4.2.) $\square$

From the denseness of S we obtain by a general extension theorem from functional analysis (see Theorem A.1) that $\mathcal{F}$ has an extension to a unitary operator from $L^2(\mathbb{R}^3)$ onto itself and (4.5) holds for all $u, v \in L^2(\mathbb{R}^3)$.

The convolution of two functions is defined by

$$(u * v)(x) = \int_{\mathbb{R}^3} u(y)\, v(x - y)\, dy, \quad x \in \mathbb{R}^3.$$

Obviously by a simple transformation we have $u * v = v * u$. The Lemma of Young clarifies the mapping properties of the convolution operator.

Lemma 4.6. *(Young) The convolution $u * v$ is well defined for $u \in L^p(\mathbb{R}^3)$ and $v \in L^1(\mathbb{R}^3)$ for any $p \geq 1$. Furthermore, in this case $u * v \in L^p(\mathbb{R}^3)$ and*

$$\|u * v\|_{L^p(\mathbb{R}^3)} \leq \|u\|_{L^p(\mathbb{R}^3)} \|v\|_{L^1(\mathbb{R}^3)} \quad \text{for all } u \in L^p(\mathbb{R}^3), \ v \in L^1(\mathbb{R}^3).$$

This lemma implies, in particular, that $L^1(\mathbb{R}^3)$ is a commutative algebra with the convolution as multiplication. The Fourier transform transforms the convolution into the pointwise multiplication of functions:

$$\mathcal{F}(u*v)(x) = (2\pi)^{3/2} \, \hat{u}(x) \, \hat{v}(x) \quad \text{for all } u \in L^p(\mathbb{R}^3), \ v \in L^1(\mathbb{R}^3), \ p \in \{1,2\}.$$
$$(4.6)$$

The following result to smoothen functions will often be used.

Theorem 4.7. *Let $\phi \in C^\infty(\mathbb{R})$ with $\phi(t) = 0$ for $|t| \geq 1$ and $\phi(t) > 0$ for $|t| < 1$ and $\int_0^1 \phi(t^2) \, t^2 dt = 1/(4\pi)$ (see Exercise 4.7 for the existence of such a function). Define $\phi_\delta \in C^\infty(\mathbb{R}^3)$ by*

$$\phi_\delta(x) = \frac{1}{\delta^3} \, \phi\left(\frac{1}{\delta^2} \, |x|^2\right), \quad x \in \mathbb{R}^3. \tag{4.7}$$

Then $\text{supp}(\phi_\delta) \subseteq B_3[0, \delta]$ and:

(a) *$u * \phi_\delta \in C^\infty(\mathbb{R}^3) \cap L^2(\mathbb{R}^3)$ for all $u \in L^2(\mathbb{R}^3)$. Let $K \subseteq \mathbb{R}^3$ be compact and $u = 0$ outside of K. Then $\text{supp}(u * \phi_\delta) \subseteq K + B_3[0, \delta] = \{x + y : x \in K, \ |y| \leq \delta\}$.*
(b) *Let $u \in L^2(\mathbb{R}^3)$ and $a \in \mathbb{R}^3$ be a fixed vector. Set $u^\delta(x) = (u * \phi_\delta)(x + \delta a)$ for $x \in \mathbb{R}^3$. Then $\|u^\delta - u\|_{L^2(\mathbb{R}^3)} \to 0$ as $\delta \to 0$.*
(c) *Let $u \in H^1(\mathbb{R}^3)$ and u^δ as in part (b). Then $u^\delta \in C^\infty(\mathbb{R}^3) \cap H^1(\mathbb{R}^3)$ and $\|u^\delta - u\|_{H^1(\mathbb{R}^3)} \to 0$ as $\delta \to 0$.*

Proof: (a) Fix any $R > 0$ and let $|x| < R$. Then

$$(u * \phi_\delta)(x) = \int_{|y| \leq R+\delta} u(y) \, \phi_\delta(x - y) \, dy$$

because $|x - y| \geq |y| - |x| \geq |y| - R \geq \delta$ for $|y| \geq R + \delta$ and thus $\phi_\delta(x - y) = 0$. The smoothness of ϕ_δ yields that this integral is infinitely often differentiable. Furthermore, if u vanishes outside some compact set K, then $(u * \phi_\delta)(x) = \int_K u(y) \, \phi_\delta(x - y) \, dy$ which vanishes for $x \notin K + B_3[0, \delta]$ because $|x - y| \geq \delta$ for these x and $y \in K$. Finally we note that $u * \phi_\delta \in L^2(\mathbb{R}^3)$ by the Lemma 4.6 of Young.

(b) Substituting $y = \delta z$ yields for the Fourier transform (note that $dy = \delta^3 dz$),

$$(2\pi)^{3/2} \hat{\phi}_\delta(x) = \frac{1}{\delta^3} \int_{\mathbb{R}^3} \phi\left(\frac{1}{\delta^2} |y|^2\right) e^{-i\,x\cdot y} \, dy$$

$$= \int_{\mathbb{R}^3} \phi(|z|^2) \, e^{-i\delta\,x\cdot z} \, dz = (2\pi)^{3/2} \, \hat{\phi}_1(\delta x).$$

Furthermore, $(2\pi)^{3/2} \hat{\phi}_1(0) = \int_{\mathbb{R}^3} \phi(|y|^2) \, dy = 4\pi \int_0^1 \phi(r^2) \, r^2 dr = 1$ by the normalization of ϕ. Therefore, the Fourier transform of u^δ is given by

$$(\mathcal{F}u^\delta)(x) = e^{-i\delta a \cdot x}(2\pi)^{3/2}\hat{u}(x)\,\hat{\phi}_\delta(x)$$

where we used (4.6) and the translation property. For $u \in L^2(\mathbb{R}^3)$ we conclude

$$\|u^\delta - u\|^2_{L^2(\mathbb{R}^3)} = \|\mathcal{F}u^\delta - \hat{u}\|^2_{L^2(\mathbb{R}^3)} = \left\| \left[\exp(-i\delta a\cdot)(2\pi)^{3/2}\hat{\phi}_\delta - 1\right]\hat{u}\right\|^2_{L^2(\mathbb{R}^3)}$$

$$= \int_{\mathbb{R}^3} \left| e^{-i\delta a\cdot x}(2\pi)^{3/2}\hat{\phi}_1(\delta x) - 1\right|^2 |\hat{u}(x)|^2 \, dx\,.$$

The integrand tends to zero as $\delta \to 0$ for every x and is bounded by the integrable function $\left[(2\pi)^{3/2}\|\hat{\phi}_1\|_\infty + 1\right]^2 |\hat{u}(x)|^2$. This proves part (b) by the theorem of dominated convergence.

(c) First we note that $\nabla(u * \phi_\delta) = \nabla u * \phi_\delta$ for $u \in H^1(\mathbb{R}^3)$. Indeed,

$$\nabla(u * \phi_\delta)(x) = \int_{\mathbb{R}^3} u(y)\,\nabla_x\phi_\delta(x-y)dy = -\int_{\mathbb{R}^3} u(y)\,\nabla_y\phi_\delta(x-y)dy$$

$$= \int_{\mathbb{R}^3} \nabla u(y)\,\phi_\delta(x-y)\,dy = (\nabla u * \phi_\delta)(x)$$

because, for fixed x, the mapping $y \mapsto \phi_\delta(x-y)$ is in $C_0^\infty(\mathbb{R}^3)$. Therefore, again by the Lemma of Young, $\nabla u^\delta \in L^2(\mathbb{R}^3)$; that is $u^\delta \in H^1(\mathbb{R}^3)$. Application of part (b) yields the assertion. $\square$

With this result we can weaken the definition of the variational derivative and prove the following product rule, compare with Lemma 4.4.

Corollary 4.8. *Let $D \subseteq \mathbb{R}^3$ be an open and bounded set.*

(a) Let $u \in L^2(D)$. Then $u \in H^1(D)$ if, and only if, there exists $F \in L^2(D, \mathbb{C}^3)$ with

$$\int_D u\,\nabla\psi\,dx = -\int_D F\,\psi\,dx \quad \text{for all } \psi \in C^1(D) \text{ with compact support in } D\,.$$

The field F coincides with ∇u.

(b) Let $u \in H^1(D)$ and $v \in C^1(\overline{D})$. Then $uv \in H^1(D)$, and the product rule holds; that is,

$$\nabla(uv) = v\,\nabla u + u\,\nabla v.$$

Proof: (a) Only one direction has to be shown. Let $u \in H^1(D)$ and $\psi \in C^1(D)$ with compact support K in D. Then the extension by zero into the outside of D belongs to $H^1(\mathbb{R}^3)$ by Lemma 4.4. Part (c) of the previous theorem yields convergence of $\psi_\delta = \psi * \phi_\delta$ to ψ in $H^1(D)$ as $\delta \to 0$. Furthermore, $\mathrm{supp}(\psi_\delta) \subseteq D$ for sufficiently small $\delta > 0$; that is, $\psi_\delta \in C_0^\infty(D)$. Therefore, by the definition of the variational derivative,

$$\int_D [u\,\nabla\psi_\delta + \psi_\delta\,\nabla u]\,dx = 0$$

for sufficiently small $\delta > 0$. Letting δ tend to zero yields

$$\int_D [u\,\nabla\psi + \psi\,\nabla u]\,dx = 0$$

which shows the assertion for $F = \nabla u$.

(b) Let $\psi \in C_0^\infty(D)$. Then $v\psi \in C^1(\overline{D})$ and vanishes in some neighborhood of ∂D. Therefore,

$$\int_D [(uv)\,\nabla\psi + (u\,\nabla v + v\,\nabla u)\,\psi]\,dx = \int_D [u\,\nabla(v\psi) + (v\psi)\,\nabla u]\,dx = 0$$

by part (a). $\square$

Lemma 4.4 implies that for bounded open sets the space $C^\infty(\overline{D})$ is contained in $H^1(D)$. By the same arguments, the space $C_0^\infty(D)$ is contained in $H^1(D)$ even for unbounded sets D. As a first denseness result we have:

Lemma 4.9. *For any open set $D \subseteq \mathbb{R}^3$ the space $C_0^\infty(D)$ is dense in $L^2(D)$.*

Proof: First we assume D to be bounded and define the open subsets D_n of D by $\{x \in D : d(x, \partial D) > 1/n\}$ where $d(x, \partial D) = \inf_{y \in \partial D} |x - y|$ denotes the distance of x to the boundary ∂D. Then $\overline{D_n} \subseteq D$, and we can find $\phi_n \in C_0^\infty(D)$ with $0 \leq \phi_n(x) \leq 1$ for all $x \in D$ and $\phi_n(x) = 1$ for all $x \in D_n$. Let now $u \in L^2(D)$. Extend u by zero into all of $\mathbb{R}^3$, then $u \in L^2(\mathbb{R}^3)$. By Theorem 4.7 there exists $\tilde{\psi}_n \in C^\infty(\mathbb{R}^3)$ with $\|\tilde{\psi}_n - u\|_{L^2(D)} \leq \|\tilde{\psi}_n - u\|_{L^2(\mathbb{R}^3)} \leq 1/n$. The functions $\psi_n = \phi_n\tilde{\psi}_n$ are in $C_0^\infty(D)$ and $\|\psi_n - u\|_{L^2(D)} \leq \|\phi_n(\tilde{\psi}_n - u)\|_{L^2(D)} + \|(\phi_n - 1)u\|_{L^2(D)}$. The first term is dominated by $\|\tilde{\psi}_n - u\|_{L^2(D)}$ which converges to zero. For the second term we apply Lebesgue's theorem of dominated convergence. Indeed, writing $\|(\phi_n - 1)u\|_{L^2(D)}^2 = \int_D |\phi_n(x) - 1|^2\,|u(x)|^2\,dx$, we observe that every $x \in D$ is element of D_n for sufficiently large n and thus $\phi_n(x) - 1$ vanishes for these n. Furthermore, $|\phi_n(x) - 1|^2 |u(x)|^2$ is bounded by the integrable function $|u(x)|^2$.

Finally, if D is not bounded, we just approximate $u \in L^2(D)$ by a function with compact support and use the previous result for bounded sets to this approximation, compare with the last part of the proof of Lemma 4.18 below. $\square$

The space $C_0^\infty(\mathbb{R}^3)$ of smooth functions with compact support is even dense in $H^1(\mathbb{R}^3)$, see Exercise 4.3. For bounded domains, however, this is not the case. Therefore, we add an other important definition.

Definition 4.10. The space $H_0^1(D)$ is defined as the closure of $C_0^\infty(D)$ with respect to $\| \cdot \|_{H^1(D)}$.

Note that this definition makes sense because $C_0^\infty(D)$ is a subspace of $H^1(D)$ by the above remark. By definition, $H_0^1(D)$ is a closed subspace of $H^1(D)$. We understand $H_0^1(D)$ as the space of differentiable (in the variational sense) functions u with $u = 0$ on ∂D. This interpretation is justified by the trace theorem, introduced in Chap. 5 (see Theorem 5.7).

Remark: Let D be an open bounded set. For fixed $u \in H^1(D)$ the left and the right-hand sides of the definition (4.3) of the variational derivative; that is,

$$\ell_1(\psi) := \int_D u \nabla \psi \, dx \quad \text{and} \quad \ell_2(\psi) := -\int_D \psi \nabla u \, dx, \quad \psi \in C_0^\infty(D),$$

are linear functionals from $C_0^\infty(D)$ into $\mathbb{C}$ which are bounded with respect to the norm $\| \cdot \|_{H^1(D)}$ by the Cauchy–Schwarz inequality:

$$|\ell_1(\psi)| \leq \|u\|_{L^2(D)} \|\nabla \psi\|_{L^2(D)} \leq \|u\|_{L^2(D)} \|\psi\|_{H^1(D)}$$

and analogously for ℓ_2. Therefore, there exist linear and bounded extensions of these functionals to the closure of $C_0^\infty(D)$ which is $H_0^1(D)$. Thus the formula of partial integration holds; that is,

$$\int_D u \nabla \psi \, dx = -\int_D \psi \nabla u \, dx \quad \text{for all } u \in H^1(D) \text{ and } \psi \in H_0^1(D). \quad (4.8)$$

The next step is to show compactness of the embedding $H_0^1(D) \rightarrow L^2(D)$ provided D is bounded. For the proof we choose a box Q of the form $Q = (-R, R)^3 \subseteq \mathbb{R}^3$ with $\overline{D} \subseteq Q$. We cite the following basic result from the theory of Fourier series (see [17] for more details and proofs).

Theorem 4.11. *Let $Q = (-R, R)^3 \subseteq \mathbb{R}^3$ be a bounded cube. For $u \in L^2(Q)$ define the Fourier coefficients $u_n \in \mathbb{C}$ by*

$$u_n = \frac{1}{(2R)^3} \int_Q u(x)\, e^{-i\frac{\pi}{R} n \cdot x} dx \quad for\ n \in \mathbb{Z}^3 . \tag{4.9}$$

Then

$$u(x) = \sum_{n \in \mathbb{Z}^3} u_n\, e^{i\frac{\pi}{R} n \cdot x}$$

in the L^2-sense; that is,

$$\int_Q \left| u(x) - \sum_{n=-N}^{M} u_n\, e^{i\frac{\pi}{R} n \cdot x} \right|^2 dx \longrightarrow 0, \quad N, M \to \infty .$$

Furthermore,

$$(u, v)_{L^2(Q)} = (2R)^3 \sum_{n \in \mathbb{Z}^3} u_n\, \overline{v_n} \quad and \quad \|u\|_{L^2(Q)}^2 = (2R)^3 \sum_{n \in \mathbb{Z}^3} |u_n|^2 .$$

Therefore, the space $L^2(Q)$ can be characterized by the space of all functions u such that $\sum_{n \in \mathbb{Z}^3} |u_n|^2$ converges. Because, for sufficiently smooth functions,

$$\nabla u(x) = i\frac{\pi}{R} \sum_{n \in \mathbb{Z}^3} u_n\, n\, e^{i\frac{\pi}{R} n \cdot x} ,$$

we observe that $i\frac{\pi}{R} n\, u_n$ are the Fourier coefficients of ∇u. The requirement $\nabla u \in L^2(D, \mathbb{C}^3)$ leads to the following definition.

Definition 4.12. We define the Sobolev space $H_{\mathrm{per}}^1(Q)$ of periodic functions by

$$H_{\mathrm{per}}^1(Q) = \left\{ u \in L^2(Q) : \sum_{n \in \mathbb{Z}^3} (1 + |n|^2)\, |u_n|^2 < \infty \right\}$$

with inner product

$$(u, v)_{H_{\mathrm{per}}^1(Q)} = (2R)^3 \sum_{n \in \mathbb{Z}^3} (1 + |n|^2)\, u_n \overline{v_n} .$$

Here, u_n and v_n are the Fourier coefficients of u and v, respectively, see (4.9).

The next theorem guarantees that the zero-extensions of functions of $H_0^1(D)$ belong to $H_{\mathrm{per}}^1(Q)$.

Theorem 4.13. *Let again $D \subseteq \mathbb{R}^3$ be a bounded open set and $Q = (-R, R)^3$ an open box which contains $\overline{D}$ in its interior. Then the extension operator*
$$\tilde{\eta} : u \mapsto \tilde{u} = \begin{cases} u \text{ on } D, \\ 0 \text{ on } Q \setminus D, \end{cases} \text{ is linear and bounded from } H_0^1(D) \text{ into } H_{\mathrm{per}}^1(Q).$$

Proof: Let first $u \in C_0^\infty(D)$. Then obviously $\tilde{u} \in C_0^\infty(Q)$. We compute its Fourier coefficients $\tilde{u}_n$ as

$$\tilde{u}_n = \frac{1}{(2R)^3} \int_Q \tilde{u}(x) \, e^{-i \frac{\pi}{R} n \cdot x} dx \;=\; \frac{iR}{\pi n_j (2R)^3} \int_Q \tilde{u}(x) \, \frac{\partial}{\partial x_j} e^{-i \frac{\pi}{R} n \cdot x} dx$$

$$= -\frac{iR}{\pi n_j (2R)^3} \int_Q \frac{\partial \tilde{u}}{\partial x_j}(x) \, e^{-i \frac{\pi}{R} n \cdot x} dx \;=\; -\frac{iR}{\pi n_j} v_n$$

where v_n are the Fourier coefficients of $v = \partial \tilde{u}/\partial x_j$. Therefore, by Theorem 4.11,

$$(2R)^3 \sum_{n \in \mathbb{Z}^3} n_j^2 |\tilde{u}_n|^2 = \frac{R^2}{\pi^2} (2R)^3 \sum_{n \in \mathbb{Z}^3} |v_n|^2$$

$$= \frac{R^2}{\pi^2} \left\| \frac{\partial \tilde{u}}{\partial x_j} \right\|_{L^2(Q)}^2 = \frac{R^2}{\pi^2} \left\| \frac{\partial u}{\partial x_j} \right\|_{L^2(D)}^2 .$$

Furthermore,

$$(2R)^3 \sum_{n \in \mathbb{Z}^3} |\tilde{u}_n|^2 = \|\tilde{u}\|_{L^2(Q)}^2 = \|u\|_{L^2(D)}^2$$

and thus adding the equations for $j = 1, 2, 3$ and for $\tilde{u}$,

$$\|\tilde{u}\|_{H_{\mathrm{per}}^1(Q)}^2 \leq \left(1 + \frac{R^2}{\pi^2}\right) \|u\|_{H^1(D)}^2 .$$

This holds for all functions $u \in C_0^\infty(D)$. Because $C_0^\infty(D)$ is dense in $H_0^1(D)$ we conclude that

$$\|\tilde{u}\|_{H_{\mathrm{per}}^1(Q)} \leq \sqrt{1 + \frac{R^2}{\pi^2}} \, \|u\|_{H^1(D)} \quad \text{for all } u \in H_0^1(D) .$$

This proves boundedness of the extension operator $\tilde{\eta}$. $\square$

As a simple application we have the compact embedding.

Theorem 4.14. *Let $D \subseteq \mathbb{R}^3$ be a bounded open set. Then the embedding $H_0^1(D) \to L^2(D)$ is compact.*

Proof: Let again $Q = (-R, R)^3 \subseteq \mathbb{R}^3$ such that $\overline{D} \subseteq Q$. First we show that the embedding $J : H^1_{\text{per}}(Q) \to L^2(Q)$ is compact. Define $J_N : H^1_{\text{per}}(Q) \to L^2(Q)$ by

$$(J_N u)(x) := \sum_{|n| \leq N} u_n \, e^{i\frac{\pi}{R} n \cdot x}, \quad x \in Q, \ N \in \mathbb{N}.$$

Then J_N is obviously bounded and finite dimensional, because the range $\mathcal{R}(J_N) = \text{span}\left\{ e^{i\frac{\pi}{R} n \cdot} : |n| \leq N \right\}$ is finite dimensional. Therefore, by a well-known result from functional analysis (see, e.g., [31, Section X.2]), J_N is compact. Furthermore,

$$\|(J_N - J)u\|^2_{L^2(Q)} = (2R)^3 \sum_{|n| > N} |u_n|^2 \leq \frac{(2R)^3}{1 + N^2} \sum_{|n| > N} (1 + |n|^2) |u_n|^2$$

$$\leq \frac{1}{1 + N^2} \|u\|^2_{H^1_{\text{per}}(Q)}.$$

Therefore, $\|J_N - J\|^2_{H^1_{\text{per}}(Q) \to L^2(Q)} \leq \frac{1}{1+N^2} \to 0$ as N tends to infinity and thus, again by a well-known result from functional analysis, also J is compact.

Now the claim of the theorem follows because the composition

$$R \circ J \circ \tilde{\eta} : \ H^1_0(D) \ \xrightarrow{\tilde{\eta}} \ H^1_{\text{per}}(Q) \ \xrightarrow{J} \ L^2(Q) \ \xrightarrow{R} \ L^2(D)$$

is compact. Here, $\tilde{\eta} : H^1_0(D) \to H^1_{\text{per}}(Q)$ denotes the extension operator from the previous theorem and $R : L^2(Q) \to L^2(D)$ is just the restriction operator. $\square$

For the space $H^1_0(D)$ we obtain a useful result which is often called an inequality of Friedrich's Type.

Theorem 4.15. *For any bounded open set $D \subseteq \mathbb{R}^3$ there exists $c > 0$ with*

$$\|u\|_{L^2(D)} \leq c \|\nabla u\|_{L^2(D)} \quad \text{for all } u \in H^1_0(D).$$

Proof: Let again first $u \in C^\infty_0(D)$, extended by zero into $\mathbb{R}^3$. Then, if again $\overline{D} \subseteq Q = (-R, R)^3$,

$$u(x) = u(x_1, x_2, x_3) = \underbrace{u(-R, x_2, x_3)}_{= 0} + \int_{-R}^{x_1} \frac{\partial u}{\partial x_1}(t, x_2, x_3) \, dt$$

and thus for $x \in Q$ by the inequality of Cauchy–Schwarz

$$|u(x)|^2 \leq (x_1 + R) \int_{-R}^{x_1} \left| \frac{\partial u}{\partial x_1}(t, x_2, x_3) \right|^2 dt \leq 2R \int_{-R}^{R} \left| \frac{\partial u}{\partial x_1}(t, x_2, x_3) \right|^2 dt \quad \text{and}$$

$$\int_{-R}^{R} |u(x)|^2 \, dx_1 \leq (2R)^2 \int_{-R}^{R} \left| \frac{\partial u}{\partial x_1}(t, x_2, x_3) \right|^2 dt.$$

Integration with respect to x_2 and x_3 yields

$$\|u\|_{L^2(D)}^2 = \|u\|_{L^2(Q)}^2 \leq (2R)^2 \int_{-R}^{R} \int_{-R}^{R} \int_{-R}^{R} \left| \frac{\partial u}{\partial x_1}(t, x_2, x_3) \right|^2 dt \, dx_2 \, dx_3$$

$$\leq (2R)^2 \|\nabla u\|_{L^2(Q)}^2 = (2R)^2 \|\nabla u\|_{L^2(D)}^2 \, .$$

By a density argument this holds for all $u \in H_0^1(D)$. □

4.1.2 Basic Properties of Sobolev Spaces of Vector Valued Functions

We follow the scalar case as closely as possible. We assume here that all functions are complex valued. First we note that in the formulation (4.1a), (4.1b) not all of the partial derivatives of the vector field E and H appear but only the curl of E and H.

Definition 4.16. Let $D \subseteq \mathbb{R}^3$ be an open set.

(a) A vector field $v \in L^2(D, \mathbb{C}^3)$ possesses a variational curl in $L^2(D, \mathbb{C}^3)$ if there exists a vector field $w \in L^2(D, \mathbb{C}^3)$ such that

$$\int_D v \cdot \operatorname{curl} \psi \, dx = \int_D w \cdot \psi \, dx \quad \text{for all vector fields } \psi \in C_0^\infty(D, \mathbb{C}^3).$$

(b) A vector field $v \in L^2(D, \mathbb{C}^3)$ possesses a variational divergence in $L^2(D)$ if there exists a scalar function $p \in L^2(D)$ such that

$$\int_D v \cdot \nabla \varphi \, dx = -\int_D p \varphi \, dx \quad \text{for all scalar functions } \varphi \in C_0^\infty(D).$$

The functions w and p are unique if they exist (compare Exercise 4.5). In view of partial integration [see (A.6) and (A.9)] we write $\operatorname{curl} v$ and $\operatorname{div} v$ for w and p, respectively.

(c) We define the space $H(\mathrm{curl}, D)$ by

$$H(\mathrm{curl}, D) = \{u \in L^2(D, \mathbb{C}^3) : u \text{ has a variational curl in } L^2(D, \mathbb{C}^3)\},$$

and equip it with the natural inner product

$$(u, v)_{H(\mathrm{curl}, D)} = (u, v)_{L^2(D)} + (\mathrm{curl}\, u, \mathrm{curl}\, v)_{L^2(D)}.$$

With this inner product the space $H(\mathrm{curl}, D)$ is a Hilbert space. The proof follows by the same arguments as in the proof of Theorem 4.3.

The following lemma corresponds to Lemma 4.4 and is proven in exactly the same way.

Lemma 4.17. *Let $D \subseteq \mathbb{R}^3$ be an open, bounded set.*

(a) *The space $C^1(\overline{D}, \mathbb{C}^3)$ is contained in $H(\mathrm{curl}, D)$, and the embedding is bounded. For $u \in C^1(\overline{D})$ the classical curl coincides with the variational curl.*
(b) *Let $u \in H(\mathrm{curl}, D)$ and $\phi \in C_0^\infty(D)$. Let v be the extension of $u\phi$ by zero into $\mathbb{R}^3$. Then $v \in H_0(\mathrm{curl}, \mathbb{R}^3)$ and the product rule holds; that is,*

$$\mathrm{curl}\, v = \begin{cases} \phi \,\mathrm{curl}\, u + \nabla\phi \times u & \text{in } D, \\ 0 & \text{in } \mathbb{R}^3 \setminus D. \end{cases}$$

We note that part (a) follows even directly from Lemma 4.4 because $H^1(D, \mathbb{C}^3) \subseteq H(\mathrm{curl}, D)$.

We continue as in the scalar case and prove denseness properties (compare with Exercise 4.3).

Lemma 4.18. *$C_0^\infty(\mathbb{R}^3, \mathbb{C}^3)$ is dense in $H(\mathrm{curl}, \mathbb{R}^3)$.*

Proof: First we show denseness of $C^\infty(\mathbb{R}^3, \mathbb{C}^3)$ in $H(\mathrm{curl}, \mathbb{R}^3)$ by applying Theorem 4.7. It is sufficient to show that

$$\mathrm{curl}(u * \phi_\delta) = (\mathrm{curl}\, u) * \phi_\delta \quad \text{for all } u \in H(\mathrm{curl}, \mathbb{R}^3)$$

with ϕ_δ from (4.7). Because $\mathrm{supp}(\phi_\delta) \subseteq B_3[0, \delta]$ we conclude for $|x| < R$ that

$$(u * \phi_\delta)(x) = \int_{|y| < R + \delta} u(y)\, \phi_\delta(x - y)\, dy$$

and for any fixed vector $a \in \mathbb{R}^3$ interchanging differentiation and integration leads to

$$a \cdot \mathrm{curl}(u * \phi_\delta)(x) = \int_{\mathbb{R}^3} a \cdot \left[\nabla_x \phi_\delta(x-y) \times u(y) \right] dy$$

$$= - \int_{\mathbb{R}^3} a \cdot \left[\nabla_y \phi_\delta(x-y) \times u(y) \right] dy$$

$$= - \int_{\mathbb{R}^3} \left[a \times \nabla_y \phi_\delta(x-y) \right] \cdot u(y)\, dy$$

$$= \int_{\mathbb{R}^3} \mathrm{curl}_y \left[a\, \phi_\delta(x-y) \right] \cdot u(y)\, dy$$

$$= \int_{\mathbb{R}^3} a \cdot \mathrm{curl}\, u(y)\, \phi_\delta(x-y)\, dy \ = \ a \cdot (\mathrm{curl}\, u * \phi_\delta)(x)$$

because, for fixed x, the mapping $y \mapsto a\, \phi_\delta(x-y)$ is in $C_0^\infty(\mathbb{R}^3, \mathbb{C}^3)$. This holds for all vectors a, thus $\mathrm{curl}(u * \phi_\delta) = (\mathrm{curl}\, u) * \phi_\delta$.

Finally, we have to approximate $u^\delta = u * \phi_\delta$ by a field with compact support. Let $\psi \in C_0^\infty(\mathbb{R}^3)$ with $\psi(x) = 1$ for $|x| \leq 1$. We define $\psi_R(x) = \psi(x/R)$ and $v^{R,\delta} = \psi_R\, u^\delta$ and have

$$\|v^{R,\delta} - u^\delta\|_{H(\mathrm{curl}, \mathbb{R}^3)}^2 = \|(\psi_R - 1) u^\delta\|_{L^2(\mathbb{R}^3)}^2$$

$$+ \|\nabla \psi_R \times u^\delta + (\psi_R - 1)\, \mathrm{curl}\, u^\delta\|_{L^2(\mathbb{R}^3)}^2$$

$$\leq \|(\psi_R - 1) u^\delta\|_{L^2(\mathbb{R}^3)}^2 + 2\,\|(\psi_R - 1)\, \mathrm{curl}\, u^\delta\|_{L^2(\mathbb{R}^3)}^2$$

$$+ 2\,\|\nabla \psi_R \times u^\delta\|_{L^2(\mathbb{R}^3)}^2 .$$

It is easily seen by the theorem of dominated convergence that all of the terms converge to zero as R tends to infinity because for every $x \in \mathbb{R}^3$ there exists R_0 with $\psi_R(x) = 1$ and $\nabla \psi_R(x) = 0$ for $R \geq R_0$. $\quad\square$

Analogously to the definition of $H_0^1(D)$ we define

Definition 4.19. For any open set $D \subseteq \mathbb{R}^3$ we define $H_0(\mathrm{curl}, D)$ as the closure of $C_0^\infty(D, \mathbb{C}^3)$ in $H(\mathrm{curl}, D)$.

Vector fields $v \in H_0(\mathrm{curl}, D)$ do *not* necessarily vanish on ∂D—in contrast to functions in $H_0^1(D)$. Only their tangential components vanish. This follows from the trace theorem in Chap. 5, see Theorem 5.21. We finish this short introduction to $H(\mathrm{curl}, D)$ and $H_0(\mathrm{curl}, D)$, respectively, by an important observation in view of the following Helmholtz decomposition and Maxwell's equations.

Lemma 4.20. *The space $\nabla H_0^1(D) = \{\nabla p : p \in H_0^1(D)\}$ is a closed subspace of $H_0(\mathrm{curl}, D)$.*

Proof By the definition of $H_0^1(D)$ there exists a sequence $p_j \in C_0^\infty(D)$ with $p_j \to p$ in $H^1(D)$. It is certainly $\nabla p_j \in C_0^\infty(D, \mathbb{C}^3)$. Because $\nabla p_j \to \nabla p$ in $L^2(D, \mathbb{C}^3)$ and $\operatorname{curl} \nabla p_j = 0$ we note that (∇p_j) is a Cauchy sequence in $H(\operatorname{curl}, D)$ and thus convergent. We conclude that $\nabla p \in H_0(\operatorname{curl}, D)$.

It remains to show closedness of $\nabla H_0^1(D)$ in $L^2(D, \mathbb{C}^3)$—and therefore, in $H(\operatorname{curl}, D)$. Let $p_j \in H_0^1(D)$ with $\nabla p_j \to F$ in $L^2(D, \mathbb{C}^3)$ for some $F \in L^2(D, \mathbb{C}^3)$. Then (∇p_j) is a Cauchy sequence in $L^2(D, \mathbb{C}^3)$. By Friedrich's inequality of Theorem 4.15 we conclude that also (p_j) is a Cauchy sequence in $L^2(D)$. Therefore, (p_j) is a Cauchy sequence in $H^1(D)$, which implies convergence. This shows that $F = \nabla p$ for some $p \in H_0^1(D)$ and ends the proof. $\square$

We recall that $H_0^1(D)$ is compactly embedded in $L^2(D)$ by Theorem 4.14, a result which is crucial for the solvability of the interior boundary value problem for the Helmholtz equation. As one can easily show (see Exercise 4.14), the space $H_0(\operatorname{curl}, D)$ fails to be compactly embedded in $L^2(D, \mathbb{C}^3)$. Therefore, we will decompose the variational problem into two problems which themselves can be treated in the same way as for the Helmholtz equation. The basis of this decomposition is set by the important Helmholtz decomposition which we discuss in the subsequent subsection.

4.1.3 The Helmholtz Decomposition

Now we turn to decompositions of vector fields, which are closely related to the Maxwell system—the Helmholtz decomposition. For a general proof we will apply the Theorem A.6 of Lax and Milgram.

There are several forms of the Helmholtz decomposition. In this section we need decompositions only in $L^2(D, \mathbb{C}^3)$ and in $H_0(\operatorname{curl}, D)$. But for later purposes, however, (see Sect. 4.3) we prove the corresponding decompositions also in $H(\operatorname{curl}, D)$. Also, we consider it for complex and matrix-valued coefficients μ and ε which we treat simultaneously by writing A.

Theorem 4.21. *Let $D \subseteq \mathbb{R}^3$ be open and bounded and $A \in L^\infty(D, \mathbb{C}^{3\times3})$ such that $A(x)$ is symmetric for almost all x. Furthermore, assume the existence of a constant $\hat{c} > 0$ such that $\operatorname{Re}(\overline{z}^\top A(x)z) \geq \hat{c}|z|^2$ for all $z \in \mathbb{C}^3$ and almost all $x \in D$. Then for the subspaces $H(\operatorname{curl} 0, D)$, $V_A \subseteq H(\operatorname{curl}, D)$, and $H_0(\operatorname{curl} 0, D)$, $V_{0,A} \subseteq H_0(\operatorname{curl}, D)$ and $\tilde{V}_{0,A} \subseteq L^2(D, \mathbb{C}^3)$ defined by*

$$H(\mathrm{curl}\,0, D) = \{u \in H(\mathrm{curl}, D) : \mathrm{curl}\,u = 0 \ in \ D\}, \tag{4.10a}$$

$$H_0(\mathrm{curl}\,0, D) = \{u \in H_0(\mathrm{curl}, D) : \mathrm{curl}\,u = 0 \ in \ D\}, \tag{4.10b}$$

$$V_A = \{u \in H(\mathrm{curl}, D) : (Au, \psi)_{L^2(D)} = 0$$
$$for \ all \ \psi \in H(\mathrm{curl}\,0, D)\}, \tag{4.10c}$$

$$V_{0,A} = \{u \in H_0(\mathrm{curl}, D) : (Au, \psi)_{L^2(D)} = 0$$
$$for \ all \ \psi \in H_0(\mathrm{curl}\,0, D)\}, \tag{4.10d}$$

$$\tilde{V}_{0,A} = \{u \in L^2(D, \mathbb{C}^3) : (Au, \psi)_{L^2(D)} = 0$$
$$for \ all \ \psi \in H_0(\mathrm{curl}\,0, D)\}, \tag{4.10e}$$

we have:

(a) $H_0(\mathrm{curl}\,0, D)$ *and* $\tilde{V}_{0,A}$ *are closed in* $L^2(D, \mathbb{C}^3)$, *and* $L^2(D, \mathbb{C}^3)$ *is the direct sum of* $\tilde{V}_{0,A}$ *and* $H_0(\mathrm{curl}\,0, D)$; *that is,*

$$L^2(D, \mathbb{C}^3) = \tilde{V}_{0,A} \oplus H_0(\mathrm{curl}\,0, D).$$

(b) $H(\mathrm{curl}\,0, D)$ *and* V_A *are closed in* $H(\mathrm{curl}, D)$, *and*

$$H(\mathrm{curl}, D) = V_A \oplus H(\mathrm{curl}\,0, D).$$

(c) $H_0(\mathrm{curl}\,0, D)$ *and* $V_{0,A}$ *are closed in* $H_0(\mathrm{curl}, D)$, *and*

$$H_0(\mathrm{curl}, D) = V_{0,A} \oplus H_0(\mathrm{curl}\,0, D).$$

Furthermore, all of the corresponding projection operators are bounded. The projection from $H_0(\mathrm{curl}, D)$ *onto* $V_{0,A}$ *is the restriction of the projection operator from* $L^2(D, \mathbb{C}^3)$ *onto* $\tilde{V}_{0,A}$.

Proof: The closedness of $H(\mathrm{curl}\,0, D)$ in $H(\mathrm{curl}, D)$ and $H_0(\mathrm{curl}\,0, D)$ in $H_0(\mathrm{curl}, D)$ is obvious. The closedness of $H_0(\mathrm{curl}\,0, D)$ in $L^2(D, \mathbb{C}^3)$ is seen as follows. Let $u_n \in H_0(\mathrm{curl}\,0, D)$ converge to some u in $L^2(D, \mathbb{C}^3)$. Then (u_n) is a Cauchy sequence in $L^2(D, \mathbb{C}^3)$. From $\mathrm{curl}\,u_n = 0$ we observe that (u_n) is also a Cauchy sequence in $H_0(\mathrm{curl}, D)$ and thus convergent in $H_0(\mathrm{curl}, D)$. This shows that $u \in H_0(\mathrm{curl}, D)$ and $\mathrm{curl}\,u = 0$; that is, $u \in H_0(\mathrm{curl}\,0, D)$.

The closedness of $\tilde{V}_{0,A}$ and $V_{0,A}$ and V_A in $L^2(D, \mathbb{C}^3)$ and $H_0(\mathrm{curl}, D)$ and $H(\mathrm{curl}, D)$, respectively, is easy to see, and we omit this part.

Furthermore it holds $\tilde{V}_{0,A} \cap H_0(\mathrm{curl}\,0, D) = \{0\}$ because $u \in \tilde{V}_{0,A} \cap H_0(\mathrm{curl}\,0, D)$ implies $(Au, u)_{L^2(D)} = 0$ and thus, taking the real part, $\hat{c}\|u\|_{L^2(D)}^2 \leq \mathrm{Re}\,(Au, u)_{L^2(D)} = 0$ which yields $u = 0$.

Next, we have to prove that $L^2(D, \mathbb{C}^3) \subseteq \tilde{V}_{0,A} + H_0(\mathrm{curl}\,0, D)$ and $H_0(\mathrm{curl}, D) \subseteq V_{0,A} + H_0(\mathrm{curl}\,0, D)$ and $H(\mathrm{curl}, D) \subseteq V_A + H(\mathrm{curl}\,0, D)$. Let $u \in$

$L^2(D, \mathbb{C}^3)$. We define the sesquilinear form $a : H_0(\mathrm{curl}\, 0, D) \times H_0(\mathrm{curl}\, 0, D)$ $\to \mathbb{C}$ by

$$a(\psi, v) = (\psi, Av)_{L^2(D)} = \int_D \psi^\top \overline{Av}\, dx \quad \text{for all } v, \psi \in H_0(\mathrm{curl}\, 0, D).$$

$$(4.11)$$

Then a is certainly bounded but also coercive in the space $H_0(\mathrm{curl}\, 0, D)$ equipped with the inner product of $H(\mathrm{curl}, D)$. Indeed, we write

$$\mathrm{Re}\, a(v, v) = \mathrm{Re}\, (v, Av)_{L^2(D)} \geq \hat{c} \|v\|^2_{L^2(D)} = \hat{c} \|v\|^2_{H(\mathrm{curl}, D)}$$

for $v \in H_0(\mathrm{curl}\, 0, D)$ because $\mathrm{curl}\, v = 0$. Therefore, the sesquilinear form a and—for every fixed $u \in L^2(D, \mathbb{C}^3)$—the bounded linear form $\ell(\psi) = (\psi, Au)_{L^2(D)}$, $\psi \in H_0(\mathrm{curl}\, 0, D)$, satisfy the assumptions of Theorem A.6 of Lax and Milgram. Therefore, for every given $u \in L^2(D, \mathbb{C}^3)$ there exists a unique $u_0 \in H_0(\mathrm{curl}\, 0, D)$ with $a(\psi, u_0) = \ell(\psi)$ for all $\psi \in H_0(\mathrm{curl}\, 0, D)$; that is, $(Au_0, \psi)_{L^2(D)} = (Au, \psi)_{L^2(D)}$ for all $\psi \in H_0(\mathrm{curl}\, 0, D)$. Therefore, $u - u_0 \in \tilde{V}_{0,A}$ and the decomposition $u = u_0 + (u - u_0) \in H_0(\mathrm{curl}\, 0, D) + \tilde{V}_{0,A}$ has been shown.

Finally, boundedness of the projection operator $u \mapsto u - u_0$ follows from general properties of direct sums. This proves the result in $L^2(D, \mathbb{C}^3)$. Analogously, the projection operator $P : H_0(\mathrm{curl}, D) \to V_{0,A}$ is defined in the same way by $Pu = u - u_0$ where $u_0 \in H_0(\mathrm{curl}\, 0, D)$ is as before. The proof of $H(\mathrm{curl}, D) = V_A + H(\mathrm{curl}\, 0, D)$ follows the same arguments. □

Remark 4.22. (a) Obviously, the space $\nabla H_0^1(D) = \{\nabla p : p \in H_0^1(D)\}$ is contained in $H_0(\mathrm{curl}\, 0, D)$. Therefore, $V_{0,A}$ and $\tilde{V}_{0,A}$ are contained in the spaces

$$H_0(\mathrm{curl}, \mathrm{div}_A\, 0, D) := \{v \in H_0(\mathrm{curl}, D) : (Av, \nabla\varphi)_{L^2(D)} = 0$$
$$\text{for all } \varphi \in H_0^1(D)\}, \qquad (4.12a)$$

$$L^2(\mathrm{div}_A\, 0, D) := \{v \in L^2(D, \mathbb{C}^3) : (Av, \nabla\varphi)_{L^2(D)} = 0$$
$$\text{for all } \varphi \in H_0^1(D)\}, \qquad (4.12b)$$

respectively, which are the spaces of vector fields with vanishing divergence $\mathrm{div}(Av) = 0$. The space $\nabla H_0^1(D)$ is a strict subspace of $H_0(\mathrm{curl}\, 0, D)$. It can be shown (see, e.g., [9, Section IX.1]) that $H_0(\mathrm{curl}\, 0, D)$ can be decomposed as a direct sum of $\nabla H_0^1(D)$ and the gradients $\nabla \mathcal{C}(D)$ of the co-homology space

$$\mathcal{C}(D) = \{p \in H^1(D) : \Delta p = 0 \text{ in } D,\ p \text{ constant}$$
$$\text{on every connected component of } \partial D\}.$$

Analogously, $\nabla H^1(D)$ is a subspace of $H(\mathrm{curl}\,0, D)$ and coincides with it for simply connected domains. Therefore, V_A is contained in

$$H(\mathrm{curl}, \mathrm{div}_A\,0, D)$$
$$= \{v \in H(\mathrm{curl}, D) : (Av, \nabla\varphi)_{L^2(D)} = 0 \text{ for all } \varphi \in H^1(D)\} \quad (4.12c)$$

which is the space of vector fields in $H(\mathrm{curl}, D)$ with vanishing divergence $\mathrm{div}(Av) = 0$ and vanishing co-normal components $(Av) \cdot v$ on ∂D.

(b) Often, as in our application, μ is a scalar, thus $A = \mu I$. In this case we write V_μ for $V_{\mu I}$ and, analogously, V_ε for scalar valued ε. In the case $\sigma = 0$; that is real-valued and symmetric and positive definite ε and μ the direct sums

$$H(\mathrm{curl}, D) = V_\mu \oplus H(\mathrm{curl}\,0, D) \quad \text{and} \quad H_0(\mathrm{curl}, D) = V_{0,\varepsilon} \oplus H_0(\mathrm{curl}\,0, D)$$

are orthogonal with respect to the inner products

$$(u, v)_{\varepsilon,\mu} = (\varepsilon^{-1}\,\mathrm{curl}\,u, \mathrm{curl}\,v)_{L^2(D)} + (\mu u, v)_{L^2(D)}, \quad (4.13a)$$
$$(u, v)_{\mu,\varepsilon} = (\mu^{-1}\,\mathrm{curl}\,u, \mathrm{curl}\,v)_{L^2(D)} + (\varepsilon u, v)_{L^2(D)}, \quad (4.13b)$$

respectively, and $L^2(D, \mathbb{C}^3) = \tilde{V}_{0,\varepsilon} \oplus H_0(\mathrm{curl}\,0, D)$ is an orthogonal decomposition with respect to the inner product $(\varepsilon u, v)_{L^2(D)}$.

By the same arguments as in the proof of the previous theorem one can show another kind of decompositions of $L^2(D, \mathbb{C}^3)$ and $H_0(\mathrm{curl}, D)$.

Theorem 4.23. *Let $L^2(\mathrm{div}_A\,0, D)$ and $H_0(\mathrm{curl}, \mathrm{div}_A\,0, D)$ be defined in (4.12a) and (4.12b), respectively, where D and A are as in Theorem 4.21. Then the decompositions*

$$L^2(D, \mathbb{C}^3) = L^2(\mathrm{div}_A\,0, D) \oplus \nabla H_0^1(D),$$
$$H_0(\mathrm{curl}, D) = H_0(\mathrm{curl}, \mathrm{div}_A\,0, D) \oplus \nabla H_0^1(D)$$

hold, and the projections are bounded.

We leave the proof to the reader (Exercise 4.15).

As mentioned before, the space $H_0(\mathrm{curl}, D)$ is not compactly embedded in $L^2(D, \mathbb{C}^3)$. However—and this is the reason for introducing the Helmholtz decomposition—the subspace $V_{0,A}$ is compactly embedded in $L^2(D, \mathbb{C}^3)$. We formulate this crucial result in the following theorem.

Theorem 4.24. *Let D be a bounded Lipschitz domain (see Definition A.7). Then the space $V_{0,A}$ from (4.10d) is compactly embedded in $L^2(D, \mathbb{C}^3)$.*

The proof of this important theorem requires more sophisticated properties of Sobolev spaces. Therefore, we postpone it to Sect. 5.1, Theorem 5.32. From this result it follows (see Corollary 4.36 below) that also V_A is compactly embedded in $L^2(D, \mathbb{C}^3)$. However, in the context of this chapter we do not need the compactness of V_A.

4.2 The Cavity Problem

4.2.1 The Variational Formulation and Existence

Before we investigate the Maxwell system (4.1a)–(4.1c) we consider the variational form of the following interior boundary value problem for the scalar inhomogeneous Helmholtz equation, namely

$$\operatorname{div}(a\,\nabla u) + k^2 b u = f \ \text{ in } D\,, \quad u = 0 \text{ on } \partial D\,, \tag{4.14}$$

where $k \geq 0$ is real. We assume that $a \in L^\infty(D)$ is real valued such that $a(x) \geq a_0$ on D for some constant $a_0 > 0$, and $b \in L^\infty(D)$, $f \in L^2(D)$ are allowed to be complex valued.

As for the derivation of the variational equation for Maxwell's equations we multiply the equation by some test function $\overline{\psi}$ with $\psi = 0$ on ∂D, integrate over D, and use Green's first formula. This yields

$$\int_D \left[a\,\nabla u \cdot \nabla \overline{\psi} - k^2 b\, u\,\overline{\psi} \right] dx \;=\; -\int_D f\,\overline{\psi}\,dx\,.$$

This equation makes perfectly sense in $H_0^1(D)$. We use this as a definition.

Definition 4.25. Let D be a bounded open set and $f \in L^2(D)$ and $b \in L^\infty(D)$ and real valued $a \in L^\infty(D)$ such that $a(x) \geq a_0$ on D for some constant $a_0 > 0$. A function $u \in H_0^1(D)$ is called a variational solution (or weak solution) of the boundary value problem (4.14) if

$$\int_D \left[a\,\nabla u \cdot \nabla \overline{\psi} - k^2 b\, u\,\overline{\psi} \right] dx \;=\; -\int_D f\,\overline{\psi}\,dx \quad \text{for all } \psi \in H_0^1(D)\,.$$

By Friedrich's inequality, the first term in this equation defines an inner product in $H_0^1(D)$.

Lemma 4.26. Let $a \in L^\infty(D)$ be real valued such that $a(x) \geq a_0$ on D for some constant $a_0 > 0$. We define a new inner product in $H_0^1(D)$ by

$$(u, v)_* \;=\; (a\,\nabla u, \nabla v)_{L^2(D)}\,, \quad u, v \in H_0^1(D)\,,$$

Then $(H_0^1(D), (\cdot, \cdot)_*)$ is a Hilbert space, and its norm is equivalent to the ordinary norm in $H_0^1(D)$; in particular,

$$\|u\|_* \leq \sqrt{\|a\|_\infty}\, \|u\|_{H^1(D)} \leq \sqrt{\frac{(1 + c^2)\|a\|_\infty}{a_0}}\, \|u\|_* \quad \text{for all } u \in H_0^1(D) \tag{4.15}$$

where c is the constant from Theorem 4.15.

Proof: From the previous theorem we conclude

$$\|u\|_*^2 = \|\sqrt{a}\,\nabla u\|_{L^2(D)}^2 \leq \|a\|_\infty \|u\|_{H^1(D)}^2 = \|a\|_\infty \left[\|\nabla u\|_{L^2(D)}^2 + \|u\|_{L^2(D)}^2\right]$$

$$\leq (1 + c^2)\|a\|_\infty \|\nabla u\|_{L^2(D)}^2 \leq \|a\|_\infty \frac{1 + c^2}{a_0}\|u\|_*^2.$$

$\square$

Therefore, we write the variational equation of Definition 4.25 in the form

$$(u, \psi)_* - k^2 (bu, \psi)_{L^2(D)} = -(f, \psi)_{L^2(D)} \quad \text{for all } \psi \in H_0^1(D). \tag{4.16}$$

The question of existence of solutions of (4.30) is again—as in the previous chapter—answered by the Riesz–Fredholm theory.

Theorem 4.27. *Let again $D \subseteq \mathbb{R}^3$ be a bounded open set and $a, b \in L^2(D)$ satisfy the above assumptions. Then the boundary value problem (4.14) has a variational solution for exactly those $f \in L^2(D)$ such that $(f, v)_{L^2(D)} = 0$ for all solutions $v \in H_0^1(D)$ of the corresponding homogeneous form of (4.14) with b replaced by $\bar{b}$; that is,*

$$\mathrm{div}(a\,\nabla v) + k^2 \bar{b} v = 0 \text{ in } D, \quad v = 0 \text{ on } \partial D. \tag{4.17}$$

If, in particular, this boundary value problem (4.17) admits only the trivial solution $v = 0$, then the inhomogeneous boundary value problem has a unique solution for every $f \in L^2(D)$.

Proof: We will use the Riesz representation theorem (Theorem A.5) from functional analysis to rewrite (4.16) in the form $u - k^2 K u = -\tilde{f}$ with some compact operator K and apply the Fredholm alternative to this equation. To carry out this idea we note that—for fixed $v \in L^2(D)$—the mapping $\psi \mapsto (\bar{b}\psi, v)_{L^2(D)}$ is linear and bounded from $(H_0^1(D), (\cdot, \cdot)_*)$ into $\mathbb{C}$. Indeed, by the inequality of Cauchy–Schwarz and Theorem 4.15 we conclude that there exists $c > 0$ such that

$$\left|(\bar{b}\psi, v)_{L^2(D)}\right| \leq \|b\|_\infty \|\psi\|_{L^2(D)} \|v\|_{L^2(D)} \leq c\,\|v\|_{L^2(D)} \|\psi\|_*$$

for all $\psi \in H_0^1(D)$. Therefore, the representation theorem of Riesz assures the existence of a unique $g_v \in H_0^1(D)$ with $(\overline{b}\psi, v)_{L^2(D)} = (\psi, g_v)_*$ for all $\psi \in H_0^1(D)$. We define the operator $\tilde{K} : L^2(D) \to H_0^1(D)$ by $\tilde{K}v = g_v$. Then $\tilde{K}$ satisfies

$$(\tilde{K}v, \psi)_* \; = \; (bv, \psi)_{L^2(D)} \quad \text{for all } v \in L^2(D) , \; \psi \in H_0^1(D) .$$

$\tilde{K}$ is linear (easy to see) and bounded because $\|\tilde{K}v\|_*^2 = (\tilde{K}v, \tilde{K}v)_* = (bv, \tilde{K}v)_{L^2(D)} \leq \|b\|_\infty \|\tilde{K}v\|_{L^2(D)} \|v\|_{L^2(D)} \leq c\|\tilde{K}v\|_* \|v\|_{L^2(D)}$; that is, $\|\tilde{K}v\|_* \leq c\|v\|_{L^2(D)}$. By the same argument there exists $\tilde{f} \in H_0^1(D)$ with $(f, \psi)_{L^2(D)} = (\tilde{f}, \psi)_*$ for all $\psi \in H_0^1(D)$. Therefore, Eq. (4.16) can be written as

$$(u, \psi)_* \; - \; k^2(\tilde{K}u, \psi)_* \; = \; -(\tilde{f}, \psi)_* \quad \text{for all } \psi \in H_0^1(D) . \tag{4.18}$$

Because this holds for all such ψ we conclude that this equation is equivalent to

$$u \; - \; k^2 \tilde{K}u \; = \; -\tilde{f} \quad \text{in } H_0^1(D) . \tag{4.19}$$

Because $H_0^1(D)$ is compactly embedded in $L^2(D)$ we conclude that the operator $K = \tilde{K} \circ J : H_0^1(D) \to H_0^1(D)$ is compact (J denotes the embedding $H_0^1(D) \subseteq L^2(D)$). To study solvability of this equation we want to apply the Fredholm alternative of Theorem A.4. We choose the bilinear form $\langle u, v \rangle = (u, \overline{v})_* = \int_D a \nabla u \cdot \nabla v \, dx$ in $H_0^1(D)$. Then K is self-adjoint because by the definition of K we have $\langle Ku, v \rangle = (bu, \overline{v})_{L^2(D)} = \int_D b \, u \, v \, dx$ and this is symmetric in u and v. Therefore, we know from the Fredholm alternative of Theorem A.4 that Eq. (4.19) is solvable if, and only if, the right-hand side $\tilde{f}$ is orthogonal with respect to $\langle \cdot, \cdot \rangle$ to the nullspace of $I - k^2 K$. Therefore, let $v \in H_0^1(D)$ be a solution of (4.17). Then $\overline{v}$ satisfies the homogeneous equation $\overline{v} - k^2 K \overline{v} = 0$ and $\langle \tilde{f}, \overline{v} \rangle = (\tilde{f}, v)_* = (f, v)_{L^2(D)}$. Therefore, we conclude that $\langle \tilde{f}, \overline{v} \rangle = 0$ if, and only if, $(f, v)_{L^2(D)} = 0$. $\quad \square$

We can also use the form (4.19) to prove existence and completeness of an *eigensystem* of

$$\operatorname{div}(a\nabla v) \; + \; k^2 b\, v \; = \; 0 \;\; \text{in } D , \quad u = 0 \;\; \text{on } \partial D , \tag{4.20}$$

for real valued $a, b \in L^\infty(D)$ such that $a(x) \geq a_0$ and $b(x) \geq b_0$ on D for some $a_0, b_0 > 0$. This equation has, of course, to be understood in the variational sense of Definition 4.25. Recall the definition of the inner product $(u, v)_* = (a \nabla u, \nabla v)_{L^2(D)}$.

Theorem 4.28. *Let D be a bounded open set and $a, b \in L^\infty(D)$ with $a(x) \geq a_0$ and $b(x) \geq b_0$ on D for some $a_0, b_0 > 0$. Then there exists an infinite number of eigenvalues $k^2 \in \mathbb{R}_{>0}$ of (4.20); that is, there exists a sequence $k_n > 0$ and corresponding eigenfunctions $\hat{u}_n \in H_0^1(D)$ with $\|\hat{u}_n\|_* = 1$ such that $\operatorname{div}(a\nabla \hat{u}_n) + k_n^2 b\, \hat{u}_n = 0$ in D and $\hat{u}_n = 0$ on ∂D in the variational* *form. Furthermore, $k_n \to \infty$ as $n \to \infty$. The sets $\{\hat{u}_n : n \in \mathbb{N}\}$ and $\{k_n\hat{u}_n : n \in \mathbb{N}\}$ form complete orthonormal systems in $\left(H_0^1(D), (\cdot, \cdot)_*\right)$ and $L^2(D, b\,dx)$, respectively. Here, we denote by $L^2(D, b\,dx)$ the space equipped with the weighted inner product $(u, v)_{L^2(D, b\,dx)} = (bu, v)_{L^2(D)}$.*

Proof: We define the operator $K : H_0^1(D) \to H_0^1(D)$ as in the previous theorem. Then K is compact and self-adjoint in $\left(H_0^1(D), (\cdot, \cdot)_*\right)$ because b is real valued. It is also positive because $(Ku, u)_* = (bu, u)_{L^2(D)} > 0$ for $u \neq 0$. Therefore, the spectral theorem for compact self-adjoint operators in Hilbert spaces (see, e.g, [16, Section 15.3]) implies the existence of a spectral system $(\mu_n, \hat{u}_n)$ of K in $\left(H_0^1(D), (\cdot, \cdot)\right)_*$. Furthermore, $\mu_n > 0$ and $\mu_n \to 0$ as $n \to \infty$ and the set $\{\hat{u}_n : n \in \mathbb{N}\}$ of orthonormal eigenfunctions $\hat{u}_n$ is complete in $\left(H_0^1(D), (\cdot, \cdot)\right)_*$. Therefore, we have $K\hat{u}_n = \mu_n\hat{u}_n$; that is, $\hat{u}_n - \frac{1}{\mu_n} K\hat{u}_n = 0$, which is the variational form of (4.20) for $k_n = \frac{1}{\sqrt{\mu_n}}$. The system $\{k_n\hat{u}_n : n \in \mathbb{N}\}$ is a system of eigenfunctions as well which is orthogonal in $L^2(D, b\,dx)$ because

$$k_n^2(b\hat{u}_n, \hat{u}_m)_{L^2(D)} = k_n^2(K\hat{u}_n \,\hat{u}_m)_* = (\hat{u}_n, \hat{u}_m)_* = \delta_{mn}.$$

This set is also complete in $L^2(D)$. This follows directly from the fact that $H_0^1(D)$ is dense in $L^2(D)$ (see Lemma 4.9). $\square$

Now we go back to the formulation (4.1a)–(4.1c) of the cavity problem for *Maxwell's equations* which we recall for the convenience of the reader.
Given $J_e \in L^2(D, \mathbb{C}^3)$ and scalar and real valued $\varepsilon, \mu, \sigma \in L^\infty(D)$ determine $E \in H_0(\text{curl}, D)$ and $H \in H(\text{curl}, D)$ such that

$$\text{curl}\, E - i\omega\mu\, H = 0 \quad \text{in } D\,, \tag{4.21a}$$

$$\text{curl}\, H + (i\omega\varepsilon - \sigma)\, E = J_e \quad \text{in } D\,. \tag{4.21b}$$

The boundary condition (4.1c) is included in the space $H_0(\text{curl}, D)$ for E.

At the beginning of the section we already computed the variational equation (4.2) for this system; that is,

$$\int_D \left[\frac{1}{\mu}\,\text{curl}\, E \cdot \text{curl}\,\psi - \omega^2\left(\varepsilon + i\frac{\sigma}{\omega}\right) E \cdot \psi\right] dx = i\omega \int_D J_e \cdot \psi\, dx\,. \tag{4.22}$$

This *variational equation* holds for all $\psi \in H_0(\text{curl}, D)$.

Analogously, we can eliminate E. Indeed, now we multiply (4.21a) by $\psi \in H(\text{curl}, D)$ and proceed as in the previous part.

$$0 = \int_D \left[\text{curl}\, E \cdot \psi - i\omega\mu\, H \cdot \psi\right] dx = \int_D \left[E \cdot \text{curl}\,\psi - i\omega\mu\, H \cdot \psi\right] dx$$

$$= \int_D \left[\frac{1}{i\omega\varepsilon - \sigma}\,(J_e - \text{curl}\, H) \cdot \text{curl}\,\psi - i\omega\mu\, H \cdot \psi\right] dx\,.$$

We note that the boundary term vanishes because of $E \in H_0(\mathrm{curl}, D)$. Multiplication with $i\omega$ yields

$$\int_D \left[\frac{1}{\varepsilon + i\sigma/\omega} \, \mathrm{curl}\, H \cdot \mathrm{curl}\, \psi - \omega^2 \mu\, H \cdot \psi \right] dx \; = \; \int_D \frac{1}{\varepsilon + i\sigma/\omega} \, J_e \cdot \mathrm{curl}\, \psi \, dx \, .$$
$$(4.23)$$

This holds for all $\psi \in H(\mathrm{curl}, D)$.

It is easy to see that the variational equations (4.22) and (4.23) are equivalent to the Maxwell system.

Lemma 4.29. Let $E \in H_0(\mathrm{curl}, D)$ satisfy the variational equation (4.22) for all $\psi \in H_0(\mathrm{curl}, D)$. Set

$$H \; = \; \frac{1}{i\omega\mu} \, \mathrm{curl}\, E \quad \text{in } D \, .$$

Then $E \in H_0(\mathrm{curl}, D)$ and $H \in H(\mathrm{curl}, D)$ satisfy the system (4.21a), (4.21b).

Furthermore, $H \in H(\mathrm{curl}, D)$ satisfies (4.23) for all $\psi \in H(\mathrm{curl}, D)$.

The same statement holds for E and H interchanged.

Proof: Let $E \in H_0(\mathrm{curl}, D)$ satisfy (4.22) for all $\psi \in H_0(\mathrm{curl}, D)$. We note that $H \in L^2(D, \mathbb{C}^3)$. Substituting the definition of H into (4.22) yields

$$\int_D \left[i\omega\, H \cdot \mathrm{curl}\, \psi - \omega^2 \left(\varepsilon + i\frac{\sigma}{\omega} \right) E \cdot \psi \right] dx \; = \; i\omega \int_D J_e \cdot \psi \, dx$$

which is the variational form of $i\omega\, \mathrm{curl}\, H$; that is,

$$i\omega\, \mathrm{curl}\, H \; = \; \omega^2 \left(\varepsilon + i\frac{\sigma}{\omega} \right) E \; + \; i\omega\, J_e \, .$$

Division by $i\omega$ yields (4.21b). □

In the following we will discuss the variational equation (4.22) in the space $H_0(\mathrm{curl}, D)$. For the remaining part of this subsection we make the following assumptions on the parameters μ, ε, σ, and on D (recall Sect. 1.3.3):

There exists $c > 0$ such that

(a) $\mu \in L^\infty(D)$ is real valued and $\mu(x) \geq c$ almost everywhere,
(b) $\varepsilon \in L^\infty(D)$ is real valued and $\varepsilon(x) \geq c$ almost everywhere on D.
(c) $\sigma \in L^\infty(D)$ is real valued and $\sigma(x) \geq 0$ almost everywhere on D.

For abbreviation we define again the complex dielectricity by

$$\varepsilon_c(x) \; = \; \varepsilon(x) \; + \; i\frac{\sigma(x)}{\omega}, \qquad x \in D \, . \qquad (4.24)$$

Also, we assume that $D \subseteq \mathbb{R}^3$ is open and bounded such that the subspace V_{0,ε_c} (see (4.10d) for $A = \varepsilon_c I$) is compactly embedded in $L^2(D, \mathbb{C}^3)$. By Theorem 4.24 this is the case for Lipschitz domains D.

Furthermore, it is convenient to define the vector field $F \in L^2(D, \mathbb{C}^3)$ by

$$F = \frac{i\omega}{\varepsilon_c} J_e .$$

Then the variational equation (4.22) takes the form

$$\int_D \left[\frac{1}{\mu} \operatorname{curl} E \cdot \operatorname{curl} \overline{\psi} - \omega^2 \varepsilon_c E \cdot \overline{\psi} \right] dx = \int_D \varepsilon_c F \cdot \overline{\psi} \, dx \qquad (4.25)$$

for all $\psi \in H_0(\operatorname{curl}, D)$. To discuss this equation in $H_0(\operatorname{curl}, D)$ we will use the Helmholtz decomposition to split this problems into two problems, one in V_{0,ε_c} and one in $H_0(\operatorname{curl} 0, D)$, where the one in $H_0(\operatorname{curl} 0, D)$ will be trivial. Recall that V_{0,ε_c} is given by (4.10d); that is,

$$V_{0,\varepsilon_c} = \{ u \in H_0(\operatorname{curl}, D) : (\varepsilon_c u, \psi)_{L^2(D)} = 0 \text{ for all } \psi \in H_0(\operatorname{curl} 0, D) \} .$$

Analogously, $\tilde{V}_{0,\varepsilon_c}$ denotes the corresponding subspace in $L^2(D, \mathbb{C}^3)$, see (4.10e).

First we use the Helmholtz decomposition to write F as $F = F_0 + f$ with $F_0 \in H_0(\operatorname{curl} 0, D)$ and $f \in \tilde{V}_{0,\varepsilon_c}$. Then we make an ansatz for E in the form $E = E_0 + u$ with $E_0 \in H_0(\operatorname{curl} 0, D)$ and $u \in V_{0,\varepsilon_c}$. Substituting this into (4.25) yields

$$\int_D \left[\frac{1}{\mu} \operatorname{curl} u \cdot \operatorname{curl} \overline{\psi} - \omega^2 \varepsilon_c (u + E_0) \cdot \overline{\psi} \right] dx = \int_D \varepsilon_c (f + F_0) \cdot \overline{\psi} \, dx \quad (4.26)$$

for all $\psi \in H_0(\operatorname{curl}, D)$. Now we take $\psi \in H_0(\operatorname{curl} 0, D)$ as test functions. Recalling that $\int_D \varepsilon_c u \cdot \overline{\psi} \, dx = 0$ and $\int_D \varepsilon_c f \cdot \overline{\psi} \, dx = 0$ for all $\psi \in H_0(\operatorname{curl} 0, D)$ yields

$$-\omega^2 \int_D \varepsilon_c E_0 \cdot \overline{\psi} \, dx = \int_D \varepsilon_c F_0 \cdot \overline{\psi} \, dx \quad \text{for all } \psi \in H_0(\operatorname{curl} 0, D) .$$

Rewriting this as

$$\int_D \varepsilon_c [F_0 + \omega^2 E_0] \cdot \overline{\psi} \, dx = 0 \quad \text{for all } \psi \in H_0(\operatorname{curl} 0, D)$$

and setting $\psi = F_0 + \omega^2 E_0 \in H_0(\operatorname{curl} 0, D)$ yields $E_0 = -\frac{1}{\omega^2} F_0$.

Therefore, (4.26) reduces to

$$\int_D \left[\frac{1}{\mu} \operatorname{curl} u \cdot \operatorname{curl} \overline{\psi} - \omega^2 \varepsilon_c u \cdot \overline{\psi} \right] dx = \int_D \varepsilon_c f \cdot \overline{\psi} \, dx \qquad (4.27)$$

for all $\psi \in H_0(\operatorname{curl}, D)$. Second, we take $\psi \in V_{0,\varepsilon_c}$ as test functions. Before we investigate (4.27) we summarize this splitting in the following lemma.

Lemma 4.30. *Let $F \in L^2(D, \mathbb{C}^3)$ have the Helmholtz decomposition $F = F_0 + f$ with $f \in \tilde{V}_{0,\varepsilon_c}$ and $F_0 \in H_0(\operatorname{curl} 0, D)$.*

(a) *Let $E \in H_0(\operatorname{curl}, D)$ be a solution of (4.25) and $E = E_0 + u$ with $u \in V_{0,\varepsilon_c}$ and $E_0 \in H_0(\operatorname{curl} 0, D)$ the Helmholtz decomposition of E. Then $E_0 = -\frac{1}{\omega^2} F_0$ and $u \in V_{0,\varepsilon_c}$ solves (4.27) for all $\psi \in V_{0,\varepsilon_c}$.*
(b) *If $u \in V_{0,\varepsilon_c}$ solves (4.27) then $E = u - \frac{1}{\omega^2} F_0 \in H_0(\operatorname{curl}, D)$ solves (4.25).*

To show solvability of (4.27) we follow the approach as in the scalar case for the Helmholtz equation and introduce the equivalent inner product $(\cdot, \cdot)_*$ on V_{0,ε_c} by

$$(v, w)_* = \int_D \frac{1}{\mu} \operatorname{curl} v \cdot \operatorname{curl} \overline{w} \, dx, \qquad v, w \in V_{0,\varepsilon_c}. \qquad (4.28)$$

Then we have the analogue of Lemma 4.26:

Lemma 4.31. *The norm $\|v\|_* = \sqrt{(v, v)_*}$ is an equivalent norm on V_{0,ε_c}.*

Proof: In contrast to the proof of Lemma 4.26 we use an indirect argument to show the existence of a constant $c > 0$ such that

$$\|v\|_{H(\operatorname{curl}, D)} \leq c \|\operatorname{curl} v\|_{L^2(D)} \quad \text{for all } v \in V_{0,\varepsilon_c}. \qquad (4.29)$$

This is sufficient because the estimates $\|\operatorname{curl} v\|_{L^2(D)}^2 \leq \|\mu\|_\infty \|v\|_*^2$ and $\|v\|_* \leq \|(1/\mu)\|_\infty \|v\|_{H(\operatorname{curl}, D)}$ hold obviously.

Assume that the inequality (4.29) is not satisfied. Then there exists a sequence (v_n) in V_{0,ε_c} with $\|v_n\|_{H(\operatorname{curl}, D)} = 1$ and $\operatorname{curl} v_n \to 0$ in $L^2(D, \mathbb{C}^3)$ as n tends to infinity. The sequence (v_n) is therefore bounded in V_{0,ε_c}. Because V_{0,ε_c} is compactly embedded in $L^2(D, \mathbb{C}^3)$ by Theorem 4.24 there exists a subsequence, denoted by the same symbol, which converges in $L^2(D, \mathbb{C}^3)$. For this subsequence, (v_n) and $(\operatorname{curl} v_n)$ are Cauchy sequences in $L^2(D, \mathbb{C}^3)$. Therefore, (v_n) is a Cauchy sequence in V_{0,ε_c} and therefore convergent: $v_n \to v$ in $H(\operatorname{curl}, D)$ for some $v \in V_{0,\varepsilon_c}$ and $\operatorname{curl} v = 0$. Therefore, $v \in V_{0,\varepsilon_c} \cap H_0(\operatorname{curl} 0, D) = \{0\}$ which shows that v vanishes. This contradicts the fact that $\|v\|_{H(\operatorname{curl}, D)} = 1$. $\quad \square$

Now we write (4.27) in the form

$$(u, \psi)_* - \omega^2 (\varepsilon_c u, \psi)_{L^2(D)} = (\varepsilon_c f, \psi)_{L^2(D)} \quad \text{for all } \psi \in V_{0,\varepsilon_c}.$$

Again, we use the representation theorem of Riesz (Theorem A.5) to show the existence of a linear and bounded operator $\tilde{K} : L^2(D, \mathbb{C}^3) \to V_{0,\varepsilon_c}$ with $(\psi, \varepsilon_c u)_{L^2(D)} = (\psi, \tilde{K} u)_*$ for all $\psi \in V_{0,\varepsilon_c}$ and $u \in L^2(D, \mathbb{C}^3)$. We carry out the arguments for the convenience of the reader although they are completely analogous to the arguments in the proof of Theorem 4.27. For fixed $u \in L^2(D, \mathbb{C}^3)$ the mapping $\psi \mapsto (\psi, \varepsilon_c u)_{L^2(D)}$ defines a linear and bounded functional on V_{0,ε_c}. Therefore, by the representation theorem of Riesz there exists a unique $g = g_u \in V_{0,\varepsilon_c}$ with $(\psi, \varepsilon_c u)_{L^2(D)} = (\psi, g_u)_*$ for all $\psi \in V_{0,\varepsilon_c}$. We set $\tilde{K} u = g_u$. Linearity of $\tilde{K}$ is clear from the uniqueness of the representation g_u. Boundedness of $\tilde{K}$ from $L^2(D, \mathbb{C}^3)$ into V_{0,ε_c} follows from the estimate

$$\|\tilde{K} u\|_*^2 = (\tilde{K} u, \tilde{K} u)_* = (\tilde{K} u, \varepsilon_c u)_{L^2(D)} \leq \|\varepsilon_c\|_\infty \|u\|_{L^2(D)} \|\tilde{K} u\|_{L^2(D)}$$
$$\leq c \|\varepsilon_c\|_\infty \|u\|_{L^2(D)} \|\tilde{K} u\|_*$$

after division by $\|\tilde{K} u\|_*$. Because V_{0,ε_c} is compactly embedded in $L^2(D, \mathbb{C}^3)$ we conclude that $K = \tilde{K} \circ J$ is even compact as an operator from V_{0,ε_c} into itself, if we denote by $J : V_{0,\varepsilon_c} \to L^2(D, \mathbb{C}^3)$ the compact embedding operator. Therefore, we can write Eq. (4.27) in the form

$$(u, \psi)_* - \omega^2 (K u, \psi)_* = (\tilde{K} f, \psi)_* \quad \text{for all } \psi \in V_{0,\varepsilon_c};$$

that is,

$$u - \omega^2 K u = \tilde{K} f \quad \text{in } V_{0,\varepsilon_c}. \tag{4.30}$$

This is again a Fredholm equation of the second kind for $u \in V_{0,\varepsilon}$. In particular, we have existence once we have uniqueness. The question of uniqueness will be discussed in the next section below.

For the general question of existence we apply again Fredholm's Theorem A.4 to $X = V_{0,\varepsilon_c}$ with bilinear form $\langle u, v \rangle = (u, \overline{v})_* = \int_D \frac{1}{\mu} \operatorname{curl} u \cdot \operatorname{curl} v \, dx$ and operator $T = \omega^2 K$. The operator K is self-adjoint. Indeed, $\langle K u, \psi \rangle = (K u, \overline{\psi})_* = (\varepsilon_c u, \overline{\psi})_{L^2(D)}$ and this is symmetric in u and ψ. Lemma 4.30 tells us that the variational form (4.25) of the cavity problem is solvable for $F \in L^2(D, \mathbb{C}^3)$ if, and only if, the variational problem (4.27) is solvable in V_{0,ε_c} for the part $f \in \tilde{V}_{0,\varepsilon_c}$ of F. By Theorem A.4 this is solvable for exactly those $f \in \tilde{V}_{0,\varepsilon_c}$ such that $\tilde{K} f$ is orthogonal to the nullspace of $I - \omega^2 K$ with respect to $\langle \cdot, \cdot \rangle$. By Lemma 4.30 $v \in V_{0,\varepsilon_c}$ solves (4.27) for $f = 0$ if, and only if, v solves (4.25) for $F = 0$. Also we note that, for $v \in V_{0,\varepsilon_c}$,

$$\langle \tilde{K} f, v \rangle = (\varepsilon_c f, \overline{v})_{L^2(D)} = (\varepsilon_c (F_0 + f), \overline{v})_{L^2(D)} = (\varepsilon_c F, \overline{v})_{L^2(D)};$$

that is $\langle \tilde{K}f, v\rangle$ vanishes if, and only if, $(\varepsilon_c F, \bar{v})_{L^2(D)}$ vanishes. Furthermore we note that v solves the homogeneous form of (4.25) if, and only if, $\bar{v}$ solves (4.25) with ε_c replaced by $\overline{\varepsilon_c}$. We have thus proven:

Theorem 4.32. *The cavity problem (4.21a), (4.21b) has a solution $E \in H_0(\mathrm{curl}, D)$ and $H \in H(\mathrm{curl}, D)$ for exactly those source terms $J_e \in L^2(D, \mathbb{C}^3)$ such that $(J_e, v)_{L^2(D)} = 0$ for all solutions $v \in H_0(\mathrm{curl}, D)$ of the corresponding homogeneous form of (4.21a), (4.21b) with σ replaced by $-\sigma$. If, in particular, the boundary value problem admits only the trivial solution $E = 0, H = 0$, then the inhomogeneous boundary value problem (4.21a), (4.21b) has a unique solution $(E, H) \in H_0(\mathrm{curl}, D) \times H(\mathrm{curl}, D)$ for all $J_e \in L^2(D, \mathbb{C}^3)$.*

For the remaining part of this subsection we assume that $\sigma = 0$; that is, $\varepsilon_c = \varepsilon$ is real valued. Then we can consider the following eigenvalue problem by the same arguments.

$$\mathrm{curl}\, E \ - \ i\omega\mu\, H = 0 \quad \text{in } D, \tag{4.31a}$$

$$\mathrm{curl}\, H \ + \ i\omega\varepsilon\, E = J \quad \text{in } D, \tag{4.31b}$$

for $J = 0$.

Definition 4.33. The resolvent set consists of all $\omega \in \mathbb{C}$ for which (4.31a), (4.31b) has a unique solution $(E, H) \in H_0(\mathrm{curl}, D) \times H(\mathrm{curl}, D)$ for all $J \in L^2(D, \mathbb{C}^3)$ and such that $J \mapsto (E, H)$ is bounded from $L^2(D, \mathbb{C}^3)$ into $H_0(\mathrm{curl}, D) \times H(\mathrm{curl}, D)$.

We have seen that this eigenvalue problem is equivalent to the corresponding variational equation (4.27) for $u = E \in V_{0,\varepsilon}$ (because the right-hand side vanishes); that is,

$$\int_D \left[\frac{1}{\mu}\, \mathrm{curl}\, u \cdot \mathrm{curl}\, \psi - \omega^2 \varepsilon u \cdot \psi\right] dx \ = \ 0 \quad \text{for all } \psi \in V_{0,\varepsilon},$$

which holds even for all $\psi \in H_0(\mathrm{curl}, D)$ because $u \in V_{0,\varepsilon}$. If u solves this equation, then also its complex conjugate. Therefore, we can assume that u is real valued which we do for the remaining part of this section for all functions. Then we write the variational equation in the form

$$(u, \psi)_{\mu,\varepsilon} \ = \ (\omega^2 + 1)\, (\varepsilon u, \psi)_{L^2(D)} \quad \text{for all } \psi \in H_0(\mathrm{curl}, D), \tag{4.32}$$

where $(\cdot, \cdot)_{\mu,\varepsilon}$ had been defined in (4.13b); that is, using the bilinear form $\langle \cdot, \cdot \rangle = (\cdot, \cdot)_*$ and the operator $K : V_{0,\varepsilon} \to V_{0,\varepsilon}$ from above,

$$(u, v)_{\mu,\varepsilon} \ = \ \langle u, v\rangle + (\varepsilon u, v)_{L^2(D)} \ = \ \langle u, v\rangle + \langle Ku, v\rangle. \tag{4.33}$$

First we consider $\omega \neq 0$. We have seen that $\omega \neq 0$ is in the resolvent set if, and only if, the operator $I - \omega^2 K$ is one-to-one in $V_{0,\varepsilon}$; that is, $1/\omega^2$ is in the resolvent set of $K : V_{0,\varepsilon} \to V_{0,\varepsilon}$. We note that K is also compact and self-adjoint with respect to $(\cdot, \cdot)_{\mu,\varepsilon}$. The reason why we take $(\cdot, \cdot)_{\mu,\varepsilon}$ instead of just $\langle \cdot, \cdot \rangle$ as the inner product will be clear in a moment.

The spectrum of K; that is, the complementary set of the resolvent set in $\mathbb{C}$, consists of 0 and eigenvalues $1/\omega_n^2$ which converge to zero. Let us now consider $\omega = 0$. From the variational equation we conclude for $\psi = u$ that $\text{curl}\, u = 0$. Therefore, $\omega = 0$ is an eigenvalue with the infinite dimensional eigenspace $H_0(\text{curl}\, 0, D)$. We summarize this in the following theorem.

Theorem 4.34. *There exists an infinite number of positive eigenvalues $\omega_n \in \mathbb{R}_{>0}$ of (4.31a), (4.31b); that is, there exists a sequence $\omega_n > 0$ for $n \in \mathbb{N}$ and corresponding real valued functions $v_n \in V_{0,\varepsilon}$ such that*

$$\text{curl}\left(\frac{1}{\mu} \text{curl}\, v_n\right) - \omega_n^2 \varepsilon\, v_n = 0 \text{ in } D, \quad \nu \times v_n = 0 \text{ on } \partial D$$

in the variational form (4.32); that is,

$$\int_D \left[\frac{1}{\mu} \text{curl}\, v_n \cdot \text{curl}\, \psi - \omega_n^2 \varepsilon\, v_n \cdot \psi\right] dx = 0 \quad \text{for all } \psi \in H_0(\text{curl}, D).$$

(4.34)

The eigenvalues $\omega_n > 0$ have finite multiplicity and tend to infinity as $n \to \infty$. We normalize v_n by $\|v_n\|_{\mu,\varepsilon} = 1$ for all $n \in \mathbb{N}$, where the norm $\|\cdot\|_{\mu,\varepsilon}$ had been defined in (4.13b). Then the sets $\{v_n : n \in \mathbb{N}\}$ and $\{\sqrt{1 + \omega_n^2}\, v_n : n \in \mathbb{N}\}$ form complete orthonormal systems in $\left(V_{0,\varepsilon}, (\cdot, \cdot)_{\mu,\varepsilon}\right)$ and in $\left(\tilde{V}_{0,\varepsilon}, (\varepsilon\cdot, \cdot)_{L^2(D)}\right)$, respectively. Furthermore, $\omega = 0$ is also an eigenvalue with infinite dimensional eigenspace $H_0(\text{curl}\, 0, D)$.

Proof: First we note that the variational equation (4.34) has been shown to hold for $\psi \in V_{0,\varepsilon}$ only. For $\psi \in H_0(\text{curl}\, 0, D)$, however, the variational equation holds as well because of $v_n \in V_{0,\varepsilon}$ and the definition of $V_{0,\varepsilon}$. From (4.34) we conclude for $\psi = v_m$ that

$$\delta_{n,m} = (v_n, v_m)_{\mu,\varepsilon} = (\omega_n^2 + 1)(\varepsilon v_n, v_m)_{L^2(D)}$$

which shows that the set $\{\sqrt{1 + \omega_n^2}\, v_n : n \in \mathbb{N}\}$ forms an orthonormal systems in $\left(\tilde{V}_{0,\varepsilon}, (\varepsilon\cdot, \cdot)_{L^2(D)}\right)$. It remains to prove completeness of this system. But this follows from the fact that $V_{0,\varepsilon}$ is dense in $\tilde{V}_{0,\varepsilon}$. To see this latter denseness result let $v \in \tilde{V}_{0,\varepsilon}$. By the denseness of $C_0^\infty(D, \mathbb{C}^3)$ in $L^2(D, \mathbb{C}^3)$ (see Lemma 4.9) there exists a sequence $v_n \in C_0^\infty(D, \mathbb{C}^3)$ with $v_n \to v$ in $L^2(D, \mathbb{C}^3)$. The decompositions of v_n with respect to the direct sums $H_0(\text{curl}, D) = V_{0,\varepsilon} + H_0(\text{curl}\, 0)$ and $L^2(D, \mathbb{C}^3) = \tilde{V}_{0,\varepsilon} + H_0(\text{curl}\, 0)$ are identical: $v_n = v_n^0 + \tilde{v}_n$ with $v_n^0 \in H_0(\mathrm{curl}\,0, D)$ and $\tilde{v}_n \in V_{0,\varepsilon}$. The bound-edness of the projections yields $v_n^0 \to 0$ in $L^2(D, \mathbb{C}^3)$ and thus $\tilde{v}_n \to v$ in $L^2(D, \mathbb{C}^3)$ which ends the proof. $\square$

As a particular result of this theorem we recall that $\{v_n : n \in \mathbb{N}\}$ forms a complete orthonormal system in the closed subspace $V_{0,\varepsilon}$ of $H_0(\mathrm{curl}, D)$ with respect to the inner product $(\cdot, \cdot)_{\mu,\varepsilon}$. The functions v_n correspond to the electric fields E. As we will see now, the corresponding magnetic fields form a complete orthonormal system in the closed subspace V_μ of $H(\mathrm{curl}, D)$, defined in (4.10c), with respect to the inner product $(\cdot, \cdot)_{\varepsilon,\mu}$, defined in (4.13a). This symmetry is the reason for using the inner product $(\cdot, \cdot)_{\mu,\varepsilon}$ instead of just $\langle \cdot, \cdot \rangle$.

Lemma 4.35. Let $\omega_n > 0$ and $\{v_n : n \in \mathbb{N}\} \subseteq V_{0,\varepsilon}$ be the complete orthonor-mal system of $V_{0,\varepsilon}$, defined by (4.34). Then $w_n := \frac{1}{\omega_n \mu} \mathrm{curl}\, v_n \in V_\mu$ for all n, and $\{w_n : n \in \mathbb{N}\}$ forms a complete orthonormal system in the closed subspace V_μ of $H(\mathrm{curl}, D)$ with respect to the inner product $(\cdot, \cdot)_{\varepsilon,\mu}$, defined in (4.13a). Furthermore, w_n satisfies

$$\int_D \left[\frac{1}{\varepsilon}\, \mathrm{curl}\, w_n \cdot \mathrm{curl}\, \psi - \omega_n^2 \mu\, w_n \cdot \psi \right] dx = 0 \quad \text{for all } \psi \in H(\mathrm{curl}, D).$$
(4.35)

Proof: Substituting the definition of w_n into (4.34) yields $\omega_n(w_n, \mathrm{curl}\, \psi)_{L^2(D)} = \omega_n^2(\varepsilon\, v_n, \psi)_{L^2(D)}$ for all $\psi \in H_0(\mathrm{curl}, D)$, that is $w_n \in H(\mathrm{curl}, D)$ and $\mathrm{curl}\, w_n = \omega_n \varepsilon v_n$ by the definition of the variational curl. Furthermore, $w_n \in V_\mu$ because for $\phi \in H(\mathrm{curl}\,0, D)$ we conclude that $\omega_n(\mu\, w_n, \phi)_{L^2(D)} = (\mathrm{curl}\, v_n, \phi)_{L^2(D)} = (v_n, \mathrm{curl}\, \phi)_{L^2(D)} = 0$ by applying Green's theorem in the form (A.16b). To show (4.35) let $\psi \in H(\mathrm{curl}, D)$. Then

$$\left(\varepsilon^{-1} \mathrm{curl}\, w_n, \mathrm{curl}\, \psi \right)_{L^2(D)} = \omega_n\, (v_n, \mathrm{curl}\, \psi)_{L^2(D)}$$
$$= \omega_n\, (\mathrm{curl}\, v_n, \psi)_{L^2(D)} = \omega_n^2\, (\mu\, w_n, \psi)_{L^2(D)}.$$

Furthermore, from

$$(w_n, w_m)_{\varepsilon,\mu} = \left(\varepsilon^{-1} \mathrm{curl}\, w_n, \mathrm{curl}\, w_m \right)_{L^2(D)} + (\mu\, w_n, w_m)_{L^2(D)}$$
$$= \omega_n \omega_m\, (\varepsilon\, v_n, v_m)_{L^2(D)} + \frac{1}{\omega_n \omega_m} \left(\mu^{-1} \mathrm{curl}\, v_n, \mathrm{curl}\, v_m \right)_{L^2(D)}$$
$$= \left(\frac{\omega_n^2}{1 + \omega_n^2} + \frac{1}{1 + \omega_n^2} \right) \delta_{nm} = \delta_{nm}$$

we see that the system $\{w_n : n \in \mathbb{N}\}$ forms an orthonormal set in V_μ. Also completeness in V_μ is seen by similar arguments: Let $\psi \in V_\mu$ such that $(w_n, \psi)_{\varepsilon,\mu} = 0$ for all $n \in \mathbb{N}$. Then

$$0 = (w_n, \psi)_{\varepsilon,\mu} = \left(\varepsilon^{-1}\operatorname{curl} w_n, \operatorname{curl}\psi\right)_{L^2(D)} + (\mu\, w_n, \psi)_{L^2(D)}$$

$$= w_n\,(v_n, \operatorname{curl}\psi)_{L^2(D)} + \frac{1}{w_n}(\operatorname{curl} v_n, \psi)_{L^2(D)}$$

$$= \left(w_n + \frac{1}{w_n}\right)(v_n, \operatorname{curl}\psi)_{L^2(D)}$$

for all $n \in \mathbb{N}$. Defining $\phi = \varepsilon^{-1}\operatorname{curl}\psi$ we note that $(\varepsilon\, v_n, \phi)_{L^2(D)} = 0$ for all $n \in \mathbb{N}$. The completeness of the system $\left\{\sqrt{1+w_n^2}\, v_n : n \in \mathbb{N}\right\}$ in $\tilde{V}_{0,\varepsilon}$ would yield $\phi = 0$ provided we had shown that $\phi \in \tilde{V}_{0,\varepsilon}$. But this is the case because for $\rho \in H_0(\operatorname{curl} 0, D)$ we have $(\varepsilon\, \phi, \rho)_{L^2(D)} = (\operatorname{curl}\psi, \rho)_{L^2(D)} = (\psi, \operatorname{curl}\rho)_{L^2(D)} = 0$. Therefore, $\operatorname{curl}\psi = 0$, thus $\psi \in H(\operatorname{curl} 0, D) \cap V_\mu = \{0\}$, which proves completeness. $\square$

As a corollary the analogue of Theorem 4.24 follows.

Corollary 4.36. *The closed subspace V_μ is compactly embedded in $L^2(D, \mathbb{C}^3)$.*

Proof: The closed subspaces V_μ and $V_{0,\varepsilon}$ are isomorphic. Indeed, the isomorphism $T : V_{0,\varepsilon} \to V_\mu$ is given by

$$Tv = \sum_{n=1}^{\infty} \alpha_n\, w_n \quad\text{for } v = \sum_{n=1}^{\infty} \alpha_n\, v_n \in V_{0,\varepsilon}$$

because of

$$\|Tv\|_{\varepsilon,\mu}^2 = \sum_{n=1}^{\infty} \alpha_n^2 = \|v\|_{\mu,\varepsilon}^2 .$$

Then compactness of V_μ follows from the compactness of $V_{0,\varepsilon}$ by Theorem 4.24. $\square$

4.2.2 Uniqueness and Unique Continuation

First we consider again the scalar boundary value problem (4.14). For conducting media; that is, complex $b \in L^\infty(D)$ with $\operatorname{Im} b > 0$ a.e. on D and real valued $a \in L^\infty(D)$ with $a(x) \geq a_0$ for some $a_0 > 0$ we have uniqueness by Green's theorem. Indeed, let $u \in H_0^1(D)$ be a solution of (4.14) in the variational form of Definition 4.25 for $f = 0$. Substituting $\psi = u$ into the integral yields

$$\int_D [a\,|\nabla u|^2 - k^2 b\,|u|] \,dx = 0 .$$

Taking the imaginary part yields $\int_D \operatorname{Im} b\,|u|^2 dx = 0$ and thus $u = 0$ because of the assumption on b.

If only Im $b \geq 0$ on D and Im $b > 0$ a.e. on some open subset U of D then, by the same argument, u vanishes on U. We want to prove that this implies $u = 0$ in all of D. This property of a differential equation is called the *unique continuation property*. If u was analytic this would follow from an analytic continuation argument which is well known from complex analysis. However, for non-analytic coefficients a and b the solution u fails to be analytic.

Before we turn to the unique continuation property we briefly consider the analogous situation for the boundary value problem (4.21a), (4.21b) for the Maxwell system. As before, we assume first that the medium is conducting; that is, $\sigma > 0$ in D. By the definition (4.24) of ε_c this is equivalent to the assumption that the imaginary part of ε_c is strictly positive on D. Let $E \in H_0(\text{curl}, D)$ be a solution of the homogeneous equation. Inserting $v = E$ yields

$$\int_D \left[\frac{1}{\mu} \left| \text{curl}\, E \right|^2 - \omega^2 \varepsilon_c \left| E \right|^2 \right] dx = 0,$$

and thus, taking the imaginary part,

$$\int_D \text{Im}\, \varepsilon_c \, |E|^2 \, dx = 0$$

from which $E = 0$ follows because Im $\varepsilon_c > 0$ on D. Again, if only $\sigma > 0$ a.e. on some open subset U of D then, by the same argument, E vanishes on U.

Our next goal is to prove the unique continuation property of the scalar equation (4.14) and the Maxwell system (4.21a), (4.21b). As a preparation we need an *interior regularity result*. We can prove it by the familiar technique of transferring the problem into the Sobolev spaces of periodic functions.

For $k \in \mathbb{N}$ and an open domain $D \subseteq \mathbb{R}^3$ the Sobolev space $H^k(D)$ is defined as before by requiring that all partial derivatives up to order k exist in the variational sense and are L^2-functions (compare with Definition 4.1). The space $C^\infty(\overline{D})$ of smooth functions is dense in $H^k(D)$. Again, $H_0^k(D)$ is the closure of $C_0^\infty(D)$ with respect to the norm in $H^k(D)$.

Let $Q = (-R, R)^3 \subseteq \mathbb{R}^3$ be a cube containing $\overline{D}$ in its interior. Then we define $H_{\text{per}}^k(Q)$ by

$$H_{\text{per}}^k(Q) = \left\{ u \in L^2(Q) : \sum_{n \in \mathbb{Z}^3} (1 + |n|^2)^k \, |u_n|^2 < \infty \right\}$$

with inner product

$$(u, v)_{H_{\text{per}}^k(Q)} = (2R)^3 \sum_{n \in \mathbb{Z}^3} (1 + |n|^2)^k \, u_n \overline{v_n},$$

compare with Definition 4.12. Here, $u_n, v_n \in \mathbb{C}$ are the Fourier coefficients of u and v, respectively, see (4.9). Then it is easy to show:

Lemma 4.37. *The following inclusions hold and are bounded (for any $k \in \mathbb{N}$):*

$$H_0^k(Q) \;\hookrightarrow\; H_{\mathrm{per}}^k(Q) \;\hookrightarrow\; H^k(Q).$$

Proof: First inclusion: We show this by induction with respect to k. For $k = 0$ there is nothing to show. Assume that the assertion is true for $k \geq 0$. Then $H_0^k(Q) \subseteq H_{\mathrm{per}}^k(Q)$ and there exists $c > 0$ such that

$$(2R)^3 \sum_{n \in \mathbb{Z}^3} (1 + |n|^2)^k \, |u_n|^2 \;=\; \|u\|_{H_{\mathrm{per}}^k(Q)}^2 \;\leq\; c \|u\|_{H^k(Q)}^2 \quad \text{for all } u \in H_0^k(Q).$$

$$(4.36)$$

Let $u \in C_0^\infty(Q)$. Then, by partial differentiation (if $n_j \neq 0$),

$$
\begin{aligned}
u_n &= \frac{1}{(2R)^3} \int_Q u(x) \, e^{-i\frac{\pi}{R} n \cdot x} dx \;=\; \frac{1}{(2R)^3} \frac{i\,R}{\pi\,n_j} \int_Q u(x) \frac{\partial}{\partial x_j} e^{-i\frac{\pi}{R} n \cdot x} dx \\
&= -\frac{1}{(2R)^3} \frac{i\,R}{\pi\,n_j} \int_Q \frac{\partial u}{\partial x_j}(x) \, e^{-i\frac{\pi}{R} n \cdot x} dx \;=\; -\frac{i\,R}{\pi\,n_j} d_n^{(j)},
\end{aligned}
$$

where $d_n^{(j)}$ are the Fourier coefficients of $\partial u/\partial x_j$. Note that the boundary term vanishes. By assumption of induction we conclude that (4.36) holds for u and for $\partial u/\partial x_j$. Therefore,

$$
\begin{aligned}
\|u\|_{H_{\mathrm{per}}^{k+1}(Q)}^2 &= (2R)^3 \sum_{n \in \mathbb{Z}^3} (1 + |n|^2)^{k+1} \, |u_n|^2 \;=\; (2R)^3 \sum_{n \in \mathbb{Z}^3} (1 + |n|^2)^k \, |u_n|^2 \\
&\quad + \frac{R^2}{\pi^2} (2R)^3 \sum_{j=1}^{3} \sum_{n \in \mathbb{Z}^3} (1 + |n|^2)^k \, |d_n^{(j)}|^2 \\
&\leq c \|u\|_{H^k(Q)}^2 \;+\; c \frac{R^2}{\pi^2} \sum_{j=1}^{3} \|\partial u/\partial x_j\|_{H^k(Q)}^2 \;\leq\; c' \|u\|_{H^{k+1}(Q)}^2.
\end{aligned}
$$

This proves boundedness of the embedding with respect to the norm of order $k + 1$. This ends the proof of the first inclusion because $C_0^\infty(Q)$ is dense in $H_0^{k+1}(Q)$.

For the second inclusion we truncate the Fourier series of $u \in H_{\mathrm{per}}^k(Q)$ into

$$u^N(x) \;=\; \sum_{|n| \leq N} u_n \, e^{i\frac{\pi}{R} n \cdot x}$$

and compute directly

$$\frac{\partial^{j_1+j_2+j_3} u^N}{\partial x_1^{j_1} \partial x_2^{j_2} \partial x_3^{j_3}}(x) = \left(i\frac{\pi}{R}\right)^{j_1+j_2+j_3} \sum_{|n|\leq N} u_n \, n_1^{j_1} n_2^{j_2} n_3^{j_3} \, e^{i\frac{\pi}{R} n \cdot x}$$

and thus for $j \in \mathbb{N}^3$ with $j_1 + j_2 + j_3 \leq k$:

$$\left\| \frac{\partial^{j_1+j_2+j_3} u^N}{\partial x_1^{j_1} \partial x_2^{j_2} \partial x_3^{j_3}} \right\|_{L^2(Q)}^2 \leq (2R)^3 \left(\frac{\pi}{R}\right)^{2k} \sum_{|n|\leq N} |u_n|^2 |n_1|^{2j_1} |n_2|^{2j_2} |n_3|^{2j_3}$$

$$\leq (2R)^3 \left(\frac{\pi}{R}\right)^{2k} \sum_{|n|\leq N} |u_n|^2 |n|^{2(j_1+j_2+j_3)}$$

$$\leq \left(\frac{\pi}{R}\right)^{2k} \|u^N\|_{H^k_{per}(Q)}^2 \, .$$

This proves the lemma by letting N tend to infinity.   □

We continue with a regularity result.

Theorem 4.38. *(Interior Regularity Property)*
Let $f \in L^2(D)$ and U be an open set with $\overline{U} \subseteq D$.

(a) Let $u \in H^1(D)$ be a solution of the variational equation

$$\int_D \nabla u \cdot \nabla \psi \, dx = \int_D f \psi \, dx \quad \text{for all } \psi \in C_0^\infty(D) \, . \qquad (4.37)$$

Then $u|_U \in H^2(U)$ and $\Delta u = -f$ in U.
(b) Let $u \in L^2(D)$ be a solution of the variational equation

$$\int_D u \, \Delta \psi \, dx = -\int_D f \psi \, dx \quad \text{for all } \psi \in C_0^\infty(D) \, . \qquad (4.38)$$

Then $u|_U \in H^2(U)$ and $\Delta u = -f$ in U.

Proof: For both parts we restrict the problem to a periodic problem in a cube by using a partition of unity. Indeed, let $\rho > 0$ such that $\rho < \text{dist}(U, \partial D)$. Then the open balls $B_3(x, \rho)$ are in D for every $x \in U$. Furthermore, $\overline{U} \subseteq \bigcup_{x \in U} B_3(x, \rho)$. Because $\overline{U}$ is compact there exist finitely many open balls $B_3(x^j, \rho) \subseteq D$ for $x^j \in U$, $j = 1, \ldots, m$, with $\overline{U} \subseteq \bigcup_{j=1}^m B_3(x^j, \rho)$. For abbreviation we set $B_j = B_3(x^j, \rho)$. We choose a partition of unity; that is, $\varphi_j \in C_0^\infty(B_j)$ with $\varphi_j \geq 0$ and $\sum_{j=1}^m \varphi_j(x) = 1$ for all $x \in \overline{U}$. Let now $u_j(x) = \varphi_j(x)u(x)$, $x \in D$. Then $\sum_{j=1}^m u_j = u$ on U and $u_j \in H_0^1(D)$ has support in B_j.

Proof of (a): For $\psi \in C_0^\infty(D)$ and any $j \in \{1, \ldots, m\}$ we have that

$$
\int_D \nabla u_j \cdot \nabla \psi \, dx = \int_D \left[\varphi_j \nabla u \cdot \nabla \psi + u \nabla \varphi_j \cdot \nabla \psi \right] dx
$$

$$
= \int_D \left[\nabla u \cdot \nabla (\varphi_j \psi) - \psi \nabla u \cdot \nabla \varphi_j - \psi u \, \Delta \varphi_j - \psi \nabla u \cdot \nabla \varphi_j \right] dx
$$

$$
= \int_D \left[\varphi_j \, f - 2 \nabla u \cdot \nabla \varphi_j - u \, \Delta \varphi_j \right] \psi \, dx
$$

$$
= \int_D g_j \, \psi \, dx
$$

with $g_j = \varphi_j \, f - 2 \nabla u \cdot \nabla \varphi_j - u \, \Delta \varphi_j \in L^2(D)$. Because the support of φ_j is contained in B_j this equation restricts to

$$
\int_{B_j} \nabla u_j \cdot \nabla \psi \, dx = \int_{B_j} g_j \, \psi \, dx \quad \text{for all } \psi \in H_0^1(D). \tag{4.39}
$$

Let now $R > 0$ such that $\overline{D} \subseteq Q = (-R, R)^3$. We fix $j \in \{1, \ldots, m\}$ and $\ell \in \mathbb{Z}$ and set $\psi_\ell(x) = \exp(-i(\pi/R)\, \ell \cdot x)$. Because $\overline{B_j} \subseteq D$ we can find a function $\tilde{\psi}_\ell \in C_0^\infty(D)$ with $\tilde{\psi}_\ell = \psi_\ell$ on B_j. Substituting $\tilde{\psi}_\ell$ into (4.39) yields

$$
\int_{B_j} \nabla u_j \cdot \nabla \psi_\ell \, dx = \int_{B_j} g_j \, \psi_\ell \, dx. \tag{4.40}
$$

We can extend the integrals to Q because u_j and g_j vanish outside of B_j. Now we expand u_j and g_j into Fourier series of the form

$$
u_j(x) = \sum_{n \in \mathbb{Z}} a_n \, e^{i \frac{\pi}{R} n \cdot x} \quad \text{and} \quad g_j(x) = \sum_{n \in \mathbb{Z}} b_n \, e^{i \frac{\pi}{R} n \cdot x}
$$

and substitute this into Eq. (4.40). This yields

$$
\left(\frac{\pi}{R} \right)^2 \sum_{n \in \mathbb{Z}} a_n \, n \cdot \ell \int_Q e^{i(n-\ell) \cdot x} \, dx = \sum_{n \in \mathbb{Z}} b_n \int_Q e^{i(n-\ell) \cdot x} \, dx.
$$

From the orthogonality of the functions $\exp(i(\pi/R)\, n \cdot x)$ we conclude that $(\pi/R)^2 a_\ell \, |\ell|^2 = b_\ell$. Because $\sum_{\ell \in \mathbb{Z}} |b_\ell|^2 < \infty$ we conclude that $\sum_{\ell \in \mathbb{Z}} |\ell|^4 |a_\ell|^2 < \infty$; that is, $u_j \in H^2_{\text{per}}(Q) \subseteq H^2(Q)$. Therefore, also $\sum_{j=1}^m u_j \in H^2(Q)$ and thus $u|_U = \sum_{j=1}^m u_j|_U \in H^2(U)$. This proves part (a).

Proof of (b): We proceed very similarly and show that $u \in H^1(D)$. Then part (a) applies and yields the assertion. For $\psi \in C_0^\infty(D)$ and any $j \in \{1, \ldots, m\}$ we have that

$$\int_D u_j \, \Delta\psi \, dx = \int_D u \, \varphi_j \, \Delta\psi \, dx$$

$$= \int_D u \left[\Delta(\varphi_j \psi) - 2\nabla\varphi_j \cdot \nabla\psi - \psi \, \Delta\varphi_j \right] dx$$

$$= \int_D \left[f\varphi_j - u \, \Delta\varphi_j \right] \psi \, dx \; - \; 2 \int_D u \, \nabla\varphi_j \cdot \nabla\psi \, dx$$

$$= \int_D g_j \, \psi \, dx \; - \; \int_D F_j \cdot \nabla\psi \, dx$$

with $g_j = f\varphi_j - u\,\Delta\varphi_j \in L^2(D)$ and $F_j = 2u\,\nabla\varphi_j \in L^2(D, \mathbb{C}^3)$. As in the proof of (a) we observe that the domain of integration is B_j. Therefore, we can again take $\psi_\ell(x) = \exp(-i(\pi/R)\,\ell \cdot x)$ for ψ and modify it outside of B_j such that it is in $C_0^\infty(D)$. With the Fourier series

$$u_j(x) = \sum_{n \in \mathbb{Z}} a_n \, e^{i\frac{\pi}{R} n \cdot x}, \quad g_j(x) = \sum_{n \in \mathbb{Z}} b_n \, e^{i\frac{\pi}{R} n \cdot x}, \text{ and } F_j(x) = \sum_{n \in \mathbb{Z}} c_n \, e^{i\frac{\pi}{R} n \cdot x}$$

we conclude that

$$-\left(\frac{\pi}{R}\right)^2 a_\ell \, |\ell|^2 \; = \; b_\ell \; + \; i \frac{\pi}{R} \ell \cdot c_\ell \quad \text{for all } \ell \in \mathbb{Z},$$

and thus $|\ell|^2 \, |a_\ell| \leq c\big[|b_\ell| + |\ell| \, |c_\ell|\big]$ which proves that $\sum_{\ell \in \mathbb{Z}} (1 + |\ell|^2) \, |a_\ell|^2 \leq \tilde{c} \sum_{\ell \in \mathbb{Z}} (|b_\ell|^2 + |c_\ell|^2) < \infty$ and thus $u_j \in H^1_{\text{per}}(Q) \subseteq H^1(Q)$. □

Remarks:

(a) If $f \in H^k(D)$ for some $k \in \mathbb{N}$, then $u|_U \in H^{k+2}(U)$ by the same arguments, applied iteratively. Indeed, we have just shown it for $k = 0$. If it is true for $k - 1$ and if $f \in H^k(D)$, then $g_j \in H^k(D)$ (note that $u_{B_j} \in H_0^{k+1}(B_j)$ by assumption of induction!) and thus $u_j \in H^{k+2}(D)$.

(b) This theorem holds without any regularity assumptions on the boundary ∂D. Without further assumptions on ∂D and the boundary data $u|_{\partial D}$ we cannot assure that $u \in H^2(D)$.

The proof of the following fundamental result is taken from [7, Section 8.3]. The proof itself goes back to Müller [22] and Protter [26].

Theorem 4.39. (*Unique Continuation Property*)

Let $D \subseteq \mathbb{R}^3$ be a domain; that is, a nonempty, open and connected set, and $u_1, \ldots, u_m \in H^2(D)$ be real valued such that

$$|\Delta u_j| \; \leq \; c \sum_{\ell=1}^m \{|u_\ell| + |\nabla u_\ell|\} \quad \text{in } D \text{ for } j = 1, \ldots, m. \tag{4.41}$$

If u_j vanish in some open set $U \subseteq D$ for all $j = 1, \ldots, m$, then u_j vanish identically in D for all $j = 1, \ldots, m$.

Proof: Let $x_0 \in U$ and $R \in (0,1)$ such that $B[x_0, R] \subseteq D$. We show that u_j vanishes in $B[x_0, R/2]$. This is sufficient because for every point $\hat{x} \in D$ we can find finitely many balls $B(x_\ell, R_\ell)$ $\ell = 0, \ldots, p$, with $R_\ell \in (0,1)$, such that $B(x_\ell, R_\ell/2) \cap B(x_{\ell-1}, R_{\ell-1}/2) \neq \emptyset$ for all $\ell = 1, \ldots, p$ and $\hat{x} \in B(x_p, R_p/2)$. Then one concludes $u_j(\hat{x}) = 0$ for all j by iteratively applying the first step. We choose the coordinate system such that $x_0 = 0$. First we fix j and write u for u_j. Let $\varphi \in C_0^\infty(B_3(0, R))$ with $\varphi(x) = 1$ for $|x| \leq R/2$ and define $\hat{u}, \hat{v} \in H_0^2(D)$ by $\hat{u}(x) = \varphi(x)u(x)$ and

$$\hat{v}(x) = \begin{cases} \exp(|x|^{-n})\,\hat{u}(x), & x \neq 0, \\ 0, & x = 0, \end{cases}$$

for some $n \in \mathbb{N}$. Note that, indeed, $\hat{v} \in H^2(D)$ because $\hat{u}$ vanishes in the neighborhood U of $x_0 = 0$. Then, with $r = |x|$,

$$\Delta\hat{u}(x) = \exp(-r^{-n})\left\{ \Delta\hat{v}(x) + \frac{2n}{r^{n+1}}\frac{\partial\hat{v}}{\partial r}(x) + \frac{n}{r^{n+2}}\left(\frac{n}{r^n} - n + 1\right)\hat{v}(x) \right\}.$$

Using the inequality $(a+b)^2 \geq 4ab$ and calling the middle term in the above expression b, we see that

$$(\Delta\hat{u}(x))^2 \geq \frac{8n\exp(-2r^{-n})}{r^{n+1}}\frac{\partial\hat{v}}{\partial r}(x)\left\{ \Delta\hat{v}(x) + \frac{n}{r^{n+2}}\left(\frac{n}{r^n} - n + 1\right)\hat{v}(x) \right\}$$

for allmost all x. From now on we drop the argument x. Multiplication with $\exp(2r^{-n})\,r^{n+2}$ and integration yields

$$\int_D \exp(2r^{-n})\,r^{n+2}(\Delta\hat{u})^2 dx \geq 8n\int_D r\frac{\partial\hat{v}}{\partial r}\left\{ \Delta\hat{v} + \frac{n}{r^{n+2}}\left(\frac{n}{r^n} - n + 1\right)\hat{v} \right\} dx.$$

$$(4.42)$$

We show that

$$\int_D r\frac{\partial\hat{v}}{\partial r}(x)\,\Delta\hat{v}(x)\,dx = \frac{1}{2}\int_D |\nabla\hat{v}(x)|^2 dx \quad \text{and} \qquad (4.43a)$$

$$\int_D \frac{1}{r^m}\hat{v}(x)\frac{\partial\hat{v}}{\partial r}(x)\,dx = \frac{m-2}{2}\int_D \frac{\hat{v}(x)^2}{r^{m+1}}\,dx \quad \text{for any integer } m. \quad (4.43b)$$

Indeed, proving the first equation we note that $r\frac{\partial\hat{v}}{\partial r} = x \cdot \nabla\hat{v}$ and $\nabla(x \cdot \nabla\hat{v}) = \sum_{j=1}^{3}\left[e^{(j)}\frac{\partial\hat{v}}{\partial x_j} + x_j\nabla\frac{\partial\hat{v}}{\partial x_j} \right]$ and thus

$$\nabla\left(x\cdot\nabla\hat{v}\right)\cdot\nabla\hat{v} = |\nabla\hat{v}|^2 + \sum_{j=1}^{3} x_j \nabla\frac{\partial\hat{v}}{\partial x_j}\cdot\nabla\hat{v} = |\nabla\hat{v}|^2 + \frac{1}{2}x\cdot\nabla\left(|\nabla\hat{v}|^2\right).$$

Since all boundary terms vanish, by partial integration we obtain

$$\int_D r\frac{\partial\hat{v}}{\partial r}\Delta\hat{v}\,dx = -\int_D \nabla\left(r\frac{\partial\hat{v}}{\partial r}\right)\cdot\nabla\hat{v}\,dx = -\int_D |\nabla\hat{v}|^2 + \frac{1}{2}x\cdot\nabla\left(|\nabla\hat{v}|^2\right)\,dx$$

$$= -\int_D |\nabla\hat{v}|^2\,dx + \frac{1}{2}\int_D \underbrace{\mathrm{div}\,x}_{=\,3}\,|\nabla\hat{v}|^2\,dx = \frac{1}{2}\int_D |\nabla\hat{v}|^2\,dx.$$

This proves Eq. (4.43a). For Eq. (4.43b) we have, using polar coordinates and partial integration,

$$\int_D \frac{1}{r^m}\hat{v}\frac{\partial\hat{v}}{\partial r}\,dx = \int_{B(0,R)} \frac{1}{r^m}\hat{v}\frac{\partial\hat{v}}{\partial r}\,dx = \int_0^R\int_{S^2} \frac{\hat{v}}{r^m}\frac{\partial\hat{v}}{\partial r}r^2\,ds\,dr$$

$$= -\int_0^R\int_{S^2} \hat{v}\frac{\partial}{\partial r}\left(\frac{1}{r^{m-2}}\hat{v}\right)ds\,dr = -\int_D \hat{v}\frac{\partial}{\partial r}\left(\frac{1}{r^{m-2}}\hat{v}\right)\frac{1}{r^2}\,dx$$

$$= -\int_D \frac{1}{r^m}\hat{v}\frac{\partial\hat{v}}{\partial r}\,dx + (m-2)\int_D \frac{\hat{v}^2}{r^{m+1}}\,dx$$

which proves Eq. (4.43b). Substituting (4.43a) and (4.43b) for $m = 2n+1$ and $m = n+1$ into the inequality (4.42) yields

$$\int_D \exp(2r^{-n})\,r^{n+2}(\Delta\hat{u})^2dx \geq 4n\int_D |\nabla\hat{v}|^2\,dx + 4n^3(2n-1)\int_D \frac{\hat{v}^2}{r^{2n+2}}\,dx$$

$$-\,4n^2(n-1)^2\int_D \frac{\hat{v}^2}{r^{n+2}}\,dx$$

$$\geq 4n\int_D |\nabla\hat{v}|^2\,dx + 4n^2(n^2+n-1)\int_D \frac{\hat{v}^2}{r^{2n+2}}\,dx$$

where in the last step we used $r \leq R \leq 1$. Now we replace the right-hand side by $\hat{u}$ again. With $\hat{v}(x) = \exp(r^{-n})\hat{u}(x)$ we have

$$\nabla\hat{v}(x) = \exp(r^{-n})\left[\nabla\hat{u}(x) - n\,r^{-n-1}\frac{x}{r}\,\hat{u}(x)\right]$$

and thus, using $|a - b|^2 \geq \frac{1}{2}|a|^2 - |b|^2$ for any vectors $a, b \in \mathbb{R}^3$,

$$|\nabla \hat{v}(x)|^2 \geq \exp(2r^{-n}) \left[\frac{1}{2} |\nabla \hat{u}(x)|^2 - n^2 r^{-2n-2} |\hat{u}(x)|^2 \right].$$

Substituting this into the estimate above yields

$$\int_D \exp(2r^{-n}) r^{n+2} (\Delta \hat{u})^2 dx \qquad\qquad (4.44)$$

$$\geq 2n \int_D \exp(2r^{-n}) |\nabla \hat{u}|^2 \, dx$$

$$+ \left[4n^2(n^2 + n - 1) - 4n^3 \right] \int_D \exp(2r^{-n}) \frac{\hat{u}^2}{r^{2n+2}} \, dx$$

$$= 2n \int_D \exp(2r^{-n}) |\nabla \hat{u}|^2 \, dx \; + \; 4n^2(n^2 - 1) \int_D \exp(2r^{-n}) \frac{\hat{u}^2}{r^{2n+2}} \, dx$$

$$\geq 2n \int_D \exp(2r^{-n}) |\nabla \hat{u}|^2 \, dx \; + \; 2n^4 \int_D \exp(2r^{-n}) \frac{\hat{u}^2}{r^{n+2}} \, dx \qquad (4.45)$$

for $n \geq 2$. Up to now we have not used the estimate (4.41). We write now u_j and $\hat{u}_j$ for u and $\hat{u}$, respectively. From this estimate, the inequality of Cauchy–Schwarz, and the estimate $(a+b)^2 \leq 2a^2 + 2b^2$ we have the following estimate

$$\left| \Delta \hat{u}_j(x) \right|^2 = \left| \Delta u_j(x) \right|^2 \leq 2m\,c^2 \sum_{\ell=1}^m \{ |u_\ell(x)|^2 + |\nabla u_\ell(x)|^2 \} \quad \text{for } |x| < \frac{R}{2}.$$

Therefore, from (4.45) for u_j we conclude that

$$2n \int_{|x| \leq R/2} \exp(2r^{-n}) |\nabla u_j|^2 \, dx \; + \; 2n^4 \int_{|x| \leq R/2} \exp(2r^{-n}) \frac{u_j^2}{r^{2n+2}} \, dx$$

$$\leq \int_D \exp(2r^{-n}) r^{n+2} (\Delta \hat{u}_j)^2 dx$$

$$\leq 2mc^2 \sum_{\ell=1}^m \int_{|x| \leq R/2} \exp(2r^{-n}) r^{n+2} \left[|u_\ell|^2 + |\nabla u_\ell|^2 \right] dx$$

$$+ \int_{R/2 \leq |x| \leq R} \exp(2r^{-n}) r^{n+2} (\Delta \hat{u}_j)^2 \, dx.$$

We set $\psi_n(r) = \frac{\exp(2r^{-n})}{r^{2n+2}}$ for $r > 0$ and note that ψ_n is monotonously decreasing. Also, because $r \leq 1$ (and thus $r^{n+2} \leq r^{-2n-2}$),

$$2n \int_{|x| \leq R/2} \exp(2r^{-n}) |\nabla u_j|^2 \, dx + 2n^4 \int_{|x| \leq R/2} \psi_n(r) \, u_j^2 \, dx$$

$$\leq 2mc^2 \sum_{\ell=1}^{m} \left\{ \int_{|x| \leq R/2} \psi_n(r) \, u_\ell^2 \, dx + \int_{|x| \leq R/2} \exp(2r^{-n}) |\nabla u_\ell|^2 \, dx \right\}$$

$$+ \psi_n(R/2) \|\Delta \hat{u}_j\|_{L^2(D)}^2 \,.$$

Now we sum with respect to $j = 1, \ldots, m$ and combine the matching terms. This yields

$$2(n - m^2 c^2) \sum_{j=1}^{m} \int_{|x| \leq R/2} \exp(2r^{-n}) |\nabla u_j|^2 \, dx + 2(n^4 - m^2 c^2) \sum_{j=1}^{m} \int_{|x| \leq R/2} \psi_n(r) \, u_j^2 \, dx$$

$$\leq \psi_n(R/2) \sum_{j=1}^{m} \|\Delta \hat{u}_j\|_{L^2(D)}^2 \,.$$

For $n \geq m^2 c^2$ we conclude that

$$2(n^4 - m^2 c^2) \psi_n(R/2) \sum_{j=1}^{m} \int_{|x| \leq R/2} u_j^2 \, dx \leq 2(n^4 - m^2 c^2) \sum_{j=1}^{m} \int_{|x| \leq R/2} \psi_n(r) \, u_j^2 \, dx$$

$$\leq \psi_n(R/2) \sum_{j=1}^{m} \|\Delta \hat{u}_j\|_{L^2(D)}^2$$

and thus

$$\sum_{j=1}^{m} \int_{|x| \leq R/2} u_j^2 \, dx \leq \frac{1}{2(n^4 - m^2 c^2)} \sum_{j=1}^{m} \|\Delta \hat{u}_j\|_{L^2(D)}^2$$

The right-hand side tends to zero as n tends to infinity. This proves $u_j = 0$ in $B(0, R/2)$. $\square$

Now we apply this result to the scalar boundary value problem (4.14) and the boundary value problem (4.21a), (4.21b) for the Maxwell system. We begin with the scalar problem.

Theorem 4.40. *Let $D \subseteq \mathbb{R}^3$ be a bounded domain and $a \in C^1(\overline{D})$ be real valued and $a \geq a_0$ on D for some $a_0 > 0$. Furthermore, let $b \in L^\infty(D)$ be complex valued with $\mathrm{Im}\, b \geq 0$ on D and $\mathrm{Im}\, b > 0$ a.e. on some open subset U of D. Then the boundary value problem (4.14) has a unique solution $u \in H_0^1(D)$ for every $f \in L^2(D)$.*

Proof: By Theorem 4.27 it suffices to prove uniqueness. Let $u \in H_0^1(D)$ be a solution for $f = 0$. We have seen above that u vanishes on U, and it remains to show that u vanishes everywhere on D. In view of Theorem 4.39 we have first to show that $u \in H^2(V)$ for every domain V with $\overline{V} \subset D$.

Let $\varphi \in C_0^\infty(D)$ and define $\psi = \frac{1}{a}\varphi$ on D. Then $\psi \in C_0^1(D)$ and $\nabla\varphi = a\nabla\psi + \psi\nabla a$. Substituting $a\nabla\psi$ and ψ into the variational equation yields

$$\int_D \left[\nabla u \cdot \nabla\varphi - \frac{1}{a}\varphi\nabla a \cdot \nabla u - k^2 \frac{b}{a} u\varphi \right] dx = 0$$

that is, $\int_D \nabla u \cdot \nabla\varphi\, dx = \int_D f\,\varphi\, dx$ for all $\varphi \in C_0^\infty(D)$ where $f = \frac{1}{a}\nabla a \cdot \nabla u + k^2 \frac{b}{a} u$. Then $f \in L^2(D)$, and by Theorem 4.38 we conclude that $u \in H^2(V)$ for any open set with $\overline{V} \subset D$. Furthermore, from the differential equation we have the estimate

$$|\Delta u(x)| \leq \frac{1}{a_0} \|\nabla a\|_\infty |\nabla u(x)| + k^2 \frac{\|b\|_\infty}{a_0} |u(x)| \quad \text{on } V.$$

From this we conclude an estimate of the type (4.41) for the real and imaginary parts of u; that is, $u_1 = \operatorname{Re} u$ and $u_2 = \operatorname{Im} u$. Application of Theorem 4.39 to any open domain V with $U \subset V$ and $\overline{V} \subset D$ yields that u vanishes in such domains V. This implies that u vanishes in D and ends the proof. $\square$

We now turn to the Maxwell case and transform Maxwell's equations to the vector Helmholtz equation. We need weaker smoothness conditions on ε_c if we work with the magnetic field—provided μ is constant which is the case for many materials. Thus, let us consider this case.

Theorem 4.41. *Let $\mu > 0$ constant, $\varepsilon_c \in C^1(\overline{D})$ with $\operatorname{Im} \varepsilon_c > 0$ on some open set $U \subset D$. Then there exists a unique solution $E \in H_0(\operatorname{curl}, D)$ of the boundary value problem (4.25) for every $F \in L^2(D, \mathbb{C}^3)$.*

Proof: Again, it suffices to prove uniqueness. Let $F = 0$ and $E \in H_0(\operatorname{curl}, D)$ be the corresponding solution of (4.25). We have shown at the beginning of this subsection that E vanishes on U. Therefore, also $H = \frac{1}{i\omega\mu}\operatorname{curl} E = 0$ in U. By Lemma 4.29, the magnetic field H satisfies

$$\int_D \left[\frac{1}{\varepsilon_c} \operatorname{curl} H \cdot \operatorname{curl}\psi - \omega^2\mu\, H \cdot \psi \right] dx = 0 \quad \text{for all } \psi \in H(\operatorname{curl}, D). \quad (4.46)$$

If we choose $\psi = \nabla\phi$ for some $\phi \in H_0^1(D)$, then we have

$$\int_D H \cdot \nabla\phi\, dx = 0 \quad \text{for all } \phi \in H_0^1(D),$$

which is the variational form of $\operatorname{div} H = 0$.

If all of the functions were sufficiently smooth, we just would rewrite the equation for H in the form

$$0 \; = \; \varepsilon_c \, \mathrm{curl}\big[\varepsilon_c^{-1} \, \mathrm{curl}\, H\big] - \omega^2 \varepsilon_c \mu \, H \; = \; \mathrm{curl}^2 \, H + \varepsilon_c \, \nabla \frac{1}{\varepsilon_c} \times \mathrm{curl}\, H - \omega^2 \varepsilon_c \mu \, H \,,$$

and thus, because $\mathrm{div}\, H = 0$ and $\mathrm{curl}^2 = \nabla\, \mathrm{div} - \Delta$,

$$\Delta H \; = \; \varepsilon_c \, \nabla \frac{1}{\varepsilon_c} \times \mathrm{curl}\, H - \omega^2 \varepsilon_c \mu \, H \; = \; \frac{1}{\varepsilon_c} \, \mathrm{curl}\, H \times \nabla \varepsilon_c - \omega^2 \varepsilon_c \mu \, H \,.$$

We derive this formula also by the variational equation. Indeed, we set $\psi = \varepsilon_c \tilde{\psi}$ for some $\tilde{\psi} \in C_0^\infty(D, \mathbb{C}^3)$. Then $\tilde{\psi} \in H_0(\mathrm{curl}, D)$ and therefore, because $\mathrm{curl}\, \psi = \varepsilon_c \, \mathrm{curl}\, \tilde{\psi} + \nabla \varepsilon_c \times \tilde{\psi}$,

$$\int_D \left[\mathrm{curl}\, H \cdot \mathrm{curl}\, \tilde{\psi} + \frac{1}{\varepsilon_c} \, \mathrm{curl}\, H \cdot (\nabla \varepsilon_c \times \tilde{\psi}) - \omega^2 \mu \varepsilon_c H \cdot \tilde{\psi} \right] dx \; = \; 0$$

for all $\tilde{\psi} \in C_0^\infty(D, \mathbb{C}^3)$, which we write as

$$\int_D \mathrm{curl}\, H \cdot \mathrm{curl}\, \tilde{\psi} \, dx \; = \; \int_D G \cdot \tilde{\psi} \, dx \quad \text{for all } \tilde{\psi} \in C_0^\infty(D, \mathbb{C}^3),$$

where $G = -\frac{1}{\varepsilon_c} \, \mathrm{curl}\, H \times \nabla \varepsilon_c + \omega^2 \mu \varepsilon_c H \in L^2(D, \mathbb{C}^3)$. Partial integration yields

$$\int_D G \cdot \tilde{\psi} \, dx = \int_D H \cdot \mathrm{curl}^2 \, \tilde{\psi} \, dx = - \int_D H \cdot \Delta \tilde{\psi} \, dx + \int_D H \cdot \nabla \, \mathrm{div}\, \tilde{\psi} \, dx$$

$$= - \int_D H \cdot \Delta \tilde{\psi} \, dx \,.$$

Here we used the fact that $\int_D H \cdot \nabla \, \mathrm{div}\, \tilde{\psi} \, dx = 0$ because $\mathrm{div}\, \tilde{\psi} \in H_0^1(D)$. This holds for all $\tilde{\psi} \in C_0^\infty(D, \mathbb{C}^3)$. By the interior regularity result of Theorem 4.38 we conclude that $H \in H^2(U, \mathbb{C}^3)$ for all domains U with $\overline{U} \subseteq D$, and $\Delta H = -G = \frac{1}{\varepsilon_c} \, \mathrm{curl}\, H \times \nabla \varepsilon_c - \omega^2 \mu \varepsilon_c H$ in U. Because every component of $\mathrm{curl}\, H$ is a combination of partial derivatives of H_ℓ for $\ell \in \{1, 2, 3\}$ we conclude the existence of a constant $c > 0$ such that

$$|\Delta H_j| \; \leq \; c \sum_{\ell=1}^{3} \big[|\nabla H_\ell| + |H_\ell| \big] \quad \text{in } D \text{ for } j = 1, 2, 3 \,.$$

Therefore, all of the assumptions of Theorem 4.39 are satisfied and thus $H = 0$ in all of U. This implies that also $E = 0$ in U. Because U is an arbitrary domain with $\overline{U} \subseteq D$ we conclude that $E = 0$ in D. $\quad\square$

Remarks:

(a) The proof of the theorem can be modified for $\mu \in C^2(\overline{D})$. Instead of div $H = 0$ we have that $0 = \operatorname{div}(\mu H) = \nabla \mu \cdot H + \mu \operatorname{div} H$, thus

$$\operatorname{curl}^2 H = -\Delta H + \nabla \operatorname{div} H = -\Delta H - \nabla \left(\frac{1}{\mu} \nabla \mu \cdot H \right)$$

$$= -\Delta H - \nabla \left[\nabla (\ln \mu) \cdot H \right]$$

$$= -\Delta H - \sum_{j=1}^{3} \left(\nabla \frac{\partial \ln \mu}{\partial x_j} H_j + \frac{\partial \ln \mu}{\partial x_j} \nabla H_j \right)$$

and this can be treated in the same way. Here we argue classically, but all of the arguments hold also in the weak case.

(b) The assumption $\varepsilon_c \in C^1(\overline{D})$ is very restrictive. One can weaken this assumption to the requirement that ε_c is piecewise continuously differentiable. We refer to [21, Section 4.6].

4.3 The Time-Dependent Cavity Problem

The spectral theorem of the previous section allows it to treat the full time-dependent system of Maxwell's equations. We begin again with the initial-boundary value problem for the scalar wave equation in some bounded Lipschitz domain $D \subseteq \mathbb{R}^3$ and some interval $(0, T)$.

$$\frac{1}{c(x)^2} \frac{\partial^2 u}{\partial t^2}(t, x) - \operatorname{div}\left(a(x) \nabla u(t, x)\right) = f(t, x), \quad (t, x) \in (0, T) \times D,$$
$$\text{(4.47a)}$$

$$u(t, x) = 0 \quad \text{for } (t, x) \in (0, T) \times \partial D, \qquad \text{(4.47b)}$$

$$u(0, x) = u_0(x) \quad \text{and} \quad \frac{\partial u}{\partial t}(0, x) = u_1(x) \quad \text{for } x \in D. \qquad \text{(4.47c)}$$

We make the following assumptions on the data:

Assumptions:

- $a, c \in L^\infty(D)$ with $a(x) \geq a_0$ and $c(x) \geq c_0$ on D for some $a_0 > 0$ and $c_0 > 0$,

- $f \in L^2\left((0, T) \times D\right)$,

- $u_0 \in H_0^1(D)$, $u_1 \in L^2(D)$.

In this section we assume that all functions are real-valued. We set $b(x) = 1/c(x)^2$ for abbreviation. Then $b \in L^\infty(D)$ and $b(x) \geq b_0 = 1/\|c\|_\infty^2$ on D.

The solution has to be understood in a variational sense. To motivate this we multiply the differential equation (4.47a) by some $\psi \in C^1([0,T] \times \overline{D})$ with $\psi(0,x) = \psi(T,x) = 0$ for all $x \in \overline{D}$ and integrate by parts with respect to t and use Green's first formula with respect to x. This yields

$$\int\limits_0^T \int\limits_D \left[b(x) \frac{\partial u}{\partial t}(t,x) \frac{\partial \psi}{\partial t}(t,x) - a(x) \nabla u(t,x) \cdot \nabla \psi(t,x) \right] dx\, dt$$

$$= -\int\limits_0^T \int\limits_D f(t,x)\, \psi(t,x)\, dx\, dt\,,$$

or, using the notation $u_t = \partial u/\partial t$ and the inner product in $L^2(D)$,

$$\int\limits_0^T \left[\big(b\, u_t(t,\cdot), \psi_t(t,\cdot)\big)_{L^2(D)} - \big(a\nabla u(t,\cdot), \nabla\psi(t,\cdot)\big)_{L^2(D)} \right] dt$$

$$= -\int\limits_0^T \big(f(t,\cdot), \psi(t,\cdot)\big)_{L^2(D)}\, dt\,.$$

We will require different smoothness properties of u with respect to t and x. This leads to the so-called *anisotropic* function spaces. In particular, the solution has to be differentiable with respect to t (in a sense to be explained in a moment). It is convenient to consider u to be a function in t with values $u(t)$ in some function space with respect to x. We have taken this point of view already when we wrote $u(t,\cdot)$. To make this idea precise, we recall the notion of a Frechét-differentiable function for this case.

Definition 4.42. Let V be a normed space (over $\mathbb{R}$) and $f : [0,T] \to V$ a function with values in V.

(a) The function f is continuous in some $t_0 \in [0,T]$ if $\lim\limits_{t \to t_0} \|f(t) - f(t_0)\|_V = 0$. The space of continuous functions on $[0,T]$ is denoted by $C[0,T;V]$.

(b) The function f is differentiable in $t_0 \in [0,T]$ with value $f'(t_0) \in V$ if $\lim\limits_{t \to t_0} \left\| \frac{f(t)-f(t_0)}{t-t_0} - f'(t_0) \right\|_V = 0$. The space of continuously differentiable functions is denoted by $C^1[0,T;V]$.

Remark: If V is a Hilbert space with inner product $(\cdot,\cdot)_V$, then the following product rule holds. For $f,g \in C^1[0,T;V]$ the scalar function $h(t) = \big(f(t), g(t)\big)_V$, $t \in [0,T]$, is differentiable and

$$h'(t) = \left(f'(t), g(t)\right)_V + \left(f(t), g'(t)\right)_V, \quad t \in [0, T].$$

The proof uses the same arguments as in the case where $V = \mathbb{R}^n$.

We define the solution space X of the initial-boundary value problem to be

$$X = C[0, T; H_0^1(D)] \cap C^1[0, T; L^2(D)] \tag{4.48a}$$

and equip X with the weighted norm

$$\|u\|_X = \max_{0 \le t \le T} \|\sqrt{a}\, \nabla u(t)\|_{L^2(D)} + \max_{0 \le t \le T} \|\sqrt{b}\, u'(t)\|_{L^2(D)}. \tag{4.48b}$$

Lemma 4.43. *X is a Banach space. The norm is equivalent to*

$$u \mapsto \max_{0 \le t \le T} \|u(t)\|_{H^1(D)} + \max_{0 \le t \le T} \|u'(t)\|_{L^2(D)}.$$

Proof: The equivalence is clear because $\|\sqrt{a}\, \nabla u\|_{L^2(D)}$ is equivalent to the ordinary norm in $H_0^1(D)$ by Lemma 4.26 and $\|\sqrt{b} \cdot \|_{L^2(D)}$ is equivalent to the ordinary norm in $L^2(D)$ by the boundedness of b and $1/b$. All properties of a normed space are very easy to see. Only the proof of completeness is a little more delicate, but follows the same arguments as in the proofs of the completeness of $C[0, T]$ and $C^1[0, T]$. $\square$

Definition 4.44. *u is a (weak) solution of the initial-boundary value problem (4.47a)–(4.47c) if $u \in X$ such that $u(0) = u_0$, $u'(0) = u_1$ and*

$$\int_0^T \left[\left(b\, u'(t), \psi'(t)\right)_{L^2(D)} - \left(a \nabla u(t), \nabla \psi(t)\right)_{L^2(D)} \right] dt$$

$$= -\int_0^T \left(f(t, \cdot), \psi(t)\right)_{L^2(D)} dt \tag{4.49}$$

for all $\psi \in X$ with $\psi(0) = \psi(T) = 0$.

We note that $t \mapsto f(t, \cdot)$ is in $L^2(D)$ by Fubini's theorem and $\int_0^T \|f(t, \cdot)\|_{L^2(D)}^2 dt = \|f\|_{L^2((0,T) \times D)}^2$. Therefore, the right-hand side of (4.49) is well defined.

The following analysis makes use of the spectral theorem (Theorem 4.28). We recall from this theorem that there exist eigenvalues $k_n \in \mathbb{R}$ and corresponding eigenfunctions $\{v_n \in H_0^1(D) : n \in \mathbb{N}\}$ such that

$$\int_D \left[a \nabla v_n \cdot \nabla \varphi - k_n^2\, b\, v_n\, \varphi \right] dx = 0 \quad \text{for all } \varphi \in H_0^1(D);$$

that is, using again the notion of the inner product $(u, v)_* = \int_D a \nabla u \cdot \nabla v \, dx$ in $H_0^1(D)$,

$$(v_n, \varphi)_* = k_n^2 (b \, v_n, \varphi)_{L^2(D)} \quad \text{for all } \varphi \in H_0^1(D). \tag{4.50}$$

Furthermore, the set $\{v_n \in H_0^1(D) : n \in \mathbb{N}\}$ forms a complete orthonormal system in $(H_0^1(D), (\cdot, \cdot)_*)$, and $\{k_n v_n : n \in \mathbb{N}\}$ forms a complete orthonormal system in $(L^2(D), (b \cdot, \cdot)_{L^2(D)})$, see Theorem 4.28.

First we prove uniqueness.

Theorem 4.45. *There exists at most one solution $u \in X$ of (4.49).*

Proof: Let u be the difference of two solutions. Then u solves the problem for $u_0 = 0$, $u_1 = 0$, and $f = 0$. We define $c_n(t)$ by $c_n(t) = (u(t), v_n)_*$ for $n \in \mathbb{N}$. Then $c_n \in C^1[0, T]$. We choose any $\varphi \in C^1[0, T]$ such that $\varphi(0) = \varphi(T) = 0$ and set $\psi(t) = \varphi(t) v_m$ for an arbitrary $m \in \mathbb{N}$. We compute the inner products in (4.49) by using the orthogonality of v_n with respect to $(\cdot, \cdot)_*$ and of $k_n v_n$ with respect to $(b \cdot, \cdot)_{L^2(D)}$. This yields

$$\big(b \, u'(t), \psi'(t)\big)_{L^2(D)} = \varphi'(t) \big(b \, u'(t), v_m\big)_{L^2(D)} = \frac{1}{k_m^2} c_m'(t) \, \varphi'(t),$$

$$\big(a \nabla u(t), \nabla \psi(t)\big)_{L^2(D)} = \big(u(t), \psi(t)\big)_* = \varphi(t) \big(u(t), v_m\big)_* = c_m(t) \, \varphi(t).$$

Inserting this into (4.49) yields

$$\int_0^T \left[\frac{1}{k_m^2} c_m'(t) \, \varphi'(t) - c_m(t) \, \varphi(t) \right] dt = 0$$

for all such φ and all m. Now we use Lemma 4.46 below which yields that $c_m \in C^2[0, T]$ and $c_m''(t) + k_n^2 c_m(t) = 0$ on $[0, T]$. Using the initial conditions $c_m(0) = 0$ and $c_m'(0) = 0$ yields $c_m(t) = 0$ for all t. This holds for all $m \in \mathbb{N}$. The completeness of the system $\{v_n : n \in \mathbb{N}\}$ implies that $u(t)$ vanishes for all t. $\square$

It remains to prove the following lemma which is sometimes called the *Fundamental Theorem of Calculus of Variations.*

Lemma 4.46. *Let $h \in C^1[0, T]$ and $g \in C[0, T]$ such that*

$$\int_0^T [h'(t) \, \varphi'(t) - g(t) \, \varphi(t)] \, dt = 0 \quad \text{for all } \varphi \in C^1[0, T] \text{ with } \varphi(0) = \varphi(T) = 0.$$

Then $h \in C^2[0, T]$ and $h''(t) + g(t) = 0$ for all $t \in [0, T]$.

Proof: Define $\tilde{g}(t) = \int_0^t g(s) ds$ for $t \in [0, T]$. Then $\tilde{g} \in C^1[0, T]$. We substitute g into the variational equation and use partial integration. This yields

$$\int_0^T \underbrace{[h'(t) + \tilde{g}(t)]}_{=:\, c(t)} \varphi'(t)\, dt \;=\; \int_0^T c(t)\, \varphi'(t)\, dt \;=\; 0$$

for all $\varphi \in C^1[0,T]$ with $\varphi(0) = \varphi(T) = 0$.

Let now $\rho \in C[0,T]$ be arbitrary. Define $\varphi(t) = \int_0^t \rho(s)ds - \frac{t}{T}\int_0^T \rho(s)ds$ for $t \in [0,T]$. Then $\varphi \in C^1[0,T]$ with $\varphi(0) = \varphi(T) = 0$. Therefore,

$$0 = \int_0^T c(t)\, \varphi'(t)\, dt \;=\; \int_0^T c(t)\, \rho(t)\, dt - \frac{1}{T}\int_0^T \rho(s)\, ds \int_0^T c(t)\, dt$$

$$= \int_0^T \rho(t)\left[c(t) - \frac{1}{T}\int_0^T c(s)\, ds\right] dt.$$

Because this holds for all $\rho \in C[0,T]$ we conclude that $c(t) = \frac{1}{T}\int_0^T c(s)\, ds$ for all t. Therefore, c is constant and thus $h' = -\tilde{g} + c \in C^1[0,T]$ and $h''(t) + g(t) = 0$ for all $t \in [0,T]$. $\square$

We can draw a second conclusion from this lemma.

Corollary 4.47. *Let $f \in C[0,T;L^2(D)]$ and $u \in X$ be a solution of (4.49). Then, for all $\psi \in H_0^1(D)$ the scalar function $t \mapsto (b\,u(t), \psi)_{L^2(D)}$ is twice continuously differentiable and*

$$\frac{d^2}{dt^2}(b\,u(t), \psi)_{L^2(D)} \;+\; (a\,\nabla u(t), \nabla\psi)_{L^2(D)} \;=\; (f(t,\cdot), \psi)_{L^2(D)} \qquad (4.51)$$

for all $t \in [0,T]$.

Proof: Let $\psi \in H_0^1(D)$ and $\varphi \in C^1[0,T]$ with $\varphi(0) = \varphi(T) = 0$. We insert $\varphi\psi$ into (4.49) which yields

$$\int_0^T \left[\varphi'(t)\frac{d}{dt}(b\,u(t), \psi)_{L^2(D)} - \varphi(t)\,(a\nabla u(t), \nabla\psi)_{L^2(D)}\right] dt$$

$$= -\int_0^T \varphi(t)\,(f(t,\cdot), \psi)_{L^2(D)}\, dt.$$

Now we apply Lemma 4.46 to $h(t) = (b\,u(t), \psi)_{L^2(D)}$ and $g(t) = (a\nabla u(t), \nabla\psi)_{L^2(D)} - (f(t,\cdot), \psi)_{L^2(D)}$ which yields the assertion because h is differentiable and g continuous. $\square$

The following simple result will be useful for the proofs of existence.

Lemma 4.48. *Let H be a Hilbert space with complete orthonormal system $\{h_n : n \in \mathbb{N}\}$. Let $c_n \in C[0,T]$ and $\gamma_n > 0$ with $|c_n(t)| \leq \gamma_n$ for all $t \in [0,T]$ and $n \in \mathbb{N}$ where $\gamma_n > 0$ with $\sum_{n=1}^{\infty} \gamma_n^2 < \infty$. Define formally*

$$u(t) = \sum_{n=1}^{\infty} c_n(t)\, h_n, \quad t \in [0,T].$$

Then $u \in C[0,T;H]$ and $\|u\|^2_{C[0,T;H]} \leq \sum_{n=1}^{\infty} \gamma_n^2$.

Proof: Let u_N be the truncated series; that is,

$$u_N(t) = \sum_{n=1}^{N} c_n(t)\, h_n, \quad t \in [0,T].$$

Then $u_N \in C[0,T;H]$. We show that (u_N) is a Cauchy sequence in $C[0,T;H]$. Indeed, for $N > M$ we have

$$\|u_N(t) - u_M(t)\|^2_H = \sum_{n=M+1}^{N} c_n(t)^2 \leq \sum_{n=M+1}^{N} \gamma_n^2.$$

Taking the maximum yields

$$\|u_N - u_M\|^2_{C[0,T;H]} \leq \sum_{n=M+1}^{N} \gamma_n^2$$

and this tends to zero as N, M tend to infinity. Therefore, (u_N) is a Cauchy sequence in $C[0,T;H]$ and thus convergent. $\square$

We will apply this result first to the case where $H = H_0^1(D)$ with inner product $(\cdot,\cdot)_*$ and complete orthonormal system $\{v_n : n \in \mathbb{N}\}$ and then to $H = L^2(D)$ with inner product $(b\cdot,\cdot)_{L^2(D)}$ and complete orthonormal system $\{k_n v_n : n \in \mathbb{N}\}$. This yields the following corollary:

Corollary 4.49. *Let $c_n \in C^1[0,T]$ and $\gamma_n > 0$ with $|c_n(t)| \leq \gamma_n$ and $|c_n'(t)| \leq k_n \gamma_n$ for all $t \in [0,T]$ and $n \in \mathbb{N}$ where $\gamma_n > 0$ with $\sum_{n=1}^{\infty} \gamma_n^2 < \infty$. Define formally*

$$u(t) = \sum_{n=1}^{\infty} c_n(t)\, v_n, \quad t \in [0,T].$$

Then $u \in X$ and $\|u\|^2_X \leq 2 \sum_{n=1}^{\infty} \gamma_n^2$.

Proof: Application of the previous lemma to c_n in $\big(H_0^1(D),(\cdot,\cdot)_*\big)$ with respect to the orthonormal system $\{v_n : n \in \mathbb{N}\}$ yields that $u \in C[0,T;H_0^1(D)]$

and $\|u\|^2_{C[0,T;H^1_0(D)]} \leq \sum_{n=1}^{\infty} \gamma_n^2$. Then we apply the lemma to u' with coefficients $\frac{c'_n}{k_n}$ in $\left(L^2(D),(b\cdot,\cdot)_{L^2(D)}\right)$ with respect to the orthonormal system $\{k_n v_n : n \in \mathbb{N}\}$. This proves that $u' \in C[0,T;L^2(D)]$ and $\|u'\|^2_{C[0,T;L^2(D)]} \leq \sum_{n=1}^{\infty} \gamma_n^2$ because also $\frac{|c'_n|}{k_n} \leq \gamma_n$ for all t and n. Adding the results yields the assertion. $\square$

To prove existence we first consider the case of no source; that is, $f = 0$.

Theorem 4.50. *For every $u_0 \in H^1_0(D)$ and $u_1 \in L^2(D)$ there exists a unique solution u of (4.49) for $f = 0$ such that $u(0) = u_0$ and $u'(0) = u_1$. The solution is given by*

$$u(t) = \sum_{n=1}^{\infty} \left[\alpha_n \cos(k_n t) + \beta_n \sin(k_n t)\right] v_n, \quad t \in [0,T],$$

where α_n and β_n are the expansion coefficients of $u_0 \in H^1_0(D)$ and $u_1 \in L^2(D)$ with respect to $\{v_n : n \in \mathbb{N}\}$ and $\{k_n v_n : n \in \mathbb{N}\}$, respectively; that is,

$$u_0 = \sum_{n=1}^{\infty} \alpha_n v_n, \qquad u_1 = \sum_{n=1}^{\infty} \beta_n k_n v_n.$$

Furthermore, the solution operator $(u_0, u_1) \mapsto u$ is bounded from $H^1_0(D) \times L^2(D)$ into X.

Proof: To show that $u \in X$ we apply the previous corollary with $c_n(t) = \alpha_n \cos(k_n t) + \beta_n \sin(k_n t)$. The assumptions are obviously satisfied because $c_n(t)^2 \leq \gamma_n^2 := 2\alpha_n^2 + 2\beta_n^2$ for all $t \in [0,T]$ and $n \in \mathbb{N}$ and, analogously, $|c'_n(t)| \leq k_n \gamma_n$ for all $t \in [0,T]$ and $n \in \mathbb{N}$ and $\sum_{n=1}^{\infty}[\alpha_n^2 + \beta_n^2] = \|u_0\|^2_* + \|u_1\|^2_{L^2(D,b\,dx)}$. Application of Corollary 4.49 yields $u \in X$. Furthermore, $u(0) = \sum_{n=1}^{\infty} \alpha_n v_n = u_0$ and $u'(0) = \sum_{n=1}^{\infty} k_n \beta_n v_n = u_1$. It remains to prove that u satisfies (4.49) for $f = 0$. Let $\psi \in X$ and expand $\psi(t)$ as $\psi(t) = \sum_{n=1}^{\infty} \psi_n(t) v_n$. Using the orthogonality of $\{v_n : n \in \mathbb{N}\}$ with respect to $(\cdot,\cdot)_*$ and of $\{k_n v_n : n \in \mathbb{N}\}$ with respect to $(b\cdot,\cdot)_{L^2(D)}$ we conclude

$$\int_0^T \left[(bu'(t),\psi'(t))_{L^2(D)} - (u(t),\psi(t))_*\right] dt$$

$$= \sum_{n=1}^{\infty} \int_0^T \left[\frac{1}{k_n^2} c'_n(t)\psi'_n(t) - c_n(t)\psi_n(t)\right] dt,$$

where again $c_n(t) = \alpha_n \cos(k_n t) + \beta_n \sin(k_n t)$ are the expansion coefficients of $u(t)$. Let $n \in \mathbb{N}$ be fixed. Using partial integration we conclude

$$\int_0^T \left[\frac{1}{k_n^2} c_n'(t) \psi_n'(t) - c_n(t) \psi_n(t) \right] dt = -\int_0^T \left[\frac{1}{k_n^2} c_n''(t) + c_n(t) \right] \psi_n(t) \, dt = 0$$

by the special form of $c_n(t)$. Therefore, u satisfies (4.49) for $f = 0$. □

Let now $f \in L^2((0,T) \times D)$ be arbitrary. We construct a particular solution of the inhomogeneous variational equation (4.49). We note that $\frac{1}{b} f \in L^2((0,T) \times D)$ and, therefore, $\frac{1}{b} f(t,\cdot)$ can be expanded with respect to the orthonormal system $\{k_n v_n : n \in \mathbb{N}\}$ in $(L^2(D), (b\cdot,\cdot)_{L^2(D)})$; that is,

$$\frac{1}{b(\cdot)} f(t,\cdot) = \sum_{n=1}^\infty f_n(t) \, k_n v_n \quad \text{with} \quad f_n(t) = (f(t,\cdot), k_n v_n)_{L^2(D)}, \quad n \in \mathbb{N}.$$
$$(4.52)$$

Theorem 4.51. *Let $f \in L^2((0,T) \times D)$ with coefficients $f_n(t)$ of (4.52). Then $\hat{u}(t) = \sum_{n=1}^\infty a_n(t) v_n$, $t \in [0,T]$, with coefficients*

$$a_n(t) = -\frac{1}{2} \int_0^T \sin(k_n|t - s|) \, f_n(s) \, ds, \quad t \in [0,T], \ n \in \mathbb{N}, \qquad (4.53)$$

is a particular solution of (4.49). Furthermore, the operator $f \mapsto \hat{u}$ is bounded from $L^2((0,T) \times D)$ into X.

Proof: First we note that $\hat{u} \in X$. Indeed, we observe that $a_n \in C^1[0,T]$ and $|a_n(t)|^2 \leq \frac{1}{4} \left[\int_0^T |f_n(s)| \, ds \right]^2 \leq \frac{T}{4} \int_0^T |f_n(s)|^2 \, ds$ for all $t \in [0,T]$ and $n \in \mathbb{N}$ and analogously $|a_n'(t)|^2 \leq \frac{k_n^2 T}{4} \int_0^T |f_n(s)|^2 \, ds$ for all $t \in [0,T]$ and $n \in \mathbb{N}$. Therefore, $\hat{u} \in X$ by Corollary 4.49 and

$$\|\hat{u}\|_X^2 \leq c \sum_n \int_0^T |f_n(s)|^2 \, ds = c \int_0^T \sum_n |f_n(s)|^2 \, ds$$

$$= c \int_0^T \|\sqrt{b} f(s,\cdot)\|_{L^2(D)}^2 \, ds \leq c' \|f\|_{L^2((0,T) \times D)}^2.$$

Let now $\psi \in X$ with $\psi(0) = \psi(T) = 0$ and expansion $\psi(t) = \sum_{n=1}^\infty \psi_n(t) v_n$. We fix n and assume first that f_n is continuous in $[0,T]$. Then it is easy to check that even $a_n \in C^2[0,T]$ and a_n satisfies the differential equation $a_n''(t) + k_n^2 a_n(t) = -k_n f_n(t)$. Multiplying this equation with $\psi_n \in C^1[0,T]$, integrating and using integration by parts (note that $\psi_n(0) = \psi_n(T) = 0$), yields

$$\int_0^T [a_n'(t) \psi_n'(t) - k_n^2 a_n(t) \, \psi_n(t)] \, dt = k_n \int_0^T f_n(t) \, \psi_n(t) \, dt.$$

By a density argument we conclude that this equation holds also if only $f_n \in L^2(0,T)$. Division by k_n^2 and summing these equations with respect to n yields

$$\sum_{n=1}^{\infty} \int_0^T \left[\frac{1}{k_n^2} a_n'(t)\psi_n'(t) - a_n(t)\psi_n(t) \right] dt = \sum_{n=1}^{\infty} \frac{1}{k_n} \int_0^T f_n(t)\psi_n(t)\, dt,$$

which can again be written as

$$\int_0^T \left[\left(b\hat{u}'(t), \psi'(t) \right)_{L^2(D)} - \left(\hat{u}(t), \psi(t) \right)_* \right] dt = \int_0^T \left(f(t), \psi(t) \right)_{L^2(D)} dt.$$

$\square$

As a corollary we have existence and uniqueness for the general inhomogeneous problem.

Corollary 4.52. *For every $u_0 \in H_0^1(D)$ and $u_1 \in L^2(D)$ and $f \in L^2((0,T) \times D)$ there exists a unique solution $u \in X$ of (4.49) such that $u(0) = u_0$ and $u'(0) = u_1$. The solution is given by the sum $u = \hat{u} + \tilde{u}$ of the particular solution $\hat{u}$ constructed in Theorem 4.51 and the solution $\tilde{u}$ of the homogeneous differential equation with initial values $\tilde{u}(0) = u_0 - \hat{u}(0)$ and $\tilde{u}'(0) = u_1 - \hat{u}'(0)$; that is,*

$$u(t) = \sum_{n=1}^{\infty} \left[(\alpha_n - a_n(0)) \cos(k_n t) + \left(\beta_n - \frac{a_n'(0)}{k_n} \right) \sin(k_n t) + a_n(t) \right] v_n$$

(4.54)

for $t \in [0,T]$ where α_n and β_n are the expansion coefficients of $u_0 \in H_0^1(D)$ and $u_1 \in L^2(D)$ with respect to $\{v_n : n \in \mathbb{N}\}$ and $\{k_n v_n : n \in \mathbb{N}\}$, respectively, and $a_n(t)$ are defined in (4.53). Furthermore, the solution $u \in X$ depends continuously on u_0, u_1, and f; that is, the solution operator $(u_0, u_1, f) \mapsto u$ is bounded from $H_0^1(D) \times L^2(D) \times L^2((0,T) \times D)$ into X.

By the same arguments we can prove the following regularity result.

Theorem 4.53. *Let $f \in C^1[0,T; L^2(D)]$ and $u_0 = 0$ and $u_1 \in H_0^1(D)$ and $u \in X$ be the solution of (4.49). Then $u' \in C(0,T; H_0^1(D))$ and $u'' \in C(0,T; L^2(D))$; that is, $u' \in X$. Furthermore, the mapping $(u_1, f) \mapsto u$ is bounded from $H_0^1(D) \times C^1[0,T; L^2(D)]$ into $\tilde{X} = C^1(0,T; H_0^1(D)) \cap C^2(0,T; L^2(D))$ equipped with the canonical norm analogously to (4.48b).*

The solution satisfies the differential equation pointwise with respect to t; that is

$$\left(b\, u''(t), \psi \right)_{L^2(D)} + \left(a\, \nabla u(t), \nabla \psi \right)_{L^2(D)} = \left(f(t, \cdot), \psi \right)_{L^2(D)}$$

(4.55)

for all $\psi \in H_0^1(D)$ and all $t \in [0,T]$, compare (4.51).

Proof: First we consider the particular solution $\hat{u}$ of Theorem 4.51 and write (4.53) as

$$
a_n(t) = -\frac{1}{2} \left[\int_0^t \sin(k_n(t-s)) \, f_n(s) \, ds \; + \; \int_t^T \sin(k_n(s-t)) \, f_n(s) \, ds \right]
$$

$$
= -\frac{1}{2k_n} \left[\cos(k_n(t-s)) \, f_n(s) \Big|_0^t \; - \; \int_0^t \cos(k_n(t-s)) \, f_n'(s) \, ds \right.
$$

$$
\left. - \; \cos(k_n(s-t)) \, f_n(s) \Big|_t^T \; + \; \int_t^T \cos(k_n(s-t)) \, f_n'(s) \, ds \right]
$$

and

$$
a_n'(t) = -\frac{k_n}{2} \left[\int_0^t \cos(k_n(t-s)) \, f_n(s) \, ds \; - \; \int_t^T \cos(k_n(s-t)) \, f_n(s) \, ds \right]
$$

$$
= -\frac{1}{2} \left[-\sin(k_n(t-s)) \, f_n(s) \Big|_0^t \; + \; \int_0^t \sin(k_n(t-s)) \, f_n'(s) \, ds \right.
$$

$$
\left. - \; \sin(k_n(s-t)) \, f_n(s) \Big|_t^T \; + \; \int_t^T \sin(k_n(s-t)) \, f_n'(s) \, ds \right]
$$

and thus

$$
|a_n(t)| \leq \frac{c_1}{k_n} \left(\|f_n\|_\infty + \|f_n'\|_{L^2(0,T)} \right) \; \leq \; \frac{c_2}{k_n} \left(\|f_n\|_{L^2(0,T)} + \|f_n'\|_{L^2(0,T)} \right),
$$

$$
|a_n'(t)| \leq c_3 \left(\|f_n\|_{L^2(0,T)} + \|f_n'\|_{L^2(0,T)} \right),
$$

$$
|a_n''(t)| = |k_n^2 a_n(t) + k_n f_n(t)| \; \leq \; c_4 \, k_n \left(\|f_n\|_{L^2(0,T)} + \|f_n'\|_{L^2(0,T)} \right)
$$

for all $t \in [0, T]$. Here we used the estimate

$$
\max_{0 \leq t \leq T} |\varphi(t)|^2 \leq 2 \max\{T, 1/T\} \left(\|\varphi\|_{L^2(0,T)}^2 + \|\varphi'\|_{L^2(0,T)}^2 \right)
$$

for any function $\varphi \in C^1[0, T]$ (see Exercise 4.10). Differentiating (4.54) for the case $\alpha_n = 0$ yields

$$
u'(t) = \sum_{n=1}^\infty \left[k_n a_n(0) \sin(k_n t) \; + \; (k_n \beta_n - a_n'(0)) \cos(k_n t) \; + \; a_n'(t) \right] v_n,
$$

$$
u''(t) = \sum_{n=1}^\infty \left[k_n^2 a_n(0) \cos(k_n t) \; + \; (k_n a_n'(0) - k_n^2 \beta_n) \sin(k_n t) \; + \; a_n''(t) \right] v_n.
$$

Now we observe from the above estimates of $a_n(t)$, $a_n'(t)$, and $a_n''(t)$ and the fact that $\sum_{n=1}^\infty k_n^2 \beta_n^2 = \|u_1\|_{H_0^1(D)}^2$ that the coefficients of this series satisfy the assumptions of Corollary 4.49. This shows that $u' \in X$.

Equation (4.51) follows from the observation that $\frac{d^2}{dt^2}\left(b\,u(t),\psi\right)_{L^2(D)} = \left(b\,u''(t),\psi\right)_{L^2(D)}$ because $u \in C^2[0,T;L^2(D)]$. □

We finish this part with an explicit example which shows that one cannot weaken the assumptions on f for deriving the C^1-regularity of u.

Example 4.54. Let $D = (0,\pi)^3 \subseteq \mathbb{R}^3$ and $a = c = 1$ and $T > 2$. The Dirichlet eigenvalues of $-\Delta$ in D are given by $k_n = |n|$, $n \in \mathbb{N}^3$, with corresponding eigenfunctions

$$v_n(x) = \frac{1}{(\pi/2)^{3/2}\,|n|} \prod_{j=1}^{3} \sin(n_j x_j), \quad x \in D, \; n \in \mathbb{N}^3.$$

They are normalized such that $\|v_n\|_* = \|\nabla v_n\|_{L^2(D)} = 1$ and $\|v_n\|_{L^2(D)} = 1/|n|$. We set $\rho_n = \sqrt{|n|^2 - |n|}$ for $n \in \mathbb{N}^3$ and define f by

$$f(t,x) = \sum_{n \in \mathbb{N}^3} f_n \sin(\rho_n t)\,|n|\,v_n(x), \quad t \in [0,T], \; x \in D,$$

where the coefficients f_n are such that $\sum_n f_n^2 < \infty$. Then

$$\|f(t,\cdot)\|_{L^2(D)}^2 = \sum_{n \in \mathbb{N}^3} f_n^2 \sin^2(\rho_n t), \quad t \in [0,T],$$

because $\{|n|v_n : n \in \mathbb{N}^3\}$ is an orthonormal system in $L^2(D)$. By $\|f(t,\cdot)\|_{L^2(D)}^2 \leq \sum_{n \in \mathbb{N}^3} f_n^2$ we observe that $f \in C[0,T;L^2(D)]$. The solution of the initial boundary value problem with $u_0 = u_1 = 0$ is given by

$$u(t,x) = \sum_{n \in \mathbb{N}^3} f_n \left[\sin(\rho_n t) - \frac{\rho_n}{|n|}\sin(|n|t)\right] v_n(x), \quad t \in [0,T], \; x \in D,$$

as one checks directly by term-by-term differentiation, which can be made rigorously by investigating the convergence. The normalization of v_n yields

$$\|u(t,\cdot)\|_*^2 = \sum_{n \in \mathbb{N}^3} f_n^2 \left[\sin(\rho_n t) - \frac{\rho_n}{|n|}\sin(|n|t)\right]^2, \quad t \in [0,T].$$

If $u \in C^1[0,T;H_0^1(D)]$, then

$$u'(t) = \sum_{n \in \mathbb{N}^3} f_n \rho_n \left[\cos(\rho_n t) - \cos(|n|t)\right] v_n$$

and

$$\|u'(t)\|_*^2 = \sum_{n \in \mathbb{N}^3} f_n^2 \rho_n^2 \left[\cos(\rho_n t) - \cos(|n|t)\right]^2$$

for $t \in [0, T]$. Integration yields

$$\int_0^T \|u'(t)\|_*^2 \, dt = \sum_{n \in \mathbb{N}^3} f_n^2 \rho_n^2 \int_0^T \left[\cos^2(\rho_n t) + \cos^2(|n|t) - 2\cos(\rho_n t)\cos(|n|t)\right] dt$$

$$= \sum_{n \in \mathbb{N}^3} f_n^2 \rho_n^2 \left[T + \frac{\sin(2\rho_n T)}{4\rho_n} + \frac{\sin(2|n|T)}{4|n|} - \frac{\sin\left((\rho_n + |n|)T\right)}{\rho_n + |n|}\right.$$

$$\left. - \frac{\sin\left((|n| - \rho_n)T\right)}{|n| - \rho_n}\right].$$

Now we observe that $|n| - \rho_n = \frac{1}{2} + \mathcal{O}(1/|n|)$. Therefore we can choose $\varepsilon > 0$ such that $2 + \varepsilon < T$ and then N so large such that

$$\frac{1}{4\rho_n} + \frac{1}{4|n|} + \frac{1}{\rho_n + |n|} + \frac{1}{|n| - \rho_n} \leq 2 + \varepsilon \quad \text{for } |n| \geq N.$$

This yields

$$\int_0^T \|u'(t)\|_*^2 \, dt \geq (T - 2 - \varepsilon) \sum_{|n| \geq N} f_n^2 \rho_n^2$$

which implies that $f \in C^1[0, T; L^2(D)]$.

After the scalar wave equation we consider now the *time-dependent Maxwell system*; that is,

$$\operatorname{curl} E(t, x) + \mu(x) \frac{\partial H}{\partial t}(t, x) = 0, \quad (t, x) \in (0, T) \times D, \tag{4.56a}$$

$$\operatorname{curl} H(t, x) - \varepsilon(x) \frac{\partial E}{\partial t}(t, x) = J_e(t, x), \quad (t, x) \in (0, T) \times D, \tag{4.56b}$$

with boundary conditions

$$\nu(x) \times E(t, x) = 0 \quad \text{for } (t, x) \in (0, T) \times \partial D, \tag{4.56c}$$

and initial conditions

$$E(0, x) = e_0(x) \quad \text{and} \quad H(0, x) = h_0(x) \quad \text{for } x \in D. \tag{4.56d}$$

Again we need some conditions on the electrical parameter. Therefore, we make the following assumptions on the data:

- $\varepsilon, \mu \in L^\infty(D)$ such that $\varepsilon(x) \geq c > 0$ and $\mu(x) \geq c > 0$ on D for some $c > 0$,

- $J_e \in L^2\left((0, T) \times D\right)$,

- $e_0 \in H_0(\operatorname{curl}, D)$ and $h_0 \in H(\operatorname{curl}, D)$.

We define the solution space X for the pair (E, H) by $X = X_e \times X_h$ where

$$X_e = C[0, T; H_0(\operatorname{curl}, D)] \cap C^1[0, T; L^2(D, \mathbb{C}^3)], \qquad (4.57a)$$
$$X_h = C[0, T; H(\operatorname{curl}, D)] \cap C^1[0, T; L^2(D, \mathbb{C}^3)], \qquad (4.57b)$$

and equip X with the product norm $\|(E, H)\|_X = \|E\| + \|H\|$ with

$$\|u\| = \max_{0 \le t \le T} \|u(t)\|_{H(\operatorname{curl}, D)} + \max_{0 \le t \le T} \|u'(t)\|_{L^2(D)} \qquad (4.57c)$$

where $u = E$ or $u = H$.

Definition 4.55. (E, H) is a solution of the initial-boundary value problem (4.56a)–(4.56d) if $(E, H) \in X$ such that $E(0) = e_0$, $H(0) = h_0$ and

$$\operatorname{curl} E(t) + \mu H'(t) = 0 \quad \text{for all } t \in [0, T], \qquad (4.58a)$$
$$\operatorname{curl} H(t) - \varepsilon E'(t) = J_e(t, \cdot) \quad \text{for all } t \in [0, T]. \qquad (4.58b)$$

We observe that the solution space is not symmetric with respect to E and H because of the boundary condition for E. The following analysis makes use of the Helmholtz decompositions of Theorem 4.21; that is,

$$L^2(D, \mathbb{C}^3) = H_0(\operatorname{curl} 0, D) \oplus \tilde{V}_{0,\varepsilon},$$
$$H_0(\operatorname{curl}, D) = H_0(\operatorname{curl} 0, D) \oplus V_{0,\varepsilon},$$
$$H(\operatorname{curl}, D) = H(\operatorname{curl} 0, D) \oplus V_{\mu},$$

where the spaces have been defined in (4.10a)–(4.10e) when we take $A = \varepsilon I$ or $A = \mu I$, respectively; that is,

$$V_{\mu} = \big\{ u \in H(\operatorname{curl}, D) : (\mu u, \psi)_{L^2(D)} = 0$$
$$\text{for all } \psi \in H(\operatorname{curl}, D) \text{ with } \operatorname{curl} \psi = 0 \big\},$$
$$V_{0,\varepsilon} = \big\{ u \in H_0(\operatorname{curl}, D) : (\varepsilon u, \psi)_{L^2(D)} = 0$$
$$\text{for all } \psi \in H_0(\operatorname{curl}, D) \text{ with } \operatorname{curl} \psi = 0 \big\},$$
$$\tilde{V}_{0,\varepsilon} = \big\{ u \in L^2(D, \mathbb{C}^3) : (\varepsilon u, \psi)_{L^2(D)} = 0$$
$$\text{for all } \psi \in H_0(\operatorname{curl}, D) \text{ with } \operatorname{curl} \psi = 0 \big\}.$$

We recall (see Remark 4.22) that $V_{0,\varepsilon}$ and V_{μ} are just the orthogonal complements of $H_0(\operatorname{curl} 0, D)$ and $H(\operatorname{curl} 0, D)$, respectively, with respect to the inner products

$$(u, v)_{\mu,\varepsilon} = (\mu^{-1} \operatorname{curl} u, \operatorname{curl} v)_{L^2(D)} + (\varepsilon u, v)_{L^2(D)},$$
$$(u, v)_{\varepsilon,\mu} = (\varepsilon^{-1} \operatorname{curl} u, \operatorname{curl} v)_{L^2(D)} + (\mu u, v)_{L^2(D)},$$

respectively, see (4.13a), (4.13b). We note that both norms are equivalent to the ordinary norm in $H(\mathrm{curl}, D)$. Analogously, $\tilde{V}_{0,\varepsilon}$ is the orthogonal complement of $H_0(\mathrm{curl}\, 0, D)$ in $L^2(D, \mathbb{R}^3)$ with respect to the inner product $(\varepsilon\, u, v)_{L^2(D)}$.

We recall the spectral theorem (Theorem 4.34) for the Maxwell system. We proved the existence of an infinite number of eigenvalues $\omega_n > 0$ and corresponding eigenfunctions $v_n \in V_{0,\varepsilon}$ such that

$$\left(\mu^{-1}\, \mathrm{curl}\, v_n, \mathrm{curl}\, \psi\right)_{L^2(D)} = \omega_n^2\, \left(\varepsilon\, v_n, \psi\right)_{L^2(D)} \quad \text{for all } \psi \in H_0(\mathrm{curl}, D)\,.$$
$$(4.59)$$

The functions v_n are normalized such that $\|v_n\|_{\mu,\varepsilon}^2 = \|\mu^{-1/2}\, \mathrm{curl}\, v_n\|_{L^2(D)}^2 + \|\varepsilon^{1/2} v_n\|_{L^2(D)}^2 = 1$ for all $n \in \mathbb{N}$. Then we saw that the system $\{v_n : n \in \mathbb{N}\}$ is a complete orthonormal system in $V_{0,\varepsilon}$ with respect to the inner product $(\cdot, \cdot)_{\mu,\varepsilon}$ and $\{\sqrt{1 + \omega_n^2}\, v_n : n \in \mathbb{N}\}$ is a complete orthonormal system in $\tilde{V}_{0,\varepsilon}$ with respect to $(\varepsilon v, w)_{L^2(D)}$. Furthermore, we saw in Lemma 4.35 that $w_n = \frac{1}{\omega_n\, \mu}\, \mathrm{curl}\, v_n$, $n \in \mathbb{N}$, form a complete orthonormal system in V_μ with respect to the inner product $(\cdot, \cdot)_{\varepsilon,\mu}$.

Now we turn to the investigation of the Maxwell system (4.56a)–(4.56d). First we prove uniqueness.

Theorem 4.56. *There exists at most one solution of (4.58a), (4.58b) with $E(0) = e_0$ and $H(0) = h_0$.*

Proof: Let (E, H) be the difference of two solutions. Then (E, H) solves the system for $J_e = 0$, $e_0 = 0$, and $h_0 = 0$. Let first $\phi \in H(\mathrm{curl}\, 0, D)$. Then, from (4.58a),

$$0 = \left(\mathrm{curl}\, E(t), \phi\right)_{L^2(D)} + \frac{d}{dt}\left(\mu\, H(t), \phi\right)_{L^2(D)} = \frac{d}{dt}\left(\mu\, H(t), \phi\right)_{L^2(D)}$$

for all t because $\left(\mathrm{curl}\, E(t), \phi\right)_{L^2(D)} = \left(E(t), \mathrm{curl}\, \phi\right)_{L^2(D)} = 0$. Therefore, $\left(\mu\, H(t), \phi\right)_{L^2(D)}$ is constant and thus zero for all t because of the initial condition $H(0) = 0$. Therefore, $H(t) \in V_\mu$ for all $t \in [0, T]$. By the same arguments for $\phi \in H_0(\mathrm{curl}\, 0, D)$ and Eq. (4.58b) one shows that $E(t) \in V_{0,\varepsilon}$ for all $t \in [0, T]$. Now we multiply Eq. (4.58a) by w_n for some n and have

$$0 = \left(\mathrm{curl}\, E(t), w_n\right)_{L^2(D)} + \frac{d}{dt}\left(\mu\, H(t), w_n\right)_{L^2(D)}$$

$$= \left(E(t), \mathrm{curl}\, w_n\right)_{L^2(D)} + \frac{d}{dt}\left(\mu\, H(t), w_n\right)_{L^2(D)}$$

$$= \omega_n\, \left(\varepsilon\, E(t), v_n\right)_{L^2(D)} + \frac{d}{dt}\left(\mu\, H(t), w_n\right)_{L^2(D)}$$

and analogously

$$
\begin{aligned}
0 &= \big(\operatorname{curl} H(t), v_n\big)_{L^2(D)} - \frac{d}{dt}\big(\varepsilon\, E(t), v_n\big)_{L^2(D)} \\
&= \big(H(t), \operatorname{curl} v_n\big)_{L^2(D)} - \frac{d}{dt}\big(\varepsilon\, E(t), v_n\big)_{L^2(D)} \\
&= \omega_n \big(\mu\, H(t), w_n\big)_{L^2(D)} - \frac{d}{dt}\big(\varepsilon\, E(t), v_n\big)_{L^2(D)} .
\end{aligned}
$$

Therefore, $\big(\varepsilon\, E(t), v_n\big)_{L^2(D)}$ and $\big(\mu\, H(t), w_n\big)_{L^2(D)}$ solve a homogeneous linear system of ordinary differential equations of first order with homogeneous initial data. Therefore, $\big(\varepsilon\, E(t), v_n\big)_{L^2(D)}$ and $\big(\mu\, H(t), w_n\big)_{L^2(D)}$ have to vanish for all t and $n \in \mathbb{N}$. The completeness of $\{v_n : n \in \mathbb{N}\}$ in $V_{0,\varepsilon}$ and of $\{w_n : n \in \mathbb{N}\}$ in V_μ implies that $E(t)$ and $H(t)$ vanish for all $t \in [0, T]$. $\square$

Next we show existence for the special case $J_e = 0$.

Theorem 4.57. *For all $e_0 \in H_0(\operatorname{curl}, D)$ and $h_0 \in H(\operatorname{curl}, D)$ there exists a unique solution of (4.58a), (4.58b) for $J_e = 0$.*

Proof: We decompose the pair (e_0, h_0) into the sum $(e_0, h_0) = (e_0^0, h_0^0) + (\tilde{e}_0, \tilde{h}_0)$ with $(e_0^0, h_0^0) \in H_0(\operatorname{curl} 0, D) \times H(\operatorname{curl} 0, D)$ and $(\tilde{e}_0, \tilde{h}_0) \in V_{0,\varepsilon} \times V_\mu$. Then the solution (E, H) is the sum of the solutions with initial data (e_0^0, h_0^0) and $(\tilde{e}_0, \tilde{h}_0)$, respectively. The solution corresponding to initial data (e_0^0, h_0^0) is just the constant (with respect to time t); that is, $E^0(t) = e_0^0$ and $H^0(t) = h_0^0$ for all $t \in [0, T]$.

We now construct the solution $(\tilde{E}, \tilde{H})$ corresponding to initial data $(\tilde{e}_0, \tilde{h}_0) \in V_{0,\varepsilon} \times V_\mu$. We expand the initial data in the forms

$$
\tilde{e}_0 = \sum_{n=1}^{\infty} \alpha_n v_n , \qquad \tilde{h}_0 = \sum_{n=1}^{\infty} \beta_n w_n
$$

where $\{v_n : n \in \mathbb{N}\}$ in $V_{0,\varepsilon}$ and $\{w_n : n \in \mathbb{N}\}$ in V_μ are the orthonormal systems studied above. Then, by Parseval's identity, $\sum_{n=1}^{\infty} \alpha_n^2 = \|\tilde{e}_0\|_{\mu,\varepsilon}^2$ and $\sum_{n=1}^{\infty} \beta_n^2 = \|\tilde{h}_0\|_{\varepsilon,\mu}^2$. We make an ansatz in the form

$$
\tilde{E}(t) = \sum_{n=1}^{\infty} a_n(t)\, v_n , \qquad \tilde{H}(t) = \sum_{n=1}^{\infty} b_n(t)\, w_n , \quad t \in [0, T] ,
$$

where $a_n, b_n \in C^1[0, T]$. Substitution of these series into Eqs. (4.58a), (4.58b) for $J_e = 0$ yields formally

$$
0 = \sum_{n=1}^{\infty} \big[a_n(t)\, \operatorname{curl} v_n + b_n'(t)\, \mu\, w_n\big] = \sum_{n=1}^{\infty} \big[\omega_n\, a_n(t) + b_n'(t)\big]\, \mu\, w_n ,
$$

$$
0 = \sum_{n=1}^{\infty} \big[b_n(t)\, \operatorname{curl} w_n - a_n'(t)\, \varepsilon\, v_n\big] = \sum_{n=1}^{\infty} \big[\omega_n\, b_n(t) - a_n'(t)\big]\, \varepsilon\, v_n .
$$

The linear independence of the systems $\{v_n : n \in \mathbb{N}\}$ in $V_{0,\varepsilon}$ and $\{w_n : n \in \mathbb{N}\}$ in V_μ yields the following system of ordinary differential equations

$$\omega_n a_n(t) + b_n'(t) = 0, \quad \omega_n b_n(t) - a_n'(t) = 0, \quad t \in [0, T], \qquad (4.60)$$

with initial conditions $a_n(0) = \alpha_n$, $b_n(0) = \beta_n$. Solving it yields

$$\tilde{E}(t) = \sum_{n=1}^\infty \left[\alpha_n \cos(\omega_n t) + \beta_n \sin(\omega_n t)\right] v_n, \quad t \in [0, T],$$

$$\tilde{H}(t) = \sum_{n=1}^\infty \left[-\alpha_n \sin(\omega_n t) + \beta_n \cos(\omega_n t)\right] w_n, \quad t \in [0, T].$$

These functions belong to X_e and X_h, respectively. Indeed, by Corollary 4.49 we have to check uniform estimate of $c(t) = \alpha_n \cos(\omega_n t) + \beta_n \sin(\omega_n t)$ and of $c(t) = -\alpha_n \sin(\omega_n t) + \beta_n \cos(\omega_n t)$, respectively, and of their derivatives. Indeed, for both cases we have that $|c(t)|^2 \leq 2\alpha_n^2 + 2\beta_n^2 =: \gamma_n^2$ and $|c'(t)| \leq \omega_n \gamma_n$ and $\sum_{n=1}^\infty \gamma_n^2 = \|\tilde{e}_0\|_{\mu,\varepsilon}^2 + \|\tilde{h}_0\|_{\varepsilon,\mu}^2$. This proves the theorem. $\square$

Now we have to determine particular solutions of the inhomogeneous differential equations; that is, for $J_e \neq 0$. We make the following additional assumption on J_e:

Assumption: Let J_e be of the form $J_e(t) = \rho(t)\,\varepsilon\,F$, $t \in [0, T]$, for some $F \in L^2(D, \mathbb{C}^3)$ and some scalar function $\rho \in C^1[0, T]$.

Again, we decompose F in the form

$$F = F^0 + \tilde{F} \quad \text{with } F^0 \in H_0(\operatorname{curl} 0, D) \text{ and } \tilde{F} \in \tilde{V}_{0,\varepsilon}.$$

A particular solution is obtained as the sum of solutions with right-hand side F^0 and with $\tilde{F}$, respectively. A solution (E^0, H^0) with right-hand side $\rho(t)\varepsilon F^0$ is simply $E^0(t) = -\int_0^t \rho(s)\,ds\,F^0$ and $H^0(t) = 0$.

Now we consider $\tilde{F} \in \tilde{V}_{0,\varepsilon}$ as right-hand side. Then $\tilde{F}$ has an expansion in the form $\tilde{F} = \sum_{n=1}^\infty f_n \sqrt{1 + \omega_n^2} v_n$ (convergence in $L^2(D)$) with $\|\sqrt{\varepsilon}\,\tilde{F}\|_{L^2(D)}^2 = \sum_{n=1}^\infty f_n^2$. We make an ansatz for $\tilde{E}$ and $\tilde{H}$ in the forms

$$\tilde{E}(t) = \sum_{n=1}^\infty a_n(t)\,v_n, \qquad \tilde{H}(t) = \sum_{n=1}^\infty b_n(t)\,w_n, \quad t \in [0, T],$$

Using the relationship between v_n and w_n a substitution into the Maxwell system yields, using the system

$$\omega_n a_n(t) + b_n'(t) = 0, \quad \omega_n b_n(t) - a_n'(t) = \sqrt{1 + \omega_n^2}\,\rho(t)\,f_n. \quad t \in [0, T]$$
$$(4.61)$$

(compare to (4.60)). We can either eliminate a_n or b_n from the system. Because of the differentiability of ρ we eliminate b_n and arrive at

$$a_n''(t) + \omega_n^2 a_n(t) = -\sqrt{1 + \omega_n^2} f_n \rho'(t), \quad t \in [0, T].$$

A solution is given by

$$a_n(t) = -\frac{\sqrt{1 + \omega_n^2}}{2\omega_n} f_n \int_0^T \sin(\omega_n|t - s|) \rho'(s) \, ds, \quad t \in [0, T].$$

Then $|a_n(t)| \leq c\|\rho'\|_\infty |f_n|$ and $|a_n'(t)| \leq c\omega_n\|\rho'\|_\infty |f_n|$ for all $t \in [0, T]$ and $n \in \mathbb{N}$ where the constant c is independent of n or t. Solving for b_n yields the estimates $|b_n(t)| \leq c(\|\rho\|_\infty + \|\rho'\|_\infty) |f_n|$ and $|b_n'(t)| \leq c\omega_n\|\rho'\|_\infty |f_n|$ for all $t \in [0, T]$ and $n \in \mathbb{N}$. Corollary 4.49 yields $(\tilde{E}, \tilde{H}) \in X = X_e \times X_h$ and ends the proof. $\square$

Collecting our results we can formulate the following theorem for the time-dependent cavity problem.

Theorem 4.58. *Let $e_0 \in H_0(\mathrm{curl}, D)$ and $h_0 \in H(\mathrm{curl}, D)$ and J_e of the form $J_e(t) = \rho(t) \varepsilon F$, $t \in [0, T]$, for some $F \in L^2(D, \mathbb{C}^3)$ and some scalar function $\rho \in C^1[0, T]$. Then there exists a unique solution $(E, H) \in X = X_e \times X_h$ of the system (4.56a), (4.56b).*

As a final remark we note that we can replace the assumption $(\rho, F) \in C^1[0, T] \times L^2(D, \mathbb{C}^3)$ in the theorem by the assumption $(\rho, F) \in C[0, T] \times H_0(\mathrm{curl}, D)$. Then $\tilde{F} \in V_{0,\varepsilon}$, and we solve the system (4.61) for b_n and proceed as before.

4.4 Exercises

Exercise 4.1. Let $D \subseteq \mathbb{R}^3$ be an open set and $u \in C(D)$. Show that the support of u as defined at the beginning of Sect. 4.1.1 can be expressed as the closure of the set $S \subseteq D$, given by $S = \{x \in D : u(x) \neq 0\}$.

Exercise 4.2. Let $U \subset \mathbb{R}^3$ be open and bounded and u the characteristic function of U; that is, $u(x) = 1$ for $x \in U$ and $u(x) = 0$ outside of U. Let $\phi_\delta \in C_0^\infty(\mathbb{R}^3)$ be as in Theorem 4.7. Show that $u_k = u * \phi_k$ converges to u in $L^2(\mathbb{R}^3)$.
Hint: Use the theorem of dominated convergence!

Exercise 4.3. Show that $C_0^\infty(\mathbb{R}^3)$ is dense in $H^1(\mathbb{R}^3)$. Therefore, $H_0^1(\mathbb{R}^3)$ coincides with $H^1(\mathbb{R}^3)$.
Hint: Use the technique of Lemma 4.18.

Exercise 4.4. Prove that $H_0^1(\mathbb{R}^3)$ is not compactly embedded in $L^2(\mathbb{R}^3)$.
Hint: Take a non-vanishing $\varphi \in C_0^\infty(\mathbb{R})$ and some $a \in \mathbb{R}^3$, $a \neq 0$, and discuss the sequence $u_n(x) = \varphi(x + na)$, $x \in \mathbb{R}^3$, $n \in \mathbb{N}$.

Exercise 4.5. Show that the vector field F in Definition 4.1 is unique if it exists.
Hint: Use Exercise 4.3!

Exercise 4.6. Let $D \subseteq \mathbb{R}^3$ be symmetric with respect to $x_3 = 0$; that is, $x \in D \Leftrightarrow x^* \in D$ where $a^* = (a_1, a_2, -a_3)^\top$ for any $a = (a_1, a_2, a_3)^\top \in \mathbb{C}^3$. Let $D^\pm = \{x \in D : x_3 \gtrless 0\}$ and $\varphi \in C^1(\overline{D^-})$ and $v \in C^1(\overline{D^-}, \mathbb{C}^3)$. Extend φ and v into D by reflection; that is,

$$\tilde{\varphi}(x) = \begin{cases} \varphi(x), & x \in D^-, \\ \varphi(x^*), & x \in D^+, \end{cases} \qquad \tilde{v}(x) = \begin{cases} v(x), & x \in D^-, \\ v^*(x^*), & x \in D^+. \end{cases}$$

Show that $\tilde{\varphi} \in H^1(D)$ and $\tilde{v} \in H(\mathrm{curl}, D)$.

Hint: Show that

$$\nabla \tilde{\varphi}(x) = \begin{cases} \nabla\varphi(x), & x \in D^-, \\ (\nabla\varphi)^*(x^*), & x \in D^+, \end{cases} \qquad \mathrm{curl}\,\tilde{v}(x) = \begin{cases} \mathrm{curl}\,v(x), & x \in D^-, \\ -(\mathrm{curl}\,v)^*(x^*), & x \in D^+, \end{cases}$$

are the gradient and curl, respectively.

Exercise 4.7. Construct a function $\phi \in C^\infty(\mathbb{R})$ such that $\phi(t) = 0$ for $|t| \geq 1$ and $\phi(t) = 1$ for $|t| \leq 1/2$.
Hint: Define φ by $\varphi(t) = \exp(-1/t)$ for $t > 0$ and $\varphi(t) = 0$ for $t \leq 0$. Discuss φ and choose ϕ as a proper rational combination of $\varphi(1-t^2)$ and $\varphi(t^2-1/4)$.

Exercise 4.8. Prove that for open and bounded $D \subseteq \mathbb{R}^3$ the space $H_0^1(D)$ is a proper subspace of $H^1(D)$.
Hint: Show that functions v with $\Delta v - v = 0$ in D are orthogonal to $H_0^1(D)$.

Exercise 4.9. Show that for any $v \in H_0^1(D)$ the extension by zero belongs to $H^1(\mathbb{R}^3)$; that is, $\tilde{v} \in H^1(\mathbb{R}^3)$ where $\tilde{v} = v$ in D and $\tilde{v} = 0$ in $\mathbb{R}^3 \setminus D$.

Exercise 4.10. Prove that $|\varphi(t)|^2 \leq 2\max\{T, 1/T\}\left(\|\varphi\|_{L^2(0,T)}^2 + \|\varphi'\|_{L^2(0,T)}^2\right)$ for all $t \in [0, T]$ and any function $\varphi \in C^1[0, T]$.
Hint: Use the Fundamental Theorem of Calculus of Variations.

Exercise 4.11. In Theorem 4.13 the existence of an extension operator is proven. Why is this not an obvious fact by our definition of $H^1(D)$ as the space of restrictions of $H^1(\mathbb{R}^3)$?

Exercise 4.12. Show that $C_0^\infty(\mathbb{R}^3, \mathbb{C}^3)$ is dense in $H(\text{curl}, \mathbb{R}^3)$.
Hint: Compare with Exercise 4.3.

Exercise 4.13. Let $Q = (-R, R)^3$. Show that the following inclusions hold and are bounded:

$$H_0(\text{curl}, Q) \;\hookrightarrow\; H_{\text{per}}(\text{curl}, Q) \;\hookrightarrow\; H(\text{curl}, Q).$$

Hint: Follow the arguments of the proof of Lemma 4.37.

Exercise 4.14. Why is, even for bounded sets D, the space $H_0(\text{curl}, D)$ not compactly embedded in $L^2(D, \mathbb{C}^3)$?
Hint: Use the Helmholtz decomposition!

Exercise 4.15. Prove Theorem 4.23; that is, show the decompositions of $L^2(D, \mathbb{C}^3)$ and $H_0(\text{curl}, D)$ in the form

$$L^2(D, \mathbb{C}^3) \;=\; L^2(\text{div}_A\, 0, D) \;\oplus\; \nabla H_0(D),$$

and

$$H_0(\text{curl}, D) \;=\; H_0(\text{curl}, \text{div}_A\, 0, D) \;\oplus\; \nabla H_0(D).$$

Hint: Use the arguments as in the proof of Theorem 4.21.

Exercise 4.16. Prove the following version of the Fundamental Theorem of Calculus of Variations.
Let $u \in L^1(0, T)$ and $v \in C[0, T]$ such that

$$\int_0^T \left[\rho'(t)\, u(t) + \rho(t)\, v(t)\right] dt \;=\; 0 \quad \text{for all } \rho \in C^1[0, T] \text{ with } \rho(T) = 0.$$

Then $u \in C^1[0, T]$ and $u(0) = 0$ and $u' = v$ on $[0, T]$. If the variational equation holds for all $\rho \in C^1[0, T]$, then also $u(T) = 0$.
Hint: Use the methods of the proof of Lemma 4.46.

Exercise 4.17. Let $D = B_3(0, R)$ be a ball. Show explicitly the Helmholtz decomposition by using the results of Chap. 2.

Chapter 5

Boundary Integral Equation Methods for Lipschitz Domains

For the boundary value problems of Chaps. 3 and 4 we made assumptions which are often not met in applications. Indeed, the classical integral equation methods discussed in Chap. 3 require smoothness of the boundary ∂D. In case of the cavity problem of Chap. 4 just a homogeneous boundary condition has been treated. Both restrictions are connected because if we like to weaken the regularity of the boundary, or if we like to allow for more general boundary conditions we have to investigate the traces of the functions or vector fields on the boundary ∂D in detail. Therefore, we continue in Sects. 5.1.1 and 5.1.2 by introducing Sobolev spaces which appear as the range spaces of the trace operators and prove denseness, trace theorems and compact embedding results. Finally we use these results to extend the boundary integral equation methods for Lipschitz domains.

5.1 Advanced Properties of Sobolev Spaces

We recall Definition A.7 for the notion of a Lipschitz domain. Let $B_2(0, \alpha) \subseteq \mathbb{R}^2$ be the two-dimensional disk of radius α.

Definition 5.1. We call a region $D \subseteq \mathbb{R}^3$ to be a Lipschitz domain, if there exists a finite number of open cylinders U_j of the form $U_j = \{R_j x + z^{(j)} : x \in B_2(0, \alpha_j) \times (-2\beta_j, 2\beta_j)\}$ with $z^{(j)} \in \mathbb{R}^3$ and rotations $R_j \in \mathbb{R}^{3 \times 3}$ and real valued Lipschitz-continuous functions $\xi_j \in C(B_2(0, \alpha_j))$ with $|\xi_j(x_1, x_2)| \leq \beta_j$ for all $(x_1, x_2) \in B_2[0, \alpha_j]$ such that $\partial D \subseteq \bigcup_{j=1}^{m} U_j$ and

© Springer International Publishing Switzerland 2015
A. Kirsch, F. Hettlich, *The Mathematical Theory of Time-Harmonic Maxwell's Equations*, Applied Mathematical Sciences 190,
DOI 10.1007/978-3-319-11086-8_5

$$\partial D \cap U_j = \left\{ R_j x + z^{(j)} : (x_1, x_2) \in B_2(0, \alpha_j), \ x_3 = \xi_j(x_1, x_2) \right\},$$
$$D \cap U_j = \left\{ R_j x + z^{(j)} : (x_1, x_2) \in B_2(0, \alpha_j), \ x_3 < \xi_j(x_1, x_2) \right\},$$
$$U_j \setminus \overline{D} = \left\{ R_j x + z^{(j)} : (x_1, x_2) \in B_2(0, \alpha_j), \ x_3 > \xi_j(x_1, x_2) \right\}.$$

We call $\{U_j, \xi_j \ : \ j = 1, \ldots, m\}$ a *local coordinate system* of ∂D. For abbreviation we denote by

$$C_j = C_j(\alpha_j, \beta_j) = B_2(0, \alpha_j) \times (-2\beta_j, 2\beta_j)$$
$$= \left\{ x = (x_1, x_2, x_3) \in \mathbb{R}^3 : x_1^2 + x_2^2 < \alpha_j^2, \ |x_3| < 2\beta_j \right\}$$

the cylinders with parameters α_j and β_j. We can assume without loss of generality that $\beta_j \geq \alpha_j$ (otherwise split the parameter region into smaller ones). Furthermore, we define the three-dimensional balls $B_j = B_3(0, \alpha_j) \subseteq C_j$ and introduce the mappings

$$\tilde{\Psi}_j(x) \ = \ R_j \begin{pmatrix} x_1 \\ x_2 \\ \xi_j(x_1, x_2) + x_3 \end{pmatrix} + z^{(j)}, \quad x = (x_1, x_2, x_3)^\top \in B_j,$$

and their restrictions Ψ_j to $B_2(0, \alpha_j)$; that is,

$$\Psi_j(\tilde{x}) \ = \ R_j \begin{pmatrix} x_1 \\ x_2 \\ \xi_j(x_1, x_2) \end{pmatrix} + z^{(j)}, \quad \tilde{x} = (x_1, x_2)^\top \in B_2(0, \alpha_j).$$

By Rademacher's result ([11, 27], see Remark A.8) we know that Ψ_j is differentiable almost everywhere on $B_2(0, \alpha_j)$. This yields a parametrization of $\partial D \cap U_j$ in the form $y = \Psi_j(\tilde{x})$ for $\tilde{x} \in B_2(0, \alpha_j)$ with $\left| \frac{\partial \Psi_j}{\partial x_1} \times \frac{\partial \Psi_j}{\partial x_2} \right| = \sqrt{1 + |\nabla \xi_j|^2}$.

We set $U_j' = \tilde{\Psi}_j(B_j)$. Then $\partial D \subseteq \bigcup_{j=1}^m U_j'$ and $B_j \cap (\mathbb{R}^2 \times \{0\}) = B_2(0, \alpha_j) \times \{0\}$, and

$$\partial D \cap U_j' = \left\{ \tilde{\Psi}_j(x) : x \in B_j, \ x_3 = 0 \right\} \ = \ \left\{ \Psi_j(\tilde{x}) : \tilde{x} \in B_2(0, \alpha_j) \right\},$$
$$D \cap U_j' = \left\{ \tilde{\Psi}_j(x) : x \in B_j, \ x_3 < 0 \right\},$$
$$U_j' \setminus \overline{D} = \left\{ \tilde{\Psi}_j(x) : x \in B_j, \ x_3 > 0 \right\}.$$

Therefore, the mappings $\tilde{\Psi}_j$ "flatten" the boundary.

We note that the Jacobian $\tilde{\Psi}_j'(x) \in \mathbb{R}^{3 \times 3}$ is given by

$$\tilde{\Psi}_j'(x) \ = \ R_j \begin{pmatrix} 1 & 0 & 0 \\ 0 & 1 & 0 \\ \partial_1 \xi_j(\tilde{x}) & \partial_2 \xi_j(\tilde{x}) & 1 \end{pmatrix}$$

where $\partial_\ell \xi_j = \partial \xi_j / \partial x_\ell$ for $\ell = 1, 2$ and $\tilde{x} = (x_1, x_2)$. Therefore, these Jacobians are regular with constant determinant $\det \tilde{\Psi}'_j(x) = 1$ and $\tilde{\Psi}_j$ are isomorphisms from B_j onto U'_j for every $j = 1, \ldots, m$.

These parametrizations allow it to transfer the notion of (periodic) Sobolev spaces on two-dimensional planar domains to the boundary ∂D. We begin with Sobolev spaces of scalar functions in Sect. 5.1.1 and continue with vector valued functions in Sect. 5.1.2.

5.1.1 Sobolev Spaces of Scalar Functions

We note first that Green's formula holds in the form of Theorem A.12 for Lipschitz domains and sufficiently smooth functions. This theorem is, e.g. used in the following simple result.

Lemma 5.2. *Let D be a bounded Lipschitz domain.*

(a) *Let $u \in C^1(\overline{D})$ with $u = 0$ on ∂D. Then the extension $\tilde{u}$ of u by zero outside of $\overline{D}$ yields $\tilde{u} \in H^1(\mathbb{R}^3)$ and $\nabla \tilde{u} = \nabla u$ in D and $\nabla \tilde{u} = 0$ outside of D.*
(b) *Let $Q \in \mathbb{R}^{3 \times 3}$ be an orthogonal matrix and $z \in \mathbb{R}^3$. For $u \in H^1(D)$ define $v(x) = u(Qx + z)$ on $U = \{x \in \mathbb{R}^3 : Qx + z \in D\}$. Then $v \in H^1(U)$ and $\nabla v(x) = Q^\top \nabla u(Qx + z)$.*

Proof: (a) Set $g = \nabla u$ in D and zero outside of D. Then $g \in L^2(\mathbb{R}^3, \mathbb{C}^3)$, and for $\psi \in C_0^\infty(\mathbb{R}^3)$ we have by partial integration:

$$\int_{\mathbb{R}^3} \tilde{u} \, \nabla \psi \, dx = \int_D u \, \nabla \psi \, dx = -\int_D \psi \, \nabla u \, dx + \underbrace{\int_{\partial D} u \, \psi \, \nu \, ds}_{= 0}$$

$$= -\int_D \psi \, \nabla u \, dx = -\int_{\mathbb{R}^3} \psi \, g \, dx .$$

(b) Let $\psi \in C_0^\infty(D)$ and let $q^{(j)}$, $j = 1, 2, 3$, be the columns of the orthogonal matrix Q. We make the substitution $y = Qx + z$ and have

$$q^{(j)} \cdot \int_D \nabla u(y)|_{y=Qx+z} \, \psi(x) \, dx$$

$$= q^{(j)} \cdot \int_U \nabla u(y) \, \psi(Q^\top(y - z)) \, dy = -q^{(j)} \cdot \int_U u(y) \, Q \nabla \psi(x)|_{x=Q^\top(y-z)} \, dy$$

$$= -\int_U u(y) \left. \frac{\partial \psi(x)}{\partial x_j} \right|_{x=Q^\top(y-z)} dy = -\int_D v(x) \frac{\partial \psi(x)}{\partial x_j} \, dx .$$

This shows that $\partial v(x)/\partial x_j = q^{(j)} \cdot \nabla u(y)|_{y=Qx+z}$. $\square$

An important property of Sobolev spaces on Lipschitz domains D is the denseness of the space $C^\infty(\overline{D})$. Its proof is technically complicated.

Theorem 5.3. *Let D be a bounded Lipschitz domain. Then $C^\infty(\overline{D})$ is dense in $H^1(D)$.*

Proof: The idea of the proof is to transform the given function $u \in H^1(D)$ onto a cylinder by using a partition of unity and local coordinates. In a cylinder we construct a smooth approximation of the transformed function. Finally, we transform back the approximation. We separate the proof into four steps.

(i) Let $u \in H^1(D)$ and $\{U_j, \xi_j : j = 1, \ldots, m\}$ be a local coordinate system. Set $U_0 = D$ and choose a partition of unity $\{\varphi_j : j = 0, \ldots, m\}$ on $\overline{D}$ corresponding to $\{U_j : j = 0, \ldots, m\}$ (see Theorem A.9). Set $u_j(y) = \varphi_j(y)u(y)$ for $j = 0, \ldots, m$ and $y \in D$. For $j = 1, \ldots, m$ we transform u_j by the definition $v_j(x) = u_j(R_j x + z^{(j)})$, $x \in C_j^-$, where $C_j = B_2(0, \alpha_j) \times (-2\beta_j, 2\beta_j)$ and

$$C_j^- := \{x = (\tilde{x}, x_3) \in C_j : x_3 < \xi_j(\tilde{x})\}$$

where again $\tilde{x} = (x_1, x_2)$. We observe that $v_j \in H^1(C_j^-)$ by the chain rule of part (b) of Lemma 5.2. Furthermore, with

$$V_j(t) := \{x \in C_j^- : |\tilde{x}| > \alpha_j - t \text{ or } x_3 < -2\beta_j + t\}$$

we note that there exists $\delta > 0$ with $v_j(x) = 0$ in the neighborhood $V_j(2\delta)$ of $\partial C_j \cap \overline{C_j^-}$ because the support of φ_j is contained in U_j. The reader should sketch the sets C_j, C_j^-, and $V_j(t)$.

(ii) Let $a = (0, 0, L+1)^\top$ where L is larger than all of the Lipschitz constants of the functions ξ_j and set

$$v_j^\varepsilon(x) = (v_j * \phi_\varepsilon)(x - \varepsilon a) = \int_{C_j^- \setminus V_j(2\delta)} v_j(y)\, \phi_\varepsilon(x - \varepsilon a - y)\, dy$$

where $\phi_\varepsilon \in C^\infty(\mathbb{R}^3)$ denotes the mollifier function from (4.7). Its most important properties had been collected in Theorem 4.7. In particular, we have that $v_j^\varepsilon \in C_0^\infty(\mathbb{R}^3)$ and $v_j^\varepsilon \to v_j$ in $L^2(\mathbb{R}^3)$ as ε tends to zero, the latter property because the zero-extension of v_j is in $L^2(\mathbb{R}^2)$.

Next we show that v_j^ε vanishes on $V_j(\delta)$ for $\varepsilon \leq \delta$. First we note that the integral in the definition of v_j^ε is taken over $C_j^- \setminus V_j(2\delta)$. Therefore, for $y \in C_j^- \setminus V_j(2\delta)$ we have $|\tilde{y}| \leq \alpha_j - 2\delta$ and $y_3 \geq -2\beta_j + 2\delta$.

Let first $x \in C_j^-$ with $|\tilde{x}| \geq \alpha_j - \delta$. Then $|x - \varepsilon a - y| \geq |\tilde{x} - \tilde{y}| \geq |\tilde{x}| - |\tilde{y}| \geq \alpha_j - \delta - (\alpha_j - 2\delta) = \delta \geq \varepsilon$. Therefore, $v_j^\varepsilon(x) = 0$ for these x.

Let second $x \in C_j^-$ with $x_3 \leq -2\beta_j + \delta$. Then $|x - \varepsilon a - y| \geq |x_3 - \varepsilon(L + 2) - y_3| \geq y_3 - x_3 + (L + 2)\varepsilon \geq -2\beta_j + 2\delta_j + 2\beta_j - \delta + (L + 2)\varepsilon \geq \delta \geq \varepsilon$. Therefore, $v_j^\varepsilon(x) = 0$ for these x.

(iii) We show now that v_j^ε converges to v_j even in $H^1(C_j^-)$. Let $\varepsilon < \delta/(L+2)$ and $x \in C_j^- \setminus V_j(\delta)$. Then

$$\nabla v_j^\varepsilon(x) = \int_{C_j^-} v_j(y) \nabla_x \phi_\varepsilon(x - \varepsilon a - y)dy = -\int_{C_j^-} v_j(y) \nabla_y \phi_\varepsilon(x - \varepsilon a - y)dy\,.$$

We show that the function $y \mapsto \phi_\varepsilon(x - \varepsilon a - y)$ belongs to $C_0^\infty(C_j^-)$. It is sufficient to prove that the ball $B_3(x - \varepsilon a, \varepsilon)$ is contained in C_j^-. Therefore, let $y \in B_3(x - \varepsilon a, \varepsilon)$.

Then, first, $|\tilde{y}| \leq |\tilde{x}| + |\tilde{y} - \tilde{x}| \leq \alpha_j - \delta + \varepsilon \leq \alpha_j$.

Second, $y_3 \geq x_3 - (L+1)\varepsilon - |x_3 - (L+1)\varepsilon - y_3| \geq -2\beta_j + \delta - (L+1)\varepsilon - \varepsilon \geq -2\beta_j$ because $(L+2)\varepsilon \leq \delta$.

Third, $y_3 \leq x_3 - (L+1)\varepsilon + |y_3 - x_3 + (L+1)\varepsilon)| \leq \xi_j(\tilde{x}) - L\varepsilon \leq \xi_j(\tilde{y}) + L|\xi_j(\tilde{y}) - \xi_j(\tilde{x})| - L\varepsilon \leq \xi_j(\tilde{y})$. This proves that $y \mapsto \phi_\varepsilon(x - \varepsilon a - y)$ belongs to $C_0^\infty(C_j^-)$.

By the definition of the weak derivative we have that

$$\nabla v_j^\varepsilon(x) = \int_{C_j^-} \nabla v_j(z)\, \phi_\varepsilon(x - \varepsilon a - z)\, dz\,. \tag{5.1}$$

For $\varepsilon < \delta/(L+2)$ and $x \in V_j(\delta)$ both sides of (5.1) vanish. Therefore, (5.1) holds for all $x \in C_j^-$ and thus $(\nabla v_j^\varepsilon)(x) = (g_j * \phi_\varepsilon)(x - \varepsilon a)$ for $x \in C_j^-$ where $g_j = \nabla v_j$ on C_j^- and $g_j = 0$ on $\mathbb{R}^3 \setminus C_j^-$. By Theorem 4.7 we conclude again that $(g_j * \phi_\varepsilon)(x - \varepsilon a)$ converges to g_j in $L^2(\mathbb{R}^3)$ and thus $\nabla v_j^\varepsilon|_{C_j^-}$ to ∇v_j in $L^2(C_j^-)$. This proves $v_j^\varepsilon \to v_j$ in $H^1(C_j^-)$ as ε tends to zero for every $j = 1, \ldots, m$.

(iv) In the last step we transform back the function and set $u_j^\varepsilon(y) = v_j^\varepsilon\big(R_j^\top(y - z^{(j)})\big)$ for $j = 1, \ldots, m$ and $y \in U_j \cap \overline{D}$. We note that $u_j^\varepsilon \in C_0^\infty(U_j \cap \overline{D})$ and $u_j^\varepsilon \to \varphi_j u$ in $H^1(U_j \cap D)$. This holds for all $j = 1, \ldots, m$. We extend u_j by zero into all of D. Finally, we note that $u_0 \in H^1(D)$ with compact support in $U_0 = D$ and thus $u_0^\varepsilon = u_0 * \phi_\varepsilon$ converges to $u_0 = \varphi_0 u$ in $H^1(D)$. To finish the proof we set $u^\varepsilon = \sum_{j=0}^m u_j^\varepsilon$ in D and conclude that $u^\varepsilon \in C_0^\infty(D)$ and u^ε converges to $\sum_{j=0}^m \varphi_j u = u$ in $H^1(D)$. $\quad\Box$

We apply this result and prove the following chain rule.

Corollary 5.4. *Let $U, D \subseteq \mathbb{R}^3$ be Lipschitz domains and $\Psi : \overline{U} \to \overline{D}$ continuous in $\overline{U}$ and differentiable at almost all points $x \in U$ such that the Jacobian $\Psi'(x) \in \mathbb{R}^{3 \times 3}$ and its inverse are essentially bounded; that is, $\Psi', (\Psi')^{-1} \in L^\infty(U, \mathbb{R}^{3 \times 3})$. For $u \in H^1(D)$ the composition $v = u \circ \Psi$ belongs to $H^1(U)$, and the chain rule holds in the form*

$$\nabla v(x) = \Psi'(x)^\top \nabla u(y)|_{y = \Psi(x)}, \quad x \in U. \tag{5.2}$$

The mapping $u \mapsto u \circ \Psi$ is bounded from $H^1(D)$ into $H^1(U)$.

Proof: For $u \in C^\infty(\overline{D})$ the formula (5.2) holds in all points where Ψ is differentiable. Also, the right-hand side of (5.2) is in $L^2(D, \mathbb{C}^3)$. Therefore, by the transformation formula the mapping $u \mapsto u \circ \Psi$ is bounded from the dense subspace $C^\infty(\overline{D})$ of $H^1(D)$ into $H^1(U)$ and thus has a bounded extension to all of $H^1(D)$. $\square$

We continue by defining Sobolev spaces of periodic functions as we have done it already in Chap. 4. Definition 4.12 of $H^1_{per}(Q_3)$ is included in the following definition. For the sake of a simpler notation we restrict ourselves to cubes in $\mathbb{R}^d$ with edge length 2π.

Definition 5.5. Let $Q = (-\pi, \pi)^d \subseteq \mathbb{R}^d$ be the cube in $\mathbb{R}^d$ for $d = 2$ or $d = 3$. For any scalar function $v : (-\pi, \pi)^d \to \mathbb{C}$ the Fourier coefficients $v_n \in \mathbb{C}$ for $n \in \mathbb{Z}^d$ are defined as

$$v_n = \frac{1}{(2\pi)^d} \int_Q v(y) \, e^{-i n \cdot y} \, dy, \quad n \in \mathbb{Z}^d.$$

Let $s \geq 0$ be any real number. The space $H^s_{per}(Q)$ is defined by

$$H^s_{per}(Q) = \left\{ v \in L^2(Q) : \sum_{n \in \mathbb{Z}^3} (1 + |n|^2)^s \, |v_n|^2 < \infty \right\}$$

with norm

$$\|v\|_{H^s_{per}(Q)} = \sqrt{\sum_{n \in \mathbb{Z}^3} (1 + |n|^2)^s \, |v_n|^2}.$$

We recall from Lemma 4.37 that $H^1_0(Q) \subseteq H^1_{per}(Q) \subseteq H^1(Q)$ with bounded inclusions. Therefore, on $H^1_0(Q)$ the norms of $H^1_{per}(Q)$ and $H^1(Q)$ are equivalent.

Such a definition of Sobolev spaces of periodic functions is useful to prove, e.g., embedding theorems as we did in Chap. 4. We have seen in Theorem 4.14 that $H^1_0(D)$ is compactly embedded in $L^2(D)$. This can be carried over to periodic Sobolev spaces of any order.

Theorem 5.6. *Let $Q = (-\pi, \pi)^d \subseteq \mathbb{R}^d$ be the cube in $\mathbb{R}^d$ for $d = 2$ or $d = 3$. For $0 \leq s < t$ the space $H_{per}^t(Q)$ is compactly embedded in $H_{per}^s(Q)$.*

Proof: see Exercise 5.1 □

The generalization of the definition of Sobolev spaces with respect to any $s \in \mathbb{R}_{>0}$, especially $s = \frac{1}{2}$, is a key ingredient and leads to our first elementary trace theorem.

Theorem 5.7. *Let again $Q_3 = (-\pi, \pi)^3 \subseteq \mathbb{R}^3$ denote the cube in $\mathbb{R}^3$ and $Q_2 = (-\pi, \pi)^2 \subseteq \mathbb{R}^2$ the corresponding square in $\mathbb{R}^2$. The trace operator $\gamma_0 : u \mapsto u|_{Q_2 \times \{0\}}$ from the spaces of trigonometric polynomials $\mathcal{P}(Q_3)$ into $\mathcal{P}(Q_2)$ has a bounded extension from $H_{per}^1(Q_3)$ into $H_{per}^{1/2}(Q_2)$. Here—and in the following—we often identify $Q_2 \times \{0\}$ with Q_2.*
Furthermore, there exists a bounded right inverse η of γ_0; that is, a bounded operator $\eta : H_{per}^{1/2}(Q_2) \to H_{per}^1(Q_3)$ with $\gamma_0 \circ \eta = \mathrm{id}$. In other words, the function $u = \eta f \in H_{per}^1(Q_3)$ coincides with $f \in H_{per}^{1/2}(Q_2)$ on $Q_2 \times \{0\}$.

Proof: Let $u \in H_{per}^1(Q_3)$ with Fourier coefficients u_n, $n \in \mathbb{Z}^3$; that is, $u(x) = \sum_{n \in \mathbb{Z}^3} u_n e^{i n \cdot x}$. We truncate the series and define $u_N(x) = \sum_{n_1^2 + n_2^2 \leq N^2} \sum_{|n_3| \leq N} u_n e^{i n \cdot x}$. For $x \in \mathbb{R}^3$ and $n \in \mathbb{Z}^3$ we set $\tilde{x} = (x_1, x_2)$ and $\tilde{n} = (n_1, n_2)$, respectively. Again $|\cdot|$ denotes the Euclidean norm of vectors. Then

$$(\gamma_0 u_N)(x) = u_N(\tilde{x}, 0) = \sum_{|\tilde{n}| \leq N} \underbrace{\left(\sum_{n_3 = -N}^{N} u_n \right)}_{=: v_{\tilde{n}}} e^{i \tilde{n} \cdot \tilde{x}} = \sum_{|\tilde{n}| \leq N} v_{\tilde{n}} e^{i \tilde{n} \cdot \tilde{x}}$$

and thus

$$\|\gamma_0 u_N\|_{H_{per}^{1/2}(Q_2)}^2 = \sum_{|\tilde{n}| \leq N} \left(1 + |\tilde{n}|^2\right)^{1/2} |v_{\tilde{n}}|^2.$$

Now we use the following elementary estimate

$$\frac{\pi}{2} \leq a \sum_{n_3 = -\infty}^{\infty} \frac{1}{a^2 + n_3^2} \leq \pi + 1 \quad \text{for all } a \geq 1 \tag{5.3}$$

which we will show at the end of the proof. Using the upper estimate for $a = \left(1 + |\tilde{n}|^2\right)^{1/2}$ and the inequality of Cauchy–Schwarz, we get

$$(1+|\tilde{n}|^2)^{1/2}\,|v_{\tilde{n}}|^2 = (1+|\tilde{n}|^2)^{1/2}\left|\sum_{n_3=-N}^{N} u_n\,(1+|n|^2)^{1/2}\,\frac{1}{(1+|n|^2)^{1/2}}\right|^2$$

$$\leq \sum_{n_3=-N}^{N}|u_n|^2\,(1+|n|^2)\left[(1+|\tilde{n}|^2)^{1/2}\sum_{n_3=-N}^{N}\frac{1}{1+|\tilde{n}|^2+x_3^2}\right]$$

$$\leq (\pi+1)\sum_{n_3=-N}^{N}|u_n|^2\,(1+|n|^2).$$

Summation with respect to $\tilde{n}$ yields

$$\|\gamma_0 u_N\|^2_{H^{1/2}_{per}(Q_2)} \leq (\pi+1)\,\|u_N\|^2_{H^1_{per}(Q_3)}.$$

This holds for all $N\in\mathbb{N}$. Letting N tend to infinity yields the boundedness of γ_0. To show the existence of a bounded extension operator η we define the extension ηf for $f(\tilde{x})=\sum_{\tilde{n}\in\mathbb{Z}^2}f_{\tilde{n}}\,e^{i\tilde{n}\cdot\tilde{x}}$ by

$$(\eta f)(x) = \sum_{n\in\mathbb{Z}^3} u_n\,e^{in\cdot x},\quad x\in Q_3,$$

where

$$u_n = f_{\tilde{n}}\,\frac{\delta_{\tilde{n}}}{1+|n|^2}\quad\text{and}\quad \delta_{\tilde{n}} = \left[\sum_{j=-\infty}^{\infty}\frac{1}{1+|\tilde{n}|^2+j^2}\right]^{-1}. \qquad (5.4)$$

We first show that $\eta f\in H^1_{per}(Q_3)$. We note that

$$\sum_{n_3=-\infty}^{\infty}|u_n|^2(1+|n|^2) = |f_{\tilde{n}}|^2\delta_{\tilde{n}}^2\sum_{n_3=-\infty}^{\infty}\frac{1}{1+|n|^2}$$

$$= \left(|f_{\tilde{n}}|^2\,\sqrt{1+|\tilde{n}|^2}\right)\frac{\delta_{\tilde{n}}}{\sqrt{1+|\tilde{n}|^2}}$$

$$\leq \frac{2}{\pi}|f_{\tilde{n}}|^2\,\sqrt{1+|\tilde{n}|^2}$$

by the lower estimate of (5.3). Summing over $\tilde{n}$ yields

$$\|\eta f\|^2_{H^1_{per}(Q_3)} = \sum_{\tilde{n}\in\mathbb{Z}^2}\sum_{n_3=-\infty}^{\infty}|u_n|^2(1+|n|^2)$$

$$\leq \frac{2}{\pi}\sum_{\tilde{n}\in\mathbb{Z}^2}|f_{\tilde{n}}|^2\,\sqrt{1+|\tilde{n}|^2} = \frac{2}{\pi}\|f\|^2_{H^{1/2}_{per}(Q_2)}.$$

Therefore, $\eta : H_{per}^{1/2}(Q_2) \to H_{per}^1(Q_3)$ is well defined and bounded. Furthermore,

$$(\gamma_0 \eta f)(\tilde{x}) = \sum_{\tilde{n} \in \mathbb{Z}^2} \sum_{n_3 = -\infty}^{\infty} u_n \, e^{i\tilde{n} \cdot \tilde{x}} = \sum_{\tilde{n} \in \mathbb{Z}^2} f_{\tilde{n}} \, e^{i\tilde{n} \cdot \tilde{x}} = f(\tilde{x}).$$

It remains to show (5.3). We note that

$$\sum_{j=-\infty}^{\infty} \frac{1}{a^2 + j^2} = \frac{1}{a^2} + 2 \sum_{j=1}^{\infty} \frac{1}{a^2 + j^2}$$

and

$$\sum_{j=1}^{N} \frac{1}{a^2 + j^2} = \sum_{j=1}^{N} \int_{j-1}^{j} \frac{dt}{a^2 + j^2}$$

$$\leq \sum_{j=1}^{N} \int_{j-1}^{j} \frac{dt}{a^2 + t^2} = \int_0^N \frac{dt}{a^2 + t^2} = \frac{1}{a} \int_0^{N/a} \frac{ds}{1 + s^2},$$

where we used the substitution $t = as$. Analogously,

$$\sum_{j=1}^{N} \frac{1}{a^2 + j^2} = \sum_{j=1}^{N} \int_j^{j+1} \frac{dt}{a^2 + j^2}$$

$$\geq \sum_{j=1}^{N} \int_j^{j+1} \frac{dt}{a^2 + t^2} = \int_1^{N+1} \frac{dt}{a^2 + t^2} = \frac{1}{a} \int_{1/a}^{(N+1)/a} \frac{ds}{1 + s^2}.$$

Letting N tend to infinity yields

$$\frac{1}{a} \left(\frac{\pi}{2} - \arctan(1/a) \right) \leq \sum_{j=1}^{\infty} \frac{1}{a^2 + j^2} \leq \frac{\pi}{2a}.$$

Noting that $a \geq 1$ implies $\arctan(1/a) \leq \arctan(1) = \pi/4$ which yields the estimates (5.3). $\square$

The motivation for the definition of the Sobolev space $H^{1/2}(\partial D)$ is given by the following transformation of the L^2-norm such that the previous lemma can be applied. Let $D \subseteq \mathbb{R}^3$ be a Lipschitz domain with local coordinate system $\{U_j, \xi_j : j = 1, \ldots, m\}$, corresponding mappings $\tilde{\Psi}_j$ from the balls B_j onto U_j', and their restrictions $\Psi_j : B_2(0, \alpha_j) \to U_j' \cap \partial D$ as in Definition 5.1. By Q_3 we denote a cube centered at the origin such that all of the balls B_j are contained in Q_3. Without loss of generality we assume that this is the cube $Q_3 = (-\pi, \pi)^3$ to make the notation simpler.

Furthermore, let $\{\phi_j : j = 1,\ldots,m\}$, be a partition of unity on ∂D corresponding to the sets U'_j (see Theorem A.9). For $f \in L^2(\partial D)$ we write $|f(y)|^2 = \sum_{j=1}^m \phi_j(y)|f(y)|^2$ for $y \in \partial D$, and thus

$$
\begin{aligned}
\|f\|^2_{L^2(\partial D)} &= \int_{\partial D} |f(y)|^2 \, ds = \sum_{j=1}^m \int_{\partial D \cap U'_j} \phi_j(y)\,|f(y)|^2 \, ds \\
&= \sum_{j=1}^m \int_{B_2(0,\alpha_j)} \phi_j(\Psi_j(x))|f(\Psi_j(x))|^2 \sqrt{1+|\nabla \xi_j(x)|^2}\, dx \\
&= \sum_{j=1}^m \int_{Q_2} |\tilde{f}_j(x)|^2 \sqrt{1+|\nabla \xi_j(x)|^2}\, dx
\end{aligned}
\tag{5.5}
$$

where

$$
\tilde{f}_j(x) := \begin{cases} \sqrt{\phi_j(\Psi_j(x))}\, f(\Psi_j(x)), & x \in B_2(0,\alpha_j), \\ 0, & x \in Q_2 \setminus B_2(0,\alpha_j). \end{cases}
\tag{5.6}
$$

From (5.5) and the estimate $1 \leq \sqrt{1+|\nabla \xi_j(x)|} \leq \max_{j=1,\ldots,m} \sqrt{1+\|\nabla \xi_j\|^2_\infty}$ we observe that $\|\cdot\|_{L^2(\partial D)}$ is equivalent to $\sqrt{\sum_{j=1}^m \|\tilde{f}_j\|^2_{L^2(Q_2)}}$. Therefore, $f \in L^2(\partial D)$ if, and only if, $\tilde{f}_j \in L^2(Q_2)$ for all $j = 1,\ldots,m$. We extend this definition to define the Sobolev space $H^{1/2}(\partial D)$.

Definition 5.8. Let $D \subseteq \mathbb{R}^3$ be a Lipschitz domain in the sense of Definition 5.1 with corresponding local coordinate system $\{U_j, \xi_j : j = 1,\ldots,m\}$ and corresponding mappings $\tilde{\Psi}_j$ from the balls B_j onto U'_j and their restrictions $\Psi_j : B_2(0,\alpha_j) \to U'_j \cap \partial D$. Furthermore, let $\{\phi_j : j = 1,\ldots,m\}$, be a partition of unity on ∂D corresponding to the sets U'_j (see Theorem A.9). Then we define

$$
H^{1/2}(\partial D) = \{f \in L^2(\partial D) : \tilde{f}_j \in H^{1/2}_{per}(Q_2) \text{ for all } j = 1,\ldots,m\}
$$

with norm

$$
\|f\|_{H^{1/2}(\partial D)} = \left[\sum_{j=1}^m \|\tilde{f}_j\|^2_{H^{1/2}_{per}(Q_2)} \right]^{1/2},
$$

where $\tilde{f}_j$ are given by (5.6), $j = 1,\ldots,m$.

Our definition of the space $H^{1/2}(\partial D)$ seems to depend on the choice of the local coordinates ξ_j and the partition of unity ϕ_j. However, we will see in Corollary 5.15 below that the norms corresponding to two such choices are equivalent.

From now on we assume always that the domain D is a Lipschitz domain in the sense of Definition 5.1.

We note the following implication of Theorem 5.6.

Corollary 5.9. *The space $H^{1/2}(\partial D)$ is compactly embedded in $L^2(\partial D)$.*

Proof: Let $(f^\ell)_\ell$ be a bounded sequence in $H^{1/2}(\partial D)$. Then, by definition, $(\tilde{f}_j^\ell)_\ell$ is bounded in $H_{per}^{1/2}(Q_2)$ for all $j = 1, \ldots, m$, where $\tilde{f}_j^\ell$ is defined by (5.6) for f^ℓ instead of f. By Theorem 5.6 for $s = 0$ and $t = 1/2$ there exists a subsequence $(\tilde{f}_j^{\ell_k})_k$ of $(\tilde{f}_j^\ell)_\ell$ which converges in $L^2(Q_2)$ for every $j = 1, \ldots, m$ as k tends to infinity. By (5.5), applied to the difference $f^{\ell_k} - f^{\ell_p}$ also $(f^{\ell_k})_k$ converges in $L^2(\partial D)$. $\square$

Definition 5.8 and Theorem 5.7 yield a trace theorem in $H^1(D)$ for Lipschitz domains.

Theorem 5.10. *The trace operator $\gamma_0 : C^1(\overline{D}) \to C(\partial D)$, $u \mapsto u|_{\partial D}$, has an extension as a bounded operator from $H^1(D)$ to $H^{1/2}(\partial D)$. Furthermore, γ_0 has a bounded right inverse $\eta : H^{1/2}(\partial D) \to H^1(D)$; that is, $\gamma_0(\eta f) = f$ for all $f \in H^{1/2}(\partial D)$.*

Proof: Let $\{U_j, \xi_j : j = 1, \ldots, m\}$ be a local coordinate system of ∂D with corresponding mappings $\tilde{\Psi}_j$ from the balls B_j onto U_j' as in Definition 5.1. Let $\{\phi_j : j = 1, \ldots, m\}$ be a partition of unity on ∂D with respect to U_j'. Set $U := \bigcup_{j=1}^m U_j'$ and let $u \in C^1(\overline{D})$. Set

$$v_j(x) := \sqrt{\phi_j(\tilde{\Psi}_j(x))} \, u(\tilde{\Psi}_j(x)), \quad x \in B_j^-$$

and extend v_j to Q_3^- by zero. Here, $B_j^\pm = \{x \in B_j : x_3 \gtrless 0\}$ and $Q_3^\pm = \{x \in Q_3 : x_3 \gtrless 0\}$. Furthermore, we extend $v_j \in C(\overline{Q_3^-})$ into Q_3 by even reflection; that is,

$$\tilde{v}_j(x) = \begin{cases} v_j(x), & x \in Q_3^-, \\ v_j(x^*), & x \in Q_3^+, \end{cases}$$

where $x^* = (x_1, x_2, -x_3)^\top$ for $x = (x_1, x_2, x_3)^\top \in \mathbb{R}^3$. Then $\tilde{v}_j \in H^1(Q_3)$ by Exercise 4.6. Also, $\tilde{v}_j$ vanishes in some neighborhood of ∂Q_3. Now we can use the arguments of Lemma 4.37 to show that $\tilde{v}_j \in H_{per}^1(Q_3)$ and that there exists $c_1 > 0$ which is independent of $\tilde{v}_j$ such that

$$\|\tilde{v}_j\|_{H_{per}^1(Q_3)} \leq c_1 \|v_j\|_{H^1(Q_3^-)} = c_1 \|v_j\|_{H^1(B_j^-)}.$$

The boundedness of the trace operator of Theorem 5.7 yields

$$\|\tilde{v}_j(\cdot, 0)\|_{H_{per}^{1/2}(Q_2)} \leq c_2 \|v_j\|_{H^1(B_j^-)} \quad \text{for all } j = 1, \ldots, m,$$

and thus by Definition 5.8

$$\|\gamma_0 u\|_{H^{1/2}(\partial D)}^2 = \sum_{j=1}^m \|\tilde{v}_j(\cdot, 0)\|_{H_{per}^{1/2}(Q_2)}^2$$

$$\leq c_2^2 \sum_{j=1}^m \|v_j\|_{H^1(B_j^-)}^2 \leq c\|u\|_{H^1(D\cap U)}^2 \leq c\|u\|_{H^1(D)}^2 .$$

This proves boundedness of γ_0.

Now we construct an extension operator $\eta : H^{1/2}(\partial D) \to H^1(D)$. For $f \in H^{1/2}(\partial D)$ we define $\tilde{f}_j \in H_{per}^{1/2}(Q_2)$ for $j = 1, \ldots, m$ as in (5.6). By Theorem 5.7 there exist extensions $v_j \in H_{per}^1(Q_3)$ of $\tilde{f}_j$ and the mappings $\tilde{f}_j \mapsto v_j$ are bounded for $j = 1, \ldots, m$. Then we set

$$u_j(y) := \sqrt{\phi_j(y)}\, v_j(\tilde{\Psi}_j^{-1}(y)), \quad y \in U_j'.$$

Then $u_j \in H^1(U_j')$ for all j and they vanish in some neighborhood of $\partial U_j'$. We extend u_j by zero into all of $\mathbb{R}^3$ and set $u = \sum_{j=1}^m u_j$. Then u is an extension of f. Indeed, for fixed $y \in \partial D$ let $J = \{j \in \{1, \ldots, m\} : y \in U_j'\}$. Then $u(y) = \sum_{j \in J} u_j(y)$. Let $j \in J$ be fixed; that is, $y \in U_j'$. Then $x = \Psi_j^{-1}(y) \in Q_2 \times \{0\}$, thus $u_j(y) = \sqrt{\phi_j(y)}\, v_j(x) = \sqrt{\phi_j(y)}\, \tilde{f}_j(x) = \phi(y)f(y)$. Therefore, $u(y) = \sum_{j \in J} u_j(y) = \sum_{j \in J} f(y)\,\phi_j(y) = f(y)$. Furthermore, all the operations in these constructions are bounded. This proves the theorem. □

Remark 5.11. We note that in the second part of the proof we have constructed an extension u of f into all of $\mathbb{R}^3$ with support in the neighborhood $U = \bigcup_{j=1}^m U_j'$ of ∂D which depends continuously on f. Applying this to $f = \gamma_0 u$ for $u \in H^1(D)$ yields the existence of a bounded extension operator $\tilde{\eta} : H^1(D) \to H^1(\mathbb{R}^3)$.

As a corollary we have the following embedding result of Rellich (compare with Theorem 4.14):

Theorem 5.12. *Let $D \subseteq \mathbb{R}^3$ be a bounded Lipschitz domain. Then the embedding $H^1(D) \to L^2(D)$ is compact.*

Proof: We can just copy the proof of Theorem 4.14 because of the existence of a bounded extension operator $\tilde{\eta} : H^1(D) \to H^1(\mathbb{R}^3)$ (see the previous Remark 5.11). □

A simple conclusion of the trace theorem in combination with the denseness result of Theorem 5.3 is formulated in the following corollary.

Corollary 5.13. *Let $D \subseteq \mathbb{R}^3$ be a Lipschitz domain.*

(a) *The formula of partial integration (see Theorem A.11) holds in the form*

$$\int_D u \, \nabla v \, dx = -\int_D v \nabla u \, dx + \int_{\partial D} (\gamma_0 u)(\gamma_0 v) \, \nu \, ds \quad \text{for all } u, v \in H^1(D).$$

(b) *Let $\Omega \subseteq \mathbb{R}^3$ be a second Lipschitz domain with $D \subseteq \Omega$ and let $u_1 \in H^1(D)$ and $u_2 \in H^1(\Omega \backslash \overline{D})$ such that $\gamma_0 u_1 = \gamma_0 u_2$ on ∂D. Then the function defined by*

$$u(x) = \begin{cases} u_1(x), & x \in D \\ u_2(x), & x \in \Omega \backslash \overline{D} \end{cases}$$

is in $H^1(\Omega)$.
If, in particular, $\gamma_0 u_1 = 0$, then the extension u of u_1 by zero outside of D belongs to $H^1(\mathbb{R}^3)$.

(c) *Let $D \subseteq \mathbb{R}^3$ be symmetric with respect to $x_3 = 0$; that is, $x \in D \Leftrightarrow x^* \in D$ where $z^* = (z_1, z_2, -z_3)^\top$ for any $z = (z_1, z_2, z_3)^\top \in \mathbb{C}^3$. Let $D^\pm = \{x \in D : x_3 \gtrless 0\}$ and $\varphi \in H^1(D^-)$. Extend φ into D by reflection; that is,*

$$\tilde{\varphi}(x) = \begin{cases} \varphi(x), & x \in D^-, \\ \varphi(x^*), & x \in D^+, \end{cases}$$

Then $\tilde{\varphi} \in H^1(D)$ and the mapping $\varphi \mapsto \tilde{\varphi}$ is bounded from $H^1(D^-)$ into $H^1(D)$.

The **proofs** are very similar to the proof of Lemma 5.2 and left as Exercise 5.3, see also Exercise 4.6.

We recall from Definition 4.10 that the subspace $H_0^1(D)$ had been defined as the closure of $C_0^\infty(D)$ in $H^1(D)$. The following equivalent characterization will be of essential importance.

Theorem 5.14. *The space $H_0^1(D)$ is the null space $\mathcal{N}(\gamma_0)$ of the trace operator; that is, $u \in H_0^1(D)$ if, and only if, $\gamma_0 u = 0$ on ∂D.*

Proof: The inclusion $H_0^1(D) \subseteq \mathcal{N}(\gamma_0)$ follows immediately because $\gamma_0 u = 0$ for all $u \in C_0^\infty(D)$, the boundedness of γ_0 and the fact that $C_0^\infty(D)$ is dense in $H_0^1(D)$ by definition. The reverse inclusion is more difficult to show. It follows closely the proof of Theorem 5.3. We separate the proof into four steps.

(i) Let $u \in H^1(D)$ with $\gamma_0 u = 0$. We extend u by zero into all of $\mathbb{R}^3$. Then $u \in H^1(\mathbb{R}^3)$ by Corollary 5.13. Let $\{U_j, \xi_j : j = 1, \ldots, m\}$ be a local coordinate system. Set $U_0 = D$ and choose a partition of unity $\{\varphi_j : j = 0, \ldots, m\}$ on $\overline{D}$ corresponding to $\{U_j : j = 0, \ldots, m\}$ (see Theorem A.9. Set $u_j(y) = \varphi_j(y)u(y)$ for $j = 0, \ldots, m$ and $y \in \mathbb{R}^3$. Then the support of u_j is contained in $U_j \cap \overline{D}$. For $j = 1, \ldots, m$ we transform u_j by the definition $v_j(x) = u_j(R_j x + z^{(j)})$, $x \in \mathbb{R}^3$, and observe that $v_j \in H^1(\mathbb{R}^3)$ by the chain rule of Corollary 5.4, applied to a large ball containing $\overline{D}$. The support of v_j is contained in the cylinder $C_j = B_2(0, \alpha_j) \times (-2\beta_j, 2\beta_j)$. Furthermore, with

$$C_j^- := \{x = (\tilde{x}, x_3) \in B_2(0, \alpha_j) \times (-2\beta_j, 2\beta_j) : x_3 < \xi_j(\tilde{x})\} \quad \text{and}$$
$$V_j(t) := \{x \in C_j^- : |\tilde{x}| > \alpha_j - t \text{ or } x_3 < -2\beta_j + t\}$$

where again $\tilde{x} = (x_1, x_2)$ we note that there exists $\delta > 0$ with $v_j(x) = 0$ in the neighborhood $V_j(2\delta)$ of $\partial C_j \cap \overline{C_j^-}$ because the support of φ_j is contained in U_j.

(ii) Let $a = (0, 0, L+2)^\top$ where L is larger than all of the Lipschitz constants of the functions ξ_j and set

$$v_j^\varepsilon(x) = (v_j * \phi_\varepsilon)(x + \varepsilon a) = \int_{C_j^- \setminus V_j(2\delta)} v_j(y) \, \phi_\varepsilon(x + \varepsilon a - y) \, dy$$

where $\phi_\varepsilon \in C^\infty(\mathbb{R}^3)$ denotes the mollifier function from (4.7). Its most important properties had been collected in Theorem 4.7. In particular, we have that $v_j^\varepsilon \in C_0^\infty(\mathbb{R}^3)$ and $v_j^\varepsilon \to v_j$ in $H^1(\mathbb{R}^3)$ as ε tends to zero.

(iii) Next we show that $\operatorname{supp}(v_j^\varepsilon) \subseteq C_j^-$ for $\varepsilon < \delta/(L+3)$. First we note that the integral in the definition of v_j^ε is taken over $C_j^- \setminus V_j(2\delta)$. Therefore, let $y \in C_j^- \setminus V_j(2\delta)$. Then $|\tilde{y}| \leq \alpha_j - 2\delta$ and $y_3 \geq -2\beta_j + 2\delta$. The boundary of C_j^- consists for three parts. We consider x being in some neighborhood of these parts separately.

For points $x \in C_j^-$ with $|\tilde{x}| \geq \alpha_j - \delta$ we have shown this in part (ii) of the proof of Theorem 5.3.

Let now $x \in C_j^-$ with $x_3 \leq -2\beta_j + \delta$. Then $|x + \varepsilon a - y| \geq |x_3 + \varepsilon(L+2) - y_3| \geq y_3 - x_3 - (L+2)\varepsilon \geq -2\beta_j + 2\delta + 2\beta_j - \delta - (L+2)\varepsilon = \delta - (L+2)\varepsilon \geq \varepsilon$. Therefore, $v_j^\varepsilon(x) = 0$ for these x.

Finally, let $x \in C_j^-$ with $x_3 \geq \xi_j(\tilde{x}) - \varepsilon$ and $|\tilde{y} - \tilde{x}| \leq \varepsilon$. Then $|x + \varepsilon a - y| \geq |x_3 + \varepsilon(L+2) - y_3| \geq x_3 + \varepsilon(L+2) - y_3 \geq \xi_j(\tilde{x}) - \varepsilon + (L+2)\varepsilon - y_3 = [\xi_j(\tilde{y}) - y_3] - [\xi_j(\tilde{y}) - \xi_j(\tilde{x})] + (L+1)\varepsilon \geq (L+1)\varepsilon - L|\tilde{y} - \tilde{x}| \geq \varepsilon$. Therefore, $v_j^\varepsilon(x) = 0$ also for these x which shows that $\operatorname{supp}(v_j^\varepsilon) \subseteq C_j^-$ for $\varepsilon < \delta/(L+3)$; that is, $v_j^\varepsilon \in C_0^\infty(C_j^-)$.

(iv) In the last step we transform back the function and set $u_j^\varepsilon(y) = v_j^\varepsilon(R_j^\top(y - z^{(j)}))$ for $j = 1, \ldots, m$ and $y \in \mathbb{R}^3$. We note that $u_j^\varepsilon \in C_0^\infty(U_j \cap D)$ and $u_j^\varepsilon \to \varphi_j u$ in $H^1(D)$. This holds for all $j = 1, \ldots, m$. Finally, we note that $u_0 \in H^1(D)$ with compact support in $U_0 = D$ and thus $u_0^\varepsilon = u_0 * \phi_\varepsilon$ converges to $u_0 = \varphi_0 u$ in $H^1(D)$. To finish the proof we set $u^\varepsilon = \sum_{j=0}^m u_j^\varepsilon$ in D and conclude that $u^\varepsilon \in C_0^\infty(D)$ and u^ε converges to $\sum_{j=0}^m \varphi_j u = u$ in $H^1(D)$. $\quad\square$

As mentioned before we still have to complete Definition 5.8 by showing that the space $H^{-1/2}(\partial D)$ is well defined.

Corollary 5.15. *Definition 5.8 is independent of the choice of the coordinate system* $\{U_j, \xi_j : j = 1, \ldots, m\}$ *and a corresponding partition of unity* $\{\phi_j : j = 1, \ldots, m\}$ *on* ∂D *with respect to* U_j'.

Proof: First we note that the space $C^\infty(\overline{D})|_{\partial D} = \{v|_{\partial D} : v \in C^\infty(\overline{D})\}$ is dense in $H^{1/2}(\partial D)$. Indeed, for $f \in H^{1/2}(\partial D)$ we conclude that $u = \eta f \in H^1(D)$ can be approximated by a sequence $u_n \in C^\infty(\overline{D})$ with respect to $\|\cdot\|_{H^1(D)}$. But then $u_n|_{\partial D} = \gamma_0 u_n$ converges to $\gamma_0 u = f$ in $H^{1/2}(\partial D)$. Therefore, $H^{1/2}(\partial D)$ is the completion of $C^\infty(\overline{D})|_{\partial D}$ with respect to $\|\cdot\|_{H^{1/2}(\partial D)}$. Furthermore, from the boundedness and surjectivity of $\gamma_0 : H^1(D) \to H^{1/2}(\partial D)$ by Theorem 5.14 we observe that the lifted operator $[\gamma_0] : H^1(D)/H_0^1(D) \to H^{1/2}(\partial D)$ is a norm-isomorphism from the factor space $H^1(D)/H_0^1(D)$ onto $H^{1/2}(\partial D)$. Therefore, $\|\cdot\|_{H^{1/2}(\partial D)}$ is equivalent to the norm $\|[v]\| = \inf\{\|v + \psi\|_{H^1(D)} : \psi \in H_0^1(D)\}$ in $H^1(D)/H_0^1(D)$; that is, there exist constants $c_1, c_2 > 0$ with $c_1\|[v]\| \leq \|v|_{\partial D}\|_{H^{1/2}(\partial D)} \leq c_2\|[v]\|$ for all $v \in C^\infty(\overline{D})$. The canonical norm in $H^1(D)/H_0^1(D)$ depends only on the norm in $H^1(D)$ and the subspace $H_0^1(D)$. Therefore, all of the norms $\|\cdot\|_{H^{1/2}(\partial D)}$ originating from different choices of the coordinate system and partitions of unity are equivalent to each other. $\square$

The proof shows also that an equivalent norm in $H^{1/2}(\partial D)$ is given by $\|f\| = \inf\{\|u\|_{H^1(D)} : \gamma_0 u = f \text{ on } \partial D\}$.

Before we turn to the vector valued case we want to discuss the *normal derivative* $\partial u/\partial \nu$ on ∂D which is well defined for $u \in C^1(\overline{D})$. It can be considered as the trace of the normal component of the gradient of u. We denote it by $\gamma_1 : C^1(\overline{D}) \to C(\partial D)$, $u \mapsto \partial u/\partial \nu$. In Exercise 5.2 we show for an example that it is not bounded from $H^1(D)$ into $L^2(\partial D)$. (It is not even well defined as one can show.) However, it defines a bounded operator on the closed subspace

$$H_D = \left\{ u \in H^1(D) : \int_D [\nabla u \cdot \nabla \psi - k^2 u \psi] \, dx = 0 \text{ for all } \psi \in H_0^1(D) \right\}$$
(5.7)

of (variational) solutions of the Helmholtz equation into the dual space of $H^{1/2}(\partial D)$ as we will see in a moment.

First we recall from Corollary 5.9 that $H^{1/2}(\partial D)$ is a subspace of $L^2(\partial D)$ with bounded—even compact—inclusion. For any $f \in L^2(\partial D)$ the linear form $\ell_f(\psi) = (f, \overline{\psi})_{L^2(\partial D)}$ defines a bounded linear functional on $H^{1/2}(\partial D)$ because $|\ell_f(\psi)| \leq \|f\|_{L^2(\partial D)}\|\psi\|_{L^2(\partial D)} \leq c\|f\|_{L^2(\partial D)}\|\psi\|_{H^{1/2}(\partial D)}$ for all $\psi \in H^{1/2}(\partial D)$. Therefore, if we identify ℓ_f with f—as done by identifying the dual space of $L^2(\partial D)$ with itself—then $L^2(\partial D)$ can be considered as a subspace of the dual space $H^{1/2}(\partial D)^*$ of $H^{1/2}(\partial D)$ which we denote by $H^{-1/2}(\partial D)$.

Definition 5.16. Let $H^{-1/2}(\partial D)$ be the dual space of $H^{1/2}(\partial D)$ equipped with the canonical norm of a dual space; that is,

$$\|\ell\|_{H^{-1/2}(\partial D)} := \sup_{\psi \in H^{1/2}(\partial D)\setminus\{0\}} \frac{|\langle \ell, \psi \rangle_*|}{\|\psi\|_{H^{1/2}(\partial D)}}, \quad \ell \in H^{-1/2}(\partial D),$$

where $\langle \ell, \psi \rangle_* = \ell(\psi)$ denotes the dual form, the evaluation of ℓ at ψ.

For real valued functions we observe that $\langle \ell, \psi \rangle_*$ is just the extension of the L^2-inner product when we identify ℓ_f with f .

Using a local coordinate system and a corresponding partition of unity one can prove that $H^{-1/2}(\partial D)$ can be characterized by the periodic Sobolev space $H_{per}^{-1/2}(Q_2)$ which is the completion of the space of trigonometric polynomials by the norm

$$\|v\|_{H_{per}^{-1/2}(Q_2)} = \left[\sum_n (1 + |n|^2)^{-1/2} |v_n|^2 \right]^{1/2} .$$

We do not need this result and, therefore, omit the details.

The definition of the trace $\partial u / \partial \nu$ is motivated by Green's first theorem: For $u \in C^2(\overline{D})$ with $\Delta u + k^2 u = 0$ in D and $\psi \in H^1(D)$ we have

$$\int_{\partial D} \frac{\partial u}{\partial \nu} \gamma_0 \psi \, ds = \int_D \left[\nabla u \cdot \nabla \psi - k^2 u \psi \right] dx .$$

Since the trace $\gamma_0 \psi$ of ψ is an element of $H^{1/2}(\partial D)$ the left-hand side is a linear functional on $H^{1/2}(\partial D)$. The right-hand side is well defined also for $u \in H^1(D)$. Therefore, it is natural to extend this formula to $u, \psi \in H^1(D)$ and replace the left-hand side by the dual form $\langle \partial u / \partial \nu, \gamma_0 \psi \rangle_*$. This is justified by the following theorem.

Definition 5.17. (and **Theorem**) The operator $\gamma_1 : H_D \to H^{-1/2}(\partial D)$, defined by Green's formula; that is,

$$\langle \gamma_1 u, \psi \rangle_* = \int_D \left[\nabla u \cdot \nabla \tilde{\psi} - k^2 u \tilde{\psi} \right] dx , \quad \psi \in H^{1/2}(\partial D), \qquad (5.8)$$

is well defined and bounded. Here, $\tilde{\psi} \in H^1(D)$ is any extension of ψ into D; that is, $\gamma_0 \tilde{\psi} = \psi$—which is possible by the surjectivity of the trace operator γ_0, see Theorem 5.10.

Proof: We have to show that this definition is independent of the choice of $\tilde{\psi}$. Indeed, if $\tilde{\psi}_1$ and $\tilde{\psi}_2$ are two extensions of ψ, then $\tilde{\psi} := \tilde{\psi}_1 - \tilde{\psi}_2 \in H_0^1(D)$ by Theorem 5.14, because $\gamma_0 \tilde{\psi} = 0$. With $u \in H_D$ we conclude that

$\int_D \left[\nabla u \cdot \nabla (\tilde{\psi}_1 - \tilde{\psi}_2) - k^2 u (\tilde{\psi}_1 - \tilde{\psi}_2) \right] dx = 0$. This shows that the definition of γ_1 is independent of the choice of $\tilde{\psi}$.

To show boundedness of γ_1 we take $\tilde{\psi} = \eta \psi$ where $\eta : H^{1/2}(\partial D) \to H^1(D)$ denotes the bounded right inverse of γ_0 of Theorem 5.10. Then, by the inequality of Cauchy–Schwarz,

$$
\begin{aligned}
|\langle \gamma_1 u, \psi \rangle_*| &\leq \|\nabla u\|_{L^2(D)} \|\nabla \tilde{\psi}\|_{L^2(D)} + k^2 \|u\|_{L^2(D)} \|\tilde{\psi}\|_{L^2(D)} \\
&\leq \max\{k^2, 1\} \|u\|_{H^1(D)} \|\tilde{\psi}\|_{H^1(D)} \\
&= \max\{k^2, 1\} \|u\|_{H^1(D)} \|\eta \psi\|_{H^1(D)} \\
&\leq \max\{k^2, 1\} \, \|\eta\| \, \|u\|_{H^1(D)} \|\psi\|_{H^{1/2}(\partial D)} \, .
\end{aligned}
$$

This proves boundedness of γ_1 with $\|\gamma_1\|_{H^{-1/2}(\partial D)} \leq \max\{k^2, 1\} \, \|\eta\| \, \|u\|_{H^1(D)}$.
□

If the region D is of the form $D = D_1 \backslash D_2$ for open sets D_j such that $\overline{D_2} \subseteq D_1$, then the boundary ∂D consists of two components ∂D_1 and ∂D_2. The spaces $H^{\pm 1/2}(\partial D)$ can be written as direct products $H^{\pm 1/2}(\partial D_1) \times H^{\pm 1/2}(\partial D_2)$, and the trace operators γ_0 and γ_1 have two components $\gamma_0|_{\partial D_j}$ and $\gamma_1|_{\partial D_j}$ for $j = 1, 2$ which are the projections onto ∂D_j. For the definition of $\gamma_0|_{\partial D_j}$ or $\gamma_1|_{\partial D_j}$ one just takes extensions $\tilde{\psi} \in H^1(D)$ which vanish on ∂D_{3-j}.

Remark: Some readers may feel perhaps unhappy with the space H_D as the domain of definition for the trace operator γ_1 because it depends on the wave number k. A more natural way is to define the trace operator on the space

$$
H^1(\Delta, D) := \left\{ u \in H^1(D) : \begin{array}{l} \exists \, v \in L^2(D) \text{ with } \int_D \nabla u \cdot \nabla \psi \, dx = - \int_D v \psi \, dx \\ \text{for all } \psi \in H_0^1(D) \end{array} \right\}
$$

as the space of functions in $H^1(D)$ for which even $\Delta u := v \in L^2(D)$. (Note that the function v is obviously unique.) The space $H^1(\Delta, D)$ is equipped with the canonical norm $\|u\|_{H^1(\Delta, D)} = \|u\|_{H^1(D)} + \|\Delta u\|_{L^2(D)}$. Definition 5.17 should be changed into

$$
\langle \gamma_1 u, \psi \rangle_* = \int_D \left[\nabla u \cdot \nabla \tilde{\psi} + \tilde{\psi} \, \Delta u \right] dx, \quad \psi \in H^{1/2}(\partial D) \, .
$$

However, we note that H_D is a subspace of $H^1(\Delta, D)$, and on this subspace the norms of $H^1(D)$ and $H^1(\Delta, D)$ are equivalent. For the following it is important that we have boundedness of γ_1 in the $H^1(D)$-norm, and this is the reason for choosing H_D as the domain of definition.

5.1.2 Sobolev Spaces of Vector Valued Functions

Now we turn to Sobolev spaces of vector functions. In Chap. 4 we have studied the space $H(\mathrm{curl}, D)$. It is the aim to prove two trace theorems for this space. To motivate these traces we consider the integral identity (A.16b); that is,

$$\int_D (v \cdot \mathrm{curl}\, u - u \cdot \mathrm{curl}\, v)\, dx = \int_{\partial D} (\nu \times u) \cdot v\, ds = \int_{\partial D} (\nu \times u) \cdot \big((\nu \times v) \times \nu\big)\, ds$$

for $u, v \in C^1(D, \mathbb{C}^3) \cap C(\overline{D}, \mathbb{C}^3)$ and a smooth domain D. Thus, the left-hand side leads to two possibilities in extending traces of the tangential components of the vector fields on ∂D namely first, for $\gamma_t u = \nu \times u$, if we consider v as a test function and, second for $\gamma_T v = (\nu \times v) \times \nu$, if we take u as a test function. Furthermore, taking gradients of the form $v = \nabla \varphi$ for scalar functions φ as test functions gives

$$\int_D \nabla \varphi \cdot \mathrm{curl}\, u\, dx = \int_{\partial D} (\nu \times u) \cdot \gamma_T \nabla \varphi\, ds = \int_{\partial D} (\nu \times u) \cdot \mathrm{Grad}\,\varphi\, ds.$$

From (A.21) we observe that the right-hand side is just $-\int_{\partial D} \mathrm{Div}(\nu \times u)\, \varphi\, ds$, see Definition 5.29 below. Thus we observe that the traces of vector fields in $H(\mathrm{curl}, D)$ have some regularity which requires more detailed investigations.

We proceed as for the scalar case and begin with the following simple lemma which corresponds to Lemma 5.2.

Lemma 5.18. *Let D be a bounded Lipschitz domain.*

(a) *Let $u \in C^1(\overline{D}, \mathbb{C}^3)$ with $\nu \times u = 0$ on ∂D. The extension $\tilde{u}$ of u by zero outside of $\overline{D}$ yields $\tilde{u} \in H(\mathrm{curl}, \mathbb{R}^3)$ and $\mathrm{curl}\,\tilde{u} = \mathrm{curl}\, u$ in D and $\mathrm{curl}\,\tilde{u} = 0$ outside of D.*

(b) *Let $R \in \mathbb{R}^{3 \times 3}$ be an orthogonal matrix and $z \in \mathbb{R}^3$. For $u \in H(\mathrm{curl}, D)$ define*

$$v(x) = R^\top u(Rx + z) \quad on\ U = \{x : Rx + z \in D\}.$$

Then $v \in H(\mathrm{curl}, U)$ and

$$\mathrm{curl}\, v(x) = (\det R)\, R^\top \mathrm{curl}\, u(y)|_{y=Rx+z} \quad on\ U.$$

Proof: The proof of (a) is very similar to the proof of Lemma 5.2 and is omitted.

(b) With $\psi \in C_0^\infty(U)$ and the substitution $x = R^\top(y - z)$ we have

$$\int_U v(x)^\top \operatorname{curl} \psi(x)\, dx = \int_U u(Rx + z)^\top R \operatorname{curl} \psi(x)\, dx$$

$$= \det R \int_D u(y)^\top R \operatorname{curl}_x \psi(x)|_{x=R^\top(y-z)}\, dy.$$

Let $q^{(j)}$, $j = 1, 2, 3$, be the columns of R. Then

$$\operatorname{curl}_y \big[R\psi(R^\top(y - z)) \big] = \operatorname{curl}_y \sum_{j=1}^3 \psi_j(R^\top(y - z))\, q^{(j)} = \sum_{j=1}^3 R\nabla \psi_j(x) \times q^{(j)}$$

$$= \sum_{j,\ell=1}^3 \frac{\partial \psi_j(x)}{\partial x_\ell}\, q^{(\ell)} \times q^{(j)}$$

where we have set $x = R^\top(y - z)$. If $\det R = 1$, then $q^{(1)} \times q^{(2)} = q^{(3)}$ and $q^{(2)} \times q^{(3)} = q^{(1)}$ and $q^{(3)} \times q^{(1)} = q^{(2)}$ and thus $\operatorname{curl}_y\big[R\psi(R^\top(y - z))\big] = R \operatorname{curl}_x \psi(x)|_{x=R^\top(y-z)}$. If $\det R = -1$, then analogous computations yield $\operatorname{curl}_y\big[R\psi(R^\top(y - z))\big] = -R \operatorname{curl}_x \psi(x)|_{x=R^\top(y-z)}$. Therefore, making again the substitution $y = Rx + z$ yields

$$\int_U v(x)^\top \operatorname{curl} \psi(x)\, dx = \int_D u(y)^\top \operatorname{curl}_y\big[R\psi(R^\top(y - z))\big]\, dy$$

$$= \int_D \operatorname{curl} u(y)^\top R\psi(R^\top(y - z))\, dy$$

$$= \det R \int_U \operatorname{curl}_y u(y)^\top|_{y=Rx+z} R\psi(x)\, dx$$

$$= \det R \int_U \big(R^\top \operatorname{curl}_y u(y)|_{y=Rx+z}\big)^\top \psi(x)\, dx.$$

This shows that $(\det R)\, R^\top \operatorname{curl}_y u(y)|_{y=Rx+z}$ is the variational curl of v. $\quad\square$

The following result corresponds to Theorem 5.3.

Theorem 5.19. *If D is a bounded Lipschitz domain, then $C^\infty(\overline{D}, \mathbb{C}^3)$ is dense in $H(\operatorname{curl}, D)$.*

Proof: We can almost copy the proof of Theorem 5.3 if we replace ∇ by curl. We omit the details. $\quad\square$

As in the scalar case we introduce first some Sobolev spaces of periodic functions. We recall that every vector valued function $v \in L^2(Q, \mathbb{C}^d)$ on a cube $Q = (-\pi, \pi)^d$ in $\mathbb{R}^d$ for $d = 2$ or $d = 3$ has an expansion in the form

$$v(x) = \sum_{n \in \mathbb{Z}^d} v_n\, e^{in \cdot x}$$

with Fourier coefficients $v_n \in \mathbb{C}^d$, given by

$$v_n = \frac{1}{(2\pi)^d} \int_{Q_d} v(y) \, e^{-i\,n\cdot y} \, dy, \quad n \in \mathbb{Z}^d. \tag{5.9}$$

For $d = 3$ formal differentiation yields

$$\operatorname{div} v(x) = i \sum_{n \in \mathbb{Z}^3} n \cdot v_n \, e^{in\cdot x}, \quad \operatorname{curl} v(x) = i \sum_{n \in \mathbb{Z}^3} n \times v_n \, e^{in\cdot x}.$$

Therefore, $\operatorname{curl} v \in L^2(Q, \mathbb{C}^3)$ if, and only if, $\sum_{n \in \mathbb{Z}^3} |n \times v_n|^2 < \infty$. This is the motivation for the following definition.

Definition 5.20. Let again $Q_3 = (-\pi, \pi)^3 \subseteq \mathbb{R}^3$ be the cube and $Q_2 = (-\pi, \pi)^2 \subseteq \mathbb{R}^2$ be the square. For any vector function $v : Q_d \to \mathbb{C}^d$, $d = 2$ or $d = 3$, the Fourier coefficients $v_n \in \mathbb{C}^d$ for $n \in \mathbb{Z}^d$ are defined as

$$v_n = \frac{1}{(2\pi)^d} \int_{Q_d} v(y) \, e^{-i\,n\cdot y} \, dy, \quad n \in \mathbb{Z}^d.$$

(a) The space $H_{per}(\operatorname{curl}, Q_3)$ is defined by

$$H_{per}(\operatorname{curl}, Q_3) = \left\{ v \in L^2(Q_3)^3 : \sum_{n \in \mathbb{Z}^3} \left[|v_n|^2 + |n \times v_n|^2 \right] < \infty \right\}$$

with norm

$$\|v\|_{H_{per}(\operatorname{curl}, Q_3)} = \sqrt{\sum_{n \in \mathbb{Z}^3} \left[|v_n|^2 + |n \times v_n|^2 \right]}.$$

(b) For any $s \in \mathbb{R}$ the space $H^s_{per}(\operatorname{Div}, Q_2)$ is defined as the completion of the space

$$\mathcal{T}(Q_2, \mathbb{C}^2) := \left\{ \sum_{|m| \leq N} v_m \, e^{i\,m\cdot x}, \ x \in Q_2, \ v_m \in \mathbb{C}^2, \ N \in \mathbb{N} \right\}$$

of all trigonometric vector polynomials with respect to the norm

$$\|v\|_{H^s_{per}(\operatorname{Div}, Q_2)} = \sqrt{\sum_{m \in \mathbb{Z}^2} (1 + |m|^2)^s \left[|v_m|^2 + |m \cdot v_m|^2 \right]}.$$

(c) For any $s \in \mathbb{R}$ the space $H^s_{per}(\operatorname{Curl}, Q_2)$ is defined as the completion of the space $\mathcal{T}(Q_2, \mathbb{C}^2)$ with respect to the norm

$$\|v\|_{H^s_{per}(\mathrm{Curl}, Q_2)} = \sqrt{\sum_{m \in \mathbb{Z}^2} (1 + |m|^2)^s \left[|v_m|^2 + |m \times v_m|^2 \right]},$$

where we have set $m \times a = m_1 a_2 - m_2 a_1$ for vectors $m, a \in \mathbb{C}^2$.

Remark: It will be convenient to consider the elements $f \in H^{-1/2}_{per}(\mathrm{Div}, Q_2)$ and $f \in H^{-1/2}_{per}(\mathrm{Curl}, Q_2)$ as vector valued functions from Q_2 into $\mathbb{C}^3$ with vanishing third component f_3. Actually, these elements are no functions anymore but distributions. For example, we will do this identification in the following lemma.

Theorem 5.21. *Let again $Q_3 = (-\pi, \pi)^3 \subseteq \mathbb{R}^3$ denote the cube in $\mathbb{R}^3$ and $Q_2 = (-\pi, \pi)^2 \subseteq \mathbb{R}^2$ the corresponding square in $\mathbb{R}^2$. Often, we will again identify $Q_2 \subseteq \mathbb{R}^2$ with $Q_2 \times \{0\} \subseteq \mathbb{R}^3$. Furthermore, let $\hat{e} = (0, 0, 1)^\top$ be the unit normal vector orthogonal to the plane $\mathbb{R}^2 \times \{0\}$ in $\mathbb{R}^3$.*

(a) *The trace operator $\gamma_t : H_{per}(\mathrm{curl}, Q_3) \to H^{-1/2}_{per}(\mathrm{Div}, Q_2)$, $u \mapsto \hat{e} \times u(\cdot, 0) = (-u_2(\cdot, 0), u_1(\cdot, 0), 0)^\top$, is well defined and bounded. Furthermore, there exists a bounded right inverse η_t of γ_t; that is, a bounded operator $\eta_t : H^{-1/2}_{per}(\mathrm{Div}, Q_2) \to H_{per}(\mathrm{curl}, Q_3)$ with $\gamma_t \circ \eta_t = \mathrm{id}$. In other words, the tangential components $\hat{e} \times u$ of $u = \eta_t f \in H_{per}(\mathrm{curl}, Q_3)$ coincide with $f \in H^{-1/2}_{per}(\mathrm{Div}, Q_2)$ on $Q_2 \times \{0\}$.*

(b) *The trace operator $\gamma_T : H_{per}(\mathrm{curl}, Q_3) \to H^{-1/2}_{per}(\mathrm{Curl}, Q_2)$, $u \mapsto (\hat{e} \times u(\cdot, 0)) \times \hat{e} = (u_1(\cdot, 0), u_2(\cdot, 0), 0)^\top$, is well defined and bounded. Furthermore, there exists a bounded right inverse η_T of γ_T.*

Proof: We restrict ourselves to part (a) and leave the proof of part (b) to the reader. Let $u \in H_{per}(\mathrm{curl}, Q_3)$ with Fourier coefficients $u_n \in \mathbb{C}^3$, $n \in \mathbb{Z}^3$; that is, $u(x) = \sum_{n \in \mathbb{Z}^3} u_n e^{i n \cdot x}$. With $\tilde{x} = (x_1, x_2)$ and $\tilde{n} = (n_1, n_2)$ we observe that

$$u(\tilde{x}, 0) = \sum_{\tilde{n} \in \mathbb{Z}^2} \left[\sum_{n_3 = -\infty}^{\infty} u_n \right] e^{i \tilde{n} \cdot \tilde{x}} \quad \text{thus} \quad (\gamma_t u)(x) = \sum_{\tilde{n} \in \mathbb{Z}^2} u_{\tilde{n}}^{\perp} e^{i \tilde{n} \cdot \tilde{x}}$$

where

$$u_{\tilde{n}}^{\perp} := \sum_{n_3 = -\infty}^{\infty} \hat{e} \times u_n = \sum_{n_3 = -\infty}^{\infty} \hat{e} \times u_{(\tilde{n}, n_3)} \in \mathbb{C}^3, \quad \tilde{n} = (n_1, n_2) \in \mathbb{Z}^2,$$

(5.10)

are the Fourier coefficients of $\gamma_t u$. Note that the third component vanishes. We decompose u_n in the form

$$u_n = \frac{1}{|n|^2} n \times (u_n \times n) + \frac{1}{|n|^2} (n \cdot u_n) n.$$

We set for the moment $\hat{n} := n/|n|$ and $v_n := u_n \times n$ and observe that

$$\hat{e} \times u_n = \frac{1}{|n|} \hat{e} \times (\hat{n} \times v_n) + \frac{1}{|n|} (\hat{n} \cdot u_n)(\hat{e} \times n), \qquad (5.11)$$

and thus for fixed $\tilde{n} \neq (0,0)$ by the inequality of Cauchy–Schwarz

$$\left| u_{\tilde{n}}^{\perp} \right|^2 = \left| \sum_{n_3=-\infty}^{\infty} (\hat{e} \times u_n) \right|^2$$

$$\leq 2 \left| \sum_{n_3=-\infty}^{\infty} \frac{1}{|n|} \hat{e} \times (\hat{n} \times v_n) \right|^2 + 2 \left| \sum_{n_3=-\infty}^{\infty} \frac{1}{|n|} (\hat{n} \cdot u_n)(\hat{e} \times n) \right|^2$$

$$\leq 2 \left[\sum_{n_3=-\infty}^{\infty} \frac{1}{|n|^2} \right] \left[\sum_{n_3=-\infty}^{\infty} |v_n|^2 + |\tilde{n}|^2 \sum_{n_3=-\infty}^{\infty} |u_n|^2 \right]$$

where we note that $|\hat{e} \times n| = |\tilde{n}|$ does not depend on n_3. We use (5.3) for $a = |\tilde{n}|$ with $\tilde{n} \neq (0,0)$ and arrive at

$$\left| u_{\tilde{n}}^{\perp} \right|^2 \leq \frac{2(\pi+1)}{|\tilde{n}|} \left[\sum_{n_3=-\infty}^{\infty} |n \times u_n|^2 + |\tilde{n}|^2 \sum_{n_3=-\infty}^{\infty} |u_n|^2 \right]$$

provided $\tilde{n} \neq (0,0)$. Now we multiply with $(1+|\tilde{n}|^2)^{-1/2}$ and sum with respect to $\tilde{n} \neq (0,0)$.

$$\sum_{\substack{\tilde{n} \in \mathbb{Z}^2 \\ \tilde{n} \neq (0,0)}} (1 + |\tilde{n}|^2)^{-1/2} \left| u_{\tilde{n}}^{\perp} \right|^2$$

$$\leq 2(\pi+1) \sum_{\substack{\tilde{n} \in \mathbb{Z}^2 \\ \tilde{n} \neq (0,0)}} \left[\frac{(1+|\tilde{n}|^2)^{-1/2}}{|\tilde{n}|} \sum_{n_3=-\infty}^{\infty} \left[|n \times u_n|^2 + |\tilde{n}|^2 |u_n|^2 \right] \right]$$

$$\leq 2(\pi+1) \sum_{n \in \mathbb{Z}^3} \left[|n \times u_n|^2 + |u_n|^2 \right].$$

Finally, we add the term with $\tilde{n} = (0,0)$. In this case $n = n_3 \hat{e}$ and thus $\hat{e} \times u_n = \frac{1}{n_3}(n \times u_n)$. Therefore,

$$\left| \sum_{n_3 \neq 0} (\hat{e} \times u_n) \right|^2 = \left| \sum_{n_3 \neq 0} \frac{1}{n_3}(n \times u_n) \right|^2$$

$$\leq \sum_{n_3 \neq 0} \frac{1}{n_3^2} \sum_{n_3 \neq 0} |n \times u_n|^2 \leq c_1 \sum_{n_3=-\infty}^{\infty} |n \times u_n|^2,$$

with $c_1 = 2 \sum_{j=1}^{\infty} \frac{1}{j^2}$. Therefore,

$$\left| \sum_{n_3=-\infty}^{\infty} \hat{e} \times u_n \right|^2 \leq 2c_1 \sum_{n_3=-\infty}^{\infty} |n \times u_n|^2 + 2\,|\hat{e} \times u_0|^2.$$

Adding this to the previous term for $\tilde{n} \neq (0,0)$ yields

$$\sum_{\tilde{n} \in \mathbb{Z}^2} (1+|\tilde{n}|^2)^{-1/2} |u_{\tilde{n}}^{\perp}|^2 = \sum_{\tilde{n} \in \mathbb{Z}^2} (1+|\tilde{n}|^2)^{-1/2} \left| \sum_{n_3=-\infty}^{\infty} (\hat{e} \times u_n) \right|^2$$

$$\leq c \sum_{n \in \mathbb{Z}^3} \left[|n \times u_n|^2 + |u_n|^2 \right] \tag{5.12}$$

for $c = 2(\pi + 1) + 2c_1 + 2$.

We study the Fourier coefficients of $\mathrm{Div}\,\gamma_t$ by the same arguments. First we note that

$$(\mathrm{Div}\,\gamma_t u)(\tilde{x}, 0) = i \sum_{n_1, n_2 \in \mathbb{Z}} \left[(n_1, n_2, 0)^{\top} \cdot u_{\tilde{n}}^{\perp} \right] e^{i\tilde{n}\cdot\tilde{x}}$$

where again $u_{\tilde{n}}^{\perp} \in \mathbb{C}^3$ are the Fourier coefficients of $\gamma_t u$ from (5.11) above. For arbitrary $n \in \mathbb{Z}^3$ we have from (5.11) with the same notations as above that

$$(n_1, n_2, 0)^{\top}\cdot(\hat{e}\times u_n) = n\cdot(\hat{e}\times u_n) = \frac{1}{|n|}\, n\cdot\big(\hat{e}\times(\hat{n}\times v_n)\big) = \frac{1}{|n|}\,(n\times\hat{e})\cdot(\hat{n}\times v_n)$$

and thus if $(n_1, n_2) \neq (0,0)$

$$|(n_1, n_2, 0)^{\top} \cdot u_{\tilde{n}}^{\perp}|^2 = \left| \sum_{n_3=-\infty}^{\infty} \frac{1}{|n|}\,(n \times \hat{e}) \cdot (\hat{n} \times v_n) \right|^2$$

$$\leq \sum_{n_3=-\infty}^{\infty} \frac{1}{|n|^2} \sum_{n_3=-\infty}^{\infty} |n \times \hat{e}|^2\,|\hat{n} \times v_n|^2$$

$$\leq \frac{\pi+1}{|\tilde{n}|} \sum_{n_3=-\infty}^{\infty} |\tilde{n}|^2\,|v_n|^2 = (\pi+1)\,|\tilde{n}| \sum_{n_3=-\infty}^{\infty} |u_n \times n|^2$$

where again $\tilde{n} = (n_1, n_2)$. Here we used $|n \times \hat{e}| = |\tilde{n}|$ because the third component of $n \times \hat{e}$ vanishes. For $\tilde{n} = 0$ we note that $n \cdot (\hat{e} \times u_n)$ vanishes. Therefore, multiplication with $(1+|\tilde{n}|^2)^{-1/2}$ and summation with respect to $\tilde{n}$ yields

$$\sum_{n_1,n_2\in\mathbb{Z}} (1+|\tilde{n}|^2)^{-1/2} |(n_1,n_2,0)^\top \cdot u_{\tilde{n}}^\perp|^2$$

$$\leq (\pi+1) \sum_{\tilde{n}\in\mathbb{Z}^2} \frac{|\tilde{n}|}{(1+|\tilde{n}|^2)^{1/2}} \sum_{n_3=-\infty}^{\infty} |u_n \times n|^2$$

$$\leq (\pi+1) \sum_{n\in\mathbb{Z}^3} |u_n \times n|^2 .$$

Adding this to (5.12) yields

$$\|\gamma_t u\|^2_{H_{per}^{-1/2}(\mathrm{Div},Q_2)} \leq (c+\pi+1) \sum_{n\in\mathbb{Z}^3} \left[|n\times u_n|^2 + |u_n|^2 \right]$$

$$= (c+\pi+1) \|u\|^2_{H_{per}(\mathrm{curl},Q_3)} .$$

Now we construct a bounded extension operator. Let $f \in H_{per}^{-1/2}(Q_2)$ be given by $f(\tilde{x}) = \sum_{\tilde{n}\in\mathbb{Z}^2} f_{\tilde{n}}\, e^{i\tilde{n}\cdot\tilde{x}}$ with Fourier coefficients $f_{\tilde{n}} \in \mathbb{C}^3$ for which the third components vanish. We define $u_n \in \mathbb{C}^3$ for $n \in \mathbb{Z}^3$ by

$$u_n = \frac{\delta_{\tilde{n}}}{1+|n|^2} (f_{\tilde{n}} \times a_n)$$

where again $\tilde{n} = (n_1,n_2)$ and

$$a_n = \begin{cases} \dfrac{1}{|\tilde{n}|^2} \left[|n|^2\,\hat{e} - n_3\,n \right], & \tilde{n} \neq (0,0), \\[2mm] \hat{e}, & \tilde{n} = (0,0), \end{cases}$$

and δ_n is given in (5.4) of the proof of Theorem 5.7; that is,

$$\delta_{\tilde{n}} = \left[\sum_{j=-\infty}^{\infty} \frac{1}{1+|\tilde{n}|^2+j^2} \right]^{-1} .$$

We easily derive the following properties of a_n:

- $\hat{e} \cdot a_n = 1$ for all $n \in \mathbb{Z}^3$,

- $n \cdot a_n = 0$ for all n with $\tilde{n} \neq (0,0)$, and $n \cdot a_n = n_3$ for all $n = n_3\hat{e}$,

- $|a_n|^2 = \frac{|n|^2}{|\tilde{n}|^2}$ for all n with $\tilde{n} \neq (0,0)$ and $|a_n| = 1$ for all $n = n_3\hat{e}$.

First we show that $(\eta_t f)(x) = u(x) = \sum_{n\in\mathbb{Z}^3} u_n\, e^{in\cdot x}$ defines a right inverse of γ_t. This follows from

$$\sum_{n_3=-\infty}^{\infty} (\hat{e} \times u_n) = \delta_{\tilde{n}} \underbrace{\sum_{n_3=-\infty}^{\infty} \frac{1}{1+|n|^2}}_{=\,1} \hat{e} \times (f_{\tilde{n}} \times a_n) = (\hat{e} \cdot a_n) f_{\tilde{n}} = f_{\tilde{n}} .$$

Now we show boundedness of η_t. For $\tilde{n} \neq (0,0)$ we conclude

$$
\sum_{n_3=-\infty}^{\infty} |u_n|^2 \leq |f_{\tilde{n}}|^2 \delta_{\tilde{n}}^2 \sum_{n_3=-\infty}^{\infty} \frac{|a_n|^2}{(1+|n|^2)^2}
$$

$$
= |f_{\tilde{n}}|^2 \delta_{\tilde{n}}^2 \frac{1}{|\tilde{n}|^2} \sum_{n_3=-\infty}^{\infty} \frac{|n|^2}{(1+|n|^2)^2} \tag{5.13}
$$

$$
\leq |f_{\tilde{n}}|^2 \frac{\delta_{\tilde{n}}}{|\tilde{n}|^2} \underbrace{\left[\delta_{\tilde{n}} \sum_{n_3=-\infty}^{\infty} \frac{1}{1+|n|^2} \right]}_{=1} = |f_{\tilde{n}}|^2 \frac{\delta_{\tilde{n}}}{|\tilde{n}|^2} .
$$

Using $\delta_{\tilde{n}} \leq \frac{2}{\pi} (1+|\tilde{n}|^2)^{1/2}$ from estimate (5.3) we arrive at

$$
\sum_{n_3=-\infty}^{\infty} |u_n|^2 \leq \frac{2}{\pi} \left[(1+|\tilde{n}|^2)^{-1/2} |f_{\tilde{n}}|^2 \right] \frac{1+|\tilde{n}|^2}{|\tilde{n}|^2} \leq \frac{4}{\pi} \left[(1+|\tilde{n}|^2)^{-1/2} |f_{\tilde{n}}|^2 \right] .
$$

This holds for $\tilde{n} \neq (0,0)$. For $\tilde{n} = (0,0)$ we conclude

$$
\sum_{n_3=-\infty}^{\infty} |u_n|^2 \leq |f_0|^2 \delta_0^2 \sum_{n_3=-\infty}^{\infty} \frac{1}{(1+n_3^2)^2} \leq c |f_0|^2 .
$$

Summing it with respect to $\tilde{n}$ yields

$$
\sum_{n \in \mathbb{Z}} |u_n|^2 \leq c' \sum_{\tilde{n} \in \mathbb{Z}^2} (1+|\tilde{n}|^2)^{-1/2} |f_{\tilde{n}}|^2 \leq c' \|f\|_{H^{-1/2}(\mathrm{Div},Q_2)}^2 .
$$

Finally, we estimate the norm of $\operatorname{curl} u(x) = i \sum_{n \in \mathbb{Z}^3} (n \times u_n) e^{in \cdot x}$. We have for $\tilde{n} \neq (0,0)$:

$$
n \times u_n = \frac{\delta_{\tilde{n}}}{1+|n|^2} n \times (f_{\tilde{n}} \times a_n)
$$

$$
= \frac{\delta_{\tilde{n}}}{1+|n|^2} \left[(n \cdot a_n) f_{\tilde{n}} - (n \cdot f_{\tilde{n}}) a_n \right] = -\frac{\delta_{\tilde{n}}}{1+|n|^2} (\tilde{n} \cdot f_{\tilde{n}}) a_n ,
$$

and thus

$$
\sum_{n_3=-\infty}^{\infty} |n \times u_n|^2 = |\tilde{n} \cdot f_{\tilde{n}}|^2 \delta_{\tilde{n}}^2 \sum_{n_3=-\infty}^{\infty} \frac{|a_n|^2}{(1+|n|^2)^2}
$$

$$
= |\tilde{n} \cdot f_{\tilde{n}}|^2 \delta_{\tilde{n}}^2 \frac{1}{|\tilde{n}|^2} \sum_{n_3=-\infty}^{\infty} \frac{|n|^2}{(1+|n|^2)^2} .
$$

This expression appeared already in (5.13) above for $|f_{\tilde{n}}|^2$ instead of $|\tilde{n} \cdot f_{\tilde{n}}|^2$. For $\tilde{n} = (0,0)$ we observe that $n \times u_n = \frac{\delta_0}{1+n_3^2} n_3 f_0$ and thus

$$\sum_{n_3=-\infty}^{\infty} |n \times u_n|^2 = \delta_0^2 |f_0|^2 \sum_{n_3=-\infty}^{\infty} \frac{n_3^2}{(1+n_3^2)^2} \leq c|f_0|^2 .$$

Therefore, we arrive at

$$\sum_{n \in \mathbb{Z}} |n \times u_n|^2 \leq c' \sum_{\tilde{n} \in \mathbb{Z}^2} (1+|\tilde{n}|^2)^{-1/2} |n \cdot f_{\tilde{n}}|^2 \leq c' \|f\|_{H^{-1/2}(\mathrm{Div},Q_2)}^2 .$$

This ends the proof of part (a). We leave the proof of part (b) to the reader.
□

It is the aim to extend the definition of the trace spaces $H_{per}^{-1/2}(\mathrm{Div}, Q_2)$ and $H_{per}^{-1/2}(\mathrm{Curl}, Q_2)$ to $H_{per}^{-1/2}(\mathrm{Div}, \partial D)$ and $H_{per}^{-1/2}(\mathrm{Curl}, \partial D)$, respectively, by the local coordinate system of Definition 5.1. The trace operators map into the tangential plane of the boundary ∂D. Therefore, $H_{per}^{-1/2}(\mathrm{Div}, \partial D)$ and $H_{per}^{-1/2}(\mathrm{Curl}, \partial D)$ are spaces of tangential vector fields. The transformation $u \mapsto \phi_j(u \circ \tilde{\Psi}_j)|_{Q_3^-}$ which we have used in Definition 5.8 is not adequate in our case because it does not map vector fields of $H(\mathrm{curl}, U_j')$ into vector fields of $H(\mathrm{curl}, B_j)$. We observe that gradients $\nabla \varphi$ are special elements of $H(\mathrm{curl}, U_j')$ for any $\varphi \in H^1(U_j')$. They satisfy

$$\nabla_x(\varphi \circ \tilde{\Psi}_j) = (\tilde{\Psi}_j')^{\top} (\nabla_y \varphi) \circ \tilde{\Psi}_j$$

where $\tilde{\Psi}_j' \in \mathbb{R}^{3 \times 3}$ denotes again the Jacobian; that is,

$$[\tilde{\Psi}_j'(x)]_{k\ell} = \frac{\partial (\tilde{\Psi}_j)_k(x)}{\partial x_\ell} , \quad k, \ell \in \{1, 2, 3\} , \ x \in B_j .$$

This will provide the correct transformation as we will see shortly. First we assume as above with loss of generality that the balls B_j are contained in Q_3. We define the subspaces of $L^2(\partial D, \mathbb{C}^3)$ and $L^2(Q_2, \mathbb{C}^3)$ of tangential vector fields by

$$L_t^2(\partial D) := \left\{ f \in L^2(\partial D, \mathbb{C}^3) : \nu(y) \cdot f(y) = 0 \text{ almost everywhere on } \partial D \right\} ,$$

$$L_t^2(Q_2) := \left\{ g \in L^2(Q_2, \mathbb{C}^3) : g_3(y) = 0 \text{ almost everywhere on } Q_2 \right\} ,$$

respectively. We note that the tangent plane at $y = \Psi_j(x)$ is spanned by the vectors $\partial \Psi_j(x)/\partial x_1$ and $\partial \Psi_j(x)/\partial x_2$. We recall that the normal vector at $y = \Psi_j(x)$ for $x \in B_2(0, \alpha_j)$ is given by

$$\nu(y) = \frac{1}{\rho_j(x)} \left(\frac{\partial \Psi_j(x)}{\partial x_1} \times \frac{\partial \Psi_j(x)}{\partial x_2} \right), \quad y = \Psi_j(x) \in \partial D \cap U_j', \quad \text{where}$$

$$\rho_j(x) = \left| \frac{\partial \Psi_j(x)}{\partial x_1} \times \frac{\partial \Psi_j(x)}{\partial x_2} \right| = \sqrt{1 + |\nabla \xi_j(x)|^2} , \quad x \in B_2(0, \alpha_j). \quad (5.14)$$

Lemma 5.22. *Let $\{U_j, \xi_j : j = 1, \ldots, m\}$ be a local coordinate system of ∂D. Let $\tilde{\Psi} : B \to U'$ be one of the isomorphism $\tilde{\Psi}_j$ from the balls B_j onto U'_j; that is, we drop the index j. For $u \in H(\mathrm{curl}, U' \cap D) \cap C(\overline{U' \cap D}, \mathbb{C}^3)$ set*

$$v(x) = \tilde{\Psi}'(x)^\top u(\tilde{\Psi}(x)), \quad x \in B^-, \tag{5.15}$$

where again B^- is the lower part of the ball B and $\tilde{\Psi}'(x)^\top \in \mathbb{R}^{3 \times 3}$ is the adjoint of the Jacobian at $x \in B$. Furthermore, define $F(x) \in \mathbb{R}^{3 \times 3}$ as

$$F(x) := \left[\frac{\partial \Psi(x)}{\partial x_1} \middle| \frac{\partial \Psi(x)}{\partial x_2} \middle| \frac{\partial \Psi(x)}{\partial x_1} \times \frac{\partial \Psi(x)}{\partial x_2}\right], \quad x \in B_2(0, \alpha_j). \tag{5.16}$$

Then, with $\hat{e} = (0, 0, 1)^\top \in \mathbb{R}^3$,

(a) $v \in H(\mathrm{curl}, B^-)$ *and*

$$\mathrm{curl}\, v(x) = \left(\tilde{\Psi}'(x)\right)^{-1} \mathrm{curl}_y\, u(y)|_{y = \tilde{\Psi}(x)}, \quad x \in B^-.$$

(b)

$$\rho(x) \left(\nu(y) \times u(y)\right)\big|_{y = \Psi(x)} = F(x) \left(\hat{e} \times v(x, 0)\right), \quad x \in B_2(0, \alpha_j),$$

where $\rho = \rho_j$ is given by (5.14).

(c)

$$\left(\nu(y) \times u(y)\right) \times \nu(y)\big|_{y = \Psi(x)} = F(x)^{-\top} \left[(\hat{e} \times v(x, 0)) \times \hat{e}\right], \quad x \in B_2(0, \alpha_j).$$

Proof: First we note that for any regular matrix $M = [a|b|c] \in \mathbb{R}^{3 \times 3}$ with column vectors a, b, c, the inverse is given by

$$M^{-1} = [a|b|c]^{-1} = \frac{1}{(a \times b) \cdot c} [b \times c | c \times a | a \times b]^\top = \frac{1}{\det M} [b \times c | c \times a | a \times b]^\top. \tag{5.17}$$

Applying this to $[a|b|c] = \tilde{\Psi}'(x)$ yields

$$[\tilde{\Psi}'(x)]^{-1} = \left[\frac{\partial \tilde{\Psi}(x)}{\partial x_2} \times \frac{\partial \tilde{\Psi}(x)}{\partial x_3} \middle| \frac{\partial \tilde{\Psi}(x)}{\partial x_3} \times \frac{\partial \tilde{\Psi}(x)}{\partial x_1} \middle| \frac{\partial \tilde{\Psi}(x)}{\partial x_1} \times \frac{\partial \tilde{\Psi}(x)}{\partial x_2}\right]^\top.$$

In the following the subscript k denotes the kth component of the vector function $\tilde{\Psi}$ or v or u.

(a) We note that the columns of the matrix $\tilde{\Psi}'(x)^\top \in \mathbb{R}^{3 \times 3}$ are just the gradients $\nabla \tilde{\Psi}_k(x)$, $k = 1, 2, 3$. Therefore,

$$v(x) = \sum_{k=1}^{3} u_k\big(\tilde{\Psi}(x)\big)\, \nabla\tilde{\Psi}_k(x),$$

and

$$\operatorname{curl} v(x) = \sum_{k=1}^{3} \nabla u_k\big(\tilde{\Psi}(x)\big) \times \nabla\tilde{\Psi}_k(x)$$

$$= \sum_{k=1}^{3} \nabla\big[u_k\big(\tilde{\Psi}(x)\big)\big] \times \nabla\tilde{\Psi}_k(x)$$

$$= \sum_{k=1}^{3} \big[\tilde{\Psi}'(x)^{\top}\nabla_y u_k(y)\big|_{y=\tilde{\Psi}(x)}\big] \times \nabla\tilde{\Psi}_k(x)$$

$$= \sum_{k=1}^{3}\sum_{\ell=1}^{3} \frac{\partial u_k(y)}{\partial y_\ell}\bigg|_{y=\tilde{\Psi}(x)} \big[\nabla\tilde{\Psi}_\ell(x) \times \nabla\tilde{\Psi}_k(x)\big]$$

$$= \big(\operatorname{curl} u(y)\big)_1 \big[\nabla\tilde{\Psi}_2(x) \times \nabla\tilde{\Psi}_3(x)\big] + \big(\operatorname{curl} u(y)\big)_2 \big[\nabla\tilde{\Psi}_3(x) \times \nabla\tilde{\Psi}_1(x)\big]$$
$$+ \big(\operatorname{curl} u(y)\big)_3 \big[\nabla\tilde{\Psi}_1(x) \times \nabla\tilde{\Psi}_2(x)\big]$$

$$= \big[\nabla\tilde{\Psi}_2(x)\times\nabla\tilde{\Psi}_3(x) \,|\, \nabla\tilde{\Psi}_3(x)\times\nabla\tilde{\Psi}_1(x) \,|\, \nabla\tilde{\Psi}_1(x)\times\nabla\tilde{\Psi}_2(x)\big]\, \operatorname{curl} u(y)\big|_{y=\tilde{\Psi}(x)}$$

$$= \big(\tilde{\Psi}'(x)\big)^{-1} \operatorname{curl}_y u(y)\big|_{y=\tilde{\Psi}(x)}$$

where we applied (5.17) to $\tilde{\Psi}(x)^{\top}$.

(b) We write $\partial_j\tilde{\Psi}$ for $\partial\tilde{\Psi}/\partial x_j$, $j = 1,2,3$, in the following and drop the argument x or $y = \tilde{\Psi}(x)$. From the definition of v we have $v = (u \cdot \partial_1\tilde{\Psi},\, u\cdot\partial_2\tilde{\Psi},\, u\cdot\partial_3\tilde{\Psi})^{\top}$ and thus $\hat{e}\times v = (-u\cdot\partial_2\Psi,\, u\cdot\partial_1\Psi,\, 0)^{\top}$ on the boundary $x_3 = 0$. Furthermore,

$$F(\hat{e}\times v) = -(u\cdot\partial_2\Psi)\,\partial_1\Psi + (u\cdot\partial_1\Psi)\,\partial_2\Psi = (\partial_1\Psi\times\partial_2\Psi)\times u = \rho\,(\nu\times u).$$

(c) We have just seen that $\rho\,(\nu \times u) = -v_2\,\partial_1\Psi + v_1\,\partial_2\Psi$, thus

$$\rho^2\,(\nu \times u) \times \nu = -v_2\,\partial_1\Psi \times (\partial_1\Psi \times \partial_2\Psi) + v_1\,\partial_2\Psi \times (\partial_1\Psi \times \partial_2\Psi)$$
$$= -v_2\big[(\partial_1\Psi \cdot \partial_2\Psi)\,\partial_1\Psi - |\partial_1\Psi|^2\,\partial_2\Psi\big]$$
$$+ v_1\big[|\partial_2\Psi|^2\,\partial_1\Psi - (\partial_1\Psi \cdot \partial_2\Psi)\,\partial_2\Psi\big]$$
$$= F \underbrace{\begin{bmatrix} |\partial_2\Psi|^2 & -\partial_1\Psi\cdot\partial_2\Psi & 0 \\ -\partial_1\Psi\cdot\partial_2\Psi & |\partial_1\Psi|^2 & 0 \\ 0 & 0 & 1 \end{bmatrix}}_{= \rho^2(F^{\top}F)^{-1} = \rho^2 F^{-1}F^{-\top}} \begin{pmatrix} v_1 \\ v_2 \\ 0 \end{pmatrix}$$
$$= \rho^2\, F^{-\top}\big[(\hat{e}\times v)\times\hat{e}\big].$$

This ends the proof. $\quad\square$

Now we are able to define the spaces $H^{-1/2}(\text{Div}, \partial D)$ and $H^{-1/2}(\text{Curl}, \partial D)$ and prove the trace theorems for $H(\text{curl}, D)$.

Definition 5.23. Let $D \subseteq \mathbb{R}^3$ be a Lipschitz domain with corresponding local coordinate system (U_j, ξ_j) and corresponding isomorphisms $\tilde{\Psi}_j$ from the balls B_j onto U'_j and their restrictions $\Psi_j : B_2(0, \alpha_j) \to U_j \cap \partial D$. Furthermore, let $\{\phi_j : j = 1, \ldots, m\}$ be a partition of unity on ∂D corresponding to the sets U'_j and $F_j(x)$ be given by (5.16). We assume without loss of generality that all of the disks $B_2(0, \alpha_j)$ are contained in the square $Q_2 = (-\pi, \pi)^2$.

Then we define the spaces $H^{-1/2}(\text{Div}, \partial D)$ and $H^{-1/2}(\text{Curl}, \partial D)$ as the completion of

$$\{f \in L_t^2(\partial D) : \tilde{f}_j^t \in H_{per}^{-1/2}(\text{Div}, Q_2), \ j = 1, \ldots, m\}$$

and

$$\{f \in L_t^2(\partial D) : \tilde{f}_j^T \in H_{per}^{-1/2}(\text{Curl}, Q_2), \ j = 1, \ldots, m\},$$

respectively, with respect to the norms

$$\|f\|_{H^{-1/2}(\text{Div}, \partial D)} = \left[\sum_{j=1}^m \|\tilde{f}_j^t\|_{H_{per}^{-1/2}(\text{Div}, Q_2)}^2 \right]^{1/2},$$

$$\|f\|_{H^{-1/2}(\text{Curl}, \partial D)} = \left[\sum_{j=1}^m \|\tilde{f}_j^T\|_{H_{per}^{-1/2}(\text{Curl}, Q_2)}^2 \right]^{1/2},$$

where

$$\tilde{f}_j^t(x) := \rho_j(x) \sqrt{\phi_j(\Psi_j(x))}\, F_j^{-1}(x) f(\Psi_j(x)), \ x \in B_2(0, \alpha_j), \quad (5.18a)$$

$$\tilde{f}_j^T(x) := \rho_j(x) \sqrt{\phi_j(\Psi_j(x))}\, F_j^\top(x) f(\Psi_j(x)), \ x \in B_2(0, \alpha_j), \quad (5.18b)$$

extended by zero into Q_2, and F_j and ρ_j are defined in (5.16) and (5.14), respectively.

Again, combining the following trace theorem and Theorem 5.25 below one shows exactly as in Corollary 5.15 that $H^{-1/2}(\text{Div}, \partial D)$ does not depend on the choice of the coordinate system.

Theorem 5.24. *The trace operators $\gamma_t : H(\text{curl}, D) \to H^{-1/2}(\text{Div}, \partial D)$, $u \mapsto \nu \times u|_{\partial D}$ and $\gamma_T : H(\text{curl}, D) \to H^{-1/2}(\text{Curl}, \partial D)$, $u \mapsto (\nu \times u|_{\partial D}) \times \nu$ are well defined and bounded and have bounded right inverses $\eta_t : H^{-1/2}(\text{Div}, \partial D) \to H(\text{curl}, D)$ and $\eta_T : H^{-1/2}(\text{Curl}, \partial D) \to H(\text{curl}, D)$, respectively.*

Proof: Boundedness of the trace operators is seen from the Definition 5.23, Lemma 5.22, and the boundedness of the trace operator of Theorem 5.21. Indeed, for $u \in H(\text{curl}, D) \cap C(\overline{D}, \mathbb{C}^3)$ we have by definition

$$\|\nu \times u\|_{H^{-1/2}(\mathrm{Div},\partial D)} = \left[\sum_{j=1}^{m} \|\tilde{f}_j^t\|^2_{H_{per}^{-1/2}(\mathrm{Div},Q_2)} \right]^{1/2}$$

with

$$\tilde{f}_j^t(x) = \rho_j(x)\sqrt{\Phi_j(y)}\, F_j^{-1}(x)\big(\nu(y) \times u(y)\big), \quad x \in B_2(0,\alpha_j), \ y = \Psi_j(x).$$

By part (b) of Lemma 5.22 we rewrite this as

$$\tilde{f}_j^t = \hat{e} \times v_j(\cdot,0) \quad \text{with} \quad v_j(x) = \sqrt{\Phi_j(y)}\, \tilde{\Psi}_j'(x)^\top u(y)\big|_{y=\tilde{\Psi}_j(x)}, \ x \in B_j^-.$$

Then $v_j \in H(\mathrm{curl}, B_j^-)$ by part (a) of Lemma 5.22 and v_j vanishes in some neighborhood of $\partial B_j \cap \overline{B_j^-}$, and we extend v_j by zero into Q_3^-. Denoting again $z^* = (z_1, z_2, -z_3)^\top$ for any vector $z = (z_1, z_2, z_3)^\top \in \mathbb{C}^3$ we extend v_j into Q_3 by

$$\tilde{v}_j(x) = \begin{cases} v_j(x), & x \in Q_3^-, \\ v_j^*(x^*), & x \in Q_3^+, \end{cases}$$

The vector field v_j fails to be continuous in Q_3^- because of the transformation into the parameter space. Therefore, we cannot directly apply Exercise 4.6 to show that $\tilde{v}_j \in H(\mathrm{curl}, Q_3)$. However, we can approximate v_j by a sequence $\psi_j^{(\ell)} \in C^\infty(\overline{Q_3^-})$ and study the analogous extension to $\tilde{\psi}_j^{(\ell)}$ which are in $H(\mathrm{curl}, Q_3)$ by Exercise 4.6. Also, they converge to $\tilde{v}_j$ in $H(\mathrm{curl}, Q_3)$ because $\mathrm{curl}_x\big(\tilde{\psi}_j^{(\ell)}\big)^*(x^*) = -\big(\mathrm{curl}\,\psi_j^{(\ell)}\big)^*(x^*)$ for $x \in Q_3^+$. Also, $\tilde{v}_j$ vanishes in some neighborhood of ∂Q_3. Now we show, using partial integration as in Lemma 4.37 (see Exercise 4.13), that $\tilde{v}_j \in H_{per}(\mathrm{curl}, Q_3)$ and that there exists $c_1 > 0$ which is independent of $\tilde{v}_j$ such that

$$\|\tilde{v}_j\|_{H_{per}(\mathrm{curl},Q_3)} \leq c_1 \|v_j\|_{H(\mathrm{curl},Q_3^-)} = c_1 \|v_j\|_{H(\mathrm{curl},B_j^-)}$$

and thus

$$\|\tilde{f}_j^t\|_{H_{per}^{-1/2}(\mathrm{Div},Q_2)} = \|\gamma_t v_j\|_{H_{per}^{-1/2}(\mathrm{Div},Q_2)} \leq c_2 \|v_j\|_{H(\mathrm{curl},B_j^-)} \leq c_3 \|u\|_{H(\mathrm{curl},U_j')}.$$

Summing these terms yields boundedness of γ_t. For γ_T one argues in the same way.

It remains to construct an extension operator $\eta_t : H^{-1/2}(\mathrm{Div}, \partial D) \to H(\mathrm{curl}, D)$. For $f \in H^{-1/2}(\mathrm{Div}, \partial D)$ we define $\tilde{f}_j^t \in H_{per}^{-1/2}(\mathrm{Div}, Q_2)$ by (5.18a). By Theorem 5.21 there exist $v_j \in H_{per}(\mathrm{curl}, Q_3)$ such that $\hat{e} \times v_j = \tilde{f}_j^t$ and the mappings $\tilde{f}_j^t \mapsto v_j$ are bounded for $j = 1, \ldots, m$. We define

$$u_j(y) := \sqrt{\phi_j(y)}\, \tilde{\Psi}_j'(x)^{-\top} v_j(x)\big|_{x=\tilde{\Psi}_j^{-1}(y)}, \quad y \in U_j',$$

extend u_j by zero into all of $\mathbb{R}^3$ and set $u = \sum_{j=1}^{m} u_j$. Then $\nu \times u = f$ on ∂D. Indeed, for fixed $y \in \partial D$ let $J = \{j \in \{1, \ldots, m\} : y \in U_j'\}$. Then $u(y) = \sum_{j \in J} u_j(y)$. Fix $j \in J$; that is, $y \in U_j'$, and $x = \tilde{\Psi}_j^{-1}(y) \in Q_2 \times \{0\}$. From the definition of u_j we observe that

$$\rho_j(x)\, \nu(y) \times u_j(y) = \sqrt{\phi_j(y)}\, F_j(x)\, (\hat{e} \times v_j(x,0)) = \sqrt{\phi_j(y)}\, F_j(x)\, \tilde{f}_j^t(x)$$
$$= \rho_j(x)\, \phi_j(y)\, f(y)\,;$$

that is, $\nu(y) \times u_j(y) = \phi_j(y)\, f(y)$. Summation with respect to j yields $\nu \times u = f$. Furthermore, all of the operations in these constructions of u are bounded. This proves the theorem. $\quad\square$

Remark: We note that in the second part of the proof we have constructed an extension u into all of $\mathbb{R}^3$ with support in the neighborhood $\bigcup_{j=1}^{m} U_j'$ of ∂D. In particular, this implies that there exists a bounded extension operator $\tilde{\eta} : H(\mathrm{curl}, D) \to H(\mathrm{curl}, \mathbb{R}^3)$.

We note that the space $H_0(\mathrm{curl}, D)$ has been defined as the closure of $C_0^\infty(D, \mathbb{C}^3)$ in $H(\mathrm{curl}, D)$, see Definition 4.19. Analogously to Theorem 5.14 it can be characterized as the nullspace of the trace operator as the following theorem shows.

Theorem 5.25. *The space $H_0(\mathrm{curl}, D)$ is the null space $\mathcal{N}(\gamma_t)$ of the trace operator γ_t; that is, $u \in H_0(\mathrm{curl}, D)$ if, and only if, $\gamma_t u = 0$ on ∂D. The same holds for γ_T; that is, $u \in H_0(\mathrm{curl}, D)$ if, and only if, $\gamma_T u = 0$ on ∂D.*

Proof: The inclusion $H_0(\mathrm{curl}, D) \subseteq \mathcal{N}(\gamma_t)$ follows again immediately from the definition of $H_0(\mathrm{curl}, D)$ as the closure of $C_0^\infty(D, \mathbb{C}^3)$. For the reverse inclusion we modify the proof of Theorem 5.14 by, essentially, replacing ∇ by curl. We omit the details. $\quad\square$

The following theorem gives a precise formulation of the fact that the spaces $H^{-1/2}(\mathrm{Div}, \partial D)$ and $H^{-1/2}(\mathrm{Curl}, \partial D)$ are dual to each other.

Theorem 5.26.

(a) *The dual space $H^{-1/2}(\mathrm{Div}, \partial D)^*$ of $H^{-1/2}(\mathrm{Div}, \partial D)$ is isomorphic to $H^{-1/2}(\mathrm{Curl}, \partial D)$. An isomorphism is given by the mapping $J_1 : H^{-1/2}(\mathrm{Curl}, \partial D) \to H^{-1/2}(\mathrm{Div}, \partial D)^*$ defined as $J_1 g = \lambda_g$ where $\lambda_g \in H^{-1/2}(\mathrm{Div}, \partial D)^*$ is given by*

$$\lambda_g(\psi) := \int_D \left[\tilde{u} \cdot \mathrm{curl}\, \tilde{\psi} - \tilde{\psi} \cdot \mathrm{curl}\, \tilde{u} \right] dx\,,$$

$g \in H^{-1/2}(\mathrm{Curl}, \partial D)$, $\psi \in H^{-1/2}(\mathrm{Div}, \partial D)$, where $\tilde{u}, \tilde{\psi} \in H(\mathrm{curl}, D)$ are any vector fields with $\gamma_T \tilde{u} = g$ and $\gamma_t \tilde{\psi} = \psi$.

(b) The dual space $H^{-1/2}(\mathrm{Curl}, \partial D)^*$ of $H^{-1/2}(\mathrm{Curl}, \partial D)$ is isomorphic to $H^{-1/2}(\mathrm{Div}, \partial D)$. An isomorphism is given by the mapping $J_2 : H^{-1/2}(\mathrm{Div}, \partial D) \to H^{-1/2}(\mathrm{Curl}, \partial D)^*$ defined as $J_2 f = \mu_f$ where $\mu_f \in H^{-1/2}(\mathrm{Curl}, \partial D)^*$ is given by

$$\mu_f(\varphi) := - \int_D \left[\tilde{v} \cdot \mathrm{curl}\, \tilde{\varphi} - \tilde{\varphi} \cdot \mathrm{curl}\, \tilde{v}\right] dx,$$

with $f \in H^{-1/2}(\mathrm{Div}, \partial D)$, $\varphi \in H^{-1/2}(\mathrm{Curl}, \partial D)$, where $\tilde{v}, \tilde{\varphi} \in H(\mathrm{curl}, D)$ are any vector fields with $\gamma_t \tilde{v} = f$ and $\gamma_T \tilde{\varphi} = \varphi$.

(c) It holds $\lambda_g(f) = \mu_f(g) =: \langle f, g \rangle_*$ for all $f \in H^{-1/2}(\mathrm{Div}, \partial D)$ and $g \in H^{-1/2}(\mathrm{Curl}, \partial D)$ and Green's formula holds in the form

$$\langle \gamma_t v, \gamma_T u \rangle_* = \int_D \left[u \cdot \mathrm{curl}\, v - v \cdot \mathrm{curl}\, u\right] dx \quad \text{for all } u, v \in H(\mathrm{curl}, D).$$
$$(5.19)$$

Proof: (a), (b) The proof is split into four parts.

(i) First we show that λ_g and μ_f are well defined; that is, the right-hand sides do not depend on the choices of the extensions. Let $\tilde{u}_j, \tilde{\psi}_j \in H(\mathrm{curl}, D)$, $j = 1, 2$, such that $\gamma_t \tilde{\psi}_j = \psi$ and $\gamma_T \tilde{u}_j = g$ for $j = 1, 2$. Then

$$\int_D \left[\tilde{u}_1 \cdot \mathrm{curl}\, \tilde{\psi}_1 - \tilde{\psi}_1 \cdot \mathrm{curl}\, \tilde{u}_1\right] dx \;-\; \int_D \left[\tilde{u}_2 \cdot \mathrm{curl}\, \tilde{\psi}_2 - \tilde{\psi}_2 \cdot \mathrm{curl}\, \tilde{u}_2\right] dx$$

$$= \int_D \left[(\tilde{u}_1 - \tilde{u}_2) \cdot \mathrm{curl}\, \tilde{\psi}_1 - \tilde{\psi}_1 \cdot \mathrm{curl}(\tilde{u}_1 - \tilde{u}_2)\right] dx$$

$$+ \int_D \left[\tilde{u}_2 \cdot \mathrm{curl}(\tilde{\psi}_1 - \tilde{\psi}_2) - (\tilde{\psi}_1 - \tilde{\psi}_2) \cdot \mathrm{curl}\, \tilde{u}_2\right] dx.$$

Because $(\tilde{u}_1 - \tilde{u}_2) \in H_0(\mathrm{curl}, D)$ and $(\tilde{\psi}_1 - \tilde{\psi}_2) \in H_0(\mathrm{curl}, D)$ by Theorem 5.25 we conclude that both integrals vanish because of the Definition 4.16 of the variational curl. Therefore, λ_g is well defined. The same arguments show that also μ_f is well defined.

(ii) λ_g and μ_f are bounded because

$$|\lambda_g(\psi)| \leq \int_D \left[|\tilde{u}| \,|\mathrm{curl}\, \tilde{\psi}| + |\tilde{\psi}| \,|\mathrm{curl}\, \tilde{u}|\right] dx \;\leq\; \|\tilde{u}\|_{H(\mathrm{curl}, D)} \|\tilde{\psi}\|_{H(\mathrm{curl}, D)}$$

$$\leq \|\eta_T\| \,\|g\|_{H(\mathrm{Curl}, \partial D)} \,\|\eta_t\| \,\|\psi\|_{H(\mathrm{Div}, \partial D)}$$

which proves that λ_g is bounded and also boundedness of the mapping J_1. Again, the arguments for μ_f and J_2 are exactly the same.

(iii) We show that J_1 is surjective. Let $\ell \in H^{-1/2}(\mathrm{Div}, \partial D)^*$. By the theorem of Riesz (Theorem A.5) there exists a unique $u \in H(\mathrm{curl}, D)$ with

$$\int_D \left[\operatorname{curl} u \cdot \operatorname{curl} \tilde{\psi} + u \cdot \tilde{\psi} \right] dx = \ell(\gamma_t \tilde{\psi}) \quad \text{for all } \tilde{\psi} \in H(\operatorname{curl}, D)$$

because the right-hand side is a bounded functional on $H(\operatorname{curl}, D)$. Taking $\tilde{\psi} \in H_0(\operatorname{curl}, D)$ we note from the definition of the variational curl (Definition 4.16) that $\operatorname{curl} u \in H(\operatorname{curl}, D)$ and $\operatorname{curl}^2 u = -u$ in D. Therefore, for $\psi \in H^{-1/2}(\operatorname{Div}, \partial D)$ and $\tilde{\psi} = \eta_t \psi$ we have that

$$\ell(\psi) = \int_D \left[\operatorname{curl} u \cdot \operatorname{curl} \tilde{\psi} - \operatorname{curl}^2 u \cdot \tilde{\psi} \right] dx$$

which shows that $g := \gamma_T \operatorname{curl} u \in H^{-1/2}(\operatorname{Curl}, \partial D)$ yields $\lambda_g = \ell$. The proof of surjectivity of J_2 follows again by the same arguments.

(iv) Finally, we prove injectivity of J_1. Let $g \in H^{-1/2}(\operatorname{Curl}, \partial D)$ with $J_1 g = \lambda_g = 0$. By a general functional analytic argument (see [31], Sect. IV.6) there exists $\ell \in H^{-1/2}(\operatorname{Curl}, \partial D)^*$ with $\|\ell\| = 1$ and $\ell(g) = \|g\|_{H^{-1/2}(\operatorname{Curl}, \partial D)}$. By the surjectivity of J_2 there exists $f \in H^{-1/2}(\operatorname{Div}, \partial D)$ with $\mu_f = \ell$. Let $\tilde{u} = \eta_T g$ and $\tilde{v} = \eta_t f$. Then, by the definitions of μ_f and λ_g,

$$\|g\|_{H^{-1/2}(\operatorname{Curl}, \partial D)} = \mu_f(g) = -\int_D \left[\tilde{v} \cdot \operatorname{curl} \tilde{u} - \tilde{u} \cdot \operatorname{curl} \tilde{v} \right] dx = \lambda_g(f) = 0.$$

Again, injectivity of J_2 is proven analogously.

(c) This is obvious by the definition of λ_g if one takes $g = \gamma_T u$ and $\psi = \gamma_t v$. Note that—by part (a)—any extensions of g and ψ can be taken in the definition of λ_g.

$\square$

The notation $H^{-1/2}(\operatorname{Div}, \partial D)$ indicates the existence of a surface divergence $\operatorname{Div} u$ for $u \in H^{-1/2}(\operatorname{Div}, \partial D)$ and that u and $\operatorname{Div} u$ belong to $H^{-1/2}(\partial D, \mathbb{C}^3)$ and $H^{-1/2}(\partial D)$, respectively. To confirm this we first show that, indeed, the space $H^{-1/2}(\operatorname{Div}, \partial D)$ can be considered as a subspace of $H^{-1/2}(\partial D, \mathbb{C}^3)$. First we note that $H^1(D, \mathbb{C}^3)$ is boundedly embedded in $H(\operatorname{curl}, D)$. Therefore, the trace operator γ_T is well defined and bounded on $H^1(D, \mathbb{C}^3)$.

Lemma 5.27. *The space $H^{-1/2}(\operatorname{Div}, \partial D)$ can be identified with a subspace of $H^{-1/2}(\partial D, \mathbb{C}^3)$. The identification is given by $a \mapsto \ell_a$ for $a \in H^{-1/2}(\operatorname{Div}, \partial D)$ where $\ell_a \in H^{-1/2}(\partial D, \mathbb{C}^3)$ is defined by*

$$\langle \ell_a, \psi \rangle_* = \langle a, \gamma_T \tilde{\psi} \rangle_*, \quad a \in H^{-1/2}(\operatorname{Div}, \partial D), \ \psi \in H^{1/2}(\partial D, \mathbb{C}^3),$$

where $\tilde{\psi} \in H^1(D, \mathbb{C}^3)$ is any extension of ψ. Here, $\langle \cdot, \cdot \rangle_*$ denotes the dual form in $\langle H^{-1/2}(\partial D, \mathbb{C}^3), H^{1/2}(\partial D, \mathbb{C}^3) \rangle$ and in $\langle H^{-1/2}(\operatorname{Div}, \partial D), H^{-1/2}(\operatorname{Curl}, \partial D) \rangle$, respectively, see part (c) of Theorem 5.26.

Proof: First we show that the definition of ℓ_a is independent of the extension. Let $\tilde{\psi}_j \in H^1(D, \mathbb{C}^3)$, $j = 1, 2$, be two extensions of ψ. Then $\tilde{\psi} := \tilde{\psi}_1 - \tilde{\psi}_2 \in H_0^1(D, \mathbb{C}^3)$ because $\gamma_0 \tilde{\psi} = 0$ and Theorem 5.14. We choose a sequence $\tilde{\phi}_n \in C_0^\infty(D, \mathbb{C}^3)$ with $\tilde{\phi}_n \to \tilde{\psi}$ in $H^1(D, \mathbb{C}^3)$. Then $\gamma_T \tilde{\phi}_n = 0$ for all n and $\gamma_T \tilde{\phi}_n \to \gamma_T \tilde{\psi}$ in $H^{-1/2}(\mathrm{Curl}, \partial D)$. This proves $\gamma_T \tilde{\psi} = 0$.

Furthermore, $\ell_a \in H^{-1/2}(\partial D, \mathbb{C}^3)$ and the mapping $a \mapsto \ell_a$ is bounded because for $\psi \in H^{1/2}(\partial D, \mathbb{C}^3)$ and the extension operator $\eta : H^{1/2}(\partial D, \mathbb{C}^3) \to H^1(D, \mathbb{C}^3)$ from the trace theorem (Theorem 5.10) we conclude that

$$\left| \langle \ell_a, \psi \rangle_* \right| = \left| \langle a, \gamma_T \eta \psi \rangle_* \right| \leq c \|a\|_{H^{-1/2}(\mathrm{Div}, \partial D)} \|\gamma_T\| \|\eta\| \|\psi\|_{H^{1/2}(\partial D)} .$$

Finally, the mapping $a \mapsto \ell_a$ is also one-to-one. Indeed, let $\ell_a = 0$ then take any $b \in H^{-1/2}(\mathrm{Curl}, \partial D)$. Choose a sequence $\tilde{\psi}_n \in C^\infty(\overline{D}, \mathbb{C}^3)$ with $\tilde{\psi}_n \to \eta_T b$ in $H(\mathrm{curl}, D)$ where $\eta_T : H^{-1/2}(\mathrm{Curl}, \partial D) \to H(\mathrm{curl}, D)$ is the extension operator of Theorem 5.24. This is possible by the denseness result of Theorem 5.19. Then $0 = \langle a, \gamma_T \tilde{\psi}_n \rangle_* \to \langle a, b \rangle_*$. This holds for all $b \in H^{-1/2}(\mathrm{Curl}, \partial D)$ which shows that a has to vanish because $H^{-1/2}(\mathrm{Curl}, \partial D)$ is the dual space of $H^{-1/2}(\mathrm{Div}, \partial D)$. □

Remark 5.28. In the following we always think of this identification when we use the identity

$$\langle a, \psi \rangle_* = \langle a, \gamma_T \tilde{\psi} \rangle_* , \quad a \in H^{-1/2}(\mathrm{Div}, \partial D), \ \psi \in H^{1/2}(\partial D, \mathbb{C}^3) .$$

Again, on the left-hand side a is considered as an element of $H^{-1/2}(\partial D, \mathbb{C}^3)$ while on the right-side a is considered as an element of $H^{-1/2}(\mathrm{Div}, \partial D)$.

The identification of the previous lemma implies in particular that for $a \in H^{-1/2}(\mathrm{Div}, \partial D)$ and scalar $\psi \in H^{1/2}(\partial D)$ the dual form $\langle a, \psi \rangle_* \in \mathbb{C}^3$ can be defined componentwise. We will use this in the definition of the vector potentials in Sect. 5.2 below.

Having in mind the definition of the variational derivative it is not a surprise that we also define the surface divergence and the surface curl by variational equations; that is, by partial integration. We take Eq. (A.21) as a definition. First we note that for $\tilde{\psi} \in H^1(D)$ it holds that $\nabla \tilde{\psi} \in H(\mathrm{curl}, D)$. Therefore, the traces $\gamma_T \nabla \tilde{\psi} \in H^{-1/2}(\mathrm{Curl}, \partial D)$ and $\gamma_t \nabla \tilde{\psi} \in H^{-1/2}(\mathrm{Div}, \partial D)$ are well defined.

Definition 5.29. Let $a \in H^{-1/2}(\mathrm{Div}, \partial D)$ and $b \in H^{-1/2}(\mathrm{Curl}, \partial D)$. Then the surface divergence $\mathrm{Div}\, a \in H^{-1/2}(\partial D)$ and surface curl $\mathrm{Curl}\, b \in H^{-1/2}(\partial D)$ are defined as the linear bounded functionals

$$\langle \mathrm{Div}\, a, \psi \rangle_* = -\langle a, \gamma_T \nabla \tilde{\psi} \rangle_*, \quad \psi \in H^{1/2}(\partial D), \tag{5.20a}$$

$$\langle \mathrm{Curl}\, b, \psi \rangle_* = -\langle \gamma_t \nabla \tilde{\psi}, b \rangle_*, \quad \psi \in H^{1/2}(\partial D), \tag{5.20b}$$

where again $\tilde{\psi} \in H^1(D)$ is any extension of ψ. On the left-hand side, $\langle \cdot, \cdot \rangle_*$ denotes the dual form in $\langle H^{-1/2}(\partial D), H^{1/2}(\partial D) \rangle$ while on the right-hand side it denotes the form in $\langle H^{-1/2}(\mathrm{Div}, \partial D), H^{-1/2}(\mathrm{Curl}, \partial D)\rangle$, see part (c) of Theorem 5.26. We note that the definitions of the surface divergence and surface curl yield $\mathrm{Curl}\, b = -\mathrm{Div}(\nu \times b)$ in a variational sense, see the remark on p. 79.

Again, we have to show that this definition is independent of the extension $\tilde{\psi}$. If $\tilde{\psi}_1$ and $\tilde{\psi}_2$ are two extension, then $\tilde{\psi} = \tilde{\psi}_1 - \tilde{\psi}_2 \in H_0^1(D)$ and thus $\nabla \tilde{\psi} \in H_0(\mathrm{curl}, D)$ which implies that $\gamma_T \nabla \tilde{\psi} = 0$ and $\gamma_t \nabla \tilde{\psi} = 0$.

We note the following applications (compare to Corollary 5.13 for the scalar case and to Exercise 4.6 for the case of sufficiently smooth functions).

Lemma 5.30. (a) Let $\Omega \subseteq \mathbb{R}^3$ be a Lipschitz domain such that $\overline{D} \subseteq \Omega$ and let $u_1 \in H(\mathrm{curl}, D)$ and $u_2 \in H(\mathrm{curl}, \Omega \setminus \overline{D})$ such that $\gamma_t u_1 = -\gamma_t u_2$ or $\gamma_T u_1 = \gamma_T u_2$ on ∂D. Then the field

$$u(x) = \begin{cases} u_1(x), \, x \in D, \\ u_2(x), \, x \in \Omega \setminus \overline{D}, \end{cases}$$

is in $H(\mathrm{curl}, \Omega)$.

(b) Let $D \subseteq \mathbb{R}^3$ be symmetric with respect to $x_3 = 0$; that is, $x \in D \Leftrightarrow x^* \in D$ where $z^* = (z_1, z_2, -z_3)^\top$ for any $z = (z_1, z_2, z_3)^\top \in \mathbb{C}^3$. Let $D^\pm = \{x \in D : x_3 \gtrless 0\}$ and $v \in H(\mathrm{curl}, D^-)$. Extend v into D by even reflection; that is,

$$\tilde{v}(x) = \begin{cases} v(x), & x \in D^-, \\ v^*(x^*), & x \in D^+, \end{cases}$$

Then $\tilde{v} \in H(\mathrm{curl}, D)$ and the mapping $v \mapsto \tilde{v}$ is bounded from $H(\mathrm{curl}, D^-)$ into $H(\mathrm{curl}, D)$.

(c) Let $D \subseteq \mathbb{R}^3$ as in part (b) and $v \in H_0(\mathrm{curl}, D^-)$. Extend v into D by odd reflection; that is,

$$\tilde{v}(x) = \begin{cases} v(x), & x \in D^-, \\ -v^*(x^*), & x \in D^+, \end{cases}$$

Then $\tilde{v} \in H_0(\mathrm{curl}, D)$ and the mapping $v \mapsto \tilde{v}$ is bounded from $H_0(\mathrm{curl}, D^-)$ into $H_0(\mathrm{curl}, D)$.

Proof: (a) Let $\gamma_T u_1 = \gamma_T u_2$ and $\psi \in H_0(\mathrm{curl}, \Omega)$ arbitrary. Using (5.19) in D for u_1 and in $\Omega \setminus \overline{D}$ for u_2 yields

$$\langle \gamma_t \psi|_-, \gamma_T u_1 \rangle_* = \int_D [u_1 \cdot \mathrm{curl}\, \psi - \psi \cdot \mathrm{curl}\, u_1]\, dx\,,$$

$$\langle \gamma_t \psi|_+, \gamma_T u_2, \rangle_* = \int_{\Omega \setminus \overline{D}} [u_2 \cdot \mathrm{curl}\, \psi - \psi \cdot \mathrm{curl}\, u_2]\, dx\,,$$

where $\gamma_t\psi|_-$ and $\gamma_t\psi|_+$ denotes the trace from D and $\Omega \setminus \overline{D}$ on ∂D, respectively. We note that $\gamma_t\psi|_- = -\gamma_t\psi|_+$ because of the orientation of ν. Now we set $v = \operatorname{curl} u_1$ in D and $v = \operatorname{curl} u_2$ in $\Omega \setminus \overline{D}$ and add both equations. Then $v \in L^2(\Omega, \mathbb{C}^3)$ and

$$\int_\Omega [u \cdot \operatorname{curl}\psi - \psi \cdot v]\, dx = 0 \quad \text{for all } \psi \in H_0(\operatorname{curl}, \Omega).$$

This is the variational form of $v = \operatorname{curl} u$ and proves the lemma for the case that $\gamma_T u_1 = \gamma_T u_2$. In the case $\gamma_t u_1 = \gamma_t u_2$ one uses (5.19) with u and v interchanged. Here we note that $\gamma_T\psi|_- = \gamma_T\psi|_+$.

(b), (c) This part is proven analogously by validating that

$$\operatorname{curl}\tilde{v}(x) = \begin{cases} \operatorname{curl} v(x), & x \in D^-, \\ \mp(\operatorname{curl} v)^*(x^*), & x \in D^+, \end{cases}$$

respectively, see the proof of the Theorem 5.24 and Exercise 4.6. □

Now we are able to prove the important compactness property of the subspace $V_{0,A}$, defined in (4.10d). First we consider the case of a ball.

Lemma 5.31. *Let $B \subseteq \mathbb{R}^3$ be a ball and $A \in L^\infty(B, \mathbb{C}^{3\times3})$ such that $A(x)$ is symmetric for almost all x and there exists $c > 0$ with $\operatorname{Re}(\bar{z}^\top A(x)z) \geq c|z|^2$ for all $z \in \mathbb{C}^3$ and almost all $x \in B$. Then the closed subspace $H_0(\operatorname{curl}, \operatorname{div}_A 0, B) \subseteq H_0(\operatorname{curl}, B)$, defined in (4.12a); that is,*

$$H_0(\operatorname{curl}, \operatorname{div}_A 0, B) = \{u \in H_0(\operatorname{curl}, B) : (Au, \nabla\varphi)_{L^2(B)} = 0 \text{ for all } \varphi \in H_0^1(B)\}.$$

is compactly embedded in $L^2(B, \mathbb{C}^3)$. For $A=I$ we just write $H_0(\operatorname{curl}, \operatorname{div} 0, B)$.

Proof: For the special case $A(x) = I$ we will prove this result in the next section (see Theorem 5.37) by expanding the functions into spherical harmonics.

Let now A be arbitrary. We recall the Helmholtz decompositions from Theorem 4.23 in the form

$$H_0(\operatorname{curl}, B) = H_0(\operatorname{curl}, \operatorname{div}_A 0, B) \oplus \nabla H_0^1(B), \tag{5.21a}$$
$$L^2(B, \mathbb{C}^3) = L^2(\operatorname{div}_A 0, B) \oplus \nabla H_0^1(B), \tag{5.21b}$$

and note that the projections onto the components are bounded. Let now (u_n) be a bounded sequence in $H_0(\operatorname{curl}, \operatorname{div}_A 0, B)$. Then we decompose u_n with respect to the Helmholtz decomposition (5.21a) for $A(x) = I$; that is,

$$u_n = v_n + \nabla p_n \quad \text{with } v_n \in H_0(\operatorname{curl}, \operatorname{div} 0, B) \text{ and } p_n \in H_0^1(B).$$

Because the projections onto the components are bounded we conclude boundedness of the sequence (v_n) in $H_0(\operatorname{curl}, \operatorname{div} 0, B)$ with respect to

$\|\cdot\|_{H(\text{curl},B)}$. Because $H_0(\text{curl}, \text{div}\,0, B)$ is compactly embedded in $L^2(B, \mathbb{C}^3)$ there exists a subsequence (denoted by the same symbol) such that (v_n) converges in $L^2(B, \mathbb{C}^3)$. On the other hand, the form $v_n = u_n - \nabla p_n$ with $u_n \in H_0(\text{curl}, \text{div}_A\,0, B) \subseteq L^2(\text{div}_A\,0, B)$ and $p_n \in H_0^1(B)$ is just the decomposition of v_n in the Helmholtz decomposition (5.21b). Again, by the boundedness of the projections in $L^2(B, \mathbb{C}^3)$ we conclude that (u_n) converges in $L^2(B, \mathbb{C}^3)$ and ends the proof. $\square$

Now we are able to prove Theorem 4.24 from the previous chapter.

Theorem 5.32. *Let D be a Lipschitz domain and let $A \in L^\infty(D, \mathbb{C}^{3\times3})$ such that $A(x)$ is symmetric for almost all x and there exists $\hat{c} > 0$ with $\mathrm{Re}\,(\overline{z}^\top A(x)z) \geq \hat{c}|z|^2$ for all $z \in \mathbb{C}^3$ and almost all $x \in D$. Then the spaces $V_{0,A}$ and $H_0(\text{curl}, \text{div}_A\,0, D)$, defined by*

$$V_{0,A} = \left\{ u \in H_0(\text{curl}, D) : (Au, \psi)_{L^2(D)} = 0 \right.$$
$$\left. \text{for all } \psi \in H_0(\text{curl}\,0, D) \right\}, \quad (5.22a)$$

$$H_0(\text{curl}, \text{div}_A\,0, D) = \left\{ u \in H_0(\text{curl}, D) : (Au, \nabla\varphi)_{L^2(D)} = 0 \right.$$
$$\left. \text{for all } \varphi \in H_0^1(D) \right\}, \quad (5.22b)$$

are compact in $L^2(D, \mathbb{C}^3)$. Here, $H_0(\text{curl}\,0, D)$ denotes the subspace of H_0 (curl, D) with vanishing curl.

Proof: As noted before (see Remark 4.22) it suffices to consider the subspace $H_0(\text{curl}, \text{div}_A\,0, D)$ because it contains $V_{0,A}$ as a closed subspace. The proof consists of several steps. First we localize the functions $u^{(\ell)}$ into $U^- := U' \cap D = \tilde{\Psi}_j(B^-)$, then we transform these functions into the half ball B_j^- in parameter space. We extend them by a suitable reflection into $H_0(\text{curl}\,\text{div}_{\tilde{M}}\,0, B_j)$ for a certain $\tilde{M} \in L^\infty(B_j, \mathbb{C}^{3\times3})$. In this space we apply Lemma 5.31 to prove L^2-convergence of a subsequence which, finally, yields convergence of the corresponding subsequence of $(u^{(\ell)})$.

Let $\{U_j, \xi_j : j = 1, \ldots, m\}$ be a local coordinate system with corresponding isomorphisms $\tilde{\Psi}_j$ from the balls B_j onto U_j' for $j_1, \ldots, m$. Furthermore, set $U_0' = D$ and choose a partition of unity $\{\phi_j : j = 0, \ldots, m\}$ on $\overline{D}$ corresponding to the sets U_j'. Before we consider bounded sequences we concentrate on one particular element.

Therefore, let $u \in H_0(\text{curl}, \text{div}_A\,0, D)$ and $j \in \{1, \ldots, m\}$ be fixed. The definition of $H_0(\text{curl}, \text{div}_A\,0, D)$ yields for any $\tilde{\varphi} \in H_0^1(U_j^-)$ where again $U_j^- := U_j' \cap D = \tilde{\Psi}_j(B_j^-)$

$$0 = \int_D u(y)^\top A(y) \nabla\big(\phi_j(y)\tilde{\varphi}(y)\big)\, dy$$
$$= \int_{U_j^-} \big(u(y)^\top A(y) \nabla\phi_j(y)\big)\, \tilde{\varphi}(y)\, dy + \int_{U_j^-} \phi_j(y)\, u(y)^\top A(y) \nabla\tilde{\varphi}(y)\, dy$$

Let now $p_j \in H_0^1(U_j^-)$ be defined by

$$\int_{U_j^-} \nabla p_j(y)^\top A(y) \nabla \tilde{\varphi}(y) \, dy = \int_{U_j^-} \left(u(y)^\top A(y) \nabla \phi_j(y) \right) \tilde{\varphi}(y) \, dy$$

for all $\tilde{\varphi} \in H_0^1(U_j^-)$. Substituting $\tilde{\varphi} = \overline{p_j}$ yields the estimate

$$\hat{c} \|\nabla p_j\|_{L^2(U_j^-)}^2 \le c_1 \|u\|_{L^2(D)} \|p_j\|_{L^2(U_j^-)} \tag{5.23a}$$

for some $c_1 > 0$ which is independent of u and j. With Friedrich's inequality we conclude that this implies

$$\|\nabla p_j\|_{L^2(U_j^-)} \le c_2 \|u\|_{L^2(D)} \tag{5.23b}$$

for some $c_2 > 0$ independent of u and j. Defining $\tilde{u}_j := \phi_j u + \nabla p_j$ we conclude that $\tilde{u}_j \in H_0(\mathrm{curl}, U_j^-)$ and even $\tilde{u}_j \in H_0(\mathrm{curl}, \mathrm{div}_A, U_j^-)$ because $\int_{U_j^-} \tilde{u}_j(y)^\top A(y) \nabla \tilde{\varphi}(y) \, dy = 0$ for all $\tilde{\varphi} \in H_0^1(U_j^-)$. This holds for all $j \in \{1, \ldots, m\}$. We transform $\tilde{u}_j$ into the parameter domain as we did it already several times. We define

$$v(x) = \tilde{\Psi}_j'(x)^\top \tilde{u}_j\big(\tilde{\Psi}_j(x)\big) \quad \text{and} \quad M(x) = \tilde{\Psi}_j'(x)^{-1} A\big(\tilde{\Psi}_j(x)\big) \tilde{\Psi}_j'(x)^{-\top}$$

for $x \in B_j^-$ where B_j^- denotes again the lower part of the ball B_j. Here we dropped the index j for v and M to keep the presentation simple. Then $v \in H_0(\mathrm{curl}, B_j^-)$ by Lemma 5.22. Also, we recall that $\nabla \varphi(x) = \tilde{\Psi}_j'(x)^\top (\nabla \tilde{\varphi})\big(\tilde{\Psi}_j(x)\big)$ for $\varphi = \tilde{\varphi} \circ \tilde{\Psi}_j$ and thus

$$0 = \int_{U_j^-} \tilde{u}_j(y)^\top A(y) \nabla \tilde{\varphi}(y) \, dy = \int_{B_j^-} v(x)^\top M(x) \nabla \varphi(x) \, dx \tag{5.24}$$

for all $\varphi \in H_0^1(B_j^-)$. Now we extend v and M into all of the ball B_j by

$$\tilde{v}(x) = \begin{cases} v(x), & x \in B_j^-, \\ -v^*(x^*), & x \in B_j^+, \end{cases}$$

and

$$\tilde{M}_{\ell k}(x) = \begin{cases} M_{\ell k}(x), & x \in B_j^-, \ \ell, k \in \{1, 2, 3\}, \\ M_{\ell k}(x^*), & x \in B_j^+, \ \ell, k \in \{1, 2\} \text{ or } \ell = k = 3, \\ -M_{\ell k}(x^*), & x \in B_j^+, \text{ else,} \end{cases}$$

where again $z^* = (z_1, z_2, -z_3)^\top$ for any $z = (z_1, z_2, z_3)^\top \in \mathbb{C}^3$. First we note that $\tilde{M} \in L^\infty(B_j, \mathbb{C}^{3 \times 3})$ and $\tilde{M}(x)$ is symmetric for almost all $x \in B_j$ and there exists $c_1 > 0$ with $\bar{z}^\top \tilde{M}(x) z \ge c_1 \tilde{M}(x) |z|^2$ for all $z \in \mathbb{C}^3$ and $x \in B_j$. Next we show that $\tilde{v} \in H_0(\mathrm{curl}, \mathrm{div}_{\tilde{M}} 0, B_j)$. The formula $\mathrm{curl}_x \, v^*(x^*) = -(\mathrm{curl} \, v)^*(x^*)$ yields $\tilde{v} \in H_0(\mathrm{curl}, B_j)$. It remains to show that

$$\int_{B_j} \tilde{v}(x)^\top \tilde{M}(x) \nabla \varphi(x) \, dx \;=\; 0 \quad \text{for all} \quad \varphi \in H_0^1(B_j) \,.$$

Let $\varphi \in H_0^1(B_j)$. Then the function $x \mapsto \varphi(x) - \varphi(x^*)$ vanishes for $x_3 = 0$ and is thus in $H_0^1(B_j^-)$. Substituting this function into (5.24) and using $\nabla_x \varphi(x^*) = (\nabla \varphi)^*(x^*)$ yields

$$
\begin{aligned}
0 &= \int_{B_j^-} v(x)^\top M(x) \nabla \big(\varphi(x) - \varphi(x^*) \big) \, dx \\
&= \int_{B_j^-} v(x)^\top M(x) \nabla \varphi(x) \, dx \;-\; \int_{B_j^-} v(x)^\top M(x) \nabla_x \varphi(x^*) \, dx \\
&= \int_{B_j^-} v(x)^\top M(x) \nabla \varphi(x) \, dx \;-\; \int_{B_j^+} v(x^*)^\top M(x^*) (\nabla \varphi)^*(x) \, dx \,.
\end{aligned}
$$

Now a careful elementary calculation shows that

$$v(x^*)^\top M(x^*)(\nabla \varphi)^*(x) \;=\; -\tilde{v}(x)^\top \tilde{M}(x) \nabla \varphi(x) \,, \quad x \in B_j^+ \,,$$

which yields

$$\int_{B_j} \tilde{v}(x)^\top \tilde{M}(x) \nabla \varphi(x) \, dx \;=\; 0 \,.$$

Since this holds for all $\varphi \in H_0^1(B_j)$ we have shown that $\tilde{v} \in H_0(\mathrm{curl}, \mathrm{div}_{\tilde{M}} \, 0, B_j)$.

Now we prove the actual compactness property. Let $(u^{(\ell)})_\ell$ be a bounded sequence in $H_0(\mathrm{curl}, \mathrm{div}_A \, 0, D)$. The scalar functions $p_j^{(\ell)} \in H_0^1(U_j^-)$ and the vector functions $\tilde{v}_j^{(\ell)} \in H_0(\mathrm{curl}, \mathrm{div}_{\tilde{M}} \, 0, B_j)$ correspond to p_j and $\tilde{v}$ above. The estimate (5.23b) implies boundedness of $(p_j^{(\ell)})_\ell$ in $H_0^1(U_j^-)$. The compact embedding of $H_0^1(U_j^-)$ in $L^2(U_j^-)$ yields L^2-convergence of a subsequence of $(p_j^{(\ell)})_\ell$. From estimate (5.23a) applied to the difference $p_j^{(\ell)} - p_j^{(k)}$ we conclude that this subsequence converges also in $H_0^1(U_j^-)$. Also, $(\tilde{u}_j^{(\ell)})_\ell$ is bounded in $H_0(\mathrm{curl}, \mathrm{div}_A \, 0, U_j^-)$ and thus also $(\tilde{v}_j^{(\ell)})_\ell$ in $H_0(\mathrm{curl}, \mathrm{div}_{\tilde{M}} \, 0, B_j)$ for every $j = 1, \ldots, m$. Now we apply Lemma 5.31 which yields convergence of a subsequence of $(\tilde{v}_j^{(\ell)})_\ell$ in $L^2(B_j, \mathbb{C}^3)$ which we denote by the same symbol. Then also the sequence $(\tilde{u}_j^{(\ell)})_\ell$ converges in $L^2(U_j', \mathbb{C}^3)$ for every $j = 1, \ldots, m$. Finally, we consider $j = 0$; that is, the bounded sequence $(\phi_0 u^{(\ell)})_\ell$ in $H_0(\mathrm{curl}, D)$. We choose a ball B containing $\overline{D}$ in its interior and extend $u_0^{(\ell)} = \phi_0 u^{(\ell)}$ by zero into B. Then we construct $p_0^{(\ell)} \in H_0^1(B)$ and $\tilde{u}_0^{(\ell)} = u_0^{(\ell)} - \nabla p_0^{(\ell)} \in H_0(\mathrm{curl}, \mathrm{div}_A \, 0, B)$ just as before (but now in B instead of U_j^-). We again have convergence of a further subsequence of $(p_0^{(\ell)})_\ell$ in $L^2(B)$. Application of Lemma 5.31 to $(\tilde{u}_0^{(\ell)})_\ell$ yields L^2-convergence of a subsequence. Altogether, we have found a common convergent subsequence of $(\tilde{u}_j^{(\ell)})_\ell$ for all $j = 0, 1, \ldots, m$. Together with the convergence of $(p_j^{(\ell)})_\ell$ in $H_0^1(U_j^-)$ this finishes the proof. $\square$

As a direct consequence we obtain the embedding of the subspaces $V_{0,\varepsilon}$ and V_μ defined also in Chap. 4 in case of Lipschitz domains.

Corollary 5.33. *The closed subspace $V_{0,\varepsilon}$ and V_μ, defined in (4.10d) and (4.10c) for $A = \varepsilon I$ and $A = \mu I$, respectively, are compactly embedded in $L^2(D, \mathbb{C}^3)$.*

Proof: For $V_{0,\varepsilon}$ this follows from the previous theorem. By Corollary 4.36 the spaces $V_{0,\varepsilon}$ and V_μ are isometric which ends the proof. $\square$

5.1.3 The Case of a Ball Revisited

In Chap. 2 we have studied the expansion of solutions of the Laplace equation and Helmholtz equation into spherical harmonics for the special case where B is a ball of radius R centered at the origin. We have seen that for L^2-boundary data the series converge in $L^2(D)$. From Chap. 4 we know that the natural solution space of the boundary value problems for the Laplace or Helmholtz equation is the Sobolev space $H^1(D)$. Therefore, according to the trace theorem, the natural space of boundary data is $H^{1/2}(\partial D)$. In this section we will complement the results of Chap. 2 and show convergence in $H^1(D)$.

In our opinion, expansions into spherical harmonics provide a clear insight into several properties of Sobolev spaces. For example, the fact that every curl-free vector field has a potential is seen very explicitly, see Theorem 5.37 below. In the same Theorem we will show the compact embedding of the spaces which appear in the Helmholtz decomposition (the special case $A = I$ and D being a ball). As an important by-product we will prove a characterization of the Sobolev space $H^{1/2}(\partial B)$ by the decay of the expansion coefficients. We remark that the purpose of this section is mainly to fill the gap between the L^2-theory of Chap. 2 and the H^1-theory of Chap. 4. The results of this section with the exception of Theorem 5.37 will not be used in the forthcoming sections.

Let $B = B(0, R)$ be the ball of radius R centered at the origin. We separate variables with respect to polar coordinates $r > 0$ and $\hat{x} \in S^2$ to expand functions into a series of spherical harmonics just as we did in Chap. 2. We begin again with spaces of scalar functions.

Theorem 5.34. *Let $B = B(0, R)$ be the ball of radius R centered at the origin. For $u \in H^1(B)$ let $u_n^m(r) \in \mathbb{C}$ be the expansion coefficients of $\hat{x} \mapsto u(r\hat{x})$ with respect to the spherical harmonics; that is,*

$$u_n^m(r) = \int_{S^2} u(r, \hat{x}) \, Y_n^{-m}(\hat{x}) \, ds(\hat{x}), \quad 0 \leq r < R, \tag{5.25}$$

for $|m| \leq n$ and $n = 0, 1, 2, \ldots$. Then

$$u_N(r, \hat{x}) := \sum_{n=0}^{N} \sum_{m=-n}^{n} u_n^m(r) Y_n^m(\hat{x}), \quad 0 \leq r < R, \; \hat{x} \in S^2,$$

converges to u in $H^1(B)$ as N tends to infinity.

Proof: First we note that, using (2.17),

$$
\begin{aligned}
u_n^m(r) &= -\frac{1}{n(n+1)} \int_{S^2} u(r, \hat{x}) \, \mathrm{Div}_{S^2} \, \mathrm{Grad}_{S^2} Y_n^{-m}(\hat{x}) \, ds(\hat{x}) \\
&= \frac{1}{n(n+1)} \int_{S^2} \mathrm{Grad}_{S^2} u(r, \hat{x}) \cdot \mathrm{Grad}_{S^2} Y_n^{-m}(\hat{x}) \, ds(\hat{x}) \\
&= \frac{1}{\sqrt{n(n+1)}} \int_{S^2} \mathrm{Grad}_{S^2} u(r, \hat{x}) \cdot U_n^{-m}(\hat{x}) \, ds(\hat{x})
\end{aligned}
$$

where $U_n^m = \frac{1}{\sqrt{n(n+1)}} \mathrm{Grad} Y_n^m$ have been defined in Theorem 2.46. Therefore, $\sqrt{n(n+1)} u_n^m(r)$ are the expansion coefficients of $\mathrm{Grad}_{S^2} u(r, \cdot)$ with respect to $\{U_n^m : |m| \leq n, \; n \in \mathbb{N}\}$. From the orthonormality of the systems $\{Y_n^m : |m| \leq n, \; n \in \mathbb{N}\}$ and $\{U_n^m : |m| \leq n, \; n \in \mathbb{N}\}$ we have by Bessel's inequality

$$\sum_{n=0}^{N} \sum_{m=-n}^{n} |u_n^m(r)|^2 \leq \int_{S^2} |u(r\hat{x})|^2 ds(\hat{x}),$$

$$\sum_{n=0}^{N} \sum_{m=-n}^{n} n(n+1) |u_n^m(r)|^2 \leq \int_{S^2} |\mathrm{Grad}_{S^2} u(r\hat{x})|^2 ds(\hat{x}), \quad \text{and}$$

$$\sum_{n=0}^{N} \sum_{m=-n}^{n} \left| \frac{du_n^m}{dr}(r) \right|^2 \leq \int_{S^2} \left| \frac{\partial u}{\partial r}(r\hat{x}) \right|^2 ds(\hat{x})$$

for almost all r. By Parseval's equation we have even equalities for $N \to \infty$. In spherical polar coordinates the gradient is given by $\nabla = \frac{\partial}{\partial r} \hat{x} + \frac{1}{r} \mathrm{Grad}_{S^2}$, thus

$$\|u\|_{H^1(B)}^2 = \int_0^R \int_{S^2} \left[|u(r\hat{x})|^2 + \left| \frac{\partial u}{\partial r}(r\hat{x}) \right|^2 + \frac{1}{r^2} |\text{Grad}_{S^2} u(r\hat{x})|^2 \right] ds(\hat{x}) \, r^2 dr$$

$$\geq \sum_{n=0}^N \sum_{m=-n}^n \int_0^R \left[\left(1 + \frac{n(n+1)}{r^2} \right) |u_n^m(r)|^2 + \left| \frac{du_n^m}{dr}(r) \right|^2 \right] r^2 dr \,.$$

$$(5.26)$$

The right-hand side converges to $\|u\|_{H^1(B)}^2$ as N tends to infinity by the theorem of monotonous convergence. Furthermore, from

$$\nabla u_N(r\hat{x}) = \sum_{n=0}^N \sum_{m=-n}^n \frac{du_n^m}{dr}(r) \, Y_n^m(\hat{x}) \, \hat{x} + \frac{\sqrt{n(n+1)}}{r} \, u_n^m(r) \, U_n^m(\hat{x}) \,, \quad (5.27)$$

$0 \leq r < R$, $\hat{x} \in S^2$, we conclude that, for $M > N$,

$$\|u_M - u_N\|_{H^1(B)}^2$$
$$= \sum_{n=N+1}^M \sum_{m=-n}^n \int_0^R \left[|u_n^m(r)|^2 \left(1 + \frac{n(n+1)}{r^2} \right) + \left| \frac{du_n^m}{dr}(r) \right|^2 \right] r^2 \, dr$$

$$(5.28)$$

tends to zero as N tends to infinity.   □

Remark: From (5.28) we conclude that $r \mapsto r u_n^m(r)$ and $r \mapsto r(u_n^m)'(r)$ are elements of $L^2(0, R)$ for all m, n.

In this case of B being a ball the space $H^{1/2}(\partial B)$ can be characterized by a proper decay of the expansion coefficients with respect to the spherical harmonics.

Theorem 5.35. *Let $B = B(0, R)$ be the ball of radius R centered at the origin. For $f \in L^2(\partial B)$ let $f_n^m \in \mathbb{C}$ be the expansion coefficients with respect to the spherical harmonics; that is,*

$$f_n^m = \int_{S^2} f(R\hat{x}) \, Y_n^{-m}(\hat{x}) \, ds(\hat{x}) \,, \quad |m| \leq n, \ n = 0, 1, 2, \dots$$

Then $f \in H^{1/2}(\partial B)$ if, and only if,

$$\|f\| := \left[\sum_{n=0}^\infty \sum_{m=-n}^n \sqrt{1 + n(n+1)} \, |f_n^m|^2 \right]^{1/2} < \infty. \quad (5.29)$$

Proof: For the moment let X be the completion of $C^\infty(\partial B)$ with respect to the norm of (5.29). Note that $C^\infty(\partial B)$ is also dense in $H^{1/2}(\partial B)$ (see Corollary 5.15 for a proof). Therefore, as in the proof of Corollary 5.15 it suffices to show that the norm $\|\cdot\|$ of (5.29) is equivalent to the norm of the factor space $H^1(B)/H_0^1(B)$. This is assured if we can prove Theorems 5.10 and 5.14 for X instead of $H^{1/2}(\partial B)$. We assume without loss of generality that $R = 1$.

To prove boundedness of the trace operator we write $u \in H^1(B)$ in polar coordinates as

$$u(r\hat{x}) = \sum_{n=0}^{\infty} \sum_{m=-n}^{n} u_n^m(r) Y_n^m(\hat{x})$$

with coefficients $u_n^m(r) \in \mathbb{C}$ and thus

$$(\tilde{\gamma}_0 u)(\hat{x}) = \sum_{n=0}^{\infty} \sum_{m=-n}^{n} u_n^m(1) Y_n^m(\hat{x}).$$

We estimate $\left|u_n^m(s)\right|^2$ for $s \in [0,1]$, using $r^3 \le r^2$ for $r \in [0,1]$, the inequality of Cauchy–Schwarz, and $2ab \le a^2 + b^2$, as

$$s^3 \left|u_n^m(s)\right|^2 = \int_0^s \frac{d}{dr}\left[r^3 |u_n^m(r)|^2\right] dr$$

$$= \int_0^s \left[3r^2 |u_n^m(r)|^2 + 2r^3 \operatorname{Re}\left(\overline{u_n^m(r)} \frac{d}{dr} u_n^m(r)\right)\right] dr$$

$$\le 3 \int_0^s |u_n^m(r)|^2 r^2 \, dr + 2\sqrt{\int_0^s |u_n^m(r)|^2 r^2 \, dr} \sqrt{\int_0^s \left|\frac{d}{dr} u_n^m(r)\right|^2 r^2 \, dr}$$

$$\le 3 \int_0^s |u_n^m(r)|^2 r^2 \, dr + \sqrt{1+n(n+1)} \int_0^s |u_n^m(r)|^2 r^2 \, dr$$

$$+ \frac{1}{\sqrt{1+n(n+1)}} \int_0^s \left|\frac{d}{dr} u_n^m(r)\right|^2 r^2 \, dr.$$

From this and the preceding remark we conclude that u_n^m is continuous on $(0,1]$. We continue with $s = 1$ and multiply this estimate by $\sqrt{1+n(n+1)}$ which yields

$$\sqrt{1+n(n+1)} \left|u_n^m(1)\right|^2 \le 4\big(1+n(n+1)\big) \int_0^1 |u_n^m(r)|^2 r^2 \, dr + \int_0^1 \left|\frac{d}{dr} u_n^m(r)\right|^2 r^2 \, dr$$

and thus

$$\|\tilde{\gamma}_0 u\|_{H^{1/2}(\partial B)}^2 = \sum_{n=0}^{\infty} \sum_{m=-n}^{n} \sqrt{1+n(n+1)} \left|u_n^m(1)\right|^2 \le 4 \|u\|_{H^1(B)}^2$$

by (5.26) for $N \to \infty$. This proves boundedness of $\tilde{\gamma}_0$ from $H^1(B)$ into X. A right inverse of $\tilde{\gamma}$ is given by

$$(\tilde{\eta} f)(r\hat{x}) = \sum_{n=0}^{\infty} \sum_{m=-n}^{n} f_n^m r^n Y_n^m(\hat{x}).$$

It suffices to prove boundedness. From (5.28) for $u_m^n(r) = f_n^m r^m$ we conclude that

$$\|\tilde{\eta} f\|_{H^1(B)}^2 = \sum_{n=0}^{\infty} \sum_{m=-n}^{n} |f_n^m|^2 \left(\frac{1}{2n+3} + \frac{n(n+1)+n^2}{2n+1} \right)$$

$$\leq c \sum_{n=0}^{\infty} \sum_{m=-n}^{n} \sqrt{1+n(n+1)} |f_n^m|^2 = c \|f\|^2 .$$

This proves the trace theorem for $H^1(B)$ and X.

To show the analogue of Theorem 5.14 it suffices again to show that the nullspace of $\tilde{\gamma}_0$ is contained in $H_0^1(B)$. Let $u \in H^1(B)$ with $\tilde{\gamma}_0 u = 0$; that is, $u_n^m(1) = 0$ for all m, n. Let u_N be the truncated sum; that is,

$$u_N(r\hat{x}) = \sum_{n=0}^{N} \sum_{m=-n}^{n} u_n^m(r) Y_n^m(\hat{x}) .$$

Then u_N is continuous in $B \setminus \{0\}$ and converges to u in $H^1(B)$ as N tends to infinity. Furthermore, let $\psi \in C^{\infty}(\mathbb{C})$ such that $\psi(z) = 0$ for $|z| \leq 1$ and $\psi(z) = z$ for $|z| \geq 2$. Define $v_{N,\varepsilon} \in H^1(D)$ by

$$v_{N,\varepsilon}(r\hat{x}) = \sum_{n=0}^{N} \sum_{m=-n}^{n} \varepsilon \psi\left(\frac{1}{\varepsilon} u_n^m(r) \right) Y_n^m(\hat{x}) .$$

Now we can argue as we did already several times. By the theorem of dominated convergence we conclude that $v_{N,\varepsilon} \to u_N$ in $H^1(B)$ as ε tends to zero. Also, we can approximate $v_{N,\varepsilon}$ by some $\tilde{v} \in C_0^{\infty}(B)$ because $v_{N,\varepsilon}$ vanishes in some neighborhood of S^2. Altogether this shows that u can be approximated arbitrarily well in $C_0^{\infty}(B)$. This ends the proof.   □

We continue by extending the previous Theorems 5.34 and 5.35 to the vector-valued case.

Theorem 5.36. *Let $B = B(0, R)$ be the ball of radius R centered at the origin. For $u \in H(\mathrm{curl}, B)$ let $u_n^m(r), v_n^m(r), w_n^m(r) \in \mathbb{C}$ be the expansion coefficients of $\hat{x} \mapsto u(r\hat{x})$ with respect to the spherical vector harmonics; that is,*

$$u_n^m(r) = \int_{S^2} u(r, \hat{x}) \cdot \hat{x} \, Y_n^{-m}(\hat{x}) \, ds(\hat{x}), \quad 0 \leq r < R, \qquad (5.30a)$$

$$v_n^m(r) = \int_{S^2} u(r, \hat{x}) \cdot U_n^{-m}(\hat{x}) \, ds(\hat{x}), \quad 0 \leq r < R, \qquad (5.30b)$$

$$w_n^m(r) = \int_{S^2} u(r, \hat{x}) \cdot V_n^{-m}(\hat{x}) \, ds(\hat{x}), \quad 0 \leq r < R, \qquad (5.30c)$$

for $|m| \leq n$ and $n = 0, 1, 2, \ldots$, where U_n^m and V_n^m have been defined in Theorem 2.46 see Corollary 2.47. Then

$$u_N(r\hat{x}) := \sum_{n=0}^{N} \sum_{m=-n}^{n} \left[u_n^m(r) Y_n^m(\hat{x}) \hat{x} + v_n^m(r) U_n^m(\hat{x}) + w_n^m(r) V_n^m(\hat{x}) \right],$$

$0 \leq r < R$, $\hat{x} \in S^2$, *converges to u in $H(\mathrm{curl}, D)$ as N tends to infinity.*

Proof: We can argue very much as in the proof of Theorem 5.34 but the formulas are technically more complicated. First we note from the completeness of the orthonormal systems $\{Y_n^m : |m| \leq n, \ n \in \mathbb{N}\}$ in $L^2(S^2)$ and $\{U_m^n, V_n^m : |m| \leq n, \ n \in \mathbb{N}\}$ in $L_t^2(S^2)$ that

$$\sum_{n=0}^{\infty} \sum_{m=-n}^{n} \left[|u_n^m(r)|^2 + |v_n^m(r)|^2 + |w_n^m(r)|^2 \right] = \|u(r, \cdot)\|_{L^2(S^2)}^2.$$

The expansion coefficients of $\mathrm{curl}\, u(r, \cdot)$ are given by (see Exercise 5.4)

$$\int_{S^2} \mathrm{curl}\, u(r\hat{x}) \cdot \hat{x}\, Y_n^{-m}(\hat{x})\, ds(\hat{x}) = -\frac{\sqrt{n(n+1)}}{r} w_n^m(r),$$

$$\int_{S^2} \mathrm{curl}\, u(r\hat{x}) \cdot U_n^{-m}(\hat{x})\, ds(\hat{x}) = -\frac{1}{r} \left(r w_n^m(r) \right)',$$

$$\int_{S^2} \mathrm{curl}\, u(r\hat{x}) \cdot V_n^{-m}(\hat{x})\, ds(\hat{x}) = -\frac{\sqrt{n(n+1)}}{r} u_n^m(r) + \frac{1}{r} \left(r v_n^m(r) \right)'.$$

This yields

$$\frac{1}{r^2} \sum_{n=0}^{\infty} \sum_{m=-n}^{n} \left[n(n+1) |w_n^m(r)|^2 + \left| \left(r w_n^m(r) \right)' \right|^2 \right.$$

$$\left. + \left| \sqrt{n(n+1)}\, u_n^m(r) - \left(r v_n^m(r) \right)' \right|^2 \right]$$

$$= \|\mathrm{curl}\, u(r, \cdot)\|_{L^2(S^2)}^2.$$

To compute $\mathrm{curl}\, u_N$ we need the formulas (see Exercise 5.4)

$$\mathrm{curl}\left[u_n^m(r) Y_n^m(\hat{x}) \hat{x} \right] = -\frac{u_n^m(r)}{r} \left(\hat{x} \times \mathrm{Grad}_{S^2} Y_n^m(\hat{x}) \right),$$

$$\mathrm{curl}\left[v_n^m(r) \mathrm{Grad}_{S^2} Y_n^m(\hat{x}) \right] = -\frac{1}{r} \left(r v_n^m(r) \right)' \left(\hat{x} \times \mathrm{Grad}_{S^2} Y_n^m(\hat{x}) \right),$$

$$\mathrm{curl}\left[w_n^m(r) \left(\hat{x} \times \mathrm{Grad}_{S^2} Y_n^m(\hat{x}) \right) \right] = -\frac{1}{r} \left(r w_n^m(r) \right)' \mathrm{Grad}_{S^2} Y_n^m(\hat{x})$$

$$- w_n^m(r) \frac{n(n+1)}{r} Y_n^m(\hat{x}) \hat{x}.$$

Therefore,

$$\operatorname{curl} u_N(x) = \sum_{n=0}^{N} \sum_{m=-n}^{n} \frac{1}{r} \left[-\sqrt{n(n+1)}\, u_n^m(r)\, V_n^m(\hat{x}) + \left(r\, v_n^m(r) \right)' V_n^m(\hat{x}) \right.$$
$$\left. - \left(r\, w_n^m(r) \right)' U_n^m(\hat{x}) - \sqrt{n(n+1)}\, w_n^m(r)\, Y_n^m(\hat{x})\, \hat{x} \right] \qquad (5.31)$$

and thus for $M > N$

$$\| u_M - u_N \|_{H(\operatorname{curl}, B)}^2$$
$$= \sum_{n=N+1}^{M} \sum_{m=-n}^{n} \int_0^1 \left[|u_n^m(r)|^2 + |v_n^m(r)|^2 + |w_n^m(r)|^2 \right] r^2 dr$$
$$+ \sum_{n=N+1}^{M} \sum_{m=-n}^{n} \int_0^1 \left[(n(n+1)) |w_n^m(r)|^2 + \left| \left(r w_n^m(r) \right)' \right|^2 + \right.$$
$$\left. + \left| \sqrt{n(n+1)}\, u_n^m(r) - \left(r v_n^m(r) \right)' \right|^2 \right] dr$$

which proves convergence of (u_N) in $H(\operatorname{curl}, B)$. The limit has to be u. $\square$

Remark: Analogously to the scalar case we note from the previous equality that the functions $r \mapsto r u_n^m(r)$, $r \mapsto r v_n^m(r)$, $r \mapsto r w_n^m(r)$, $r \mapsto (r w_n^m(r))'$, and $r \mapsto (r v_n^m(r))'$ are in $L^2(0, R)$. This yields in particular that v_n^m and w_n^m are continuous in $(0, R]$.

As a simple consequence of this result we show that for balls the subspace $H(\operatorname{curl} 0, B)$ of vector fields with vanishing curl coincides with the space $\nabla H^1(B)$ of gradients and also compactness of the subspaces appearing in the Helmholtz decompositions (see Theorem 4.21 and Remark 4.22) in $L^2(B, \mathbb{C}^3)$.

Theorem 5.37. *Let B be a ball.*

(a) *The space $H(\operatorname{curl} 0, B)$, defined by $H(\operatorname{curl} 0, B) = \{ u \in H(\operatorname{curl}, B) : \operatorname{curl} u = 0 \text{ in } B \}$ coincides with the space $\nabla H^1(B) = \{ \nabla \varphi : \varphi \in H^1(B) \}$.*
(b) *The space $H(\operatorname{curl}, \operatorname{div} 0, B) = \{ u \in H(\operatorname{curl}, B) : \int_B u \cdot \nabla \varphi \, dx = 0 \text{ for all } \varphi \in H^1(B) \}$ is compactly embedded in $L^2(B, \mathbb{C}^3)$.*
(c) *The space $H_0(\operatorname{curl}, \operatorname{div} 0, B) = \{ u \in H_0(\operatorname{curl}, B) : \int_B u \cdot \nabla \varphi \, dx = 0 \text{ for all } \varphi \in H_0^1(B) \}$ is compactly embedded in $L^2(B, \mathbb{C}^3)$.*

Proof: (a) The inclusion $\nabla H^1(B) \subseteq H(\operatorname{curl} 0, B)$ is obvious and holds for any domain. To show the reverse inclusion let $u \in H(\operatorname{curl} 0, B)$. Then u has an expansion in the form

$$u(r\hat{x}) = \sum_{n=0}^{\infty} \sum_{m=-n}^{n} \left[u_n^m(r) Y_n^m(\hat{x})\,\hat{x} + v_n^m(r) U_n^m(\hat{x}) + w_n^m(r) V_n^m(\hat{x}) \right].$$

$$(5.32)$$

Since $\operatorname{curl} u = 0$ we conclude from (5.31) that

$$w_n^m(r) = 0 \quad \text{and} \quad \sqrt{n(n+1)}\, u_n^m(r) = \left(r v_n^m(r) \right)'$$

for all r and all $n = 0, 1, 2, \ldots$, $|m| \leq n$. Now we set

$$\varphi(r\hat{x}) = \sum_{n=1}^{\infty} \sum_{m=-n}^{n} \frac{1}{\sqrt{n(n+1)}}\, r\, v_n^m(r)\, Y_n^m(\hat{x}).$$

From $\sum_{n=1}^{\infty} \sum_{m=-n}^{n} |v_n^m|^2 < \infty$ we conclude that $\varphi \in H^1(B)$ and, by (5.27),

$$\nabla\varphi(r\hat{x}) = \sum_{n=0}^{N} \sum_{m=-n}^{n} \frac{1}{\sqrt{n(n+1)}} \left[\left(r\, v_n^m(r) \right)' Y_n^m(\hat{x}) + \sqrt{n(n+1)}\, v_n^m(r) U_n^m(\hat{x}) \right]$$

$$= \sum_{n=0}^{N} \sum_{m=-n}^{n} \left[u_n^m(r) Y_n^m(\hat{x}) + v_n^m(r) U_n^m(\hat{x}) \right] = u(r\hat{x}).$$

(b) Without loss of generality let B be the unit ball. Every $u \in H(\operatorname{curl}, \operatorname{div} 0, B)$ has an expansion in the form (5.32). Let $\varphi(r\hat{x}) = \rho(r) Y_n^{-m}(\hat{x})$ for some $n \in \mathbb{N}$ and $|m| \leq n$ and $\rho \in C^1[0,1]$. Then $\nabla\varphi(r\hat{x}) = \rho'(r) Y_n^{-m}(\hat{x})\,\hat{x} + \frac{\sqrt{n(n+1)}}{r} U_n^{-m}(\hat{x})$ and thus

$$0 = \int_B u(x)\cdot\nabla\varphi(x)\,dx = \int_0^1 \left[\rho'(r)\, u_n^m(r) + \frac{\sqrt{n(n+1)}}{r}\, \rho(r)\, v_n^m(r) \right] r^2\, dr.$$

$$(5.33)$$

This holds for all such ρ. Note that $r \mapsto r v_n^m(r)$ is continuous and $r \mapsto r^2 u_n^m(r)$ is in $L^1(0,1)$. A modification of the Fundamental Theorem of Calculus (Lemma 4.46, see Exercise 4.16) yields $u_n^m \in C^1(0,1]$ and $u_n^m(1) = 0$ and $\left(r^2 u_n^m(r) \right)' = \sqrt{n(n+1)}\, r v_n^m(r)$ for all $r \in (0,1]$. From (5.31) we conclude that $\sum_{n=0}^{\infty} \sum_{m=-n}^{n} \int_0^1 |q_n^m(r)|^2 dr$ converges where $q_n^m(r) = \sqrt{n(n+1)}\, u_n^m(r) - \left(r v_n^m(r) \right)'$. We estimate, using these two relationships between u_n^m and v_n^m and q_n^m and integration by parts,

$$\sqrt{n(n+1)} \int_0^1 r^2 |v_n^m(r)|^2 dr$$

$$= \int_0^1 \left(r^2 u_n^m(r) \right)'\, r\, \overline{v_n^m(r)}\, dr = -\int_0^1 r^2 u_n^m(r) \left(r\, \overline{v_n^m(r)} \right)' dr$$

$$= -\sqrt{n(n+1)} \int_0^1 r^2 |u_n^m(r)|^2 dr + \int_0^1 r^2 u_n^m(r)\, \overline{q_n^m(r)}\, dr.$$

Note that no boundary terms appear because $u_n^m(1) = 0$. Therefore,

$$
\sqrt{n(n+1)} \int_0^1 r^2 \big[|u_n^m(r)|^2 + |v_n^m(r)|^2\big]\, dr
$$

$$
\leq \sqrt{\int_0^1 r^2 |u_n^m(r)|^2\, dr} \sqrt{\int_0^1 r^2 |q_n^m(r)|^2\, dr}
$$

$$
\leq \frac{1}{2} \int_0^1 r^2 |u_n^m(r)|^2\, dr \; + \; \frac{1}{2} \int_0^1 r^2 |q_n^m(r)|^2\, dr\,.
$$

From this we easily derive the estimate

$$
\int_0^1 r^2 \big[|u_n^m(r)|^2 + |v_n^m(r)|^2\big]\, dr \;\leq\; \frac{1}{n} \int_0^1 |q_n^m(r)|^2 dr\,.
$$

This holds for all m and n.
Now we prove compactness of the embedding operator $I : H(\mathrm{curl}, \mathrm{div}\, 0, B) \to L^2(B, \mathbb{C}^3)$. Define the operator $I_N : H(\mathrm{curl}, \mathrm{div}\, 0, B) \to L^2(B, \mathbb{C}^3)$ by

$$
(I_N u)(r\hat{x}) = \sum_{n=0}^{N} \sum_{m=-n}^{n} \big[u_n^m(r)\, Y_n^m(\hat{x})\, \hat{x} + v_n^m(r)\, U_n^m(\hat{x}) + w_n^m(r)\, V_n^m(\hat{x})\big]\,.
$$

Then

$$
\|I_N u - I u\|_{L^2(B)} = \sum_{n=N+1}^{\infty} \sum_{m=-n}^{n} \int_0^1 r^2 \big[|u_n^m(r)|^2 + |v_n^m(r)|^2 + |w_n^m(r)|^2\big]\, dr
$$

$$
\leq \frac{1}{N+1} \sum_{n=N+1}^{\infty} \sum_{m=-n}^{n} \int_0^1 |q_n^m(r)|^2 dr
$$

$$
+ \frac{1}{(N+2)(N+1)} \sum_{n=N+1}^{\infty} \sum_{m=-n}^{n} \int_0^1 n(n+1)|w_n^m(r)|^2 r^2\, dr\,.
$$

This yields $\|I_N u - I u\|_{L^2(B)} \leq \frac{1}{N} \|u\|_{H(\mathrm{curl}, B)}$; that is, I_N converges to I in the operator norm. I_N is compact as a bounded finite dimensional operator, therefore also I is compact.

(c) We can very much follow the arguments of part (b) and only indicate the differences. We choose $\rho \in C^1[0,1]$ with $\rho(1) = 0$ in order to have $\varphi \in H_0^1(B)$ by Theorem 5.14. Then we derive the variational equation (5.33) for $\rho \in C^1[0,1]$ with $\rho(1) = 0$. Again this shows that $u_n^m \in C^1(0,1]$ and $\big(r^2 u_n^m(r)\big)' = \sqrt{n(n+1)}\, r v_n^m(r)$ for all $r \in (0,1]$ by Exercise 4.16 but no longer $u_n^m(1) = 0$. Nevertheless, one can continue with the proof because $v_n^m(1) = 0$ by the boundary condition $\nu \times u = 0$ on ∂B (and Theorem 5.25). We leave the details to the reader. $\square$

Now we turn to the characterization of the boundary spaces $H^{-1/2}(\mathrm{Div}, \partial B)$ and $H^{-1/2}(\mathrm{Curl}, \partial B)$ by the decay of the expansion coefficients, compare with Theorem 5.35 for the scalar case.

Theorem 5.38. *Let again $B = B(0, R)$ be the ball of radius R centered at the origin. For $f \in L_t^2(\partial B)$ let $a_n^m, b_n^m \in \mathbb{C}$ be the expansion coefficients with respect to the spherical vector-harmonics; that is,*

$$a_n^m = \int_{S^2} f(R\hat{x}) \cdot U_n^{-m}(\hat{x})\, ds(\hat{x}), \quad b_n^m = \int_{S^2} f(R\hat{x}) \cdot V_n^{-m}(\hat{x}))\, ds(\hat{x}),$$

$|m| \leq n$, $n = 0, 1, 2, \ldots$. *Then $f \in H^{-1/2}(\mathrm{Div}, \partial B)$ or $f \in H^{-1/2}(\mathrm{Curl}, \partial B)$ if, and only if,*

$$\|f\|_D^2 := \sum_{n=0}^{\infty} \sum_{m=-n}^{n} \left[\left(1 + n(n+1)\right)^{1/2} |a_n^m|^2 + \left(1 + n(n+1)\right)^{-1/2} |b_n^m|^2 \right. < \infty,$$

(5.34)

or

$$\|f\|_C^2 := \sum_{n=0}^{\infty} \sum_{m=-n}^{n} \left[\left(1 + n(n+1)\right)^{-1/2} |a_n^m|^2 + \left(1 + n(n+1)\right)^{+1/2} |b_n^m|^2 \right. < \infty,$$

(5.35)

respectively.

Proof: Let X be the completion of $C_t^{\infty}(\partial B)$ with respect to the norm of (5.34). The assertion follows as in Corollary 5.15 (see also Theorem 5.35) once we have proven the trace theorem in X and that $H_0^1(B)$ coincides with the null space of γ_0; that is, Theorems 5.24 and 5.25. First we compute $\|u\|_{H(\mathrm{curl}, B)}$. For u of the form

$$u(x) = \sum_{n=0}^{\infty} \sum_{m=-n}^{n} \left[u_n^m(r)\, Y_n^m(\hat{x})\, \hat{x} + v_n^m(r)\, U_n^m(\hat{x}) + w_n^m(r)\, V_n^m(\hat{x}) \right],$$

we have seen in the previous theorem that

$$\|u\|_{L^2(B)}^2 = \sum_{n=0}^{\infty} \sum_{m=-n}^{n} \int_0^R \left[|u_n^m(r)|^2 + |v_n^m(r)|^2 + |w_n^m(r)|^2 \right] r^2\, dr, \quad (5.36a)$$

$$\|\operatorname{curl} u\|_{L^2(B)}^2 = \sum_{n=0}^{\infty} \sum_{m=-n}^{n} \int_0^R \left[\left| \sqrt{n(n+1)}\, |u_n^m(r)|^2 - \left(r\, v_n^m(r) \right)' \right|^2 \right.$$
$$\left. + \left| \left(r\, w_n^m(r) \right)' \right|^2 + n(n+1)\, |w_n^m(r)|^2 \right] dr. \quad (5.36b)$$

Without loss of generality we take $R = 1$. Then we observe that

$$\hat{x} \times u(1, \hat{x}) = \sum_{n=0}^{\infty} \sum_{m=-n}^{n} \left[v_n^m(1) \, \frac{\hat{x} \times \mathrm{Grad}_{S^2} Y_n^m(\hat{x})}{\sqrt{n(n+1)}} \; - \; w_n^m(1) \, \frac{\mathrm{Grad}_{S^2} Y_n^m(\hat{x})}{\sqrt{n(n+1)}} \right].$$

Note that v_n^m and w_n^m are continuous in $(0, 1]$ by the previous remark. Now we have to estimate $\left| v_n^m(1) \right|^2$ and $\left| w_n^m(1) \right|^2$. With $2|a||b| \leq a^2 + b^2$ we conclude

$$\left| v_n^m(1) \right|^2 = \int_0^1 \left(r \left| r \, v_n^m(r) \right|^2 \right)' dr$$

$$= 2 \, \mathrm{Re} \int_0^1 \left(r \, v_n^m(r) \right)' \, \overline{v_n^m(r)} \, r^2 \, dr \; + \; \int_0^1 \left| v_n^m(r) \right|^2 r^2 \, dr$$

$$= 2 \, \mathrm{Re} \int_0^1 \left[\left(r \, v_n^m(r) \right)' - \sqrt{n(n+1)} \, u_n^m(r) \right] \overline{v_n^m(r)} \, r^2 \, dr$$

$$+ \, 2 \sqrt{n(n+1)} \, \mathrm{Re} \int_0^1 u_n^m(r) \, \overline{v_n^m(r)} \, r^2 \, dr \; + \; \int_0^1 \left| v_n^m(r) \right|^2 r^2 \, dr$$

$$\leq \int_0^1 \left| \left(r \, v_n^m(r) \right)' - \sqrt{n(n+1)} \, u_n^m(r) \right|^2 dr \; + \; \int_0^1 \left| v_n^m(r) \right|^2 r^4 \, dr$$

$$+ \, \sqrt{n(n+1)} \int_0^1 \left| u_n^m(r) \right|^2 r^2 \, dr \; + \; (1 + \sqrt{n(n+1)}) \int_0^1 \left| v_n^m(r) \right|^2 r^2 \, dr \, .$$

We divide by $\sqrt{1 + n(n+1)}$, observe that $r^4 \leq r^2$ for $r \in [0, 1]$ and sum. Comparing this with (5.36a), (5.36b) yields

$$\sum_{n=0}^{\infty} \sum_{m=-n}^{n} \left(1 + n(n+1) \right)^{-1/2} \left| v_n^m(1) \right|^2 \; \leq \; 2 \|u\|_{H(\mathrm{curl}, B)}^2 \, .$$

For $w_n^m(1)$ we argue analogously:

$$\left| w_n^m(1) \right|^2 = \int_0^1 \left(r \left| r \, w_n^m(r) \right|^2 \right)' dr$$

$$= 2 \, \mathrm{Re} \int_0^1 \left(r \, w_n^m(r) \right)' \, \overline{w_n^m(r)} \, r^2 \, dr \; + \; \int_0^1 \left| w_n^m(r) \right|^2 r^2 \, dr$$

$$\leq \frac{1}{\sqrt{1 + n(n+1)}} \int_0^1 \left| \left(r \, w_n^m(r) \right)' \right|^2 dr$$

$$+ \sqrt{1 + n(n+1)} \int_0^1 \left| w_n^m(r) \right|^2 r^4 \, dr$$

$$+ \int_0^1 \left| w_n^m(r) \right|^2 r^2 \, dr \, ,$$

and thus

$$\sum_{n=0}^{\infty} \sum_{m=-n}^{n} \left(1 + n(n+1)\right)^{1/2} \left|w_n^m(1)\right|^2 \leq 2\|u\|_{H(\mathrm{curl},B)}^2 \, .$$

This proves boundedness of the trace operator $H(\mathrm{curl}, B) \to X$. A right inverse is given by

$$(\eta f)(r\hat{x}) = \sum_{n=0}^{\infty} \sum_{m=-n}^{n} \left[b_n^m \, U_n^m(\hat{x}) \, - \, a_n^m \, V_n^m(\hat{x})\right] r^n \, ,$$

where a_m^n, b_n^m are the Fourier coefficients of f. Boundedness follows from (5.36a), (5.36b) for $u_n^m(r) = 0$, $v_n^m(r) = b_n^m \, r^n$, and $w_n^m(r) = -a_n^m \, r^n$. Finally, we have to show that the kernel of γ_t coincides with $H_0(\mathrm{curl}, B)$. We follow the arguments as in the proof of Theorem 5.35 and approximate

$$u_N(r\hat{x}) = \sum_{n=0}^{N} \sum_{m=-n}^{n} \left[u_n^m(r) \, Y_n^m(\hat{x}) \, \hat{x} \, + \, v_n^m(r) \, U_n^m(\hat{x}) \, + \, w_n^m(r) \, V_n^m(\hat{x})\right]$$

by

$$u_{N,\varepsilon}(r\hat{x}) = \sum_{n=0}^{\infty} \sum_{m=-n}^{n} \left[\phi\left(\frac{1-r}{\varepsilon}\right) u_n^m(r) \, Y_n^m(\hat{x}) \, \hat{x} \, + \, \frac{\varepsilon}{r} \, \psi\left(\frac{r}{\varepsilon} v_n^m(r)\right) U_n^m(\hat{x}) \right.$$
$$\left. + \, \frac{\varepsilon}{r} \, \psi\left(\frac{r}{\varepsilon} w_n^m(r)\right) V_n^m(\hat{x})\right]$$

where again $\psi \in C^{\infty}(\mathbb{C})$ with $\psi(z) = 0$ for $|z| \leq 1$ and $\psi(z) = z$ for $|z| \geq 2$ and $\phi \in C^{\infty}(\mathbb{R})$ with $\phi(t) = 0$ for $|t| \leq 1$ and $\phi(t) = 1$ for $|t| \geq 2$. Then $u_{N,\varepsilon}$ vanishes in some neighborhood of ∂D because the continuous functions v_n^m and w_n^m vanish for $r = 1$. Estimating the difference $\|u_N - u_{N,\varepsilon}\|_{H(\mathrm{curl},B)}$ requires (see (5.36a), (5.36b)) to consider expressions of the form

$$\frac{\varepsilon}{r} \, \psi\left(\frac{r}{\varepsilon} v_n^m(r)\right) - v_n^m(r) \quad \text{and} \quad \left[\phi\left(\frac{1-r}{\varepsilon}\right) - 1\right] u_n^m(r)$$

and

$$\frac{d}{dr} \left[r\left(\frac{\varepsilon}{r} \, \psi\left(\frac{r}{\varepsilon} v_n^m(r)\right) - v_n^m(r)\right)\right] = \left[\psi'\left(\frac{r}{\varepsilon} v_n^m(r)\right) - 1\right] (r v_n^m(r))' \, .$$

These expressions converge to zero pointwise almost everywhere as ε tends to zero. Also there exist integrable bounds. Then we can continue as in the proof of Theorem 5.35.

For γ_T one argues analogously. $\quad\square$

5.2 Surface Potentials

It is the aim of this section to study the mapping properties of the single- and double layer potentials. First we recall—and rename—the notion of the traces. In this section D is always a bounded Lipschitz domain with boundary ∂D which separates the interior D from the exterior $\mathbb{R}^3 \setminus \overline{D}$. We fix the normal vector $\nu(x)$, which exists for almost all points $x \in \partial D$ by the differentiability assumption on the parametrization, and let it direct into the exterior of D. Then $\gamma_0 u|_{\pm}$ is the trace of u from the exterior $(+)$ and interior $(-)$, respectively. The traces of the normal derivatives $\gamma_1 u|_{\pm}$ for (variational) solutions u of the Helmholtz equation are defined by

$$\langle \gamma_1 u|_-, \psi \rangle_{\partial D} = \int_D [\nabla u \cdot \nabla \hat{\psi} - k^2 u \hat{\psi}] \, dx \,, \quad u \in H_D \,, \tag{5.37a}$$

$$\langle \gamma_1 u|_+, \psi \rangle_{\partial D} = - \int_{\mathbb{R}^3 \setminus \overline{D}} [\nabla u \cdot \nabla \hat{\psi} - k^2 u \hat{\psi}] \, dx \,, \quad u \in H_{\mathbb{R}^3 \setminus \overline{D}} \,, \tag{5.37b}$$

where $\hat{\psi} \in H^1(D)$ or $\hat{\psi} \in H^1(\mathbb{R}^3 \setminus \overline{D})$, respectively, are extensions of $\psi \in H^{1/2}(\partial D)$ with bounded support in the latter case, and

$$H_D = \left\{ u \in H^1(D) : \int_D [\nabla u \cdot \nabla \psi - k^2 u \, \psi] \, dx = 0 \text{ for all } \psi \in H_0^1(D) \right\},$$

$$H_{\mathbb{R}^3 \setminus \overline{D}} = \left\{ u \in H_{loc}^1(\mathbb{R}^3 \setminus \overline{D}) : \int_{\mathbb{R}^3 \setminus \overline{D}} [\nabla u \cdot \nabla \psi - k^2 u \, \psi] \, dx = 0 \right.$$

$$\left. \text{for all } \psi \in H_0^1(\mathbb{R}^3 \setminus \overline{D}) \right\}$$

denote the spaces of variational solutions of the Helmholtz equation in D and in $\mathbb{R}^3 \setminus \overline{D}$, respectively, compare with (5.7). Here, $H_{loc}^1(\mathbb{R}^3 \setminus \overline{D}) = \{u : \mathbb{R}^3 \setminus \overline{D} \to \mathbb{C} : u|_B \in H^1(B) \text{ for all balls } B\}$.

We denoted the dual form in $\langle H^{-1/2}(\partial D), H^{1/2}(\partial D) \rangle$ by $\langle \cdot, \cdot \rangle_{\partial D}$ instead of $\langle \cdot, \cdot \rangle_*$ which we will do from now on. Note, that we have changed the sign in $\gamma_1 u|_+$ because the normal vector ν is directed into the *interior* of $\mathbb{R}^3 \setminus \overline{D}$.

In the first part of this section where we study scalar potentials we let $k \in \mathbb{C}$ with $\text{Re } k \geq 0$ and $\text{Im } k \geq 0$ be arbitrary. When we consider vector potentials we let $k = \omega \sqrt{\varepsilon \mu}$ for any constant $\varepsilon, \mu \in \mathbb{C}$.

We begin with the representation theorem for solutions of the Helmholtz equation, compare with Theorem 3.3.

Theorem 5.39. *(Green's Representation Theorem)*

Let $\Phi(x,y)$ be the fundamental solution of the Helmholtz equation; that is,

$$\Phi(x,y) = \frac{e^{ik|x-y|}}{4\pi|x-y|} \,, \quad x \neq y \,. \tag{5.38}$$

For any solution $u \in H^1(D)$ of the Helmholtz equation; that is, $u \in H_D$, we have the representation

$$\int_{\partial D} (\gamma_0 u)(y) \frac{\partial \Phi}{\partial \nu(y)}(x,y)\, ds(y) \;-\; \langle \gamma_1 u, \gamma_0 \Phi(x,\cdot)\rangle_{\partial D} \;=\; \begin{cases} -u(x), \; x \in D, \\ \quad 0, \quad x \notin \overline{D}. \end{cases}$$

Proof: First we note that elements of H_D are smooth solutions of the Helmholtz equation in D. Fix $x \in D$. As in the classical case we apply Green's formula (5.8) to u and $\Phi(x,\cdot)$ in $D \setminus B[x,\varepsilon]$:

$$\langle \gamma_1 u, \gamma_0 \Phi(x,\cdot)\rangle_{\partial D} \;-\; \int_{|y-x|=\varepsilon} \Phi(x,y) \frac{\partial u(y)}{\partial \nu}\, ds(y)$$

$$= \int_{D \setminus B[x,\varepsilon]} \left[\nabla u(y) \cdot \nabla_y \Phi(x,y) - k^2 u(y)\,\Phi(x,y)\right] dy$$

$$= \int_{D \setminus B[x,\varepsilon]} \left[\nabla u(y) \cdot \nabla_y \Phi(x,y) + u(y)\, \Delta_y \Phi(x,y)\right] dy$$

$$= -\int_{|y-x|=\varepsilon} u(y) \frac{\partial \Phi(x,y)}{\partial \nu(y)}\, ds(y) \;+\; \int_{\partial D} (\gamma_0 u)(y) \frac{\partial \Phi(x,y)}{\partial \nu(y)}\, ds(y);$$

that is,

$$\langle \gamma_1 u, \Phi(x,\cdot)\rangle_{\partial D} \;-\; \int_{\partial D} (\gamma_0 u)(y) \frac{\partial \Phi(x,y)}{\partial \nu(y)}\, ds(y)$$

$$= \int_{|y-x|=\varepsilon} \left[\Phi(x,y) \frac{\partial u(y)}{\partial \nu} - u(y) \frac{\partial \Phi(x,y)}{\partial \nu(y)} -\right] ds(y) \;=\; u(x)$$

where we applied Theorem 3.3 in the last step. For $x \notin \overline{D}$ we argue in the same way. $\quad\square$

Corollary 5.40. *Every $u \in H_D$ is infinitely often differentiable; that is, $u \in C^\infty(D)$. Therefore, u is even analytic by Corollary 3.4.*

Proof: Green's representation theorem shows that u can be expressed as a difference of a double layer potential with density in $L^2(\partial D)$ (even in $H^{1/2}(\partial D)$) and a single layer potential with the density $\varphi = \gamma_1 u \in H^{-1/2}(\partial D)$. Only the latter one has to be considered. Let A be a differential operator. It suffices to show that $A\langle \varphi, \Phi(x,\cdot)\rangle_{\partial D} = \langle \varphi, (A_x \Phi)(x,\cdot)\rangle_{\partial D}$ in every ball B such that $\overline{B} \subseteq D$. Using an argument by induction it suffices to show this for $A = \partial/\partial x_j$ and every function $\hat{\Phi} \in C^\infty(B \times (\mathbb{R}^3 \setminus B))$ instead of Φ. Set $f(x) = \langle \varphi, \hat{\Phi}(x,\cdot)\rangle_{\partial D}$ for $x \in B$. Choose an open neighborhood U of ∂D such that $d := \mathrm{dist}(U,B) > 0$. Then $\hat{\Phi}_j := \partial \hat{\Phi}(x,\cdot)/\partial x_j \in H^1(U)$ and

$$\left| f(x + h\hat{e}^{(j)}) - f(x) - h\left\langle \varphi, \hat{\Phi}_j \right\rangle_{\partial D} \right|$$
$$= \left| \left\langle \varphi, \hat{\Phi}(x + h\hat{e}^{(j)}, \cdot) - \hat{\Phi}(x, \cdot) - h\hat{\Phi}_j \right\rangle_{\partial D} \right|$$
$$\leq c \, \|\varphi\|_{H^{-1/2}(\partial D)} \|\hat{\Phi}(x + h\hat{e}^{(j)}, \cdot) - \hat{\Phi}(x, \cdot) - h\hat{\Phi}_j\|_{H^{1/2}(\partial D)}$$
$$\leq c' \, \|\varphi\|_{H^{-1/2}(\partial D)} \|\hat{\Phi}(x + h\hat{e}^{(j)}, \cdot) - \hat{\Phi}(x, \cdot) - h\hat{\Phi}_j\|_{H^1(U)} \, .$$

The differentiability of $\hat{\Phi}$ yields that

$$\|\hat{\Phi}(x + h\hat{e}^{(j)}, \cdot) - \hat{\Phi}(x, \cdot) - h\hat{\Phi}_j\|_{H^1(U)} \; = \; \mathcal{O}(h^2)$$

which proves the assertion. $\square$

In the following we drop the symbol γ_0, thus we write just v for $\gamma_0 v$. If ∂D is the interface of two domains, then we write $v|_\pm$ and $\gamma_1 v|_\pm$ to indicate the trace from the interior $(-)$ or exterior $(+)$ of D, compare with the remark at the beginning of this section.

Let now u and v be the single and double layer potentials with densities $\varphi \in H^{-1/2}(\partial D)$ and $\varphi \in H^{1/2}(\partial D)$, respectively; that is,

$$(\tilde{S}\varphi)(x) = \langle \varphi, \Phi(x, \cdot) \rangle_{\partial D}, \quad x \notin \partial D, \quad \varphi \in H^{-1/2}(\partial D), \tag{5.39a}$$

$$(\tilde{D}\varphi)(x) = \int_{\partial D} \varphi(y) \frac{\partial \Phi}{\partial \nu(y)}(x, y)\, ds(y), \quad x \notin \partial D, \quad \varphi \in H^{1/2}(\partial D), \tag{5.39b}$$

compare with (3.4a), (3.4b).

It is the aim to prove that $\tilde{S}$ and $\tilde{D}$ are bounded maps into $H^1(D)$ and $H^1(B \setminus \overline{D})$ for any ball B containing $\overline{D}$ in its interior.

From the jump conditions of Theorems 3.12 and 3.16 we recall that for smooth domains D and smooth densities φ the single layer $u = \tilde{S}\varphi$ solves the transmission problem

$$\Delta u + k^2 u = 0 \quad \text{in } \mathbb{R}^3 \setminus \partial D,$$

$$u|_+ = u|_- \text{ on } \partial D, \quad \left.\frac{\partial u}{\partial \nu}\right|_- - \left.\frac{\partial u}{\partial \nu}\right|_+ = \varphi \text{ on } \partial D,$$

and u satisfies also the radiation condition (3.2). This interpretation of the single layer potential, together with the trace theorems, will allow us to prove the required properties of the boundary operators. A problem is that the region of this transmission problem is all of $\mathbb{R}^3$; that is, unbounded. Therefore, we will restrict the transmission problem to a bounded region B containing $\overline{D}$ in its interior and add the additional boundary condition $\partial u/\partial \nu - iku = 0$ on ∂B. The solution of this transmission problem in the ball B will lead to a compact perturbation of $\tilde{S}$.

The variational form is studied in the following theorem.

Lemma 5.41. *Let $k \in \mathbb{C} \setminus \{0\}$ with $\operatorname{Im} k \geq 0$ and B be an open ball with $\overline{D} \subseteq B$. For every $\varphi \in H^{-1/2}(\partial D)$ there exists a unique solution $v \in H^1(B)$ such that*

$$\int_B \left[\nabla v \cdot \nabla \psi - k^2 v \psi \right] dx \; - \; ik \int_{\partial B} v \psi \, ds \; = \; \langle \varphi, \psi \rangle_{\partial D} \quad \text{for all } \psi \in H^1(B).$$
$$(5.40)$$

Furthermore, the operator $\varphi \mapsto v$ is bounded from $H^{-1/2}(\partial D)$ into $H^1(B)$.

Proof: We write (5.40) in the form

$$(v, \psi)_{H^1(B)} \; - \; a(v, \psi) \; = \; \langle \varphi, \overline{\psi} \rangle_{\partial D} \quad \text{for all } \psi \in H^1(B),$$

where a denotes the sesquilinear form

$$a(v, \psi) \; = \; (k^2 + 1) \int_B v \, \overline{\psi} \, dx \; + \; ik \int_{\partial B} v \, \overline{\psi} \, ds, \quad v, \psi \in H^1(B).$$

The mapping $\ell : \psi \mapsto \langle \varphi, \overline{\psi} \rangle_{\partial D}$ is a bounded linear functional on $H^1(B)$ with $\|\ell\| \leq c' \|\varphi\|_{H^{-1/2}(\partial D)}$ because

$$\left| \langle \varphi, \overline{\psi} \rangle_{\partial D} \right| \leq c \, \|\varphi\|_{H^{-1/2}(\partial D)} \|\psi\|_{H^{1/2}(\partial D)} \; \leq \; c' \|\varphi\|_{H^{-1/2}(\partial D)} \|\psi\|_{H^1(D)}$$
$$\leq c' \|\varphi\|_{H^{-1/2}(\partial D)} \|\psi\|_{H^1(B)}.$$

The boundedness of the sesquilinear form a is shown as follows.

$$|a(v, \psi)| \; \leq \; (k^2 + 1) \|v\|_{L^2(B)} \|\psi\|_{L^2(B)} \; + \; k \, \|v\|_{L^2(\partial B)} \|\psi\|_{L^2(\partial B)}. \quad (5.41)$$

Using the boundedness of the trace operator from $H^1(B)$ into $L^2(\partial B)$ we conclude that a is bounded. The theorem of Riesz (Theorem A.5) assures the existence of $r \in H^1(B)$ and a bounded operator K from $H^1(B)$ into itself such that $\langle \varphi, \overline{\psi} \rangle_{\partial D} = (r, \psi)_{H^1(B)}$ and $a(v, \psi) = (Kv, \psi)_{H^1(B)}$ for all $v, \psi \in H^1(B)$. Furthermore, $\|r\|_{H^1(B)} = \|\ell\| \leq c' \|\varphi\|_{H^{-1/2}(\partial D)}$. Then we write (5.40) as $v - Kv = r$ in $H^1(B)$ and show that K is compact. From (5.41) for $\psi = Kv$ we note that

$$\|Kv\|_{H^1(B)}^2 = (Kv, Kv)_{H^1(B)} \; = \; a(v, Kv)$$
$$\leq (k^2 + 1) \|v\|_{L^2(B)} \|Kv\|_{L^2(B)} \; + \; k \, \|v\|_{L^2(\partial B)} \|Kv\|_{L^2(\partial B)}$$
$$\leq c \left[\|v\|_{L^2(B)} \; + \; \|v\|_{L^2(\partial B)} \right] \|Kv\|_{H^1(B)},$$

thus

$$\|Kv\|_{H^1(B)} \; \leq \; c \left[\|v\|_{L^2(B)} \; + \; \|v\|_{L^2(\partial B)} \right].$$

Since in a Hilbert space an operator K is compact if, and only if, it maps weakly convergent sequences into norm-convergent sequences we consider any such sequence $v_j \rightharpoonup 0$ weakly in $H^1(B)$. From the boundedness of the trace operator γ_0 and the compactness of the embeddings $H^1(B)$ in $L^2(B)$ and $H^{1/2}(\partial B)$ in $L^2(\partial B)$ we conclude that the right-hand side of the previous estimate for v_j instead of v converges to zero. Therefore, $\|Kv_j\|_{H^1(B)}$ converges to zero which proves compactness of K.

Therefore, we can apply the Riesz–Fredholm theory to (5.40). To show existence and boundedness of the solution operator $r \mapsto v$ it suffices to prove uniqueness. Therefore, let $v \in H^1(B)$ be a solution of (5.40) for $\varphi = 0$. Substituting $\psi = \overline{kv}$ into (5.40) yields

$$\int_B \left[\overline{k}|\nabla v|^2 - k|k|^2|v|^2\right] dx \; - \; i|k|^2 \int_{\partial B} |v|^2\, ds \; = \; 0\,.$$

Taking the imaginary part and noting that Im $k \geq 0$ yields $v = 0$ on ∂B. Extending v by zero into the exterior of B yields $v \in H^1(\mathbb{R}^3)$ (see Exercise 4.9) and

$$\int_{\mathbb{R}^3} \left[\nabla v \cdot \nabla \psi - k^2 v\,\psi\right] dx \; = \; 0 \quad \text{for all } \psi \in H^1(\mathbb{R}^3)\,;$$

that is, $\Delta v + k^2 v = 0$ in $\mathbb{R}^3$. The regularity result of Corollary 5.40 and unique continuation (see Theorem 4.39) yields $v = 0$ in $\mathbb{R}^3$. $\;\square$

We are now able to prove the following basic properties of the single layer potential.

Theorem 5.42. *Let* $u := \tilde{S}\varphi$ *in* $\mathbb{R}^3 \setminus \partial D$ *be the single layer potential with density* $\varphi \in H^{-1/2}(\partial D)$, *defined by* $u(x) = (\tilde{S}\varphi)(x) = \langle \varphi, \Phi(x, \cdot)\rangle_{\partial D}$, $x \in \mathbb{R}^3 \setminus \partial D$. *Then:*

(a) $u \in C^\infty(\mathbb{R}^3 \setminus \partial D)$ *and* u *satisfies the Helmholtz equation* $\Delta u + k^2 u = 0$ *in* $\mathbb{R}^3 \setminus \partial D$ *and Sommerfeld's radiation condition (3.2) for* $|x| \to \infty$; *that is,*

$$\frac{\partial u^s(r\hat{x})}{\partial r} \; - \; ik\,u^s(r\hat{x}) \; = \; \mathcal{O}\left(\frac{1}{r^2}\right) \quad \text{for } r \to \infty, \qquad (5.42)$$

uniformly with respect to $\hat{x} \in S^2$.

(b) Let Q *be open and bounded such that* $\overline{D} \subseteq Q$. *The operator* $\tilde{S}$ *is well defined and bounded from* $H^{-1/2}(\partial D)$ *into* $H^1(Q)$.

(c) $u|_D \in H_D$ *and* $u|_{\mathbb{R}^3 \setminus \overline{D}} \in H_{\mathbb{R}^3 \setminus \overline{D}}$ *and* $\gamma_1 u|_- - \gamma_1 u|_+ = \varphi$.

(d) The traces

$$S = \gamma_0 \tilde{S} : H^{-1/2}(\partial D) \; \to \; H^{1/2}(\partial D), \qquad (5.43a)$$

$$D' = \frac{1}{2}\left(\gamma_1 \tilde{S}|_+ + \gamma_1 \tilde{S}|_-\right) : H^{-1/2}(\partial D) \; \to \; H^{-1/2}(\partial D), \quad (5.43b)$$

are well defined and bounded.

(e) *With these notations the jump conditions for* $u = \tilde{\mathcal{S}}\varphi$ *and* $\varphi \in H^{-1/2}(\partial D)$ *take the form*

$$\gamma_0 u|_\pm = \mathcal{S}\varphi, \quad \gamma_1 u|_\pm = \mp \frac{1}{2}\varphi + \mathcal{D}'\varphi. \tag{5.44}$$

Proof: (a) follows from Corollary 5.40.

(b) Let B be a ball which contains $\overline{Q}$ in its interior. We prove the following representation of $\tilde{\mathcal{S}}\varphi$ for $\varphi \in H^{-1/2}(\partial D)$.

$$\tilde{\mathcal{S}}\varphi = v + w \quad \text{in } Q, \tag{5.45}$$

where $v \in H^1(B)$ is the solution of (5.40), and $w \in H^1(Q)$ is explicitly given by

$$w(x) = \int_{\partial B} v(y) \left[\frac{\partial \Phi}{\partial \nu(y)}(x,y) - ik\,\Phi(x,y) \right] ds(y), \quad x \in Q.$$

Indeed, (5.40) implies that the restrictions satisfy $v|_D \in H_D$ and $v|_{B \setminus \overline{D}} \in H_{B \setminus \overline{D}}$ if we choose $\psi \in H_0^1(D)$ and $\psi \in H_0^1(B \setminus \overline{D})$, respectively, and extend them by zero into the remaining parts of B. Two applications of Green's representation theorem (Theorem 5.39 in D and in $B \setminus \overline{D}$) for $x \in D$ yield

$$v(x) = -\int_{\partial D} v(y)|_- \frac{\partial \Phi}{\partial \nu(y)}(x,y)\,ds(y) + \langle \gamma_1 v|_-, \Phi(x,\cdot)\rangle_{\partial D}, \quad x \in D,$$

$$0 = \int_{\partial D} v(y)|_+ \frac{\partial \Phi}{\partial \nu(y)}(x,y)\,ds(y) - \langle \gamma_1 v|_+, \Phi(x,\cdot)\rangle_{\partial D}$$
$$- \int_{\partial B} v(y) \frac{\partial \Phi}{\partial \nu(y)}(x,y)\,ds(y) + \langle \gamma_1 v, \Phi(x,\cdot)\rangle_{\partial B}, \quad x \in D,$$

where we dropped the symbol γ_0 for the trace operator. Adding both equation yields

$$v(x) = \langle \gamma_1 v|_- - \gamma_1 v|_+, \Phi(x,\cdot)\rangle_{\partial D} - \tilde{w}(x), \quad x \in D, \tag{5.46}$$

where

$$\tilde{w}(x) := \int_{\partial B} v(y) \frac{\partial \Phi}{\partial \nu(y)}(x,y)ds(y) - \langle \gamma_1 v, \Phi(x,\cdot)\rangle_{\partial B}$$
$$= \int_{\partial B} v(y) \frac{\partial \Phi}{\partial \nu(y)}(x,y)ds(y) - \int_{B \setminus \overline{D}} \left[\nabla v \cdot \nabla \Phi_x - k^2 v \Phi_x \right] dx$$
$$= \int_{\partial B} v(y) \frac{\partial \Phi}{\partial \nu(y)}(x,y)ds(y) - ik \int_{\partial B} v(y)\Phi_x(y)ds = w(x), \; x \in Q.$$

Here, $\Phi_x \in H^1(B)$ is chosen such that $\Phi_x = \Phi(x, \cdot)$ on ∂B and $\Phi_x = 0$ on $\overline{D}$. Then Φ_x is an extension of $\Phi(\cdot, x) \in H^{1/2}(\partial B)$ into B, and the last equation holds by the definition of v.

For fixed $x \in D$ we choose $\tilde{\Phi}_x \in H^1(B)$ such that $\tilde{\Phi}_x = \Phi(x, \cdot)$ on ∂D and $\tilde{\Phi}_x = 0$ on ∂B. Now we recall the definition (5.37a), (5.37b) of the traces $\gamma_1 v|_{\pm}$ and rewrite the first term of (5.46) as

$$\langle \gamma_1 v|_- - \gamma_1 v|_+ \,, \Phi(x, \cdot) \rangle_{\partial D} = \langle \gamma_1 v|_- - \gamma_1 v|_+ \,, \tilde{\Phi}_x \rangle_{\partial D}$$

$$= \int_B \left[\nabla v \cdot \nabla \tilde{\Phi}_x - k^2 v \, \tilde{\Phi}_x \right] dx$$

$$= \langle \varphi, \tilde{\Phi}_x \rangle_{\partial D} = \langle \varphi, \Phi(x, \cdot) \rangle_{\partial D} = (\tilde{S}\varphi)(x)$$

by (5.40) for $\psi = \tilde{\Phi}_x$. Thus $v = \tilde{S}\varphi - w$ in D. For $x \in B \setminus \overline{D}$ we argue exactly in the same way to show (5.46) which proves (5.45).

Now we observe that $v|_{\partial B} \mapsto w|_Q$ is bounded from $L^2(\partial B)$ into $H^1(Q)$. Furthermore, $\varphi \mapsto v$ is bounded from $H^{-1/2}(\partial D)$ into $H^1(B)$. Combining this with the boundedness of the trace operator $v \mapsto v|_{\partial B}$ yields boundedness of $\tilde{S}$ from $H^{-1/2}(\partial D)$ into $H^1(Q)$.

(c) From the representation (5.45) it suffices to show that $\gamma_1 v|_- - \gamma_1 v|_+ = \varphi$ on ∂D because w is a classical solution of the Helmholtz equation in B. By the definitions of $\gamma_1|_{\pm}$ and of v we have that $\langle \gamma_1 v|_- - \gamma_1 v|_+, \psi \rangle_{\partial D} = \int_Q [\nabla v \cdot \nabla \tilde{\psi} - k^2 v \, \tilde{\psi}] \, dx = \langle \varphi, \psi \rangle_{\partial D}$ for all extensions $\tilde{\psi}$ of ψ with compact support. This proves part (c).

(d) This follows from (b) and the boundedness of the trace operator γ_0.

(e) This follows directly from (c) and the definition of $\mathcal{D}'$. $\square$

Corollary 5.43. *For every $\varphi \in H^{-1/2}(\partial D)$ the single layer potential $u = \tilde{S}\varphi$ is the only variational solution of the transmission problem*

$$\Delta u + k^2 u = 0 \ \text{in} \ \mathbb{R}^3 \setminus \partial D, \quad u|_- = u|_+ \ \text{on} \ \partial D, \quad \frac{\partial u}{\partial \nu}\bigg|_- - \frac{\partial u}{\partial \nu}\bigg|_+ = \varphi \ \text{on} \ \partial D,$$

and u satisfies the Sommerfeld radiation condition (5.42); that is, $u \in H^1_{loc}(\mathbb{R}^3)$ is the unique radiating solution of

$$\int_{\mathbb{R}^3} \left[\nabla u \cdot \nabla \psi - k^2 u \, \psi \right] dx = \langle \varphi, \gamma_0 \psi \rangle_{\partial D} \quad \text{for all } \psi \in H^1(\mathbb{R}^3) \text{ with compact support.}$$

$$(5.47)$$

Proof: The previous theorem implies that $u|_D \in H_D$ and $u|_{\mathbb{R}^3 \setminus \overline{D}} \in H_{\mathbb{R}^3 \setminus \overline{D}}$ and u satisfies the radiation condition. Let now $\psi \in H^1(\mathbb{R}^3)$ with compact support which is contained in some ball B. The definitions of $\gamma_1|_{\pm}$ yield again

$$\langle \varphi, \psi \rangle_{\partial D} = \langle \gamma_1 u|_- - \gamma_1 u|_+, \psi \rangle_{\partial D} = \int_B \left[\nabla u \cdot \nabla \psi - k^2 u \, \psi \right] dx .$$

This proves that $u = \tilde{S}\varphi$ solves (5.47). To prove uniqueness let $\varphi = 0$. Then

$$\int_{\mathbb{R}^3} \left[\nabla u \cdot \nabla \psi - k^2 u \psi\right] dx = 0 \quad \text{for all } \psi \in H^1(\mathbb{R}^3) \text{ with compact support};$$

that is, u is a radiating variational solution of the Helmholtz equation in all of $\mathbb{R}^3$. The regularity result from Corollary 5.40 yields $u \in C^\infty(\mathbb{R}^3)$. Also, u satisfies the radiation condition which implies $u = 0$ in $\mathbb{R}^3$ by Theorem 3.23. $\square$

The following properties are helpful for using the boundary integral equation method for solving the interior or exterior boundary value problems.

Theorem 5.44. *Let S_i be the single layer boundary operator for the special value $k = i$. Then:*

(a) S_i is symmetric and coercive in the sense that there exists $c > 0$ with

$$\langle \varphi, S_i \phi \rangle_{\partial D} = \langle \phi, S_i \varphi \rangle_{\partial D}, \quad \langle \varphi, S_i \overline{\varphi} \rangle_{\partial D} \geq c\|\varphi\|^2_{H^{-1/2}(\partial D)}$$

for all $\phi, \varphi \in H^{-1/2}(\partial D)$. In particular, $\langle \varphi, S_i \overline{\varphi} \rangle_{\partial D}$ is real valued.
(b) S_i is an isomorphism from $H^{-1/2}(\partial D)$ onto $H^{1/2}(\partial D)$.
(c) $S - S_i$ is compact from $H^{-1/2}(\partial D)$ into $H^{1/2}(\partial D)$ for any $k \in \mathbb{C}$ with $\operatorname{Im} k \geq 0$.

Proof: (a) Define $u = \tilde{S}_i \varphi$ and $v = \tilde{S}_i \psi$ in $\mathbb{R}^3 \setminus \partial D$. As shown before, u is a classical solution of $\Delta u - u = 0$ in $\mathbb{R}^3 \setminus \partial D$ and satisfies the radiation condition. From the representation theorem (Theorem 3.3 for $k = i$) we observe that u and v and their derivatives decay exponentially for $|x| \to \infty$. Let $\phi \in C^\infty(\mathbb{R}^3)$ with $\phi(x) = 1$ for $|x| \leq R$ $\phi(x) = 0$ for $|x| \geq R + 1$ where R is chosen such that $\overline{D} \subseteq B(0, R)$. By the definitions (5.37a), (5.37b) of the normal derivatives (note that $u|_D, v|_D \in H_D$ and $u|_{\mathbb{R}^3 \setminus \overline{D}}, v|_{\mathbb{R}^3 \setminus \overline{D}} \in H_{\mathbb{R}^3 \setminus \overline{D}}$ for $k = i$) we conclude that

$$\langle \varphi, S_i \psi \rangle_{\partial D} = \langle \gamma_1 u|_- - \gamma_1 u|_+, \gamma_0 v \rangle_{\partial D}$$

$$= \int_D [\nabla u \cdot \nabla v + uv]\, dx + \int_{\mathbb{R}^3 \setminus \overline{D}} [\nabla u \cdot \nabla(\phi v) + \phi uv]\, dx$$

$$= \int_{B(0,R)} [\nabla u \cdot \nabla v + uv]\, dx + \int_{R < |x| < R+1} [\nabla u \cdot \nabla(\phi v) + \phi uv]\, dx$$

$$= \int_{B(0,R)} [\nabla u \cdot \nabla v + uv]\, dx - \int_{|x|=R} v \frac{\partial u}{\partial \nu}\, ds$$

where we used the classical Green's theorem in last step to the smooth functions u and ϕv. Now we use the fact that v and ∇u decay exponentially to zero as R tends to infinity. Therefore we arrive at

$$\langle \varphi, \mathcal{S}_i \psi \rangle_{\partial D} = \int_{\mathbb{R}^3} \left[\nabla u \cdot \nabla v + uv \right] dx$$

which is symmetric in u and v. To prove coercivity we choose $\psi = \overline{\varphi}$. Then $v = \overline{u}$ and thus

$$\langle \varphi, \mathcal{S}_i \overline{\varphi} \rangle_{\partial D} = \int_{\mathbb{R}^3} \left[|\nabla u|^2 + |u|^2 \right] dx = \|u\|_{H^1(\mathbb{R}^3)}^2 .$$

Now we recall from Theorem 5.10 the existence of a bounded extension operator $\eta : H^{1/2}(\partial D) \to H^1\big(B(0,R)\big)$; that is, $\gamma_0 \eta \psi|_{\pm} = \psi$ and $\eta \psi$ has compact support in $B(0,R)$. Therefore, for $\psi \in H^{1/2}(\partial D)$ and $\tilde{\psi} = \eta \psi \in H^1\big(B(0,R)\big)$ we can estimate as above

$$\langle \varphi, \psi \rangle_{\partial D} = \langle \gamma_1 u|_- - \gamma_1 u|_+, \psi \rangle_{\partial D} = \int_{\mathbb{R}^3} \left[\nabla u \cdot \nabla \tilde{\psi} + u \tilde{\psi} \right] dx$$

$$\leq \|u\|_{H^1(\mathbb{R}^3)} \|\tilde{\psi}\|_{H^1(B(0,R))} \leq \|\eta\| \, \|u\|_{H^1(\mathbb{R}^3)} \|\psi\|_{H^{1/2}(\partial D)} ,$$

and thus $\|\varphi\|_{H^{-1/2}(\partial D)} = \sup\limits_{\|\psi\|_{H^{1/2}(\partial D)}=1} \langle \varphi, \psi \rangle_{\partial D} \leq \|\eta\| \, \|u\|_{H^1(\mathbb{R}^3)}$. Therefore, $\langle \varphi, \mathcal{S}_i \overline{\varphi} \rangle_{\partial D} = \|u\|_{H^1(\mathbb{R}^3)}^2 \geq \frac{1}{\|\eta\|^2} \|\varphi\|_{H^{-1/2}(\partial D)}^2$. This finishes the proof of part (a).

(b) Injectivity of $\mathcal{S}_i$ follows immediately from the coerciveness property of part (a). Furthermore, from part (a) we observe that $(\varphi, \psi)_S := \langle \varphi, \mathcal{S}_i \overline{\psi} \rangle_{\partial D)}$ defines an inner product in $H^{-1/2}(\partial D)$ such that its corresponding norm is equivalent to the ordinary norm in $H^{-1/2}(\partial D)$. Surjectivity of $\mathcal{S}_i$ is now an immediate consequence of the Riesz Representation Theorem A.5. Indeed, let $f \in H^{1/2}(\partial D)$. It defines a linear and bounded functional ℓ on $H^{-1/2}(\partial D)$ by $\ell(\varphi) := \langle \varphi, \overline{f} \rangle_{\partial D)}$ for $\varphi \in H^{-1/2}(\partial D)$. By the theorem of Riesz (Theorem A.5) there exists $\psi \in H^{1/2}(\partial D)$ such that $(\varphi, \psi)_S = \ell(\varphi)$ for all $\varphi \in H^{-1/2}(\partial D)$; that is, $\langle \varphi, \mathcal{S}_i \overline{\psi} \rangle_{\partial D)} = \langle \varphi, \overline{f} \rangle_{\partial D)}$ for all $\varphi \in H^{-1/2}(\partial D)$; that is, $\mathcal{S}_i \psi = f$.

(c) Choose open balls Q and B such that $\overline{D} \subseteq Q$ and $\overline{Q} \subseteq B$. We use that facts that $\tilde{\mathcal{S}}\varphi$ and $\tilde{\mathcal{S}}_i\varphi$ have representations of the form (5.45); that is,

$$\tilde{\mathcal{S}}\varphi = v + w \quad \text{and} \quad \tilde{\mathcal{S}}_i\varphi = v_i + w_i \quad \text{in } Q$$

where $v, v_i \in H^1(B)$ are the solutions of (5.40) for k and $k = i$, respectively, and $w, w_i \in H^1(Q)$ are explicitly given by

$$w(x) = \int_{\partial B} v(y) \left[\frac{\partial \Phi_k}{\partial \nu(y)}(x, y) - ik \, \Phi_k(x, y) \right] ds(y), \quad x \in Q,$$

and analogously for w_i (see proof of Theorem 5.42). We note that the mapping $\varphi \mapsto v$ is bounded from $H^{-1/2}(\partial D)$ into $H^1(B)$. Therefore, by the trace theorem and the compact embedding of $H^{1/2}(\partial B)$ into $L^2(\partial B)$ the mapping $\varphi \mapsto v|_{\partial B}$ is compact from $H^{-1/2}(\partial D)$ into $L^2(\partial B)$. This proves compactness of the mappings $\varphi \mapsto w|_{\partial D}$ from $H^{-1/2}(\partial D)$ into $H^{1/2}(\partial D)$. It remains to consider $v - v_i$. Taking the difference of the equations for v and v_i; that is,

$$\int_B \left[\nabla v \cdot \nabla \psi - k^2 v\,\psi\right] dx \; - \; ik \int_{\partial B} v\,\psi\,ds = \langle \varphi, \psi \rangle_{\partial D} ,$$

$$\int_B \left[\nabla v_i \cdot \nabla \psi + v_i\,\psi\right] dx \; + \; \int_{\partial B} v_i\,\psi\,ds = \langle \varphi, \psi \rangle_{\partial D} ,$$

yields

$$\int_B \left[\nabla(v - v_i) \cdot \nabla \psi + (v - v_i)\,\psi\right] dx \; + \; \int_{\partial B} (v - v_i)\,\psi\,ds$$
$$= (k^2 + 1) \int_B v\,\psi\,dx \; + \; (ik + 1) \int_{\partial B} v\,\psi\,ds$$

for all $\psi \in H^1(B)$ which is of the form

$$\int_B \left[\nabla(v - v_i) \cdot \nabla \psi + (v - v_i)\,\psi\right] dx + \int_{\partial B} (v - v_i)\,\psi\,ds = \int_B f\,\psi\,dx + \int_{\partial B} g\,\psi\,ds$$
$$\tag{5.48}$$

with $f \in L^2(B)$ and $g \in L^2(\partial B)$. As mentioned before, the mapping $\varphi \mapsto v$ is bounded from $H^{-1/2}(\partial D)$ into $H^1(B)$ and thus compact as a mapping into $L^2(B)$ as well as the mapping $\varphi \mapsto v|_{\partial B}$ from $H^{-1/2}(\partial D)$ into $L^2(\partial B)$. Finally, the mapping (f, g) into the solution $v - v_i$ of (5.48) is bounded from $L^2(B) \times L^2(\partial B)$ into $H^1(B)$. This proves compactness of the mapping $\varphi \mapsto v - v_i$ from $H^{-1/2}(\partial D)$ into $H^1(B)$. The trace theorem yields compactness of $\mathcal{S} - \mathcal{S}_i$ and ends the proof. $\quad\square$

Remarks 5.45 (i) The proof of part (b) implies in particular that the dual space of $H^{-1/2}(\partial D)$ can be identified with $H^{1/2}(\partial D)$; that is, the spaces $H^{1/2}(\partial D)$ and $H^{-1/2}(\partial D)$ are reflexive Banach spaces.

(ii) The symmetry of $\mathcal{S}$ holds for any $k \in \mathbb{C} \setminus \{0\}$ with $\operatorname{Im} k \geq 0$. Indeed, we can just copy the previous proof of part (a) and arrive at

$$\langle \varphi, \mathcal{S}\psi \rangle_{\partial D} = \int_{B(0,R)} \left[\nabla u \cdot \nabla v - k^2 uv\right] dx \; - \; \int_{|x|=R} v\,\frac{\partial u}{\partial \nu}\,ds$$

and thus

$$\langle \varphi, \mathcal{S}\psi \rangle_{\partial D} - \langle \psi, \mathcal{S}\varphi \rangle_{\partial D} = -\int_{|x|=R} \left(v\,\frac{\partial u}{\partial \nu} - u\,\frac{\partial v}{\partial \nu}\right) ds .$$

It suffices to show that the last integral tends to zero as $R \to \infty$. To see this we just write

$$\int_{|x|=R} \left(v \frac{\partial u}{\partial \nu} - u \frac{\partial v}{\partial \nu} \right) ds = \int_{|x|=R} v \left(\frac{\partial u}{\partial \nu} - iku \right) ds$$

$$- \int_{|x|=R} u \left(\frac{\partial v}{\partial \nu} - ikv \right) ds ,$$

and this tends to zero by Sommerfeld's radiation condition as in the proof of Theorem 3.6.

We finish this part of the section by formulating the corresponding theorems for the double layer potential without detailed proofs.

Theorem 5.46. *Let Q be open and bounded such that $\overline{D} \subseteq Q$. The double layer operator $\tilde{\mathcal{D}}$, defined in (5.39b), is well defined and bounded from $H^{1/2}(\partial D)$ into $H^1(D)$ and into $H^1(Q \setminus \overline{D})$.*
Furthermore, with $u := \tilde{\mathcal{D}}\varphi$ in $\mathbb{R}^3 \setminus \partial D$ we have that $u|_D \in H_D$ and $u|_{\mathbb{R}^3 \setminus \overline{D}} \in H_{\mathbb{R}^3 \setminus \overline{D}}$ and $\gamma_0 u|_+ - \gamma_0 u|_- = \varphi$ and $\gamma_1 u|_- - \gamma_1 u|_+ = 0$. In particular, $u \in C^\infty(\mathbb{R}^3 \setminus \partial D)$ and satisfies the Helmholtz equation $\Delta u + k^2 u = 0$ in $\mathbb{R}^3 \setminus \partial D$ and the Sommerfeld radiation condition. The traces

$$\mathcal{T} = \gamma_1 \tilde{\mathcal{D}} : H^{1/2}(\partial D) \to H^{-1/2}(\partial D) , \tag{5.49a}$$

$$\mathcal{D} = \frac{1}{2} \left(\gamma_0 \tilde{\mathcal{D}}|_+ + \gamma_0 \tilde{\mathcal{D}}|_- \right) : H^{1/2}(\partial D) \to H^{1/2}(\partial D) \tag{5.49b}$$

are well defined and bounded. Furthermore, we have the jump conditions for $u = \tilde{\mathcal{D}}\varphi$ and $\varphi \in H^{1/2}(\partial D)$:

$$\gamma_0 u|_\pm = \pm \frac{1}{2} \varphi + \mathcal{D}\varphi , \quad \gamma_1 u|_\pm = \mathcal{T}\varphi . \tag{5.50}$$

Proof (Only Sketch): We follow the proof of Theorem 5.42 and choose a ball B with $\overline{D} \subseteq B$ and prove a decomposition of $u = \tilde{\mathcal{D}}\varphi$ in the form $u = v + w$ where the pair $\left(v|_D, v|_{B \setminus \overline{D}} \right) \in H^1(D) \times H^1(B \setminus \overline{D})$ is the variational solution of the transmission problem

$$\Delta v + k^2 v = 0 \text{ in } B \setminus \partial D, \quad \frac{\partial v}{\partial \nu} - ikv = 0 \text{ on } \partial B ,$$

$$\left. \frac{\partial v}{\partial \nu} \right|_+ = \left. \frac{\partial v}{\partial \nu} \right|_- , \quad v|_+ - v|_- = \varphi \text{ on } \partial D .$$

With this v we define w just as in the proof of Theorem 5.42. The mapping properties of $\mathcal{D}$ and $\mathcal{T}$ follow now directly from this representation and the trace theorems. Finally, (5.50) follows directly from $\gamma_0 u|_+ - \gamma_0 u|_- = \varphi$ and the definition of $\mathcal{D}\varphi$. $\square$

We note that—in contrast to the case of a smooth boundary—the operator $\mathcal{D}$ is, in general, not compact anymore. Similarly to Theorem 5.44 the properties of the operator $\mathcal{T}$ with respect to the special wave number $k = i$ are useful. We leave the proof of the following result to the reader because it follows the same arguments as in the proof of Theorem 5.44.

Theorem 5.47. *Let $\mathcal{T}_i$ be the normal derivative of the double layer boundary operator for the special value $k = i$. Then:*

(a) *$\mathcal{T}_i$ is symmetric and coercive in the sense that there exists $c > 0$ with*

$$\langle \mathcal{T}_i \varphi, \phi \rangle_{\partial D} \; = \; \langle \mathcal{T}_i \phi, \varphi \rangle_{\partial D}\,, \quad \langle \mathcal{T}_i \varphi, \overline{\varphi} \rangle_{\partial D} \; \leq \; -c \|\varphi\|^2_{H^{1/2}(\partial D)}$$

  for all $\phi, \varphi \in H^{1/2}(\partial D)$. In particular, $\langle \mathcal{T}_i \varphi, \overline{\varphi} \rangle_{\partial D}$ is real valued.
(b) *$\mathcal{T}_i$ is an isomorphism from $H^{1/2}(\partial D)$ onto $H^{-1/2}(\partial D)$.*
(c) *$\mathcal{T} - \mathcal{T}_i$ is compact from $H^{1/2}(\partial D)$ into $H^{-1/2}(\partial D)$ for any $k \in \mathbb{C}$ with Im $k \geq 0$.*

We continue with the *vector valued case*. Let now $\varepsilon, \mu \in \mathbb{C} \setminus \{0\}$ be constant and $\omega > 0$. Define the wave number $k \in \mathbb{C}$ with Re $k \geq 0$ and Im $k \geq 0$ by $k^2 = \omega^2 \varepsilon \mu$. We recall the trace operators $\gamma_t : H(\mathrm{curl}, D) \rightarrow H^{-1/2}(\mathrm{Div}, \partial D)$ and $\gamma_T : H(\mathrm{curl}, D) \rightarrow H^{-1/2}(\mathrm{Curl}, \partial D)$. As in the scalar case we fix the direction of the unit normal vector to point into the exterior of D and distinguish in the following between the traces from the exterior $(+)$ and interior $(-)$ by writing $\gamma_t u|_{\pm}$ and $\gamma_T u|_{\pm}$, respectively. Due to the direction of ν we have the following forms of Green's formula 5.19.

$$\langle \gamma_t v|_-, \gamma_T u|_- \rangle_{\partial D} \; = \; \int_D \left[u \cdot \mathrm{curl}\, v - v \cdot \mathrm{curl}\, u \right] dx \quad \text{for all } u, v \in H(\mathrm{curl}, D)\,,$$

and

$$\langle \gamma_t v|_+, \gamma_T u|_+ \rangle_{\partial D} \; = \; - \int_{\mathbb{R}^3 \setminus \overline{D}} \left[u \cdot \mathrm{curl}\, v - v \cdot \mathrm{curl}\, u \right] dx$$

for all $u, v \in H(\mathrm{curl}, \mathbb{R}^3 \setminus \overline{D})$, where $\langle \cdot, \cdot \rangle_{\partial D}$ denotes the dual form in $\langle H^{-1/2}(\mathrm{Div}, \partial D), H^{-1/2}(\mathrm{Curl}, \partial D) \rangle$, see Theorem 5.26. Again we changed the notation by writing $\langle \cdot, \cdot \rangle_{\partial D}$ instead of $\langle \cdot, \cdot \rangle_*$. In the following we want to discuss vector potentials of the form

$$\mathrm{curl} \int_{\partial D} a(y)\, \Phi(x, y)\, ds(y) \quad \text{and} \quad \mathrm{curl}\, \mathrm{curl} \int_{\partial D} a(y)\, \Phi(x, y)\, ds(y) \quad \text{for } x \notin \partial D\,,$$

see Sect. 3.2.2. From the trace theorem we know that in general we have to allow a to be in $H^{-1/2}(\mathrm{Div}, \partial D)$. Therefore, we have to give meaning to the boundary integral. We recall from Lemma 5.27 that $H^{-1/2}(\mathrm{Div}, \partial D)$ can be considered as a subspace of $H^{-1/2}(\partial D, \mathbb{C}^3)$ which is the dual space of $H^{1/2}(\partial D, \mathbb{C}^3)$. The dual form is denoted by $\langle \cdot, \cdot \rangle_{\partial D}$, see the beginning of this section. We define the bilinear mapping

$$\langle \cdot, \cdot \rangle_{\partial D} \, : \, H^{-1/2}(\mathrm{Div}, \partial D) \times H^{1/2}(\partial D) \, \longrightarrow \, \mathbb{C}^3$$

componentwise; that is, $\big(\langle a, \varphi \rangle_{\partial D} \big)_j = \langle a_j, \varphi \rangle_{\partial D}$, $j = 1, 2, 3$, for $a \in H^{-1/2}(\mathrm{Div}, \partial D)$ and $\psi \in H^{1/2}(\partial D)$. We refer to the remark following Lemma 5.27. For smooth tangential fields a and φ this is exactly the integral $\int_{\partial D} \varphi(y)\, a(y)\, ds(y)$. With this bilinear mapping we have the following form of Green's theorem.

Lemma 5.48. *For $v \in H(\mathrm{curl}, D)$ and $\psi \in H^1(D)$ we have*

$$\langle \gamma_t v, \gamma_0 \psi \rangle_{\partial D} \; = \; \int_D \left[\psi\, \mathrm{curl}\, v + \nabla \psi \times v \right] dx \,.$$

Proof: For any $z \in \mathbb{C}^3$ we have $\langle \gamma_t v, \gamma_0 \psi \rangle_{\partial D} \cdot z = \langle \gamma_t v, \gamma_T(\psi z) \rangle_{\partial D}$ which can be seen by approximating v and ψ by smooth functions. Therefore, by (5.19),

$$\langle \gamma_t v, \gamma_0 \psi \rangle_{\partial D} \cdot z = \langle \gamma_t v, \gamma_T(\psi z) \rangle_{\partial D} = \int_D \left[\psi\, z \cdot \mathrm{curl}\, v - v \cdot \mathrm{curl}(\psi z) \right] dx$$

$$= \int_D \left[\psi\, z \cdot \mathrm{curl}\, v - v \cdot (\nabla \psi \times z) \right] dx$$

$$= z \cdot \int_D \left[\psi\, \mathrm{curl}\, v - v \times \nabla \psi \right] dx \,.$$

$\square$

Now we are able to generalize the definition of the vector potentials of Sect. 3.2.2 for any $a \in H^{-1/2}(\mathrm{Div}, \partial D)$. We define

$$v(x) = \mathrm{curl}\, \langle a, \Phi(x, \cdot) \rangle_{\partial D} \,, \qquad x \in \mathbb{R}^3 \setminus \partial D, \tag{5.51}$$

$$u(x) = \mathrm{curl}^2 \langle a, \Phi(x, \cdot) \rangle_{\partial D} \,, \qquad x \in \mathbb{R}^3 \setminus \partial D, \tag{5.52}$$

where we dropped the symbol γ_0 for the scalar trace operator.

The starting point for our discussion of the vector potentials is the following version of the representation theorem (compare to the Stratton–Chu formula of Theorem 3.27).

Theorem 5.49. *(Stratton–Chu Formulas)*

Let again $\Phi(x, y)$ be the fundamental solution from (5.38) where $k = \omega \sqrt{\varepsilon \mu}$ with $\mathrm{Im}\, k > 0$ or $k > 0$.
(a) For any solutions $E, H \in H(\mathrm{curl}, D)$ of $\mathrm{curl}\, E - i\omega\mu H = 0$ and $\mathrm{curl}\, H + i\omega\varepsilon E = 0$ in D we have the representation

$$- \operatorname{curl}\langle \gamma_t E, \Phi(x, \cdot)\rangle_{\partial D} \;+\; \frac{1}{i\omega\varepsilon}\,\operatorname{curl}^2\langle \gamma_t H, \Phi(x, \cdot)\rangle_{\partial D} \;=\; \begin{cases} E(x)\,, & x \in D\,, \\ 0\,, & x \notin \overline{D}\,. \end{cases}$$

$$\text{(5.53a)}$$

(b) For $\omega, \varepsilon, \mu \in \mathbb{R}_{>0}$ and any solutions $E, H \in H_{loc}(\operatorname{curl}, \mathbb{R}^3 \setminus \overline{D})$ of $\operatorname{curl} E - i\omega\mu H = 0$ and $\operatorname{curl} H + i\omega\varepsilon E = 0$ in $\mathbb{R}^3 \setminus \overline{D}$ which satisfy the Silver–Müller radiation condition (3.37a); that is,

$$\sqrt{\varepsilon}\,E(x) \;-\; \sqrt{\mu}\,H(x) \times \frac{x}{|x|} \;=\; \mathcal{O}\!\left(\frac{1}{|x|^2}\right), \qquad |x| \to \infty, \qquad \text{(5.53b)}$$

uniformly with respect to $\hat{x} = x/|x|$, we have the representation

$$\operatorname{curl}\langle \gamma_t E, \Phi(x, \cdot)\rangle_{\partial D} \;-\; \frac{1}{i\omega\varepsilon}\,\operatorname{curl}^2\langle \gamma_t H, \Phi(x, \cdot)\rangle_{\partial D} \;=\; \begin{cases} 0\,, & x \in D\,, \\ E(x)\,, & x \notin \overline{D}\,. \end{cases}$$

$$\text{(5.53c)}$$

Proof: We prove only part (a). First we note that E and H are smooth solutions of $\operatorname{curl}^2 u - k^2 u = 0$ in D. We fix $z \in D$ and choose a ball $B_r = B(z, r)$ such that $\overline{B_r} \subseteq D$. We apply Green's formula of Lemma 5.48 in $D_r = D \setminus B[z, r]$ to E and $\Phi(x, \cdot)$ for any $x \in B_r$ and obtain

$$\langle \gamma_t E, \Phi(x, \cdot)\rangle_{\partial D} - \int_{\partial B_r} (\nu \times E)\,\Phi(x, \cdot)\,ds \;=\; \int_{D_r} \big[\Phi(x, \cdot)\,\operatorname{curl} E + \nabla_y \Phi(x, \cdot) \times E\big]\,dy\,.$$

Analogously, we have for H instead of E:

$$\langle \gamma_t H, \Phi(x, \cdot)\rangle_{\partial D} - \int_{\partial B_r} (\nu \times H)\,\Phi(x, \cdot)\,ds \;=\; \int_{D_r} \big[\Phi(x, \cdot)\,\operatorname{curl} H + \nabla_y \Phi(x, \cdot) \times H\big]\,dy$$

and thus

$$\begin{aligned} I_r(x) := &-\operatorname{curl}\left[\langle \gamma_t E, \Phi(x, \cdot)\rangle_{\partial D} - \int_{\partial B_r} (\nu \times E)\,\Phi(x, \cdot)\,ds\right] \\ &+ \frac{1}{i\omega\varepsilon}\,\operatorname{curl}^2\left[\langle \gamma_t H, \Phi(x, \cdot)\rangle_{\partial D} - \int_{\partial B_r} (\nu \times H)\,\Phi(x, \cdot)\,ds\right] \\ = &-\operatorname{curl}\int_{D_r} \big[\Phi(x, \cdot)\,\operatorname{curl} E + \nabla_y \Phi(x, \cdot) \times E\big]\,dy \\ &+ \frac{1}{i\omega\varepsilon}\,\operatorname{curl}^2 \int_{D_r} \big[\Phi(x, \cdot)\,\operatorname{curl} H + \nabla_y \Phi(x, \cdot) \times H\big]\,dy \end{aligned}$$

Now we use that $\nabla_y \Phi(x, \cdot) \times E = -\operatorname{curl}_x\big(\Phi(x, \cdot)E\big)$, $\operatorname{curl} E = i\omega\mu H$, and $\operatorname{curl} H = -i\omega\varepsilon E$. This yields

$$I_r(x) = \text{curl}^2 \int_{D_r} \Phi(x, \cdot) \, E \, dy \; - \; i\omega\mu \, \text{curl} \int_{D_r} \Phi(x, \cdot) \, H \, dy$$

$$- \frac{1}{i\omega\varepsilon} \text{curl}^3 \int_{D_r} \Phi(x, \cdot) \, H \, dy \; - \; \text{curl}^2 \int_{D_r} \Phi(x, \cdot) \, E \, dy$$

$$= - i\omega\mu \, \text{curl} \left[\int_{D_r} \Phi(x, \cdot) \, H \, dy - \frac{1}{k^2} \text{curl}^2 \int_{D_r} \Phi(x, \cdot) \, H \, dy \right] = 0$$

by using $\text{curl}^2 = \nabla \, \text{div} - \Delta$ and the Helmholtz equation for Φ. Therefore,

$$- \text{curl}[\langle \gamma_t E, \Phi(x, \cdot) \rangle_{\partial D} \; + \; \frac{1}{i\omega\varepsilon} \text{curl}^2 \langle \gamma_t H, \Phi(x, \cdot) \rangle_{\partial D}$$

$$= - \text{curl} \int_{\partial B_r} (\nu \times E) \, \Phi(x, \cdot) \, ds \; + \; \frac{1}{i\omega\varepsilon} \text{curl}^2 \int_{\partial B_r} (\nu \times H) \, \Phi(x, \cdot) \, ds$$

$$= E(x)$$

by the classical Stratton–Chu formula of Theorem 3.27. The case $x \notin \overline{D}$ is treated in the same way by applying Lemma 5.48 in all of D. $\square$

Corollary 5.50. *For any variational solution $E \in H(\text{curl}, D)$ of $\text{curl}^2 E - k^2 E = 0$ in D we have the representation*

$$- \text{curl}\langle \gamma_t E, \Phi(x, \cdot) \rangle_{\partial D} \; - \; \frac{1}{k^2} \text{curl}^2 \langle \gamma_t \, \text{curl} \, E, \Phi(x, \cdot) \rangle_{\partial D} \; = \; \begin{cases} E(x), \; x \in D, \\ 0, \quad x \notin \overline{D}. \end{cases}$$

Proof: We define $H = \frac{1}{i\omega\mu} \text{curl} \, E$ and observe that also $H \in H(\text{curl}, D)$, and E, H satisfy the assumptions of the previous theorem. Substituting the form of H into (5.53a) yields the assertion. $\square$

For our proof of the boundedness of the vector potentials and the corresponding boundary operators we need the (unique) solvability of a certain boundary value problem of transmission type—just as in the scalar case. As a familiar technique we will use the Helmholtz decomposition in the form

$$H(\text{curl}, B) \; = \; H(\text{curl}, \text{div} \, 0, B) \; \oplus \; \nabla H^1(B) \tag{5.54}$$

in some ball B where $H(\text{curl}, \text{div} \, 0, B) = \{u \in H(\text{curl}, B) : \int_B u \cdot \nabla\varphi \, dx = 0 \text{ for all } \varphi \in H^1(B)\}$. We have seen in Theorem 5.37 that $H(\text{curl}, \text{div} \, 0, B)$ is compactly embedded in $L^2(B, \mathbb{C}^3)$.

Theorem 5.51. *Let B_1 and B_2 be two open balls such that $\overline{B_1} \subseteq D$ and $\overline{D} \subseteq B_2$. Let $\eta \in \mathbb{C}$ with $\text{Im}\,(\eta \overline{k}) > 0$ and $K_j : H^{-1/2}(\text{Div}, \partial B_j) \to H^{-1/2}(\text{Curl}, \partial B_j)$ for $j \in \{1, 2\}$ be linear and compact operators such that $\langle \overline{\psi}, K_j\psi \rangle_{\partial B_j}$ are real valued and $\langle \overline{\psi}, K_j\psi, \rangle_{\partial B_j} > 0$ for all $\psi \in H^{-1/2}(\text{Div}, \partial B_j), \psi \neq 0, j = 1, 2$. Then, for every $a \in H^{-1/2}(\text{Div}, \partial D)$*

the following boundary value problem is uniquely solvable in $H(\mathrm{curl}, B)$ where we set $B = B_2 \setminus \overline{B}_1$ for abbreviation.

$$\mathrm{curl}^2 v - k^2 v = 0 \ \ in \ B \setminus \partial D \,,$$

$$\nu \times v|_- = \nu \times v|_+ \ \ on \ \partial D \,, \quad \nu \times \mathrm{curl}\, v|_+ - \nu \times \mathrm{curl}\, v|_- = a \ \ on \ \partial D \,,$$

$$\nu \times \mathrm{curl}\, v - \eta\, \nu \times K_2(\nu \times v) = 0 \ \ on \ \partial B_2 \,, \quad \nu \times \mathrm{curl}\, v + \eta\, \nu \times K_1(\nu \times v) = 0 \ \ on \ \partial B_1 \,;$$

that is, in variational form:

$$\int_B \left[\mathrm{curl}\, v \cdot \mathrm{curl}\, \psi - k^2 v \cdot \psi\right] dx \ - \ \eta \sum_{j=1}^{2} \langle \gamma_t \psi, K_j \gamma_t v \rangle_{\partial B_j} \ = \ \langle a, \gamma_T \psi \rangle_{\partial D} \quad (5.55)$$

for all $\psi \in H(\mathrm{curl}, B)$. The operator $a \mapsto \gamma_t v|_{\partial D}$ is an isomorphism from $H^{-1/2}(\mathrm{Div}, \partial D)$ onto itself.

Proof: We make use of the Helmholtz decomposition (5.54). Setting $v = v_0 - \nabla p$ and $\psi = \psi_0 + \nabla \varphi$ with $v_0, \psi_0 \in H(\mathrm{curl}, \mathrm{div}\, 0, B)$ and $p, \varphi \in H^1(B)$, the variational equation (5.55) is equivalent to

$$\int_B \left[\mathrm{curl}\, v_0 \cdot \mathrm{curl}\, \psi_0 - k^2 v_0 \cdot \psi_0 + k^2 \nabla p \cdot \nabla \varphi\right] dx$$

$$- \eta \sum_{j=1}^{2} \langle \gamma_t(\psi_0 + \nabla \varphi), K_j \gamma_t(v_0 - \nabla p) \rangle_{\partial B_j}$$

$$= \langle a, \gamma_T(\psi_0 + \nabla \varphi) \rangle_{\partial D} \quad (5.56)$$

for all $(\psi_0, \varphi) \in X := H(\mathrm{curl}, \mathrm{div}\, 0, B) \times H^1(B)$. We equip X with the norm $\|(\psi_0, \varphi)\|_X^2 = \|\psi_0\|_{H(\mathrm{curl}, B)}^2 + k^2 \|\varphi\|_{H^1(B)}^2$ and denote the corresponding inner product by $(\cdot, \cdot)_X$. Then the variational equation (5.56) can be written as

$$\big((v_0, p), (\psi_0, \varphi)\big)_X \ - \ \int_B \left[(k^2 + 1) v_0 \overline{\psi_0} + k^2 p \overline{\varphi}\right] dx$$

$$- \eta \sum_{j=1}^{2} \langle \gamma_t(\overline{\psi_0} + \nabla \overline{\varphi}), K_j \gamma_t(v_0 - \nabla p) \rangle_{\partial B_j}$$

$$= \langle a, \gamma_T(\overline{\psi_0} + \nabla \overline{\varphi}) \rangle_{\partial D} \quad \text{for all } (\psi_0, \varphi) \in X \,. \quad (5.57)$$

The representation theorem of Riesz (Theorem A.5) guarantees the existence of $(g, q) \in X$ and a bounded operator A from X into itself such that

$$\big((g,q),(\psi_0,\varphi)\big)_X = \langle a, \gamma_T(\overline{\psi_0} + \nabla\overline{\varphi})\rangle_{\partial D}$$

$$\big(A(v_0,p),(\psi_0,\varphi)\big)_X = \int_B \big[(k^2+1)v_0\,\overline{\psi_0} + k^2 p\,\overline{\varphi}\big]\,dx$$

$$+\,\eta\sum_{j=1}^2 \langle \gamma_t(\overline{\psi_0}+\nabla\overline{\varphi}), K_j\gamma_t(v_0-\nabla p)\rangle_{\partial B_j}$$

for all $(v_0,p),(\psi_0,\varphi) \in X$. Therefore, (5.57) can be written as

$$(v_0,p) \;-\; A(v_0,p) \;=\; (g,q) \quad \text{in } X\,.$$

From the estimate

$$\big|\big(A(v_0,p),(\psi_0,\varphi)\big)_X\big| \;\le\; (k^2+1)\|(v_0,p)\|_{L^2(B)\times L^2(B)}\,\|(\psi_0,\varphi)\|_{L^2(B)\times L^2(B)}$$

$$+\,c_1\sum_{j=1}^2 \|K_j\gamma_t(v_0-\nabla p)\|_{H^{-1/2}(\mathrm{Curl},\partial B_j)}\,\|\gamma_t(\psi_0+\nabla\varphi)\|_{H^{-1/2}(\mathrm{Div},\partial B_j)}$$

$$\le c\left[\|(v_0,p)\|_{L^2(B)\times L^2(B)} + \sum_{j=1}^2 \|K_j\gamma_t(v_0-\nabla p)\|_{H^{-1/2}(\mathrm{Curl},\partial B_j)}\right]\|(\psi_0,\varphi)\|_X$$

we conclude (set $(\psi_0,\varphi)=A(v_0,p)$) that

$$\|A(v_0,p)\|_X \;\le\; c\big[\|(v_0,p)\|_{L^2(B)\times L^2(B)} + \sum_{j=1}^2 \|K_j\gamma_t(v_0-\nabla p)\|_{H^{-1/2}(\mathrm{Curl},\partial B_j)}\big]\,.$$

From this and the compact embedding of X in $L^2(B,\mathbb{C}^3) \times L^2(B)$ and the compactness of K_j we conclude that A is compact from X into itself. Therefore, existence of a solution of (5.57) holds once uniqueness has been shown. To prove uniqueness, let $a = 0$ and $v \in H(\mathrm{curl}, B)$ a corresponding solution. Substituting $\psi = \overline{k}v$ in (5.55) and taking the imaginary part yields $\sum_{j=1}^2 \langle \overline{\gamma_t v}, K_j\gamma_t v\rangle_{\partial B_j} = 0$ and thus $\gamma_t v = 0$ on $\partial B_1 \cup \partial B_2$. Now we extend v by zero into all of $\mathbb{R}^3$. Then $v \in H(\mathrm{curl}, \mathbb{R}^3)$. Furthermore, (5.55) takes the form

$$\int_{\mathbb{R}^3}\big[\mathrm{curl}\,v \cdot \mathrm{curl}\,\psi - k^2 v \cdot \psi\big]\,dx \;=\; 0 \quad \text{for all } \psi \in H(\mathrm{curl},\mathbb{R}^3)\,;$$

that is, v solves $\mathrm{curl}^2 v - k^2 v = 0$ in $\mathbb{R}^3$ and vanishes outside of B. The unique continuation principle yields $v = 0$ in $\mathbb{R}^3$.

It remains to show that the operator $a \mapsto \gamma_t v|_{\partial D}$ is an isomorphism from $H^{-1/2}(\mathrm{Div},\partial D)$ onto itself. This follows immediately from the unique solvability of the following two boundary value problems for any given $b \in H^{-1/2}(\mathrm{Div},\partial D)$:

$$\text{curl}^2 v_i - k^2 v_i = 0 \text{ in } D \setminus \overline{B}_1, \quad \nu \times v_i = b \text{ on } \partial D, \qquad (5.58\text{a})$$

$$\nu \times \text{curl} \, v_i + \eta \, \nu \times K_1(\nu \times v_i) = 0 \text{ on } \partial B_1, \qquad (5.58\text{b})$$

and

$$\text{curl}^2 v_e - k^2 v_e = 0 \text{ in } B_2 \setminus \overline{D}, \quad \nu \times v_e = b \text{ on } \partial D,$$

$$\nu \times \text{curl} \, v_e - \eta \, \nu \times K_2(\nu \times v_e) = 0 \text{ on } \partial B_2.$$

We leave the proof of this fact to the reader (see Exercise 5.7). □

The previous theorem requires the existence of linear and compact operators $K_j : H^{-1/2}(\text{Div}, \partial B_j) \to H^{-1/2}(\text{Curl}, \partial B_j)$ such that $\langle \overline{\psi}, K_j \psi \rangle_{\partial B_j} > 0$ for all $\psi \in H^{-1/2}(\text{Div}, \partial B_j)$, $\psi \neq 0$. Such operators exist. For example, consider the case that $\partial B_j = S^2$ and the operator K_j is defined by

$$K_j \psi = \sum_{n=0}^{\infty} \sum_{m=-n}^{n} \frac{1}{1 + n(n+1)} \left[a_n^m U_n^m + b_n^m V_n^m \right]$$

for $\psi = \sum_{n=0}^{\infty} \sum_{m=-n}^{n} \left[a_n^m U_n^m + b_n^m V_n^m \right] \in H^{-1/2}(\text{Div}, S^2)$. This K_j has the desired properties, see Exercise 5.6.

Now we are able to prove all of the desired properties of the vector potentials and their behaviors at the boundary.

Theorem 5.52. *Let $k \in \mathbb{C} \setminus \{0\}$ with $\text{Im} \, k \geq 0$ and Q a bounded domain such that $\partial D \subseteq Q$.*

(a) The operators $\tilde{\mathcal{L}}$ and $\tilde{\mathcal{M}}$, defined by

$$(\tilde{\mathcal{L}} a)(x) = \text{curl}^2 \langle a, \Phi(x, \cdot) \rangle_{\partial D} \quad \text{for } x \in Q,$$
$$(\tilde{\mathcal{M}} a)(x) = \text{curl} \langle a, \Phi(x, \cdot) \rangle_{\partial D} \quad \text{for } x \in Q,$$

are well defined and bounded from $H^{-1/2}(\text{Div}, \partial D)$ into $H(\text{curl}, Q)$.

(b) For $a \in H^{-1/2}(\text{Div}, \partial D)$ the fields $u = \tilde{\mathcal{M}} a$ and $\text{curl} \, u = \tilde{\mathcal{L}} a$ satisfy $u|_D, \text{curl} \, u|_D \in H(\text{curl}, D)$ and $u|_{Q \setminus \overline{D}}, \text{curl} \, u|_{Q \setminus \overline{D}} \in H(\text{curl}, Q \setminus \overline{D})$ and $\gamma_t u|_- - \gamma_t u|_+ = a$ and $\gamma_t \, \text{curl} \, u|_- - \gamma_t \, \text{curl} \, u|_+ = 0$. In particular, $u \in C^{\infty}(\mathbb{R}^3 \setminus \partial D, \mathbb{C}^3)$ and u satisfies the equation $\text{curl}^2 u - k^2 u = 0$ in $\mathbb{R}^3 \setminus \partial D$. Furthermore, u and $\text{curl} \, u$ satisfy the Silver–Müller radiation condition (3.42); that is,

$$\text{curl} \, u \times \hat{x} \; - \; i k u \; = \; \mathcal{O}\left(|x|^{-2} \right), \quad |x| \to \infty,$$

uniformly with respect to $\hat{x}$.

(c) The traces

$$\mathcal{L} = \gamma_t \tilde{\mathcal{L}} \quad on \ \partial D, \tag{5.59a}$$

$$\mathcal{M} = \frac{1}{2} \left(\gamma_t \tilde{\mathcal{M}}|_- + \gamma_t \tilde{\mathcal{M}}|_+ \right) \quad on \ \partial D, \tag{5.59b}$$

are bounded from $H^{-1/2}(\mathrm{Div}, \partial D)$ into itself. With these notations the jump conditions hold for $u = \tilde{\mathcal{M}} a$ and $a \in H^{-1/2}(\mathrm{Div}, \partial D)$ in the form

$$\gamma_t u|_\pm = \mp \frac{1}{2} a + \mathcal{M} a, \quad \gamma_t \operatorname{curl} u|_\pm = \mathcal{L} a. \tag{5.60}$$

(d) $\mathcal{L}$ is the sum $\mathcal{L} = \hat{\mathcal{L}} + \hat{\mathcal{K}}$ of an isomorphism $\hat{\mathcal{L}}$ from $H^{-1/2}(\mathrm{Div}, \partial D)$ onto itself and a compact operator $\hat{\mathcal{K}}$.

(e) $\tilde{\mathcal{L}} a$ can be written as

$$\tilde{\mathcal{L}} a = \nabla \tilde{S} \operatorname{Div} a + k^2 \tilde{S} a, \quad a \in H^{-1/2}(\mathrm{Div}, \partial D), \tag{5.61}$$

with the single layer $\tilde{S} a$ from (5.39a) (in the second occurrence $\tilde{S} a$ is taken componentwise).

We note that some authors call $\tilde{\mathcal{L}} a$ the *Maxwell single layer* and $\tilde{\mathcal{M}} a$ the *Maxwell double layer*. The matrix operator

$$\mathcal{C} = \begin{pmatrix} \frac{1}{2} + \mathcal{M} & \mathcal{L} \\ \mathcal{L} & \frac{1}{2} + \mathcal{M} \end{pmatrix}$$

is called the electromagnetic *Calderon operator*.

Proof: (a)–(d) We argue similarly as in the proof of Theorem 5.42. It is sufficient to prove this for Q being a neighborhood of ∂D. Therefore, let B_1 and B_2 be two open balls such that $\overline{B_1} \subseteq D$ and $\overline{D} \subseteq B_2$ and let Q be an open set with $\partial D \subseteq Q$. Choose $\eta \in \mathbb{C}$ with $\mathrm{Im}\,(\eta \overline{k}) > 0$ and $K_j : H^{-1/2}(\mathrm{Div}, \partial B_j) \to H^{-1/2}(\mathrm{Curl}, \partial B_j)$ for $j \in \{1, 2\}$ as in the previous theorem. For $a \in H^{-1/2}(\mathrm{Div}, \partial D)$ let $v \in H(\mathrm{curl}, B)$ be the solution of (5.55). We prove the following representation of $u = \tilde{\mathcal{L}} f$.

$$\tilde{\mathcal{L}} a = k^2 (v + w) \quad in \ B \tag{5.62}$$

where

$$w(x) = \sum_{j=1}^{2} (-1)^j \left[\operatorname{curl} \langle \gamma_t v, \Phi(x, \cdot) \rangle_{\partial B_j} + \frac{1}{k^2} \operatorname{curl}^2 \langle \gamma_t \operatorname{curl} v, \Phi(x, \cdot) \rangle_{\partial B_j} \right]$$

for $x \in B_2 \setminus \overline{B}_1$. To prove this we fix $x \in D \setminus \overline{B}_1$ and apply Corollary 5.50 of the Stratton–Chu formula in $D \setminus \overline{B}_1$ and in $B_2 \setminus \overline{D}$, respectively, and obtain (note the different signs because of the orientation of ν)

$$v(x) = -\operatorname{curl}\langle \gamma_t v|_-, \Phi(x, \cdot)\rangle_{\partial D} \; - \; \frac{1}{k^2}\operatorname{curl}^2\langle \gamma_t \operatorname{curl} v|_-, \Phi(x, \cdot)\rangle_{\partial D}$$
$$+ \; \operatorname{curl}\langle \gamma_t v, \Phi(x, \cdot)\rangle_{\partial B_1} \; + \; \frac{1}{k^2}\operatorname{curl}^2\langle \gamma_t \operatorname{curl} v, \Phi(x, \cdot)\rangle_{\partial B_1},$$
$$0 = \operatorname{curl}\langle \gamma_t v|_+, \Phi(x, \cdot)\rangle_{\partial D} \; + \; \frac{1}{k^2}\operatorname{curl}^2\langle \gamma_t \operatorname{curl} v|_+, \Phi(x, \cdot)\rangle_{\partial D}$$
$$- \; \operatorname{curl}\langle \gamma_t v, \Phi(x, \cdot)\rangle_{\partial B_2} \; - \; \frac{1}{k^2}\operatorname{curl}^2\langle \gamma_t \operatorname{curl} v, \Phi(x, \cdot)\rangle_{\partial B_2}.$$

Adding both equations and using the transmission condition yields

$$v(x) \;=\; \frac{1}{k^2}\operatorname{curl}^2\langle a, \Phi(x, \cdot)\rangle_{\partial D} \;-\; w(x) \;=\; \frac{1}{k^2}(\tilde{\mathcal{L}}a)(x) \;-\; w(x),$$

which proves (5.62) for $x \in D$. For $x \in B_2 \setminus \overline{D}$ we argue analogously. Taking the trace in (5.62) yields

$$\mathcal{L}a \;=\; k^2\,(\gamma_t v \;+\; \gamma_t w) \;=\; \hat{\mathcal{L}}a \;+\; \hat{\mathcal{K}}a.$$

$\hat{\mathcal{L}}$ is an isomorphism by Theorem 5.51, and $\hat{\mathcal{K}}$ is compact by the smoothness of w. Finally, the properties of $\tilde{\mathcal{M}}$ and $\mathcal{M}$ follow from the relation $\operatorname{curl}\tilde{\mathcal{L}}a = \operatorname{curl}^2 \tilde{\mathcal{M}}a = k^2\tilde{\mathcal{M}}a$, thus $\tilde{\mathcal{M}}a = \operatorname{curl} v + \operatorname{curl} w$, and the trace theorem.

(e) Using $\operatorname{curl}\operatorname{curl} = \nabla\operatorname{div} - \Delta$ and the Helmholtz equation for Φ and the fact that we can differentiate with respect to the parameter in the dual form (see the proof of Corollary 5.40 for the arguments) yields for $x \notin \partial D$

$$\tilde{\mathcal{L}}a(x) = \nabla\operatorname{div}\langle a, \Phi(x, \cdot)\rangle_{\partial D} \;-\; \Delta\langle a, \Phi(x, \cdot)\rangle_{\partial D}$$
$$= \nabla\sum_{j=1}^{3}\left\langle a_j, \frac{\partial}{\partial x_j}\Phi(x, \cdot)\right\rangle_{\partial D} \;+\; k^2\langle a, \Phi(x, \cdot)\rangle_{\partial D}$$
$$= \nabla\langle a, \nabla_x\Phi(x, \cdot)\rangle_{\partial D} \;+\; k^2\langle a, \Phi(x, \cdot)\rangle_{\partial D}$$
$$= -\nabla\langle a, \nabla_y\Phi(x, \cdot)\rangle_{\partial D} \;+\; k^2\langle a, \Phi(x, \cdot)\rangle_{\partial D}.$$

By the identification of $\langle a, \nabla_y\Phi(x, \cdot)\rangle_{\partial D}$ with $\langle a, \gamma_T\nabla_y\Phi(x, \cdot)\rangle_{\partial D}$ (see Remark 5.28) and the definition of the surface divergence (Definition 5.29) we conclude that $\langle a, \nabla_y\Phi(x, \cdot)\rangle_{\partial D} = -\langle\operatorname{Div} a, \Phi(x, \cdot)\rangle_{\partial D}$ which proves the representation (5.61). □

5.3 Boundary Integral Equation Methods

We begin again with the scalar case and formulate the interior and exterior boundary value problems. We assume that $D \subseteq \mathbb{R}^3$ is a Lipschitz domain in the sense of Definition 5.1. Furthermore, we assume that the exterior $\mathbb{R}^3 \setminus \overline{D}$ of D is connected. Let $f \in H^{1/2}(\partial D)$ be given boundary data.

Interior Dirichlet Problem: Find $u \in H^1(D)$ such that $\gamma_0 u = f$ and $\Delta u + k^2 u = 0$ in D; that is, in variational form

$$\int_D \left[\nabla u \cdot \nabla \psi - k^2 u\,\psi \right] dx \;=\; 0 \quad \text{for all } \psi \in H_0^1(D). \tag{5.63a}$$

To formulate the exterior problem we define the local Sobolev space by

$$H_{loc}^1(\mathbb{R}^3 \setminus \overline{D}) \;:=\; \left\{ u : \mathbb{R}^3 \setminus \overline{D} \to \mathbb{C} : u|_B \in H^1(B) \text{ for all balls } B \right\}.$$

Exterior Dirichlet Problem: Find $u \in H_{loc}^1(\mathbb{R}^3 \setminus \overline{D})$ such that $\gamma_0 u = f$ and $\Delta u + k^2 u = 0$ in $\mathbb{R}^3 \setminus \overline{D}$; that is, in variational form

$$\int_{\mathbb{R}^3 \setminus D} \left[\nabla u \cdot \nabla \psi - k^2 u\,\psi \right] dx \;=\; 0 \quad \text{for all } \psi \in H_0^1(\mathbb{R}^3 \setminus \overline{D}) \text{ with compact support,}$$

$$\tag{5.63b}$$

and u satisfies Sommerfeld's radiation condition (5.42). Note that u is a smooth solution of the Helmholtz equation in the exterior of $\overline{D}$.

First we consider the question of uniqueness.

Theorem 5.53. *(a) There exists at most one solution of the exterior Dirichlet boundary value problem (5.63b).*

(b) The interior Dirichlet boundary value problem (5.63a) has at most one solution if, and only if, the single layer boundary operator S is one-to-one. More precisely, the null space $\mathcal{N}(S)$ of S is given by all Neumann traces $\gamma_1 u$ of solutions $u \in H_0^1(D)$ of the Helmholtz equation with vanishing Dirichlet boundary data; that is,

$$\mathcal{N}(S) = \left\{ \gamma_1 u : u \in H_0^1(D), \int_D \left[\nabla u \cdot \nabla \psi - k^2 u\,\psi \right] dx = 0 \right.$$

$$\left. \text{for all } \psi \in H_0^1(D) \right\}.$$

Proof: (a) Let $u \in H_{loc}^1(\mathbb{R}^3 \setminus \overline{D})$ be a solution of the exterior Dirichlet boundary value problem for $f = 0$. Choose a ball $B(0, R)$ which contains $\overline{D}$ in its interior and a function $\phi \in C^\infty(\mathbb{R}^3)$ such that $\phi = 1$ on

$B[0, R]$ and $\phi = 0$ in the exterior of $B(0, R+1)$. Application of Green's formula (5.19) in the region $B(0, R+1) \setminus \overline{D}$ to u and $\psi = \phi \overline{u}$ yields

$$0 = -\langle \gamma_1 u, \psi \rangle_{\partial D} + \int_{|x|=R+1} \frac{\partial u}{\partial \nu} \psi \, ds = \int_{B(0,R+1) \setminus \overline{D}} \left[\nabla u \cdot \nabla \psi - k^2 u \psi \right] dx$$

$$= \int_{B(0,R) \setminus \overline{D}} \left[|\nabla u|^2 - k^2 |u|^2 \right] dx + \int_{R < |x| < R+1} \left[\nabla u \cdot \nabla \psi - k^2 u \psi \right] dx$$

$$= \int_{B(0,R) \setminus \overline{D}} \left[|\nabla u|^2 - k^2 |u|^2 \right] dx - \int_{|x|=R} \frac{\partial u}{\partial \nu} \overline{u} \, ds.$$

Note that ψ vanishes on ∂D and outside of $B(0, R+1)$. Now we proceed similarly to the proof of Theorem 3.23. Indeed,

$$\int_{|x|=R} \left| \frac{\partial u}{\partial r} - iku \right|^2 ds = \int_{|x|=R} \left\{ \left| \frac{\partial u}{\partial r} \right|^2 + |ku|^2 \right\} ds - 2 \, \text{Im} \left[\overline{k} \int_{|x|=R} \overline{u} \frac{\partial u}{\partial r} ds \right]$$

$$= \int_{|x|=R} \left\{ \left| \frac{\partial u}{\partial r} \right|^2 + |ku|^2 \right\} ds$$

$$+ 2 \, \text{Im} \, k \int_{B(0,R) \setminus \overline{D}} \left[|\nabla u|^2 + |ku|^2 \right] dx.$$

Now we have to distinguish between two cases. If $\text{Im} \, k > 0$, then u vanishes in $B(0, R) \setminus \overline{D}$. If k is real valued, then the left-hand side of the previous equation converge to zero as R tends to infinity by the radiation condition. Rellich's lemma (Lemma 3.21 or 3.22) implies that u vanishes in the (connected) exterior of D. This proves part (a).

(b) Let first $u \in H^1(D)$ satisfy (5.63a) and $\gamma_0 u = 0$. Then, by the definition of $\tilde{S}$ and Green's representation formula of Theorem 5.39

$$(\tilde{S} \gamma_1 u)(x) = \langle \gamma_1 u, \Phi(x, \cdot) \rangle_{\partial D}$$

$$= \langle \gamma_1 u, \Phi(x, \cdot) \rangle_{\partial D} - \int_{\partial D} (\gamma_0 u)(y) \frac{\partial \Phi}{\partial \nu(y)}(x, y) \, ds(y)$$

$$= u(x), \quad x \in D.$$

Taking the trace yields $S \gamma_1 u = \gamma_0 \tilde{S} \gamma_1 u = \gamma_0 u = 0$ on ∂D.
Second, let $\varphi \in H^{-1/2}(\partial D)$ such that $S\varphi = 0$. Define $u = \tilde{S}\varphi$ in $\mathbb{R}^3 \setminus \partial D$. Then $u|_D \in H_0^1(D)$ and $\gamma_0 u|_+ = 0$ on ∂D. Therefore, $u|_{\mathbb{R}^3 \setminus \overline{D}}$ satisfies the homogeneous exterior boundary value problem and therefore vanishes by part (a). The jump condition of Theorem 5.42 implies $\varphi = \gamma_1 u|_- - \gamma_1 u|_+ = \gamma_1 u|_-$ which ends the proof. $\square$

Remark: This theorem relates the homogeneous interior Dirichlet problem to the null space of the boundary operator S. Therefore, k^2 is a Dirichlet eigenvalue of $-\Delta$ in the sense of Sect. 4.2.1, see Theorem 4.28, if and only if S fails to be one-to-one.

In particular, there exist only a countable number of values k for which S fails to be one-to-one. Furthermore, the null space $\mathcal{N}(S)$ is finite dimensional.

The following theorem studies the question of existence for the case that k^2 is not a Dirichlet eigenvalue.

Theorem 5.54. *Assume in addition to the assumptions at the beginning of this section that k^2 is not an eigenvalue of $-\Delta$ in D with respect to Dirichlet boundary conditions; that is, the only solution of the variational equation (5.63a) in $H_0^1(D)$ is the trivial one $u = 0$. Then there exist (unique) solutions of the exterior and the interior Dirichlet boundary value problems for every $f \in H^{1/2}(\partial D)$. The solutions can be represented as single layer potentials in the form*

$$u(x) \;=\; (\tilde{S}\varphi)(x) \;=\; \langle \varphi, \Phi(x,\cdot)\rangle_{\partial D}\,, \quad x \notin \partial D\,,$$

where the density $\varphi \in H^{-1/2}(\partial D)$ satisfies $S\varphi = f$.

Proof: By the mapping properties of $\tilde{S}$ of Theorem 5.42 it suffices to study solvability of the equation $S\varphi = f$. Because S is a compact perturbation of the isomorphism S_i (the operator corresponding to $k = i$) by Theorem 5.42, the well-known—and already often used—result by Riesz (Theorem A.5) guarantees surjectivity of this operator S provided injectivity holds. But this is assured by the previous theorem. Indeed, if $S\varphi = 0$, then the corresponding single layer potential u solves both, the exterior and the interior boundary value problems with homogeneous boundary data $f = 0$. The uniqueness result implies that u vanishes in all of $\mathbb{R}^3$. The jumps of the normal derivatives (Theorem 5.42 again) yields $\varphi = 0$. $\square$

Furthermore we consider the case of k^2 being a Dirichlet eigenvalue of $-\Delta$ in D. This corresponds to the case where S fails to be one-to-one. In this case we have to apply the Fredholm theory of Theorem A.4 to the boundary equation $S\varphi = f$.

Theorem 5.55. *The interior and exterior boundary value problems are solvable as single layer potentials $u = \tilde{S}\varphi$ for exactly those $f \in H^{1/2}(\partial D)$ which are orthogonal to all Neumann traces $\gamma_1 v \in H^{-1/2}(\partial D)$ of eigenfunctions $v \in H_0^1(D)$ of $-\Delta$ in D; that is, $\langle \gamma_1 v, f\rangle_{\partial D} = 0$ for all $v \in H_0^1(D)$ with $\Delta v + k^2 v = 0$ in the variational sense.*

Proof: We have to study solvability of the equation $\mathcal{S}\varphi = f$ and want to apply Theorem A.4. Therefore, we set $X_1 = Y_2 = H^{-1/2}(\partial D)$, $X_2 = Y_1 = H^{1/2}(\partial D)$, $\langle \varphi, f \rangle_1 = \langle \varphi, f \rangle_{\partial D}$ for $\varphi \in H^{-1/2}(\partial D)$, $f \in H^{1/2}(\partial D)$ and $\langle g, \psi \rangle_2 = \langle \psi, g \rangle_{\partial D}$ for $\psi \in H^{-1/2}(\partial D)$, $g \in H^{1/2}(\partial D)$. The operator $\mathcal{S}$ is self-adjoint in this dual system, see Remarks 5.45. Application of Theorem A.4 yields solvability of the equation $\mathcal{S}\varphi = f$ for exactly those $f \in H^{1/2}(\partial D)$ with $\langle \psi, f \rangle_{\partial D} = 0$ for all $\psi \in \mathcal{N}(\mathcal{S}^*) = \mathcal{N}(\mathcal{S})$. The characterization of the nullspace $\mathcal{N}(\mathcal{S})$ in Theorem 5.53 yields the assertion. $\square$

Remark: For scattering problems by plane waves we have to solve the exterior Dirichlet problem with boundary data $f(x) = -\exp(ik\,\hat{\theta} \cdot x)$ on ∂D. This boundary data satisfies the orthogonality condition. Indeed, by Green's formula (5.8) we conclude for every eigenfunction $v \in H_0^1(D)$ that

$$\langle \gamma_1 v, \exp(ik\,\hat{\theta}\cdot) \rangle_{\partial D} = \int_D \left[\nabla v(x) \cdot \nabla e^{ik\,\hat{\theta}\cdot x} - k^2 v(x)\, e^{ik\,\hat{\theta}\cdot x} \right] dx$$

$$= \int_{\partial D} v(x)\, \frac{\partial}{\partial \nu} e^{ik\,\hat{\theta}\cdot x} ds = 0\,.$$

Therefore, the scattering problem for plane waves by the obstacle D can always be solved by a single layer ansatz.

From the uniqueness result (and Chap. 3) we expect that the exterior boundary value problem is always uniquely solvable, independently of the wave number. However, this can't be done by just one single layer potential. There are several ways to modify the ansatz. In our context perhaps the simplest possibility is to choose an ansatz as the following combination of a single and a double layer ansatz

$$u = \tilde{\mathcal{S}}\varphi + \eta \tilde{\mathcal{D}}\mathcal{S}_i\varphi \quad \text{in the exterior of } D$$

for some $\varphi \in H^{-1/2}(\partial D)$ and some $\eta \in \mathbb{C}$ to be chosen in a moment. The boundary operator $\mathcal{S}_i : H^{-1/2}(\partial D) \to H^{1/2}(\partial D)$ corresponds to $\mathcal{S}$ for $k = i$. Then u is a solution of the exterior boundary value problem provided $\varphi \in H^{-1/2}(\partial D)$ solves

$$\mathcal{S}\varphi + \frac{\eta}{2} \mathcal{S}_i\varphi + \eta \mathcal{D}\mathcal{S}_i\varphi = f \text{ on } \partial D\,. \tag{5.64}$$

By Theorem 5.44 the operator $\mathcal{S}$ is a compact perturbation of the isomorphism $\mathcal{S}_i$ from $H^{-1/2}(\partial D)$ onto $H^{1/2}(\partial D)$ and the operator $\mathcal{D}\mathcal{S}_i$ is bounded. Therefore, for sufficiently small $\eta_0 > 0$ also $\mathcal{S} + \frac{\eta}{2}\mathcal{S}_i + \eta \mathcal{D}\mathcal{S}_i$ is a compact perturbation of an isomorphism for all $\eta \in \mathbb{C}$ with $|\eta| \leq \eta_0$. We choose $\eta \in \mathbb{C}$ with $|\eta| \leq \eta_0$ and $\text{Im}\,(k\eta) > 0$ and show that for this choice of η the boundary equation (5.64) is uniquely solvable for every $f \in H^{1/2}(\partial D)$. By the previous remarks it is sufficient to show that the homogeneous equations admit only the trivial solution. Therefore, let $\varphi \in H^{-1/2}(\partial D)$ be a solution of (5.64) for

$f = 0$ and define u by $u = \tilde{\mathcal{S}}\varphi + \eta\,\tilde{\mathcal{D}}\mathcal{S}_i\varphi$ in all of $\mathbb{R}^3 \setminus \partial D$. Then $u|_+$ vanishes by the jump conditions and (5.64). The uniqueness result of the exterior Dirichlet boundary value problem yields that u vanishes in the exterior of D. The jump conditions yield

$$\gamma_0 u|_- = \gamma_0 u|_- - \gamma_0 u|_+ = -\eta\,\mathcal{S}_i\varphi, \quad \gamma_1 u|_- = \gamma_1 u|_- - \gamma_1 u|_+ = \varphi\,.$$

Elimination of φ yields $\eta\mathcal{S}_i\gamma_1 u|_- + \gamma_0 u|_- = 0$. Now we apply Green's theorem in D; that is,

$$\int_D \left[\nabla u \cdot \nabla\psi - k^2 u\,\psi\right] dx \;=\; \langle\gamma_1 u|_-, \gamma_0\psi\rangle_{\partial D}\,.$$

Substituting $\psi = \overline{ku}$ yields

$$\int_D \left[\overline{k}\,|\nabla u|^2 - k\,|ku|^2\right] dx \;=\; \overline{k}\,\langle\gamma_1 u|, \gamma_0\overline{u}|_-\rangle_{\partial D} \;=\; -\overline{k\eta}\,\langle\gamma_1 u|_-, \mathcal{S}_i\gamma_1\overline{u}|_-\rangle_{\partial D}\,.$$

Now we note that $\langle\gamma_1 u|_-, \mathcal{S}_i\gamma_1\overline{u}|_-\rangle_{\partial D}$ is real and non-negative by Theorem 5.44 and $\mathrm{Im}\,(k\eta) > 0$ by assumption. Therefore, taking the imaginary part yields $\langle\gamma_1 u|_-, \mathcal{S}_i\gamma_1\overline{u}|_-\rangle_{\partial D} = 0$. The property of $\mathcal{S}_i$ from Theorem 5.44 yields $\gamma_1 u|_- = 0$ and thus also $\varphi = \gamma_1 u|_- - \gamma_1 u|_+ = 0$. We formulate the result as a theorem.

Theorem 5.56. *The exterior Dirichlet boundary value problem is uniquely solvable for every $f \in H^{1/2}(\partial D)$. The solution can be expressed in the form*

$$u \;=\; \tilde{\mathcal{S}}\varphi \;+\; \eta\,\tilde{\mathcal{D}}\mathcal{S}_i\varphi \quad in\;\mathbb{R}^3 \setminus D$$

for some sufficiently small η such that $\mathrm{Im}\,(k\eta) > 0$ and the density $\varphi \in H^{-1/2}(\partial D)$ solves (5.64).

We note that for real (and positive) values of k we can choose η with $\mathrm{Im}\,\eta > 0$ independent of k. Indeed, one can show by the same arguments as in the proof of Theorem 5.44 that $\mathcal{D} - \mathcal{D}_i$ is compact. Therefore, we have to choose $|\eta|$ small enough such that $(1 + \eta/2)\mathcal{S}_i + \eta\mathcal{D}_i\mathcal{S}_i$ is an isomorphism from $H^{-1/2}(\partial D)$ onto $H^{1/2}(\partial D)$.

Exactly the same results hold for the interior and exterior *Neumann problems*. We formulate the results in two theorems but leave the proofs to the reader.

Theorem 5.57. *The interior Neumann problem is solvable exactly for those $f \in H^{-1/2}(\partial D)$ with $\langle f, \gamma_0 v\rangle_{\partial D} = 0$ for all solutions $v \in H^1(D)$ of $\Delta v + k^2 v = 0$ in D and $\gamma_1 v = 0$ on ∂D. In this case the solution can be represented as a double layer potential $u = \tilde{\mathcal{D}}\varphi$ in D with $\varphi \in H^{1/2}(\partial D)$ satisfying $\mathcal{T}\varphi = f$ on ∂D where $\mathcal{T} : H^{1/2}(\partial D) \to H^{-1/2}(\partial D)$ denotes the trace of the normal derivative of the double layer potential, see Theorem 5.46.*

Theorem 5.58. *The exterior Neumann problem is always uniquely solvable for every $f \in H^{-1/2}(\partial D)$. The solution can be represented in the form*

$$u = \tilde{D}\varphi + \eta \tilde{S} T_i \varphi \quad in \ \mathbb{R}^3 \setminus \overline{D},$$

with $\varphi \in H^{1/2}(\partial D)$ for some sufficiently small $\eta \in \mathbb{C}$ with $\mathrm{Im}\,(\eta \overline{k}) > 0$. The density $\varphi \in H^{1/2}(\partial D)$ satisfies the boundary equation

$$T\varphi - \frac{\eta}{2} T_i \varphi + D' T_i \varphi = f \quad on \ \partial D.$$

We now turn to the *electromagnetic case* and formulate the boundary value problems. We assume again that $D \subseteq \mathbb{R}^3$ is a bounded Lipschitz domain with connected exterior $\mathbb{R}^3 \setminus \overline{D}$. Furthermore, let $f \in H^{-1/2}(\mathrm{Div}, \partial D)$ be given boundary data. We recall that the trace operator $\gamma_t : H(\mathrm{curl}, D) \to H^{-1/2}(\partial D)$ extends the mapping $u \mapsto \nu \times u|_{\partial D}$ (and analogously for exterior domains). Furthermore, let $\omega > 0$, $\mu > 0$, and $\varepsilon \in \mathbb{C}$ with $\mathrm{Im}\,\varepsilon \geq 0$ and $k = \omega\sqrt{\mu\varepsilon}$ with $\mathrm{Re}\,k > 0$ and $\mathrm{Im}\,k \geq 0$.

The *Interior Boundary Value Problem:* Find $E, H \in H(\mathrm{curl}, D)$ such that $\gamma_t E = f$ and

$$\mathrm{curl}\,E - i\omega\mu H = 0 \text{ in } D \quad \text{and} \quad \mathrm{curl}\,H + i\omega\varepsilon E = 0 \text{ in } D; \qquad (5.65a)$$

that is, in variational form for the field E (see, e.g., (4.2) or (4.22)).

$$\int_D \left[\mathrm{curl}\,E \cdot \mathrm{curl}\,\psi - k^2 E \cdot \psi\right] dx = 0 \quad \text{for all } \psi \in H_0(\mathrm{curl}, D). \qquad (5.65b)$$

As in the scalar case define the local Sobolev space as

$$H_{loc}(\mathrm{curl}, \mathbb{R}^3 \setminus \overline{D}) := \left\{ u : \mathbb{R}^3 \setminus \overline{D} \to \mathbb{C}^3 : u|_B \in H(\mathrm{curl}, B) \text{ for all balls } B \right\}.$$

The *Exterior Boundary Value Problem:* Find $E, H \in H_{loc}(\mathrm{curl}, \mathbb{R}^3 \setminus \overline{D})$ such that $\gamma_t E = f$ and

$$\mathrm{curl}\,E - i\omega\mu H = 0 \text{ in } \mathbb{R}^3 \setminus \overline{D} \quad \text{and} \quad \mathrm{curl}\,H + i\omega\varepsilon E = 0 \text{ in } \mathbb{R}^3 \setminus \overline{D}; \quad (5.66a)$$

that is, in variational form for the field E,

$$\int_{\mathbb{R}^3 \setminus \overline{D}} \left[\mathrm{curl}\,E \cdot \mathrm{curl}\,\psi - k^2 E \cdot \psi\right] dx = 0 \qquad (5.66b)$$

for all $\psi \in H_0(\mathrm{curl}, \mathbb{R}^3 \setminus \overline{D})$ with compact support. Furthermore, (E, H) have to satisfy the Silver–Müller radiation condition (5.53b); that is,

$$\sqrt{\varepsilon}\,E(x) - \sqrt{\mu}\,H(x) \times \frac{x}{|x|} = \mathcal{O}\left(\frac{1}{|x|^2}\right) \qquad (5.66c)$$

uniformly with respect to $x/|x| \in S^2$. Note that E and H are a smooth solutions of $\mathrm{curl}^2 u - k^2 u = 0$ in the exterior of $\overline{D}$.

The question of uniqueness and existence are treated in a very analogous way to the scalar problems and are subject of the following theorems.

Theorem 5.59. *(a) There exists at most one solution of the exterior bound-ary value problem (5.66a)–(5.66c).*

(b) The interior boundary value problem (5.65a)–(5.65b) has at most one solution if, and only if, the boundary operator $\mathcal{L}$ is one-to-one. More pre-cisely, the null space $\mathcal{N}(\mathcal{L})$ of $\mathcal{L}$ is given by all traces $\gamma_t \, \mathrm{curl} \, u$ of solutions $u \in H_0(\mathrm{curl}, D)$ of $\mathrm{curl}^2 u - k^2 u = 0$ with vanishing boundary data $\gamma_t u$; that is,

$$\mathcal{N}(\mathcal{L}) = \left\{ \gamma_t \, \mathrm{curl} \, u \; : \; u \in H_0(\mathrm{curl}, D), \; \begin{array}{l} \int_D [\mathrm{curl} \, u \cdot \mathrm{curl} \, \psi - k^2 u \cdot \psi] \, dx = 0 \\ \text{for all } \psi \in H_0(\mathrm{curl}, D) \end{array} \right\}.$$

Proof: (a) Let E, H be a solution of (5.65a)–(5.65b) corresponding to $f = 0$. Choose again balls such that $\overline{D} \subseteq B(0, R) \subseteq B(0, R+1)$ and a function $\phi \in C^\infty(\mathbb{R}^3)$ such that $\phi = 1$ on $B[0, R]$ and $\phi = 0$ outside of $B(0, R+1)$. Green's formula (5.19) applied to $u = \phi E$ and $v = \overline{H}$ in the region $B(0, R+1) \setminus \overline{D}$ yields (note that $\gamma_T u$ vanishes on ∂D and on $\partial B(0, R+1)$)

$$0 = -\langle \gamma_t v, \gamma_T u \rangle_{\partial D} + \int_{|x|=R+1} u \cdot (\nu \times v) \, ds$$

$$= \int_{B(0,R+1) \setminus \overline{D}} \left[\phi E \cdot \mathrm{curl} \, \overline{H} - \overline{H} \cdot \mathrm{curl}(\phi E) \right] dx$$

$$= i\omega \int_{B(0,R) \setminus \overline{D}} \left[\overline{\varepsilon} \, |E|^2 - \mu \, |H|^2 \right] dx + \int_{R < |x| < R+1} \left[u \cdot \mathrm{curl} \, v - v \cdot \mathrm{curl} \, u \right] d_{\cdot}$$

$$= i\omega \int_{B(0,R) \setminus \overline{D}} \left[\overline{\varepsilon} \, |E|^2 - \mu \, |H|^2 \right] dx - \int_{|x|=R} E \cdot (\hat{x} \times \overline{H}) \, ds.$$

We consider again the case $\mathrm{Im} \, \varepsilon > 0$ and $\mathrm{Im} \, \varepsilon = 0$ separately. If $\mathrm{Im} \, \varepsilon > 0$, then the fields decay exponentially as R tends to infinity. This follows again from the Stratton–Chu formula of Theorem 5.49. Taking the imag-inary part of the previous formula and letting R tend to infinity yields that E vanishes.

Let now ε be real valued. Taking the real part of the previous formula yields

$$\mathrm{Re} \int_{|x|=R} E \cdot (\hat{x} \times \overline{H}) \, ds \geq 0.$$

This yields

$$0 \geq 2\sqrt{\mu}\sqrt{\varepsilon} \, \text{Re} \int_{|x|=R} E \cdot (\overline{H} \times \hat{x}) \, ds$$

$$= \int_{|x|=R} \left\{ |\sqrt{\varepsilon}E|^2 + |\sqrt{\mu}H \times \hat{x}|^2 \right\} ds \; - \; \int_{|x|=R} |\sqrt{\varepsilon}E - \sqrt{\mu}H \times \hat{x}|^2 ds \, .$$

The Silver–Müller radiation condition implies that the second integral tends to zero as R tends to infinity. This implies that $\int_{|x|=R} |E|^2 ds$ tends to zero. Now we proceed as in the proof of Theorem 3.35 and apply Rellich's lemma to conclude that E and H vanish in the exterior of $\overline{D}$.

(b) Let first $u \in H_0(\text{curl}, D)$ with $\text{curl}^2 u - k^2 u = 0$ in D. By Corollary 5.50 we conclude for $x \in D$, because $\gamma_t u = 0$,

$$(\tilde{\mathcal{L}}\gamma_t \, \text{curl} \, u)(x) = \text{curl}^2 \langle \gamma_t \, \text{curl} \, u, \Phi(x, \cdot) \rangle_{\partial D}$$
$$= \text{curl}^2 \langle \gamma_t \, \text{curl} \, u, \Phi(x, \cdot) \rangle_{\partial D} \; + \; k^2 \, \text{curl} \langle \gamma_t u, \Phi(x, \cdot) \rangle_{\partial D}$$
$$= -k^2 u(x) \, .$$

Taking the trace yields $\mathcal{L}(\gamma_t \, \text{curl} \, u) = -k^2 \gamma_t u = 0$ on ∂D.

Second, let $a \in H^{-1/2}(\text{Div}, \partial D)$ with $\mathcal{L}a = 0$. Define u by $u(x) = (\tilde{\mathcal{L}}a)(x) = \text{curl}^2 \langle a, \Phi(x, \cdot) \rangle_{\partial D}$ for $x \in \mathbb{R}^3 \setminus \partial D$. Then $u \in H_{loc}(\text{curl}, \mathbb{R}^3)$ solves the exterior and the interior boundary value problem with homogeneous boundary data $\gamma_t u = \mathcal{L}a = 0$. The uniqueness result for the exterior problem implies that u vanishes in the exterior. The jump conditions of Theorem 5.52 yield $a = \gamma_t \, \text{curl} \, u|_- - \gamma_t \, \text{curl} \, u|_+ = \gamma_t \, \text{curl} \, u|_-$. This proves part (b). □

Theorem 5.60. *Assume in addition to the assumptions at the beginning of this section that k^2 is not an eigenvalue of curl^2 in D with respect to the boundary condition $\nu \times u = 0$; that is, the only solution of the variational equation (5.65b) in $H_0(\text{curl}, D)$ is the trivial one $u = 0$. Then there exist (unique) solutions of the exterior and the interior boundary value problems for every $f \in H^{-1/2}(\text{Div}, \partial D)$. The solutions can be represented as boundary layer potentials in the form*

$$E(x) \; = \; (\tilde{\mathcal{L}}a)(x) \; = \; \text{curl}^2 \langle a, \Phi(x, \cdot) \rangle_{\partial D} \, , \quad H = \frac{1}{i\omega\mu} \, \text{curl} \, E \, , \quad x \notin \partial D \, ,$$

where the density $a \in H^{-1/2}(\text{Div} \, \partial D)$ satisfies $\mathcal{L}a = f$.

Proof: This is clear from the uniqueness result of both, the interior and the exterior boundary value problem and the fact that $\mathcal{L}$ is a compact perturbation of an isomorphism. □

We want to study the equation $\mathcal{L}a = f$ for the case when $\mathcal{L}$ fails to be one-to-one. It is the aim to apply the abstract Fredholm result of Theorem A.4 and have to find the proper dual system. The following result will provide the adjoint of $\mathcal{L}$.

Lemma 5.61.

(a) The bilinear form $\langle \cdot, \cdot \rangle$ on $H^{-1/2}(\mathrm{Div}, \partial D) \times H^{-1/2}(\mathrm{Div}, \partial D)$, defined by $\langle a, b \rangle = \langle a, \nu \times b \rangle_{\partial D}$ is well defined and a dual system.

(b) For $a, b \in H^{-1/2}(\mathrm{Div}, \partial D)$ we have

$$\langle \mathcal{L}a, \nu \times b \rangle_{\partial D} = \langle \mathcal{L}b, \nu \times a \rangle_{\partial D},$$

and the adjoint of $\mathcal{L}$ with respect to $\langle \cdot, \cdot \rangle$ is given by $\mathcal{L}^* = -\mathcal{L}$.

(c) Let $\mathcal{L}_i$ be the operator $\mathcal{L}$ for $k = i$. Then there exists $c > 0$ with

$$\langle \mathcal{L}_i a, \overline{a} \rangle \geq c \, \|a\|^2_{H^{-1/2}(\mathrm{Div}, \partial D)} \quad \text{for all } a \in H^{-1/2}(\mathrm{Div}, \partial D).$$

In particular, the left-hand side is real valued.

(d) $\mathcal{L}$ satisfies the variational equation

$$\langle \mathcal{L}a, b \rangle = -\langle \mathrm{Div}\, b, \mathcal{S}\, \mathrm{Div}\, a \rangle_{\partial D} + k^2 \langle b, \mathcal{S}a \rangle_{\partial D}$$

for all $a, b \in H^{-1/2}(\mathrm{Div}, \partial D)$ where $\mathcal{S} : H^{-1/2}(\partial D) \to H^{1/2}(\partial D)$ denotes the single layer boundary operator. In the second occurrence it is considered as a bounded operator from $H^{-1/2}(\mathrm{Div}, \partial D)$ into $H^{-1/2}(\mathrm{Curl}, \partial D)$, see Lemma 5.27.

Proof: (a) On the dense subspace $\{\nu \times u|_{\partial D} : u \in C^\infty(\overline{D}, \mathbb{C}^3)\}$ of $H^{-1/2}(\mathrm{Div}, \partial D)$ the mapping $a \mapsto a \times \nu$ is expressed as $a \times \nu = \gamma_T \eta_t a$ with the extension operator $\eta_t : H^{-1/2}(\mathrm{Div}, \partial D) \to H(\mathrm{curl}, D)$ and the trace operator $\gamma_T : H(\mathrm{curl}, D) \to H^{-1/2}(\mathrm{Curl}, \partial D)$ from Theorem 5.24. This operator $\gamma_T \eta_t a$ has a bounded extension from $H^{-1/2}(\mathrm{Div}, \partial D)$ into $H^{-1/2}(\mathrm{Curl}, \partial D)$. The bilinear form $\langle \cdot, \cdot \rangle$ is nondegenerated because $H^{-1/2}(\mathrm{Curl}, \partial D)$ is the dual of $H^{-1/2}(\mathrm{Div}, \partial D)$.

(b) Define $u, v \in H_{loc}(\mathrm{curl}, \mathbb{R}^3)$ by

$$u(x) = \mathrm{curl}^2 \langle a, \Phi(x, \cdot) \rangle_{\partial D}, \quad v(x) = \mathrm{curl}^2 \langle b, \Phi(x, \cdot) \rangle_{\partial D}, \quad x \notin \partial D.$$

Then, by Theorem 5.52, $\gamma_t \mathrm{curl}\, v|_- - \gamma_t \mathrm{curl}\, v|_+ = k^2 b$ and thus $k^2 b \times \nu = \gamma_T \mathrm{curl}\, v|_- - \gamma_T \mathrm{curl}\, v|_+$. Therefore, by Green's theorem in the form (5.19), applied in D and in $B(0, R) \setminus \overline{D}$, respectively, and adding the results yields

$$k^2 \langle \mathcal{L}a, b \times \nu \rangle_{\partial D} = \langle \gamma_t u, \gamma_T \operatorname{curl} v|_- \rangle_{\partial D} - \langle \gamma_t u, \gamma_T \operatorname{curl} v|_+ \rangle_{\partial D}$$

$$= \int_{|x|<R} [\operatorname{curl} v \cdot \operatorname{curl} u - u \cdot \operatorname{curl}^2 v] \, dx$$

$$- \int_{|x|=R} (\nu \times u) \cdot \operatorname{curl} v \, ds$$

$$= \int_{|x|<R} [\operatorname{curl} v \cdot \operatorname{curl} u - k^2 v \cdot u] \, dx$$

$$- \int_{|x|=R} (\nu \times u) \cdot \operatorname{curl} v \, ds \,.$$

Changing the roles of a and b and subtracting the results yield

$$k^2 \langle \mathcal{L}a, b \times \nu \rangle_{\partial D} - k^2 \langle \mathcal{L}b, a \times \nu \rangle_{\partial D} = \int_{|x|=R} [(\nu \times v) \cdot \operatorname{curl} u - (\nu \times u) \cdot \operatorname{curl} v] \, ds \,.$$

The last term converges to zero as R tends to infinity by the radiation condition. For $a, b \in H^{-1/2}(\operatorname{Div}, \partial D)$ we have with respect to the dual system

$$\langle \mathcal{L}a, b \rangle = \langle \mathcal{L}a, \nu \times b \rangle_{\partial D} = \langle \mathcal{L}b, \nu \times a \rangle_{\partial D} = -\langle a, \nu \times \mathcal{L}b \rangle_{\partial D} = -\langle a, \mathcal{L}b \rangle \,.$$

(c) Let now $k = i$. Taking $b = \bar{a}$ in part (b) yields

$$\langle \mathcal{L}_i a, \bar{a} \rangle = -\langle \mathcal{L}_i a, \bar{a} \times \nu \rangle_{\partial D}$$

$$= \int_{|x|<R} [|\operatorname{curl} u|^2 + |u|^2] \, dx - \int_{|x|=R} (\nu \times u) \cdot \operatorname{curl} \bar{u} \, ds \,.$$

u and $\operatorname{curl} u$ decay exponentially to zero as R tends to infinity. Therefore, we arrive at

$$\langle \mathcal{L}_i a, \bar{a} \rangle = \|u\|^2_{H(\operatorname{curl}, \mathbb{R}^3)} \,.$$

Finally, the boundedness of the trace operator yields

$$\|a\|_{H^{-1/2}(\operatorname{Div}, \partial D)} = \|\gamma_t \operatorname{curl} u|_+ - \gamma_t \operatorname{curl} u|_- \|_{H^{-1/2}(\operatorname{Div}, \partial D)}$$

$$\leq c' \|u\|_{H(\operatorname{curl}, \mathbb{R}^3)} \,.$$

(d) Taking the trace in (5.61) yields

$$\mathcal{L}a = \gamma_t \nabla \tilde{S} \operatorname{Div} a + k^2 \gamma_t \tilde{S} a$$

and thus for $a, b \in H^{-1/2}(\operatorname{Div}, \partial D)$:

$$\langle \mathcal{L}a, b \rangle = \langle \mathcal{L}a, \nu \times b \rangle_{\partial D} = \langle \gamma_t \nabla \tilde{S} \operatorname{Div} a, \nu \times b \rangle_{\partial D} + k^2 \langle \nu \times Sa, \nu \times b \rangle_{\partial D}$$

$$= \langle b, \gamma_T \nabla \tilde{S} \operatorname{Div} a \rangle_{\partial D} + k^2 \langle b, \gamma_T \tilde{S} a \rangle_{\partial D}$$

$$= -\langle \operatorname{Div} b, S \operatorname{Div} a \rangle_{\partial D} + k^2 \langle b, Sa \rangle_{\partial D}$$

by the definition of the surface divergence. $\quad\square$

Theorem 5.62. *The interior and exterior boundary value problems are solvable by a layer potential $u = \tilde{\mathcal{L}}a$ for exactly those $f \in H^{-1/2}(\mathrm{Div}, \partial D)$ which are orthogonal to all traces $\gamma_T \operatorname{curl} v \in H^{-1/2}(\mathrm{Curl}, \partial D)$ of eigenfunctions $v \in H_0(\mathrm{curl}, D)$ of curl^2 in D; that is,*

$$\langle f, \gamma_T \operatorname{curl} v \rangle_{\partial D} = 0$$

for all $v \in H_0(\mathrm{curl}, D)$ with $\operatorname{curl}^2 v - k^2 v = 0$ in the variational sense.

Proof: We have to discuss solvability of the equation $\mathcal{L}a = f$. As mentioned above we want to apply Theorem A.4 and define the dual system by $X_1 = X_2 = Y_1 = Y_2 = H^{-1/2}(\mathrm{Div}, \partial D)$ and $\langle a, b \rangle = \langle a, \nu \times b \rangle_{\partial D}$ for $a, b \in H^{-1/2}(\mathrm{Div}, \partial D)$ as in the previous Lemma. The adjoint $\mathcal{L}^*$ of $\mathcal{L}$ is given by $-\mathcal{L}$.
Let now $f \in H^{-1/2}(\mathrm{Div}, \partial D)$ and $g \in \mathcal{N}(\mathcal{L})$; that is, $g = \gamma_t \operatorname{curl} v$ for some $v \in H_0(\mathrm{curl}, D)$ with $\operatorname{curl}^2 v - k^2 v = 0$ in D. The solvability condition $\langle f, g \rangle = 0$ reads as $0 = \langle f, \nu \times \gamma_t \operatorname{curl} v \rangle_{\partial D} = -\langle f, \gamma_T \operatorname{curl} v \rangle_{\partial D}$ which proves the theorem. □

For our reference scattering problem from the introduction (see Sect. 1.5) we obtain a final conclusion.

Corollary 5.63. *For any Lipschitz domain $D \subseteq \mathbb{R}^3$ the scattering problem (3.36a)–(3.37b) has a unique solution. The scattered field E^s can be represented by a layer potential $E^s = \tilde{\mathcal{L}}a$ where the density $a \in H^{-1/2}(\mathrm{Div} \, \partial D)$ satisfies $\mathcal{L}a = -\gamma_t E^{inc}$.*

Proof: For the scattering problem by a perfect conductor we have to solve the exterior problem with boundary data $f(x) = -\gamma_t E^{inc}$ on ∂D. This boundary data satisfies the orthogonality condition of the previous theorem. Indeed, by Green's formula (5.19) we conclude for every eigenfunction $v \in H_0(\mathrm{curl}, D)$ that

$$\langle \gamma_t E^{inc}, \gamma_T \operatorname{curl} v \rangle_{\partial D} = \int_D \left[\operatorname{curl} v \cdot \operatorname{curl} E^{inc} - k^2 v \, E^{inc} \right] dx$$
$$= \langle \gamma_t v, \gamma_T \operatorname{curl} E^{inc} \rangle_{\partial D} = 0.$$

Therefore, for this choice of f the boundary equation $\mathcal{L}a = f$ is solvable. □

As in the scalar case we expect that the exterior boundary value problem is uniquely for all boundary data f, independently of the wave number. By the previous results it is clear that we have to modify the ansatz $u = \tilde{\mathcal{L}}a$. We proceed just as in the scalar case and propose an ansatz in the form

$$u = \tilde{\mathcal{L}}a + \eta \tilde{\mathcal{M}}\mathcal{L}_i a \quad \text{in the exterior of } D.$$

By the jump conditions of Theorem 5.52 the vector field u solves the exterior boundary value problem for $f \in H^{-1/2}(\text{Div}, \partial D)$ if, and only if, $a \in H^{-1/2}(\text{Div}, \partial D)$ satisfies the equation

$$\mathcal{L}a - \frac{\eta}{2}\mathcal{L}_i a + \eta \mathcal{M}\mathcal{L}_i a = f \quad \text{on } \partial D. \tag{5.67}$$

As shown in Theorem 5.51 the operator $\mathcal{L}$ is the sum of an isomorphism and a compact operator. Therefore, there exists $\eta_0 > 0$ such that also $\mathcal{L} - \frac{\eta}{2}\mathcal{L}_i + \eta \mathcal{M}\mathcal{L}_i$ is a compact perturbation of an isomorphism for every $\eta \in \mathbb{C}$ with $|\eta| \leq \eta_0$. We choose $\eta \in \mathbb{C}$ with $|\eta| \leq \eta_0$ and $\text{Im}\,(\overline{k}\eta) > 0$ and have to prove uniqueness of (5.67). Therefore, let $a \in H^{-1/2}(\text{Div}, \partial D)$ satisfy (5.67) for $f = 0$ and define u in $\mathbb{R}^3 \setminus \partial D$ by $u = \tilde{\mathcal{L}}a + \eta \tilde{\mathcal{M}}\mathcal{L}_i a$. Then $\gamma_t u|_+ = 0$ on ∂D and thus $u = 0$ in the exterior of D by the uniqueness theorem. The jump conditions yield

$$\gamma_t u|_- = \gamma_t u|_- - \gamma_t u|_+ = \eta\mathcal{L}_i a, \quad \gamma_t \,\text{curl}\, u|_- = \gamma_t \,\text{curl}\, u|_- - \gamma_t \,\text{curl}\, u|_+ = k^2 a$$

because $\text{curl}\,\tilde{\mathcal{L}}a = k^2\tilde{\mathcal{M}}a$. Elimination of a yields $\eta\mathcal{L}_i\gamma_t \,\text{curl}\, u|_- = k^2\gamma_t u|_-$. Now we apply Green's theorem in D; that is,

$$\int_D \left[\text{curl}\, u \cdot \text{curl}\, \psi - k^2 u \cdot \psi\right] dx = \langle \gamma_t \psi, \gamma_T \,\text{curl}\, u|_-\rangle_{\partial D}.$$

Substituting $\psi = \overline{k u}$ yields

$$\int_D \left[\overline{k}\,|\text{curl}\, u|^2 - k\,|ku|^2\right] dx = -\overline{k}\,\langle \gamma_t \overline{u}|_-, \nu \times \gamma_t \,\text{curl}\, u|_-\rangle_{\partial D}$$

$$= -\frac{k\overline{\eta}}{|k|^2}\,\langle \mathcal{L}_i\gamma_t \,\text{curl}\, \overline{u}|_-, \nu \times \gamma_t \,\text{curl}\, u|_-\rangle_{\partial D}.$$

Now we observe that $\langle \mathcal{L}_i\gamma_t \,\text{curl}\, \overline{u}|_-, \nu \times \gamma_t \,\text{curl}\, u|_-\rangle_{\partial D}$ is non-negative by Lemma 5.61 and $\text{Im}\,(k\overline{\eta}) < 0$ by assumption. Therefore, the imaginary part of the right-hand side is non-negative while the imaginary part of the left-hand side is non-positive. Taking the imaginary part yields $\langle \mathcal{L}_i\gamma_t \,\text{curl}\, \overline{u}|_-, \nu \times \gamma_t \,\text{curl}\, u|_-\rangle_{\partial D} = 0$ and thus $\gamma_t \,\text{curl}\, u|_- = 0$ by Lemma 5.61. This ends the proof because $k^2 a = \gamma_t \,\text{curl}\, u|_- - \gamma_t \,\text{curl}\, u|_+ = 0$.

We formulate the result as a theorem.

Theorem 5.64. *The exterior boundary value problem is uniquely solvable for every $f \in H^{-1/2}(\text{Div}, \partial D)$. The solution can be expressed in the form*

$$u = \tilde{\mathcal{L}}a + \eta \tilde{\mathcal{M}}\mathcal{L}_i a \quad \text{in } \mathbb{R}^3 \setminus D$$

for some sufficiently small η such that $\text{Im}\,(\overline{k}\eta) > 0$ and the density $a \in H^{-1/2}(\text{Div}, \partial D)$ solves (5.67).

It is not clear to the authors whether or not the parameter η can be chosen independently of the wave number k. This is different from the scalar case (see remark following Theorem 5.56) because the difference $\mathcal{L} - \mathcal{L}_i$ fails to be compact.

5.4 Exercises

Exercise 5.1. Prove Theorem 5.6; that is, compactness of the mapping J : $u \mapsto u$ from $H^t_{per}(Q)$ into $H^s_{per}(Q)$ for $s < t$.
Hint: Show that the mapping $(J_N u)(x) = \sum_{|n| \leq N} u_n \exp(in \cdot x)$ is compact and converges in the operator norm to J.

Exercise 5.2. Show that the mapping $u \mapsto \partial u / \partial r$ fails to be bounded from $H^1(D)$ into $L^2(\partial D)$ for $D = B_2(0, 1)$ being the unit disk in $\mathbb{R}^2$ by

(a) computing α_n for $n \in \mathbb{N}$ such that $\|u_n\|_{H^1(D)} = 1$ where $u_n(r, \varphi) = \alpha_n r^n$, and

(b) showing that $\|\partial u_n / \partial r\|_{L^2(\partial D)} \to \infty$ as $n \to \infty$.

Exercise 5.3. Prove Corollary 5.13
Hints: (a) Approximate u and v by smooth functions. (b), (c) Argue analogously to the proof of Lemma 5.2, part (a).

Exercise 5.4. Show the formulas from the proof of Theorem 5.4; that is,

$$\int_{S^2} \operatorname{curl} u(r\hat{x}) \cdot \hat{x}\, Y_n^{-m}(\hat{x})\, ds(\hat{x}) = -\frac{\sqrt{n(n+1)}}{r}\, w_n^m(r)\,,$$

$$\int_{S^2} \operatorname{curl} u(r\hat{x}) \cdot U_n^{-m}(\hat{x})\, ds(\hat{x}) = -\frac{1}{r}\left(r w_n^m(r)\right)'\,,$$

$$\int_{S^2} \operatorname{curl} u(r\hat{x}) \cdot V_n^{-m}(\hat{x})\, ds(\hat{x}) = -\frac{\sqrt{n(n+1)}}{r}\, u_n^m(r) + \frac{1}{r}\left(r v_n^m(r)\right)'\,.$$

$$\operatorname{curl}\left[u_n^m(r)\, Y_n^m(\hat{x})\, \hat{x}\right] = -\frac{u_n^m(r)}{r}\left(\hat{x} \times \operatorname{Grad}_{S^2} Y_n^m(\hat{x})\right),$$

$$\operatorname{curl}\left[v_n^m(r)\, \operatorname{Grad}_{S^2} Y_n^m(\hat{x})\right] = -\frac{1}{r}\left(r\, v_n^m(r)\right)'\left(\hat{x} \times \operatorname{Grad}_{S^2} Y_n^m(\hat{x})\right),$$

$$\operatorname{curl}\left[w_n^m(r)\left(\hat{x} \times \operatorname{Grad}_{S^2} Y_n^m(\hat{x})\right)\right] = -\frac{1}{r}\left(r\, w_n^m(r)\right)' \operatorname{Grad}_{S^2} Y_n^m(\hat{x})$$

$$- w_n^m(r)\, \frac{n(n+1)}{r}\, Y_n^m(\hat{x})\, \hat{x}\,.$$

Exercise 5.5. Let $B = B(0, R)$ be a ball and $f \in C_0^\infty(B)$.

(a) Show that the volume potential

$$\tilde{v}(x) = \int_B f(y) \frac{1}{4\pi|x - y|} \, dy, \quad x \in B,$$

is in $C^\infty(\overline{B})$ and solves $\Delta \tilde{v} = -f$ in $\overline{B}$.
(b) Prove that there exists a unique solution $v \in C^\infty(\overline{B})$ of the boundary value problem $\Delta v = -f$ in B, $v = 0$ on ∂B.

Hints: For part (a) use the proof of Theorem 3.9. For part (b) make a proper ansatz in the form $v = \tilde{v} + u$.

Exercise 5.6. Define the operator K by

$$K\psi = \sum_{n=0}^{\infty} \sum_{m=-n}^{n} \frac{1}{1 + n(n+1)} \left[a_n^m U_n^m + b_n^m V_n^m \right] \quad \text{for}$$

$$\psi = \sum_{n=0}^{\infty} \sum_{m=-n}^{n} \left[a_n^m U_n^m + b_n^m V_n^m \right] \in H^{-1/2}(\mathrm{Div}, S^2).$$

Show that K is well defined and compact from $H^{-1/2}(\mathrm{Div}, S^2)$ into $H^{-1/2}(\mathrm{Curl}, S^2)$. Show, furthermore, that $\langle K\psi, \overline{\psi} \rangle_{S^2}$ is real valued and $\langle K\psi, \overline{\psi} \rangle_{S^2} > 0$ for all $\psi \in H^{-1/2}(\mathrm{Div}, S^2)$, $\psi \neq 0$.

Exercise 5.7. Show that the boundary value problem (5.58a), (5.58b) is uniquely solvable for all $b \in H^{-1/2}(\mathrm{Div}, \partial D)$.
Hints: Transform first the boundary value problem to homogeneous boundary data by choosing a proper extension $\hat{v}$ of b and make the ansatz $v_i = \hat{v} + v$. Second, use the Helmholtz decomposition as in the proof of Theorem 5.51.

Appendix A

Some notations and basic results are essential throughout the whole monograph. Therefore we present a brief collection in this appendix for convenience.

A.1 Differential Operators

Maxwell's equations have to be considered in the Euclidian space $\mathbb{R}^3$, where we denote the common inner product and the cross product by

$$x \cdot y = \sum_{j=1}^{3} x_j y_j \quad \text{and} \quad x \times y = \begin{pmatrix} x_2 y_3 - x_3 y_2 \\ x_3 y_1 - x_1 y_3 \\ x_1 y_2 - x_2 y_1 \end{pmatrix}.$$

Frequently we apply the elementary formulas

$$x \cdot (y \times z) = y \cdot (z \times x) = z \cdot (x \times y) \qquad (A.1)$$
$$x \times (y \times z) = (x \cdot z)y - (x \cdot y)z \qquad (A.2)$$

Furthermore, we have to fix the notation of some basic differential operators. We will use the gradient ∇. Additionally we introduce $\operatorname{div} F = \nabla \cdot F$ for the divergence of a vector field F and $\operatorname{curl} F = \nabla \times F$ for its rotation. For scalar fields we denote the Laplacian operator by $\Delta = \operatorname{div} \nabla = \nabla \cdot \nabla$. Let $u : \mathbb{R}^3 \to \mathbb{C}$ and $F : \mathbb{R}^3 \to \mathbb{C}^3$ be sufficiently smooth functions then in a *cartesian coordinate system* we have

© Springer International Publishing Switzerland 2015
A. Kirsch, F. Hettlich, *The Mathematical Theory of Time-Harmonic Maxwell's Equations*, Applied Mathematical Sciences 190,
DOI 10.1007/978-3-319-11086-8

$$\nabla u = \begin{pmatrix} \dfrac{\partial u}{\partial x_1} \\ \dfrac{\partial u}{\partial x_2} \\ \dfrac{\partial u}{\partial x_3} \end{pmatrix} \qquad \operatorname{curl} F = \nabla \times F = \begin{pmatrix} \dfrac{\partial F_3}{\partial x_2} - \dfrac{\partial F_2}{\partial x_3} \\ \dfrac{\partial F_1}{\partial x_3} - \dfrac{\partial F_3}{\partial x_1} \\ \dfrac{\partial F_2}{\partial x_1} - \dfrac{\partial F_1}{\partial x_2} \end{pmatrix}$$

$$\operatorname{div} F = \nabla \cdot F = \sum_{j=1}^{3} \frac{\partial F_j}{\partial x_j}, \qquad \Delta u = \operatorname{div} \nabla u = \sum_{j=1}^{3} \frac{\partial^2 u}{\partial x_j^2}$$

The following frequently used formulas can be obtained from straightforward calculations. With sufficiently smooth scalar valued functions $u, \lambda : \mathbb{R}^3 \to \mathbb{C}$ and vector valued functions $F, G : \mathbb{R}^3 \to \mathbb{C}^3$ we have

$$\operatorname{curl} \nabla u = 0 \qquad\qquad\qquad\qquad\qquad\qquad \text{(A.3)}$$
$$\operatorname{div} \operatorname{curl} F = 0 \qquad\qquad\qquad\qquad\qquad\qquad \text{(A.4)}$$
$$\operatorname{curl} \operatorname{curl} F = \nabla \operatorname{div} F \; - \; \Delta F \qquad\qquad\qquad \text{(A.5)}$$

$$\operatorname{div}(\lambda F) = F \cdot \nabla \lambda \; + \; \lambda \operatorname{div} F \qquad\qquad\qquad \text{(A.6)}$$
$$\operatorname{curl}(\lambda F) = \nabla \lambda \times F \; + \; \lambda \operatorname{curl} F \qquad\qquad\qquad \text{(A.7)}$$
$$\nabla(F \cdot G) = (F')^\top G + (G')^\top F \qquad\qquad\qquad \text{(A.8)}$$
$$\operatorname{div}(F \times G) = G \cdot \operatorname{curl} F \; - \; F \cdot \operatorname{curl} G \qquad\qquad \text{(A.9)}$$
$$\operatorname{curl}(F \times G) = F \operatorname{div} G \; - \; G \operatorname{div} F \; + \; F'G \; - \; G'F, \qquad \text{(A.10)}$$

where $F'(x), G'(x) \in \mathbb{C}^{3 \times 3}$ are the Jacobian matrices of F and G, respectively, at x; that is, $F'_{ij} = \partial F_i / \partial x_j$.

For completeness we add the expressions of the differential operators ∇, div, curl, and Δ also in other coordinate systems. Let $f : \mathbb{R}^3 \to \mathbb{C}$ be a scalar function and $F : \mathbb{R}^3 \to \mathbb{C}^3$ a vector field and consider *cylindrical coordinates*

$$x = \begin{pmatrix} r \cos \varphi \\ r \sin \varphi \\ z \end{pmatrix}, \quad r > 0, \; 0 \leq \varphi < 2\pi, \; z \in \mathbb{R}.$$

With the coordinate unit vectors $\hat{z} = (0,0,1)^\top$, $\hat{r} = (\cos \varphi, \sin \varphi, 0)^\top$, and $\hat{\varphi} = (-\sin \varphi, \cos \varphi, 0)^\top$ and the representation $F = F_r \hat{r} + F_\varphi \hat{\varphi} + F_z \hat{z}$ we obtain

$$\nabla f(r, \varphi, z) = \frac{\partial f}{\partial r} \hat{r} + \frac{1}{r} \frac{\partial f}{\partial \varphi} \hat{\varphi} + \frac{\partial f}{\partial z} \hat{z}, \qquad \text{(A.11a)}$$

$$\operatorname{div} F(r, \varphi, z) = \frac{1}{r} \frac{\partial (r F_r)}{\partial r} + \frac{1}{r} \frac{\partial F_\varphi}{\partial \varphi} + \frac{\partial F_z}{\partial z}, \qquad \text{(A.11b)}$$

$$\operatorname{curl} F(r, \varphi, z) = \left(\frac{1}{r} \frac{\partial F_z}{\partial \varphi} - \frac{\partial F_\varphi}{\partial z} \right) \hat{r} + \left(\frac{\partial F_r}{\partial z} - \frac{\partial F_z}{\partial r} \right) \hat{\varphi}$$
$$+ \frac{1}{r} \left(\frac{\partial (r F_\varphi)}{\partial \theta} - \frac{\partial F_\theta}{\partial \varphi} \right) \hat{z}, \qquad \text{(A.11c)}$$

$$\Delta f(r, \varphi, z) = \frac{1}{r} \frac{\partial}{\partial r} \left(r \frac{\partial f}{\partial r} \right) + \frac{1}{r^2} \frac{\partial^2 f}{\partial \varphi^2} + \frac{\partial^2 f}{\partial z^2}. \qquad \text{(A.11d)}$$

Essential for the second chapter are *spherical coordinates*

$$x = \begin{pmatrix} r \sin \theta \cos \varphi \\ r \sin \theta \sin \varphi \\ r \cos \theta \end{pmatrix}, \quad r > 0, 0 \le \theta \le \pi, \ 0 \le \varphi < 2\pi$$

and the coordinate unit vectors $\hat{r} = (\sin \theta \cos \varphi, \sin \theta \sin \varphi, \cos \theta)^\top$, $\hat{\theta} = (\cos \theta \cos \varphi, \cos \theta \sin \varphi, -\sin \theta)^\top$, and $\hat{\varphi} = (-\sin \varphi, \cos \varphi, 0)^\top$. With $F = F_r \hat{r} + F_\theta \hat{\theta} + F_\varphi \hat{\varphi}$ we have the equations

$$\nabla f(r, \theta, \varphi) = \frac{\partial f}{\partial r} \hat{r} + \frac{1}{r} \frac{\partial f}{\partial \theta} \hat{\theta} + \frac{1}{r \sin \theta} \frac{\partial f}{\partial \varphi} \hat{\varphi}, \qquad \text{(A.12a)}$$

$$\operatorname{div} F(r, \theta, \varphi) = \frac{1}{r^2} \frac{\partial (r^2 F_r)}{\partial r} + \frac{1}{r \sin \theta} \frac{\partial (\sin \theta \, F_\theta)}{\partial \theta} + \frac{1}{r \sin \theta} \frac{\partial F_\varphi}{\partial \varphi}, \qquad \text{(A.12b)}$$

$$\operatorname{curl} F(r, \theta, \varphi) = \frac{1}{r \sin \theta} \left(\frac{\partial (\sin \theta \, F_\varphi)}{\partial \theta} - \frac{\partial F_\theta}{\partial \varphi} \right) \hat{r}$$
$$+ \frac{1}{r} \left(\frac{1}{\sin \theta} \frac{\partial F_r}{\partial \varphi} - \frac{\partial (r F_\varphi)}{\partial r} \right) \hat{\theta} \qquad \text{(A.12c)}$$
$$+ \frac{1}{r} \left(\frac{\partial (r F_\theta)}{\partial r} - \frac{\partial F_r}{\partial \theta} \right) \hat{\varphi}, \qquad \text{(A.12d)}$$

$$\Delta f(r, \theta, \varphi) = \frac{1}{r^2} \frac{\partial}{\partial r} \left(r^2 \frac{\partial f}{\partial r} \right) + \frac{1}{r^2 \sin \theta} \frac{\partial}{\partial \theta} \left(\sin \theta \frac{\partial f}{\partial \theta} \right)$$
$$+ \frac{1}{r^2 \sin^2 \theta} \frac{\partial^2 f}{\partial \varphi^2}. \qquad \text{(A.12e)}$$

A.2 Results from Linear Functional Analysis

We assume that the reader is familiar with the basic facts from linear functional analysis: in particular with the notions of normed spaces, Banach- and Hilbert spaces, linear, bounded, and compact operators. We list some results which are needed often in this monograph.

Theorem A.1. *Let X, Y be Banach spaces, $V \subseteq X$ a linear subspace, and $T : V \to Y$ a linear and bounded operator; that is, there exists $c > 0$ with $\|Tx\|_Y \leq c\|x\|_X$ for all $x \in V$. Then there exists a unique extension $\tilde{T} : \overline{V} \to Y$ to the closure $\overline{V}$ of V; that is, $\tilde{T}x = Tx$ for all $x \in V$ and $\|\tilde{T}x\|_Y \leq c\|x\|_X$ for all $x \in \overline{V}$. Furthermore, $\|\tilde{T}\| = \|T\|$.*

This theorem is often applied to the case where V is a dense subspace of X. For example, if X is a Sobolev space, then V can be taken to be the space of infinitely often differentiable functions. The proof of boundedness of an operator (e.g., a trace operator) is usually easier for smooth functions.

The next two theorems are the functional analytic basis of many existence theorems for boundary value problems.

Theorem A.2. *Let $T : X \to Y$ be a linear and bounded operator between the normed spaces X and Y. Let T be of the form $T = A + K$ such that A is an isomorphism from X onto Y and $K : X \to Y$ is compact. If T is one-to-one then also onto, and T^{-1} is bounded from Y onto X. In other words, if the homogeneous equation $Tx = 0$ admits only the trivial solution $x = 0$, then the inhomogeneous equation $Tx = y$ is uniquely solvable for all $y \in Y$, and the solution x depends continuously on y.*

By writing $T = A(I + A^{-1}K)$ it is obvious that it is sufficient to consider the case $Y = X$ and $A = I$. For this case the result is proven in, e.g., [16, Chapter 3]. It is also a special case of Theorem A.4 below.

The case where $I - K$ fails to be one-to-one is answered by the following theorem which is called *Fredholm's alternative*. We need the notions of a dual system and adjoint operators (see [16, Chapter 4]).

Definition A.3. Two normed spaces X, Y, equipped with a bilinear form $\langle \cdot, \cdot \rangle : X \times Y \to \mathbb{C}$ is called a *dual system*, if $\langle \cdot, \cdot \rangle$ is nondegenerated; that is, for every $x \in X$, $x \neq 0$, the linear form $y \mapsto \langle x, y \rangle$ does not vanish identically and vice versa, for every $y \in Y$, $y \neq 0$, the linear form $x \mapsto \langle x, y \rangle$ does not vanish identically.
Let $\langle X_1, Y_1 \rangle$ and $\langle X_2, Y_2 \rangle$ be two dual forms. Two operators $T : X_1 \to X_2$ and $S : Y_2 \to Y_1$ are called *adjoint* to each other if

$$\langle Tx, y \rangle_2 = \langle x, Sy \rangle_1 \quad \text{for all } x \in X_1, \, y \in Y_2.$$

Theorem A.4 (Fredholm). *Let $\langle X_j, Y_j \rangle_j$, $j = 1, 2$, be two dual systems, $T : X_1 \to X_2$ a bounded operator with bounded adjoint operator $T^* : Y_2 \to Y_1$ such that $T = \hat{T} + K$ and $T^* = \hat{T}^* + K^*$ with isomorphisms $\hat{T}$ and $\hat{T}^*$ and compact operators K and K^*. Then the following holds:*

(a) *The dimensions of the null spaces of T and T^* are finite and coincide; that is, $\dim \mathcal{N}(T) = \dim \mathcal{N}(T^*) < \infty$.*
(b) *The equations $Tx = u$, $T^*y = v$ are solvable for exactly those $u \in X_2$ and $v \in Y_1$ for which*

$$\langle u, \psi \rangle_2 = 0 \text{ for all } \psi \in \mathcal{N}(T^*) \subset Y_2$$

and

$$\langle \varphi, v \rangle_1 = 0 \text{ for all } \varphi \in \mathcal{N}(T) \subset X_1 .$$

Proof: We can easily reduce the problem to the case that $X_1 = X_2$ and $Y_1 = Y_2$ and $\hat{T} = \hat{T}^* = id$ which is of particular importance in itself. Indeed, the equation $Tx = u$ is equivalent to $x + \hat{T}^{-1}Kx = \hat{T}^{-1}u$ and the operator $\tilde{K} = \hat{T}^{-1}K$ maps X_1 into itself. On the other hand, the equation $T^*y = v$ is equivalent to $z + K^*(\hat{T}^*)^{-1}z = v$ for $z = \hat{T}^*y$. The operator $K^*(\hat{T}^*)^{-1}$ from Y_1 into itself is just the adjoint of $\tilde{K}$. Also, $\langle u, y \rangle_2 = \langle \hat{T}^{-1}u, z \rangle_1$ for $z = \hat{T}^*y \in Y_1$, $u \in X_2$. For this special case we refer to Theorem 4.15 of [16]. $\square$

Also the following two results are used for proving existence of solutions of boundary value problems, in particular for those formulated by variational equations.

Theorem A.5. (Representation Theorem of Riesz)
Let X be a Hilbert space with inner product $(\cdot, \cdot)_X$ and $\ell : X \to \mathbb{C}$ a linear and bounded functional. Then there exists a unique $z \in X$ with $\ell(x) = (x, z)_X$ for all $x \in X$. Furthermore, $\|\ell\| = \|x\|_X$.

For a proof we refer to any book on functional analysis as, e.g., [31, Section III.6].

An extension is given by the theorem of Lax–Milgram.

Theorem A.6. *(Lax and Milgram)*
Let X be a Hilbert space over $\mathbb{C}$, $\ell : X \to \mathbb{C}$ linear and bounded, $a : X \times X \to \mathbb{C}$ sesquilinear and bounded and coercive; that is, there exist $c_1, c_2 > 0$ with

$$|a(u, v)| \leq c_1 \|u\|_X \|v\|_X \quad \text{for all } u, v \in X ,$$
$$\text{Re } a(u, u) \geq c_2 \|u\|_X^2 \quad \text{for all } u \in X .$$

Then there exists a unique $u \in X$ with

$$a(\psi, u) \;=\; \ell(\psi) \quad \text{for all } \psi \in X \,.$$

Furthermore, there exists $c > 0$, independent of u, such that $\|u\|_X \geq c\|\ell\|_{X^*}$.

For a proof we refer to, e.g., [11, Section 6.2].

A.3 Elementary Facts from Differential Geometry

Before we recall the basic integral identity of Gauss and Green we have to define rigorously the notion of domain with a C^n-boundary or Lipschitz boundary (see Evans [11]). We denote by $B_j(x, r) := \{y \in \mathbb{R}^j : |y - x| < r\}$ and $B_j[x, r] := \{y \in \mathbb{R}^j : |y - x| \leq r\}$ the open and closed ball, respectively, of radius $r > 0$ centered at x in $\mathbb{R}^j$ for $j = 2$ or $j = 3$.

Definition A.7. We call a region $D \subset \mathbb{R}^3$ to be C^n-smooth (that is, $D \in C^n$), if there exists a finite number of open cylinders U_j of the form $U_j = \{R_j x + z^{(j)} : x \in B_2(0, \alpha_j) \times (-2\beta_j, 2\beta_j)\}$ with $z^{(j)} \in \mathbb{R}^3$ and rotations[1] $R_j \in \mathbb{R}^{3 \times 3}$ and real valued functions $\xi_j \in C^n(B[0, \alpha_j])$ with $|\xi_j(x_1, x_2)| \leq \beta_j$ for all $(x_1, x_2) \in B_2[0, \alpha_j]$ such that $\partial D \subset \bigcup_{j=1}^m U_j$ and

$$\partial D \cap U_j = \left\{R_j x + z^{(j)} : (x_1, x_2) \in B_2(0, \alpha_j),\; x_3 = \xi_j(x_1, x_2)\right\},$$

$$D \cap U_j = \left\{R_j x + z^{(j)} : (x_1, x_2) \in B_2(0, \alpha_j),\; x_3 < \xi_j(x_1, x_2)\right\},$$

$$U_j \setminus \overline{D} = \left\{R_j x + z^{(j)} : (x_1, x_2) \in B_2(0, \alpha_j),\; x_3 > \xi_j(x_1, x_2)\right\}.$$

We call D to be a *Lipschitz domain* if the functions ξ_j which describe the boundary locally are Lipschitz continuous; that is, there exists a constant $L > 0$ such that $|\xi_j(z) - \xi_j(y)| \leq L|z - y|$ for all $z, y \in B_2[0, \alpha_j]$ and all $j = 1, \ldots, m$.

We call $\{U_j, \xi_j : j = 1, \ldots, m\}$ a *local coordinate system* of ∂D. For abbreviation we denote by

$$\begin{aligned} C_j \;&=\; C_j(\alpha_j, \beta_j) = B_2(0, \alpha_j) \times (-2\beta_j, 2\beta_j) \\ &=\; \left\{x = (x_1, x_2, x_3) \in \mathbb{R}^3 : x_1^2 + x_2^2 < \alpha_j^2,\; |x_3| < 2\beta_j\right\} \end{aligned}$$

the cylinders with parameters α_j and β_j. We can assume without loss of generality that $\beta_j \geq \alpha_j$ (otherwise split the parameter region into smaller ones)

---

[1] That is, $R_j^\top R_j = I$ and $\det R_j = 1$.

Furthermore, we set $B_j := B_3(0, \alpha_j) \subset C_j$, $j = 1, \ldots, m$, and introduce the mappings $\tilde{\Psi}_j : B_j \to \mathbb{R}^3$ defined by

$$\tilde{\Psi}_j(x) = R_j \begin{pmatrix} x_1 \\ x_2 \\ \xi_j(x_1, x_2) + x_3 \end{pmatrix} + z^{(j)}, \quad x = (x_1, x_2, x_3)^\top \in B_j,$$

and its restriction Ψ_j to $B_2(0, \alpha_j)$; that is,

$$\Psi_j(\tilde{x}) = R_j \begin{pmatrix} x_1 \\ x_2 \\ \xi_j(x_1, x_2) \end{pmatrix} + z^{(j)}, \quad \tilde{x} = (x_1, x_2)^\top \in B_2(0, \alpha_j),$$

which yields a parametrization of $\partial D \cap U_j$ in the form $y = \Psi_j(\tilde{x})$ for $\tilde{x} \in B_2(0, \alpha_j)$ with $\left| \frac{\partial \Psi_j}{\partial x_1} \times \frac{\partial \Psi_j}{\partial x_2} \right| = \sqrt{1 + |\nabla \xi_j|^2}$ provided the functions ξ_j are differentiable.

We set $U'_j = \tilde{\Psi}_j(B_j)$. Then $\partial D \subset \bigcup_{j=1}^m U'_j$ and $B_j \cap (\mathbb{R}^2 \times \{0\}) = B_2(0, \alpha_j) \times \{0\}$, and

$$\partial D \cap U'_j = \{\tilde{\Psi}_j(x) : x \in B_j, \, x_3 = 0\} = \{\Psi_j(\tilde{x}) : \tilde{x} \in B_2(0, \alpha_j)\},$$
$$D \cap U'_j = \{\tilde{\Psi}_j(x) : x \in B_j, \, x_3 < 0\},$$
$$U'_j \setminus \overline{D} = \{\tilde{\Psi}_j(x) : x \in B_j, \, x_3 > 0\}.$$

Therefore, the mappings $\tilde{\Psi}_j$ "flatten" the boundary. For C^1-domains D we note that the Jacobian is given by

$$\tilde{\Psi}'_j(x) = R_j \begin{pmatrix} 1 & 0 & 0 \\ 0 & 1 & 0 \\ \partial_1 \xi_j(x) & \partial_2 \xi_j(x) & 1 \end{pmatrix},$$

where $\partial_\ell \xi_j = \partial \xi_j / \partial x_\ell$ for $\ell = 1, 2$. We note that its determinant is one. The *tangential vectors* at $y = \Psi_j(x) \in \partial D \cap U_j$ are computed as

$$\frac{\partial \Psi_j}{\partial x_1}(x) = R_j \begin{pmatrix} 1 \\ 0 \\ \partial_1 \xi_j(x) \end{pmatrix}, \quad \frac{\partial \Psi_j}{\partial x_2}(x) = R_j \begin{pmatrix} 0 \\ 1 \\ \partial_2 \xi_j(x) \end{pmatrix}.$$

They span the *tangent plane* at $y = \Psi_j(x)$. The vector

$$\frac{\partial \Psi_j}{\partial x_1}(x) \times \frac{\partial \Psi_j}{\partial x_2}(x) = R_j \begin{pmatrix} -\partial_1 \xi_j(x) \\ -\partial_2 \xi_j(x) \\ 1 \end{pmatrix}$$

is orthogonal to the tangent plane and is directed into the exterior of D. The corresponding unit vector

$$\nu(y) \;=\; \frac{\frac{\partial \Psi_j}{\partial x_1}(x) \times \frac{\partial \Psi_j}{\partial x_2}(x)}{\left|\frac{\partial \Psi_j}{\partial x_1}(x) \times \frac{\partial \Psi_j}{\partial x_2}(x)\right|} \;=\; \frac{1}{\sqrt{1+|\nabla \xi_j(x)|^2}}\, R_j \begin{pmatrix} -\partial_1 \xi_j(x) \\ -\partial_2 \xi_j(x) \\ 1 \end{pmatrix}$$

is called the *exterior unit normal vector*.

Remark A.8. For *Lipschitz domains* the functions ξ_j are merely Lipschitz continuous. Therefore, $\tilde{\Psi}_j$ and its inverse $\tilde{\Psi}_j^{-1}$, given by

$$\tilde{\Psi}_j^{-1}(y) \;=\; \begin{pmatrix} \hat{y}_1 \\ \hat{y}_2 \\ \hat{y}_3 - \xi_j(\hat{y}_1, \hat{y}_2) \end{pmatrix}, \quad \hat{y} = R_j^\top (y - z^{(j)}), \quad y \in U_j' = \tilde{\Psi}_j(B_j),$$

are also Lipschitz continuous. A celebrated result of Rademacher [27] (see also [11, Section 5.8]) states that every Lipschitz continuous function ξ_j is differentiable at almost every point $x \in B_2(0, \alpha_j)$ with $|\nabla \xi_j(x)| \leq L$ for almost all $x \in B_2(0, \alpha_j)$ where L is the Lipschitz constant. Therefore, for Lipschitz domains the exterior unit normal vector $\nu(x)$ exists at almost all points $x \in \partial D$. Furthermore, $u \in L^1(U_j')$ if, and only if, $u \circ \tilde{\Psi}_j \in L^1(B_j)$ and the transformation formula holds in the form

$$\int_{U_j'} u(y)\, dy \;=\; \int_{B_j} u\big(\tilde{\Psi}_j(x)\big)\, dx. \tag{A.13}$$

Application of this result to $|u|^2$ shows that the operator $u \mapsto u \circ \tilde{\Psi}_j$ is bounded from $L^2(B_j)$ into $L^2(C_j')$.

For such domains and continuous functions $f : \partial D \to \mathbb{C}$ the *surface integral* $\int_{\partial D} f\, ds$ exists. Very often in the following we need the following tool (see, e.g., [11]):

Theorem A.9. *(Partition of Unity)*
Let $K \subset \mathbb{R}^3$ be a compact set. For every finite set $\{U_j : j = 1, \ldots, m\}$ of open domains with $K \subset \bigcup_{j=1}^m U_j$ there exist $\phi_j \in C^\infty(\mathbb{R}^3)$ with $\mathrm{supp}(\phi_j) \subset U_j$ for all j and $\sum_{j=1}^m \phi_j(y) = 1$ for all $y \in K$. We call (U_j, ϕ_j) a partition of unity on K.

Using a local coordinate system $\{U_j, \xi_j : j = 1, \ldots, m\}$ of ∂D with corresponding mappings $\tilde{\Psi}_j$ from the balls B_j onto U_j' and their restrictions $\Psi_j : B_2(0, \alpha_j) \to U_j' \cap \partial D$ as in Definition A.7 and a corresponding partition of unity ϕ_j on ∂D with respect to U_j' we write $\int_{\partial D} f\, ds$ in the form

$$\int_{\partial D} f\, ds = \sum_{j=1}^{m} \int_{\partial D \cap U'_j} \phi_j\, f\, ds = \sum_{j=1}^{m} \int_{\partial D \cap U'_j} f_j\, ds$$

with $f_j(y) = \phi_j(y) f(y)$. The integral over the surface patch $U'_j \cap \partial D$ is given by

$$\int_{U'_j \cap \partial D} f_j\, ds = \int_{B_2(0,\alpha_j)} f_j\big(\Psi_j(x)\big) \sqrt{1 + |\nabla \xi_j(x)|^2}\, dx\,.$$

We collect important properties of the smooth domain D in the following lemma.

Lemma A.10. *Let $D \in C^2$. Then there exists $c_0 > 0$ such that*

(a) $\big|\nu(y) \cdot (y - z)\big| \le c_0 |z - y|^2$ *for all $y, z \in \partial D$,*
(b) $\big|\nu(y) - \nu(z)\big| \le c_0 |y - z|$ *for all $y, z \in \partial D$.*
(c) Define
$$H_\eta := \big\{ z + t\nu(z) : z \in \partial D,\ |t| < \eta \big\}\,.$$

Then there exists $\eta_0 > 0$ such that for all $\eta \in (0, \eta_0]$ and every $x \in H_\eta$ there exist unique (!) $z \in \partial D$ and $|t| \le \eta$ with $x = z + t\nu(z)$. The set H_η is an open neighborhood of ∂D for every $\eta \le \eta_0$. Furthermore, $z - t\nu(z) \in D$ and $z + t\nu(z) \notin \overline{D}$ for $0 < t < \eta$ and $z \in \partial D$.
One can choose η_0 such that for all $\eta \le \eta_0$ the following holds:

* $|z - y| \le 2|x - y|$ *for all $x \in H_\eta$ and $y \in \partial D$, and*
* $|z_1 - z_2| \le 2|x_1 - x_2|$ *for all $x_1, x_2 \in H_\eta$.*

If $U_\delta := \big\{ x \in \mathbb{R}^3 : \inf_{z \in \partial D} |x - z| < \delta \big\}$ denotes the strip around ∂D, then there exists $\delta > 0$ with

$$\overline{U}_\delta \subset H_{\eta_0} \subset U_{\eta_0} \tag{A.14}$$

(d) There exists $r_0 > 0$ such that the surface area of $\partial B(z, r) \cap D$ for $z \in \partial D$ can be estimated by

$$\big| |\partial B(z, r) \cap D| - 2\pi r^2 \big| \le 4\pi c_0\, r^3 \quad \text{for all } r \le r_0\,. \tag{A.15}$$

Proof: We use a local coordinate system $\{U_j, \xi_j : j = 1, \ldots, m\}$ which yields the parametrization $\Psi_j : B_2(0, \alpha_j) \to \partial D \cap U'_j$. First, it is easy to see (proof by contradiction) that there exists $\delta > 0$ with the property that for every pair $(z, x) \in \partial D \times \mathbb{R}^3$ with $|z - x| < \delta$ there exists U'_j with $z, x \in U'_j$. Let $\operatorname{diam}(D) = \sup\{|x_1 - x_2| : x_1, x_2 \in D\}$ be the diameter of D.

(a) Let $x, y \in \partial D$ and assume first that $|y - x| \geq \delta$. Then

$$\left| \nu(y) \cdot (y - x) \right| \leq |y - x| \leq \frac{\mathrm{diam}(D)}{\delta^2} \delta^2 \leq \frac{\mathrm{diam}(D)}{\delta^2} |y - x|^2 \,.$$

Let now $|y - x| < \delta$. Then $y, x \in U'_j$ for some j. Let $x = \Psi_j(u)$ and $y = \Psi_j(v)$. Then

$$\nu(x) = \frac{\frac{\partial \Psi_j}{\partial u_1}(u) \times \frac{\partial \Psi_j}{\partial u_2}(u)}{\left| \frac{\partial \Psi_j}{\partial u_1}(u) \times \frac{\partial \Psi_j}{\partial u_2}(u) \right|}$$

and, by the definition of the derivative,

$$y - x = \Psi_j(v) - \Psi_j(u) = \sum_{k=1}^{2} (v_k - u_k) \frac{\partial \Psi_j}{\partial u_k}(u) + a(v, u)$$

with $|a(v, u)| \leq c|u - v|^2$ for all $u, v \in U'_j$ and some $c > 0$. Therefore,

$$\left| \nu(x) \cdot (y - x) \right|$$

$$\leq \frac{1}{\left| \frac{\partial \Psi_j}{\partial u_1}(u) \times \frac{\partial \Psi_j}{\partial u_2}(u) \right|} \sum_{k=1}^{2} (v_k - u_k) \underbrace{\left| \left(\frac{\partial \Psi_j}{\partial u_1}(u) \times \frac{\partial \Psi_j}{\partial u_2}(u) \right) \cdot \frac{\partial \Psi_j}{\partial u_k}(u) \right|}_{= 0}$$

$$+ \frac{1}{\left| \frac{\partial \Psi_j}{\partial u_1}(u) \times \frac{\partial \Psi_j}{\partial u_2}(u) \right|} \left| \left(\frac{\partial \Psi_j}{\partial u_1}(u) \times \frac{\partial \Psi_j}{\partial u_2}(u) \right) \cdot a(v, u) \right|$$

$$\leq c|u - v|^2 = c \left| \Psi_j^{-1}(x) - \Psi_j^{-1}(y) \right|^2 \leq c_0 |x - y|^2 \,.$$

This proves part (a). The proof of (b) follows analogously from the differentiability of $u \mapsto \nu$.

(c) Choose $\eta_0 > 0$ such that

(i) $\eta_0 c_0 < 1/16$ and
(ii) $\nu(x_1) \cdot \nu(x_2) \geq 0$ for $x_1, x_2 \in \partial D$ with $|x_1 - x_2| \leq 2\eta_0$ and
(iii) $H_{\eta_0} \subset \bigcup_{j=1}^{m} U'_j$.

Assume that $x \in H_\eta$ for $\eta \leq \eta_0$ has two representation as $x = z_1 + t_1 \nu_1 = z_2 + t_2 \nu_2$ where we write ν_j for $\nu(z_j)$. Then

$$|z_1 - z_2| = \left| (t_2 - t_1) \nu_2 + t_1 (\nu_2 - \nu_1) \right| \leq |t_1 - t_2| + \eta c_0 |z_1 - z_2|$$

$$\leq |t_1 - t_2| + \frac{1}{16} |z_1 - z_2| \,,$$

thus $|z_1 - z_2| \le \frac{16}{15}|t_1 - t_2| \le 2|t_1 - t_2|$. Furthermore, because $\nu_1 \cdot \nu_2 \ge 0$,

$$(\nu_1 + \nu_2) \cdot (z_1 - z_2) = (\nu_1 + \nu_2) \cdot (t_2\nu_2 - t_1\nu_1) = (t_2 - t_1)\underbrace{(\nu_1 \cdot \nu_2 + 1)}_{\ge 1},$$

thus

$$|t_2 - t_1| \le |(\nu_1 + \nu_2) \cdot (z_1 - z_2)| \le 2c_0|z_1 - z_2|^2 \le 8c_0|t_1 - t_2|^2 \,;$$

that is, $|t_2 - t_1|(1 - 8c_0|t_2 - t_1|) \le 0$. This yields $t_1 = t_2$ because $1 - 8c_0|t_2 - t_1| \ge 1 - 16c_0\eta > 0$ and thus also $z_1 = z_2$.

Let U' be one of the sets U_j' and $\Psi : \mathbb{R}^2 \supset B_2(0, \alpha) \to U' \cap \partial D$ the corresponding bijective mapping. We define the new mapping $F : \mathbb{R}^2 \supset B_2(0, \alpha) \times (-\eta, \eta) \to H_\eta$ by

$$F(u, t) = \Psi(u) + t\nu(u), \quad (u, t) \in B_2(0, \alpha) \times (-\eta, \eta).$$

For sufficiently small η the mapping F is one-to-one and satisfies $\left|\det F'(u, t)\right| \ge \tilde{c} > 0$ on $B_2(0, \alpha) \times (-\eta, \eta)$ for some $\tilde{c} > 0$. Indeed, this follows from

$$F'(u, t) = \left(\frac{\partial\Psi}{\partial u_1}(u) + t\frac{\partial\nu}{\partial u_1}(u),\ \frac{\partial\Psi}{\partial u_2}(u) + t\frac{\partial\nu}{\partial u_2}(u),\ \nu(u)\right)^\top$$

and the fact that for $t = 0$ the matrix $F'(u, 0)$ has full rank 3. Therefore, F is a bijective mapping from $B_2(0, \alpha) \times (-\eta, \eta)$ onto $U' \cap H_\eta$. Therefore, $H_\eta = \bigcup(H_\eta \cap U_j)$ is an open neighborhood of ∂D. This proves also that $x = z - t\nu(z) \in D$ and $x = z + t\nu(z) \notin \overline{D}$ for $0 < t < \eta$.

For $x = z + t\nu(z)$ and $y \in \partial D$ we have

$$\begin{aligned}
|x - y|^2 &= \left|(z - y) + t\nu(z)\right|^2 \ge |z - y|^2 + 2t(z - y) \cdot \nu(z) \\
&\ge |z - y|^2 - 2\eta c_0|z - y|^2 \\
&\ge \frac{1}{4}|z - y|^2 \quad \text{because} \quad 2\eta c_0 \le \frac{3}{4}.
\end{aligned}$$

Therefore, $|z - y| \le 2|x - y|$. Finally,

$$\begin{aligned}
|x_1 - x_2|^2 &= \left|(z_1 - z_2) + (t_1\nu_1 - t_2\nu_2)\right|^2 \\
&\ge |z_1 - z_2|^2 - 2\left|(z_1 - z_2) \cdot (t_1\nu_1 - t_2\nu_2)\right| \\
&\ge |z_1 - z_2|^2 - 2\eta\left|(z_1 - z_2) \cdot \nu_1\right| - 2\eta\left|(z_1 - z_2) \cdot \nu_2\right| \\
&\ge |z_1 - z_2|^2 - 4\eta c_0|z_1 - z_2|^2 = (1 - 4\eta c_0)|z_1 - z_2|^2 \\
&\ge \frac{1}{4}|z_1 - z_2|^2
\end{aligned}$$

because $1 - 4\eta c_0 \ge 1/4$.

The proof of (A.14) is simple and left as an exercise.

(d) Let c_0 and η_0 as in parts (a) and (c). Choose r_0 such that $B[z,r] \subset H_{\eta_0}$ for all $r \leq r_0$ [which is possible by (A.14)] and $\nu(z_1) \cdot \nu(z_2) > 0$ for $|z_1 - z_2| \leq 2r_0$. For fixed $r \leq r_0$ and arbitrary $z \in \partial D$ and $\sigma > 0$ we define

$$Z(\sigma) = \left\{ x \in \partial B(z,r) : (x - z) \cdot \nu(z) \leq \sigma \right\}$$

We show that

$$Z(-2c_0 r^2) \subset \partial B(z,r) \cap D \subset Z(+2c_0 r^2)$$

Let $x \in Z(-2c_0 r^2)$ have the form $x = x_0 + t\nu(x_0)$. Then

$$(x - z) \cdot \nu(z) = (x_0 - z) \cdot \nu(z) + t\,\nu(x_0) \cdot \nu(z) \leq -2c_0 r^2 \, ;$$

that is,

$$t\,\nu(x_0) \cdot \nu(z) \leq -2c_0 r^2 + \left| (x_0 - z) \cdot \nu(z) \right| \leq -2c_0 r^2 + c_0 |x_0 - z|^2$$
$$\leq -2c_0 r^2 + 2c_0 |x - z|^2 = 0 \, ;$$

that is, $t \leq 0$ because $|x_0 - z| \leq 2r$ and thus $\nu(x_0) \cdot \nu(z) > 0$. This shows $x = x_0 + t\nu(x_0) \in D$.

Analogously, for $x = x_0 - t\nu(x_0) \in \partial B(z,r) \cap D$ we have $t > 0$ and thus

$$(x-z) \cdot \nu(z) = (x_0 - z) \cdot \nu(z) - t\,\nu(x_0) \cdot \nu(z) \leq c_0 |x_0 - z|^2 \leq 2c_0 |x - z|^2 = 2c_0 r^2 \, .$$

Therefore, the surface area of $\partial B(z,r) \cap D$ is bounded from below and above by the surface areas of $Z(-2c_0 r^2)$ and $Z(+2c_0 r^2)$, respectively. Since the surface area of $Z(\sigma)$ is $2\pi r (r + \sigma)$ we have

$$-4\pi c_0\, r^3 \leq |\partial B(z,r) \cap D| - 2\pi r^2 \leq 4\pi c_0\, r^3 \, .$$

$\square$

A.4 Integral Identities

Now we can formulate the mentioned integral identities. We do it only in $\mathbb{R}^3$. By $C^n(D, \mathbb{C}^3)$ we denote the space of vector fields $F : D \to \mathbb{C}^3$ which are n-times continuously differentiable. By $C^n(\overline{D}, \mathbb{C}^3)$ we denote the subspace of $C^n(D, \mathbb{C}^3)$ that consists of those functions F which, together with all derivatives up to order n, have continuous extensions to the closure $\overline{D}$ of D.

Theorem A.11. *(Theorem of Gauss, Divergence Theorem)*

Let $D \subset \mathbb{R}^3$ be a bounded Lipschitz domain. For $F \in C(\overline{D}, \mathbb{C}^3)$ with $\operatorname{div} F \in C(D)$ the identity

$$\int_D \operatorname{div} F(x) \, dx = \int_{\partial D} F(x) \cdot \nu(x) \, ds$$

holds. In particular, the integral on the left-hand side exists in the sense of an improper integral.

Furthermore, application of this formula to $F = uve^{(j)}$ for $u, v \in C^1(\overline{D})$ and the j-th unit vector $e^{(j)}$ yields the formula of partial integration in the form

$$\int_D u \, \nabla v \, dx = - \int_D v \, \nabla u \, dx + \int_{\partial D} u \, v \, \nu \, ds \,.$$

For a **proof** for Lipschitz domains we refer to [20]. For smooth domains a proof can be found in [11]. As a conclusion one derives the theorems of Green.

Theorem A.12. *(Green's First and Second Theorem)*

Let $D \subset \mathbb{R}^3$ be a bounded Lipschitz domain. Furthermore, let $u, v \in C^2(D) \cap C^1(\overline{D})$. Then

$$\int_D (u \, \Delta v + \nabla u \cdot \nabla v) \, dx = \int_{\partial D} u \, \frac{\partial v}{\partial \nu} \, ds \,,$$

$$\int_D (u \, \Delta v - \Delta u \, v) \, dx = \int_{\partial D} \left(u \, \frac{\partial v}{\partial \nu} - v \, \frac{\partial u}{\partial \nu} \right) ds \,.$$

Here, $\partial u(x)/\partial \nu = \nu(x) \cdot \nabla u(x)$ denotes the normal derivative of u at $x \in \partial D$.

Proof: The first identity is derived from the divergence theorem be setting $F = u \nabla v$. Then F satisfies the assumption of Theorem A.11 and $\operatorname{div} F = u \, \Delta v + \nabla u \cdot \nabla v$.
The second identity is derived by interchanging the roles of u and v in the first identity and taking the difference of the two formulas. $\quad \square$

We will also need their vector valued analogies.

Theorem A.13. *(Integral Identities for Vector Fields)*
Let $D \subset \mathbb{R}^3$ be a bounded Lipschitz domain. Furthermore, let $A, B \in C^1(D, \mathbb{C}^3) \cap C(\overline{D}, \mathbb{C}^3)$ and let $u \in C^2(D) \cap C^1(\overline{D})$. Then

$$\int_D \operatorname{curl} A \, dx = \int_{\partial D} \nu \times A \, ds \,, \qquad \text{(A.16a)}$$

$$\int_D (B \cdot \operatorname{curl} A - A \cdot \operatorname{curl} B) \, dx = \int_{\partial D} (\nu \times A) \cdot B \, ds \,, \qquad \text{(A.16b)}$$

$$\int_D (u \operatorname{div} A + A \cdot \nabla u) \, dx = \int_{\partial D} u \, (\nu \cdot A) \, ds \,. \qquad \text{(A.16c)}$$

Proof: For the first identity we consider the components separately. For the first one we have

$$\int_D (\operatorname{curl} A)_1 \, dx = \int_D \left(\frac{\partial A_3}{\partial x_2} - \frac{\partial A_2}{\partial x_3} \right) dx = \int_D \operatorname{div} \begin{pmatrix} 0 \\ A_3 \\ -A_2 \end{pmatrix} dx$$

$$= \int_{\partial D} \nu \cdot \begin{pmatrix} 0 \\ A_3 \\ -A_2 \end{pmatrix} ds = \int_{\partial D} (\nu \times A)_1 \, ds \,.$$

For the other components it is proven in the same way.
For the second equation we set $F = A \times B$. Then $\operatorname{div} F = B \cdot \operatorname{curl} A - A \cdot \operatorname{curl} B$ and $\nu \cdot F = \nu \cdot (A \times B) = (\nu \times A) \cdot B$.
For the third identity we set $F = uA$ and have $\operatorname{div} F = u \operatorname{div} A + A \cdot \nabla u$ and $\nu \cdot F = u(\nu \cdot A)$. $\square$

A.5 Surface Gradient and Surface Divergence

We have to introduce two more notions from differential geometry, the *surface gradient* and the *surface divergence* which are differential operators on the boundary ∂D. We assume throughout this section that $D \subseteq \mathbb{R}^3$ is a C^2-smooth domain in the sense of Definition A.7. First we define the spaces of differentiable functions and vector fields on ∂D.

Definition A.14. Let $D \subseteq \mathbb{R}^3$ be a C^2-smooth domain with boundary ∂D. Let $\{U_j, \xi_j : j = 1, \ldots, m\}$ be a local coordinate system and $\{\phi_j : j = 1, \ldots, m\}$ be a corresponding partition of unity on ∂D. We set again $\Psi_j(x) = R_j(x_1, x_2, \xi_j(x))^\top + z^{(j)}$ for $x = (x_1, x_2) \in B_2(0, \alpha_j)$ and define

$$C^1(\partial D) := \left\{ f \in C(\partial D) : (\phi_j f) \circ \Psi_j \in C^1\big(B_2(0, \alpha_j)\big) \text{ for all } j = 1, \ldots, m \right\}$$
$$C^1(\partial D, \mathbb{C}^3) := \left\{ F \in C(\partial D, \mathbb{C}^3) : F_j \in C^1(\partial D) \text{ for } j = 1, 2, 3 \right\},$$
$$C_t(\partial D) := \left\{ F \in C(\partial D, \mathbb{C}^3) : F \cdot \nu = 0 \text{ on } \partial D \right\},$$
$$C_t^1(\partial D) := C_t(\partial D) \cap C^1(\partial D, \mathbb{C}^3).$$

There exist several different—but equivalent—approaches to define the surface gradient and surface divergence. We decided to choose one which uses the ordinary gradient and divergence, respectively, on a neighborhood of the boundary ∂D. To do this we need to extend functions and vector fields. We point out that the same technique is used to construct extension operators for Sobolev spaces, see, e.g., Theorem 4.13.

Lemma A.15. *Let $D \subseteq \mathbb{R}^3$ be a C^2-smooth domain with boundary ∂D.*

(a) For every $f \in C^1(\partial D)$ there exists $\tilde{f} \in C^1(\mathbb{R}^3)$ with compact support and $\tilde{f} = f$ on ∂D.

(b) For every $F \in C_t^1(\partial D)$ there exists $\tilde{F} \in C^1(\mathbb{R}^3, \mathbb{C}^3)$ with compact support and $\tilde{F} = F$ on ∂D.

Proof: (a) Using a local coordinate system $\{U_j, \xi_j : j = 1, \ldots, m\}$ and a corresponding partition of unity $\{\phi_j : j = 1, \ldots, m\}$ on ∂D as in Definition A.14 we note that $f = \sum_{j=1}^{m} f_j$ on ∂D where $f_j = f\phi_j$. The functions $f_j \circ \Psi_j$ are continuously differentiable functions from $B_2(0, \alpha_j)$ into $\mathbb{C}$ with support in $B_2(0, \alpha_j)$. We extend $f_j \circ \Psi_j$ into the cylinder $\tilde{C}_j = C_j(\alpha_j, \beta_j)$ by setting $g_j(x) := \rho(x_3)(f_j \circ \Psi_j)(x_1, x_2)$ for $x = (x_1, x_2, x_3) \in \tilde{C}_j$ where $\rho \in C_0^\infty(-\beta_j, \beta_j)$ is such that $\rho = 1$ in a neighborhood of 0. Then $g_j : \tilde{C}_j \to \mathbb{C}$ is continuously differentiable, has compact support, and $g_j = f_j \circ \tilde{\Psi}_j$ on $B_2(0, \alpha_j) \times \{0\}$. Therefore, $\tilde{f}_j := g_j \circ \tilde{\Psi}_j^{-1}$ has compact support in $\tilde{U}_j = \tilde{\Psi}_j(\tilde{C}_j)$. We extend $\tilde{f}_j$ by zero into all of $\mathbb{R}^3$ and set $\tilde{f} := \sum_{j=1}^{m} \tilde{f}_j$ in $\mathbb{R}^3$. Then $\tilde{f} \in C^1(\mathbb{R}^3)$ with support in $\bigcup_{j=1}^{m} \tilde{U}_j$ such that $\tilde{f} = f$ on ∂D.

The proof of (b) is identical by using the argument for every component. $\square$

Definition A.16. Let $f \in C^1(\partial D)$ and $\tilde{f} \in C^1(U)$ be an extension of f into a neighborhood U of the boundary ∂D of the domain $D \in C^2$. Furthermore, let $F \in C_t^1(\partial D)$ be a tangential vector field and $\tilde{F} \in C^1(U, \mathbb{C}^3)$ be an extension into U.

(a) The *surface gradient* of f is defined as the orthogonal projection of $\nabla \tilde{f}$ onto the tangent plane; that is,

$$\text{Grad } f = \nu \times (\nabla \tilde{f} \times \nu) = \nabla \tilde{f} - \frac{\partial \tilde{f}}{\partial \nu} \nu \quad \text{on } \partial D, \qquad (A.17)$$

where $\nu = \nu(x)$ denotes the exterior unit normal vector at $x \in \partial D$.

(b) The *surface divergence* of F is given by

$$\text{Div } F = \text{div } \tilde{F} - \nu \cdot (\tilde{F}' \nu) \quad \text{on } \partial D \qquad (A.18)$$

where $\tilde{F}'(x) \in \mathbb{C}^{3 \times 3}$ denotes the Jacobian matrix of $\tilde{F}$ at x.

We will see in Lemma A.19 that the definitions are independent of the choices of the extensions.

Example A.17. As an example we consider the sphere of radius $R > 0$; that is, $D = B(0, R)$. We parametrize the boundary of this ball by spherical coordinates

$$\Psi(\theta, \phi) = R \left(\sin\theta\cos\phi, \sin\theta\sin\phi, \cos\theta \right)^\top .$$

Then the surface gradient and surface divergence, respectively, on the sphere ∂D are given by

$$\text{Grad } f(\theta, \phi) = \frac{1}{R} \frac{\partial f}{\partial \theta}(\theta, \phi)\, \hat{\theta} + \frac{1}{R\sin\theta} \frac{\partial f}{\partial \phi}(\theta, \phi)\, \hat{\phi}, \qquad (A.19)$$

$$\text{Div } F(\theta, \phi) = \frac{1}{R\sin\theta} \frac{\partial}{\partial \theta}\left(\sin\theta\, F_\theta(\theta, \phi) \right) + \frac{1}{R\sin\theta} \frac{\partial F_\phi}{\partial \phi}(\theta, \phi), \quad (A.20)$$

where $\hat{\theta} = (\cos\theta\cos\phi, \cos\theta\sin\phi, -\sin\theta)^\top$ and $\hat{\phi} = (-\sin\phi, \cos\phi, 0)^\top$ are the tangential unit vectors which span the tangent plane and F_θ, F_ϕ are the components of F with respect to these vectors; that is, $F = F_\theta\hat{\theta} + F_\phi\hat{\phi}$.

In case of calculations in spherical coordinates, $x = r\hat{x} \in \mathbb{R}^3$, often the surface differential operators with respect to the unit sphere are used instead of the operators on $\partial D = \{x \in \mathbb{R}^3 : |x| = r\}$. Therefore, we indicate this by using the index S^2 for the differential operators with respect to the unit sphere. Thus on a sphere of radius r it is

$$\text{Grad}_{S^2} f(r, \hat{x}) = r \text{ Grad } f(r\hat{x}), \quad\text{and}\quad \text{Div}_{S^2} F(r, \hat{x}) = r \text{ Div } F(r\hat{x}),$$

where we understand $f(r, \cdot)$ as a function on the unit sphere on the left side and f as a function on the sphere ∂D on the right side.

Furthermore, in general the differential operator

$$\Delta_{\partial D} = \text{Div Grad}$$

on a surface ∂D is called *Laplace–Beltrami operator*. From the previous example on $|x| = R$ we note that in spherical coordinates it holds

$$\Delta_{\partial D} f(R\hat{x}) = \text{Div Grad } f(\theta, \phi)$$

$$= \frac{1}{R^2\sin\theta} \frac{\partial}{\partial \theta}\left(\sin\theta \frac{\partial f}{\partial \theta}(\theta, \phi) \right) + \frac{1}{R^2\sin^2\theta} \frac{\partial^2 f}{\partial \phi^2}(\theta, \phi).$$

Using the corresponding operators for the unit sphere we obtain with the spherical Laplace–Beltrami operator introduced in Definition 2.2 by the different views on the function f that

$$\Delta_{S^2} f(r\hat{x}) = \text{Div}_{S^2} \text{Grad}_{S^2} f(r\hat{x}) = r^2\Delta_{\partial D} f(x) = r^2 \text{Div Grad } f(x)$$

where we read the function f on the left as a function on the unit sphere.

We collect some important properties in the following lemma. It will be necessary to extend also the vector field ν into a neighborhood U of ∂D such that $|\tilde{\nu}(x)| = 1$ on U. This is possible because by Lemma A.15 there exists an extension $\hat{\nu} \in C^1(\mathbb{R}^3, \mathbb{C}^3)$ of ν. Certainly, $\hat{\nu} \neq 0$ in a neighborhood U of ∂D because $|\hat{\nu}| = 1$ on ∂D. Therefore, $\tilde{\nu} = \hat{\nu}/|\hat{\nu}|$ will be the required extension into U.

Lemma A.18. *Let $D \in C^2$ and $F \in C_t^1(\partial D)$ and $f \in C^1(\partial D)$ with extensions $\tilde{F} \in C^1(\mathbb{R}^3, \mathbb{C}^3)$ and $\tilde{f} \in C^1(\mathbb{R}^3)$, respectively. Then:*

(a) $\operatorname{Div} F = \nu \cdot \operatorname{curl}(\tilde{\nu} \times \tilde{F})$ *on ∂D where $\tilde{\nu} \in C^1(U)$ is an extension of ν into a neighborhood U of ∂D such that $|\tilde{\nu}(x)| = 1$ on U.*

(b) *Let $\Gamma \subset \partial D$ be a relatively open subset[2] such that the (relative) boundary $C = \partial \Gamma$ is a closed curve with continuously differentiable tangential unit vector $\tau(x)$ for $x \in C$. The orientation of τ is chosen such that (Γ, C) is mathematically positively orientated; that is, the vector $\tau(x) \times \nu(x)$ (which is a tangential vector to the boundary ∂D) is directed "outwards" of Γ for all $x \in C$. Then*

$$\int_\Gamma \operatorname{Div} F \, ds = \int_C F \cdot (\tau \times \nu) \, d\ell.$$

In particular, $\int_{\partial D} \operatorname{Div} F \, ds = 0$.

(c) *Partial integration holds in the following form:*

$$\int_{\partial D} f \operatorname{Div} F \, ds = -\int_{\partial D} F \cdot \operatorname{Grad} f \, ds. \qquad (A.21)$$

Proof: (a) The product rule for the curl of a vector product (see A.10) yields

$$\operatorname{curl}(\tilde{\nu} \times \tilde{F}) = \tilde{\nu} \operatorname{div} \tilde{F} - \tilde{F} \operatorname{div} \tilde{\nu} + \tilde{\nu}' \tilde{F} - \tilde{F}' \tilde{\nu}$$
$$= \nu \operatorname{div} \tilde{F} - F \operatorname{div} \tilde{\nu} + \tilde{\nu}' F - \tilde{F}' \nu$$

and thus

$$\nu \cdot \operatorname{curl}(\tilde{\nu} \times \tilde{F}) = \operatorname{div} \tilde{F} - \underbrace{\nu \cdot F}_{=\,0} \operatorname{div} \tilde{\nu} + \nu^\top \tilde{\nu}' F - \nu^\top \tilde{F}' \nu.$$

From $\tilde{\nu}^\top \tilde{\nu} = 1$ in U we have by differentiation that $\tilde{\nu}^\top \tilde{\nu}' = 0$ in U and thus

$$\nu \cdot \operatorname{curl}(\tilde{\nu} \times \tilde{F}) = \operatorname{div} \tilde{F} - \nu^\top \tilde{F}' \nu = \operatorname{Div} F.$$

---

[2] That is, $\Gamma = \partial D \cap U$ for some open set $U \subset \mathbb{R}^3$.

(b) Let $\tilde{\nu}$ and $\tilde{F}$ as in part (a) and set $\tilde{G} = \tilde{\nu} \times \tilde{F}$ in U. Then $\tilde{G} \in C^1(U, \mathbb{C}^3)$. By part (a) we conclude that $\int_\Gamma \operatorname{Div} F \, ds = \int_\Gamma \nu \cdot \operatorname{curl} \tilde{G} \, ds$. Choose a sequence $\tilde{G}_n \in C^\infty(U, \mathbb{C}^3)$ with $\tilde{G}_n \to \tilde{G}$ in $C^1(U, \mathbb{C}^3)$. Then, by the Theorem of Stokes on Γ, we conclude that $\int_\Gamma \nu \cdot \operatorname{curl} \tilde{G}_n \, ds = \int_C \tilde{G}_n \cdot \tau \, d\ell$. The convergence $\operatorname{curl} \tilde{G}_n \to \operatorname{curl} \tilde{G}$ in $C(\partial D, \mathbb{C}^3)$ and $\tilde{G}_n \to \nu \times F$ in $C(\partial D, \mathbb{C}^3)$ yields $\int_\Gamma \operatorname{Div} F \, ds = \int_C (\nu \times F) \cdot \tau \, d\ell = \int_C F \cdot (\tau \times \nu) \, d\ell$.

(c) By part (b) it suffices to prove the product rule

$$\operatorname{Div}(f\,F) \;=\; \operatorname{Grad} f \cdot F \;+\; f \operatorname{Div} F\,. \tag{A.22}$$

Indeed, using the definitions yields

$$
\begin{aligned}
\operatorname{Div}(f\,F) &= \operatorname{div}(\tilde{f}\,\tilde{F}) \;-\; \nu \cdot \big((\tilde{f}\,\tilde{F})'\nu\big) \\
&= \nabla \tilde{f} \cdot \tilde{F} \;+\; \tilde{f} \operatorname{div} \tilde{F} \;-\; \nu^\top \tilde{F} \tilde{f}'\nu \;-\; \tilde{f}\nu^\top \tilde{F}'\nu \\
&= \operatorname{Grad} f \cdot F \;+\; f \operatorname{div} \tilde{F} \;-\; \underbrace{\nu^\top F \tilde{f}'\nu}_{=\,0} \;-\; f\nu^\top \tilde{F}'\nu \\
&= \operatorname{Grad} f \cdot F \;+\; f \operatorname{Div} F\,.
\end{aligned}
$$

This ends the proof. $\square$

Lemma A.19. *The tangential gradient and the tangential divergence depend only on the values of f and the tangential field F on ∂D, respectively.*

Proof Let $\tilde{f}_1, \tilde{f}_2 \in C^1(\overline{D})$ be two extensions of $f \in C^1(\partial D)$. Then $\tilde{f} = \tilde{f}_1 - \tilde{f}_2$ vanishes on ∂D. The identity (A.21) shows that

$$\int_{\partial D} \big[\nu \times (\nabla \tilde{f} \times \nu)\big] \cdot F \, ds \;=\; 0 \quad \text{for all } F \in C_t^1(\partial D)\,.$$

The space $C_t^1(\partial D)$ is dense in the space $L_t^2(\partial D)$ of all tangential vector fields with L^2-components. Therefore, $\nu \times (\nabla \tilde{f} \times \nu)$ vanishes which shows that the definition of $\operatorname{Grad} f$ is independent of the extension. Similarly the assertion for the tangential divergence is obtained. $\square$

Corollary A.20. *Let $w \in C^2(D, \mathbb{C}^3)$ such that $w, \operatorname{curl} w \in C(\overline{D}, \mathbb{C}^3)$. Then the surface divergence of $\nu \times w$ exists and is given by*

$$\operatorname{Div}(\nu \times w) \;=\; -\nu \cdot \operatorname{curl} w \quad \text{on } \partial D\,. \tag{A.23}$$

Proof: For any $\varphi \in C^2(\overline{D})$ we have by the divergence theorem

$$
\int_{\partial D} \varphi \, \nu \cdot \operatorname{curl} w \, ds = \int_D \operatorname{div}(\varphi \, \operatorname{curl} w) \, dx = \int_D \nabla \varphi \cdot \operatorname{curl} w \, dx
$$

$$
= \int_D \operatorname{div}(w \times \nabla \varphi) \, dx = \int_{\partial D} \nu \cdot (w \times \nabla \varphi) \, ds
$$

$$
= \int_{\partial D} (\nu \times w) \cdot \operatorname{Grad} \varphi \, ds = - \int_{\partial D} \operatorname{Div}(\nu \times w) \, \varphi \, ds \, .
$$

The assertion follows because the traces $\{\varphi|_{\partial D} : \varphi \in C^2(\overline{D})\}$ are dense in $L^2(\partial D)$. $\square$

References

1. A. Buffa, P. Ciarlet Jr., On traces for functional spaces related to Maxwell's equations. Part I: An integration by parts formula in Lipschitz polyhedra. Math. Methods Appl. Sci. **24**, 9–30 (2001)
2. A. Buffa, P. Ciarlet Jr., On traces for functional spaces related to Maxwell's equations. Part II: Hodge decompositions on the boundary of Lipschitz polyhedra and applications. Math. Methods Appl. Sci. **24**, 31–48 (2001)
3. A. Buffa, M. Costabel, D. Sheen, On traces for $H(\mathrm{curl}, \Omega)$ in Lipschitz domains. J. Math. Anal. Appl. **276**, 845–876 (2002)
4. A. Buffa, R. Hiptmair, Galerkin boundary element methods for electromagnetic scattering, in *Computational Methods in Wave Propagation*, ed. by M. Ainsworth et al. Lecture Notes in Computational Science and Engineering, vol. 31 (Springer, Berlin, Heidelberg, 2003), pp. 83–124
5. M. Cessenat, *Mathematical Methods in Electromagnetism* (World Scientific, Singapore, 1996)
6. D. Colton, R. Kress, *Integral Equation Methods in Scattering Theory*. Classics of Applied Mathematics (SIAM, New York, 2013)
7. D. Colton, R. Kress, *Inverse Acoustic and Electromagnetic Scattering Theory*, 3rd edn. (Springer, New York, 2013)
8. M. Costabel, Boundary integral operators on Lipschitz domains: elementary results. SIAM J. Math. Anal. **19**, 613–626 (1988)
9. R. Dautray, J.-L. Lions, *Mathematical Analysis and Numerical Methods for Science and Technology* (Springer, Berlin, 1990)
10. D.W. Dearholt, W.R. McSpadden, *Electromagnetic Wave Propagation* (McGraw-Hill, New York, 1973)
11. L.C. Evans, *Partial Differential Equations* (Springer, New York, 1998)
12. G. Hsiao, W. Wendland, *Boundary Integral Equations* (Springer, Berlin, 2008)
13. A. Ishimaru, *Electromagnetic Wave Propagation, Radiation, and Scattering* (Prentice Hall, London, 1991)
14. J.D. Jackson, *Classical Electrodynamics*, 3rd edn. (Wiley, New York, 1998)
15. D.S. Jones, *Methods of Electromagnetic Wave Propagation* (Clarendon Press, Oxford, 1979)
16. R. Kress, *Linear Integral Equations*, 2nd edn. (Springer, New York, 1999)
17. A. Kufner, J. Kadlec, *Fourier Series* (Lliffe Books, London, 1971)
18. R. Leis, *Vorlesungen über Partielle Differentialgleichungen zweiter Ordnung* (Bibliographisches Institut, Mannheim, 1967)

© Springer International Publishing Switzerland 2015   333
A. Kirsch, F. Hettlich, *The Mathematical Theory of Time-Harmonic Maxwell's Equations*, Applied Mathematical Sciences 190,
DOI 10.1007/978-3-319-11086-8

19. E. Martensen, *Potentialtheorie* (Teubner, Stuttgart, 1968)
20. W. McLean, *Strongly Elliptic Systems and Boundary Integral Equations* (Cambridge University Press, Cambridge, 2000)
21. P. Monk, *Finite Element Methods for Maxwell's Equations* (Oxford Science Publications, Oxford, 2003)
22. C. Müller, On the behavior of solutions of the differential equation $\Delta u = F(x, u)$ in the neighborhood of a point. Commun. Pure Appl. Math. **7**, 505–515 (1954)
23. C. Müller, *Grundprobleme der mathematischen Theorie elektromagnetischer Schwingungen* (Springer, Berlin, 1957)
24. J.-C. Nédélec, *Acoustic and Electromagnetic Equations* (Springer, New York, 2001)
25. C.H. Papas, *Theory of Electromagnetic Wave Propagation* (Dover, New York, 1988)
26. M.H. Protter, Unique continuation for elliptic equations. Trans. Am. Math. **95**, 81–90 (1960)
27. H. Rademacher, Über partielle und totale Differenzierbarkeit von Funktionen mehrerer Variablen und über die Transformation der Doppelintegrale. Math. Ann. **79**, 340–359 (1919)
28. F. Rellich, Über das asymptotische Verhalten von Lösungen von $\Delta u + \lambda u = 0$ in unendlichen Gebieten. Jber. Deutsch. Math. Verein. **53**, 57–65 (1943)
29. J.A. Stratton, *Electromagnetic Theory* (Wiley, New York, 2007)
30. J. Van Bladel, *Electromagnetic Fields*, 2nd edn. (Wiley, New York, 2007)
31. K. Yosida, *Functional Analysis* (Springer, New York, 1978)

Index

© Springer International Publishing Switzerland 2015
A. Kirsch, F. Hettlich, *The Mathematical Theory of Time-Harmonic Maxwell's Equations*, Applied Mathematical Sciences 190,
DOI 10.1007/978-3-319-11086-8

Printed in the United States
by Bookmasters

Printed in the United States
By Bookmasters